Industrializing Knowledge

Industrializing Knowledge

University-Industry Linkages in Japan and the United States

edited by Lewis M. Branscomb, Fumio Kodama, and Richard Florida

The MIT Press, Cambridge, Massachusetts, and London, England

This book was printed and bound in the United States of America.

Library of Congress Cataloging-in-Publication Data

Industrializing knowledge : university-industry linkages in Japan and the United States / edited by Lewis M. Branscomb, Fumio Kodama, and Richard Florida.
p. cm.
Includes bibliographical references and index.
ISBN 0-262-02465-9 (hc: alk. paper)
1. Research—United States. 2. Research—Japan. 3. Research, Industrial—United States. 4. Research, Industrial—Japan. 5. Universities and colleges—Research. I. Branscomb, Lewis M., 1926– . II. Kodama, Fumio. III. Florida, Richard L.
Q180.U5I43 1999
607'.052—dc21 99-15447
CIP

Contents

Preface

Research-intensive universities in all the industrialized democracies are actively pursuing opportunities to commercialize the creative ideas of their faculties, and a supportive relationship between higher education and economic performance is an old story. Before the onset of heavy government investments in academic research after the Second World War, industry and foundations were almost the only source of external support of university research in the United States. Only in the last decade or so have we seen so much experimentation in the new ways in which high technology firms and universities can relate to one another.

The motives for this new entrepreneurial trend in the United States are varied—the search for recognition as institutions of societal importance, the demonstration of new paths to research employment for students, a desire to see the fruits of science brought to public benefit, a response to pressure from governments—and, yes, the search for new sources of revenue. This project does not explore the appropriateness of the university-industry links, about which many academics have concerns, but rather seeks to describe what is actually taking place, what motivates it, and whether expectations are being fulfilled.

Recognizing that our best universities are hallowed institutions with long histories, it would be surprising if all of them pursued the same paths. To the extent that their traditions are different, producing quite different institutional responses to public expectations, we concluded that a comparison of American and Japanese experience would help illuminate those differences, permitting some sorting out of the effects of a global economy, through which a common set of forces acts on all high-tech institutions. In essence we wanted to ask, "Is it reasonable to expect research universities to make substantial and direct contributions to industrial innovation, and if so, to what extent has this happened in the United States and in Japan?"

This book is a culmination of a project entitled "Universities and Science-Based Industrial Development: A Comparative USA and Japan Dialogue on Public Policy for Economic Development," funded by a grant from the Japan Foundation's Center for Global Partnership for Collaborative Policy-Oriented Research. We are grateful to the Foundation and to Mr. Junichi Chano for their help and encouragement.

Two teams of researchers studied the situation in the United States and in Japan. They deserve the credit for whatever contribution this volume may be judged to make to the understanding of industry-university relations in Japan and in the United States. In September 1998, these two teams and a number of invited experts were brought together in a conference held at Harvard University's John F. Kennedy School of Government to compare experiences on the two sides and draw such conclusions as the facts warranted. As a result of this interaction, papers by both teams were revised and edited. The U.S. team was headed by Lewis M. Branscomb, Aetna Professor, emeritus, in Public Policy and Corporate Management, John F. Kennedy School of Government, Harvard University, in collaboration with Richard Florida, H. John Heinz III School of Public Policy and Management, Carnegie Mellon University. The Japanese team was headed by Professor Fumio Kodama, Professor of Science, Technology and Policy, Research Center for Advanced Science and Technology, University of Tokyo, aided by Professor Fujio Niwa, of the National Graduate Institute for Policy Studies, Saitama University; these four constituted the Project Steering Committee. We were aided by a distinguished Project Advisory Board, to whom we are grateful for many helpful suggestions and most especially for their role in the September 1998 workshop conference, where their experience and wisdom were of great value. The advisors were: Professors Harvey Brooks of Harvard

University, Leo Esaki, President, Tsukuba University, Ezra Vogel, Director for the Center for Asian Studies, Harvard University; Dr. John Armstrong, former vice president for research of IBM; Dr. Hiroshi Inose, Director of the National Center for Science and Information Services and former Dean of Engineering, University of Tokyo; and Dr. Michiyuki Uenohara, Executive Advisor, NEC Corporation.

The operations of the project were coordinated in Japan by Mr. Kenneth Pechter, project director for the Japanese studies and faculty research associate to Professor Kodama. Mr. Pechter made important contributions, not only through his own paper in this book but to the overall conception of the project and to the resolution of many issues that arose during the project. His tireless efforts are much appreciated; they were critical to keeping the project on track and of good quality. At Harvard, Dr. Lucien Randazzese began as Project Manager and was followed by Mr. Masamichi Ishii, who performed most of his service pro bono, for which we are especially grateful. The fact that both Pechter and Ishii are bilingual and know the cultures and academic traditions of both countries was critical in the ultimate melding of Japanese and American approaches in this project.

We also owe a debt of gratitude to Mr. Andrew Russell, assistant to Professor Branscomb, for his many contributions to the project, especially the establishment and maintenance of the project Web site <www.ksg.harvard.edu/iip/usbid/usbid.html>. The executive editor for this book is Ms. Lois Malone, an expert in the editing of collaborative Japanese and American scholarly work. Her editorial judgment and attention to detail contributed much to the quality of this volume. We are equally appreciative of MIT Press and its acquisitions editor, Mr. Larry Cohen, who has encouraged this book project and made it real.

About the Authors

Lewis M. Branscomb is Aetna Professor of Public Policy and Corporate Management (emeritus) in Harvard University's John F. Kennedy School of Government, and former director of its Science, Technology and Public Policy Program. His research focuses on domestic and international technology policy, the management of science-based innovation, and comparative studies of technology policy in Korea, Japan, and the United States. A physicist, Branscomb was director of the U.S. National Bureau of Standards from 1951 to 1969, vice president and chief scientist of IBM Corporation from 1972 to 1986, and chairman of the National Science Board from 1980 to 1984. His most recent book, edited with James Keller, is *Investing in Innovation* (MIT Press, 1998).

Amy Bertin Candell is a senior economist at The Economics Resource Group, Inc. Her area of expertise is applied micro-economic analysis relating to public policy and economic regulation. She has worked with corporate and government clients on intellectual property issues, including the analysis and design of licensing systems for copyrighted materials and the modeling of contributions to the local economy from federally funded R&D. Her other consulting work focuses on antitrust and regulatory policy issues.

Yi-Tzuu (Y.T.) Chien is director of the Division of Information and Intelligent Systems at the National Science Foundation (NSF). In 1998–99, he was a research fellow in the Kennedy School of Government at Harvard University. His current research interest lies at the intersection of information technology and public policy, with a focus on managing the Internet as a research tool and as public space. Before moving to NSF, Chien was a professor and later an assistant dean of engineering for computer research at the University of Connecticut. He has published many articles and several books in computing, pattern recognition, and knowledge-based systems and is a fellow of the Institute of Electrical and Electronics Engineers.

Wesley M. Cohen is a professor of economics and social science in the Department of Social and Decision Sciences at Carnegie Mellon University, a research associate of the National Bureau of Economic Research in Cambridge, and currently serves as a main editor of *Research Policy*. His research has focused on the economics of technological change, considering topics such as industry- and firm-level determinants of R&D, firm learning, the nature and impact on R&D of patent and other appropriability mechanisms, and the relationship between university research and technical advance.

Henry Etzkowitz is an associate professor of sociology and director of the Science Policy Institute at the State University of New York at Purchase. He is author of *The Second Academic Revolution: MIT and the Rise of Entrepreneurial Science* (Gordon and Breach, in press) and co-editor of *The Capitalization of Knowledge: New Intersections of Industry and Academia* (State University of New York Press, 1998). He is also co-organizer of the international biannual conference series on "The Triple Helix of University-Industry-Government Relations." He is currently involved in studies of university-industry relations and knowledge-based economic development in Sweden, Brazil, Portugal, and New York City.

Irwin Feller is director of the Institute for Policy Research and Evaluation and professor of economics at The Pennsylvania State University, where he has been on the faculty since 1963. His current research interests include the economics of academic research, the university's role in technology-based economic development, and the evaluation of federal and state technology programs. He is the author of *Universities and State Governments: A Study in Policy Analysis* (Praeger, 1986). He is a member of the Panel on International Benchmarking of U.S. Research Fields—Immunology, National Academy of Sciences/National Research Council; Transportation Research Board, Research and Technology Coordinating Committee, National Research Council; and NIST—Manufacturing Extension Partnership

National Advisory Board, and formerly chaired the American Association for the Advancement of Science's Committee on Science, Engineering, and Public Policy.

Richard Florida is the H. John Heinz III Professor of Regional Economic Devlopment at Carnegie Mellon University's H. John Heinz III School of Public Policy and Management, and an adjunct scholar with the American Enterprise Institute. He has been a visiting professor at Harvard University's John F. Kennedy School of Government and a visiting scholar at the Massachusetts Institute of Technology. He has written five books, including *Beyond Mass Production* (Oxford University Press, 1993) and *The Breakthrough Illusion* (Basic Books, 1990), both with Martin Kenney.

Michael S. Fogarty is a professor of economics and a research fellow at the Center for Regional Economic Issues, Case Western Reserve University, where he is director of the Program on Invention and Regions. His research focuses on regional growth and development through investments in science and technology. This work uses patent data and a new systems methodology to study R&D spillovers and the regional innovation systems of both high-tech and older industrial cities. A current project surveys U.S. inventors to validate the use of patent citations as a measure of R&D spillovers.

Gerald J. Hane is a special assistant for policy and plans for national security and international affairs at the Office of Science and Technology Policy of the Executive Office of the President. His responsibilities include the interaction of trade and security in science and technology policies, technology transfer, sustainable economic development, and regional concerns such as U.S. relationships with Japan, the Asia Pacific Economic Cooperation (APEC) forum, and the Summit of the Americas. He has managed the development of an administration strategy for strengthening science and technology relations with Japan and the effort that produced the National Security Science and Technology Strategy. Prior to his appointment to this position, he was a professional staff member of the Committee on Science, Space and Technology of the U.S. House of Representatives.

Takehiko Hashimoto is an associate professor of science and technology studies at the University of Tokyo. He has written on the histories of aeronautical engineering, the physical sciences, and engineering education. His current research interest lies in the relationship between scientific knowledge and technical skills and in the role of science and technology in the modernization of Japan.

Adam B. Jaffe is a professor of economics at Brandeis University and coordinator of the Project on Industrial Technology and Productivity of the National Bureau of Economic Research (NBER). His research has focused on such topics as industrial research and development, the economics of basic research and universities, incentive regulation and regulatory reform, and the determinants of the diffusion of new technologies.

Sumio Kakinuma is an associate professor at the National Center for Science Information Systems (NACSIS) in Tokyo. His research interests include the bibliometric study of Japanese science and technology and the university-industry research linkage in Japan. Before moving to NACSIS in 1996, he was with the Ministry of Education, where his responsibilities included editing the "White Papers on Education, Science, Sports and Culture."

Shingo Kano is a consultant to The Nomura Securities Co., Ltd., where he is responsible for high-tech-related investment banking activities, including origination of new companies, venture capital, and company evaluation for IPOs, mainly in the health care sector. Previously at Nomura Research Institute and the Research Center for Advanced Science and Technology (RCAST), he has recently been a doctoral candidate at the University of Tokyo, studying innovation policies.

Robert Kneller worked in the U.S. National Institutes of Health from 1988 to 1997, first in cancer epidemiology, then in science policy at the Fogarty International Center, and finally in technology transfer at the National Cancer Institute's Technology Development and Commercialization Branch. In 1997 he received an Abe Fellowship to study technology transfer in Japan. In 1998, he became a professor in the Department of Intellectual Property of the Research Center for Advanced Science and Technology (RCAST) of the University of Tokyo, where he continues research and teaching on technology transfer and the development of biomedical technologies.

Fumio Kodama is a professor of science, technology and policy at the Research Center for Advanced Science and Technology (RCAST), director of the Department of Advanced Interdisciplinary Studies at the Graduate School of Engineering at the University of Tokyo, and an adjunct professor at the National Center for Science and Information

Systems (NACSIS). He is a member of the Engineering Academy of Japan, an editor of the *Journal of Research Policy*, and an associate editor of *Technology Analysis & Strategic Management*. The Japanese version of his book *Analyzing Japanese High Technology: The Techno-paradigm Shift* (Pinter, 1991) received the 1991 Sakuzo Yoshino Prize. He is also a recipient of the 1991 Science and Technology Minister's Award for Research Excellence. His most recent book is *Emerging Patterns of Innovation: Sources of Japan's Technological Edge* (Harvard Business School Press, 1995).

Hiroto Kotake is a lecturer on public policy at Gunma University. His research interests have been mainly in local government behavior and intergovernment relationships. He is an adviser to the Ministry of Construction and the Ministry of Agriculture on various grant-in-aid programs. With Seiritsu Ogura he coauthored "A Simulation Analysis of the Higher Education Market by an Econometric Model," in *Proposals on the Reform of the Japanese Higher Education System: Proposals of Economists* (Economic Planning Agency of Japan, 1998).

Josh Lerner is a professor at Harvard Business School, with a joint appointment in the areas of finance and entrepreneurial management. Much of his research focuses on the structure of venture capital organizations and was collected in *The Venture Capital Cycle* (with Paul Gompers, MIT Press, 1999). He also examines the impact of intellectual property protection, particularly patents, on the competitive strategies of firms in high-technology industries. He is a faculty research fellow in the National Bureau of Economic Research's Corporate Finance and Productivity Programs. He has developed an elective course for second-year MBAs on private equity finance, "Venture Capital and Private Equity," whose collected course materials have been published as *Venture Capital and Private Equity: A Casebook* (John Wiley, 1999).

David C. Mowery is the Milton W. Terrill Professor of Business Administration at the Walter A. Haas School of Business, University of California, Berkeley, and the director of the Haas School Ph.D. Program. He has published widely on the economics of technological innovation.

Masamitsu Negishi is a professor in the Research Trend Division at the National Center for Science Information Systems (NACSIS) and the director of its Developmental Research Division. His research topics cover scientific information databases, electronic library systems, and the bibliometric study of research trends. He has been an Expert Member of the Science Council to the Minister of Education, Science, Sports and Culture, and of the Council for Science and Technology of the Prime Minister's Office. He chairs the editorial committee for "The Database White Paper" published by the Database Promotion Center, Japan.

Richard R. Nelson is the George Blumenthal Professor of International and Public Affairs at Columbia University. He was formerly a professor of economics at Yale University, a senior staff member of the President's Council of Economic Advisors, and a researcher at the RAND Corporation. His research has been largely on long-run economic change, with a special interest in technological change. His books include *An Evolutionary Theory of Economic Change* (with Sidney Winter, Belknap Press, 1982) and *National Innovation Systems* (Oxford University Press, 1993).

Fujio Niwa is a professor of science and technology policy at the National Graduate Institute for Policy Studies, formerly the Graduate School of Policy Science, of Saitama University. His research focuses on domestic science and technology policy and on technology management mainly based on quantitative analysis. He developed the Japanese Science and Technology Indicator System. An engineering scientist by training, Niwa was a researcher at the Studiengruppe für Systemforschung at Heidelberg, a professor at Tsukuba University from 1975 to 1992, and director-in-research of the National Institute of Science and Technology Policy (NISTEP) of the Science and Technology Agency (STA) from 1988 to 1990.

Hiroyuki Odagiri is a professor at the Graduate School of Economics, Hitotsubashi University. He has also taught at Oberlin College and the University of Tsukuba, and was a senior research fellow at the International Institute of Management in Berlin and the Center for Business Strategy, London Business School. His research has covered the theory of the firm, industrial organization, and the economics of innovation. His books include *The Theory of Growth in a Corporate Economy* (Oxford University Press, 1981); *Growth Through Competition, Competition Through Growth* (Oxford University Press, 1992); and *Technology and Industrial Development in Japan* (Oxford University Press, 1996).

Yoshiyuki Ohtawa is director of the Private Education Institution Management Division, Higher Education Bureau of the Ministry of Education, Science, Sports and Culture of the Government of Japan (Monbusho). He was an associate professor and professor in the R&D Department, National Center for Science Information Systems, from 1994 to 1997. His work focuses on science policy, research statistics and indicators, trends in research, and university policy.

Seiritsu Ogura is a professor of economics at Hosei University, Tokyo, and a chief economist for the Japan Center of Economic Research. He has worked on both theoretical and empirical issues in many fields, including public finance, environmental economics, industrial policies, population economics, health economics, the economics of aging, and education. He is a member of the Expert Committee of the Tax Commission of Japan.

Kenneth Pechter is an associate professor of innovation management at the University of Tokyo's Institute of Social Science, researching the role of networks in industrial innovation. An electrical engineer by training, he has been a fellow at the University of California, Irvine, at the Tokyo Institute of Technology, at Stanford's Inter-University Center in Japan, and at MIT; a faculty research associate at the University of Tokyo's Research Center for Advanced Science and Technology; and a member of science and technology policy-oriented commissions in the United States and Japan, including those of the U.S. Congress, National Research Council, and the Japan Society for the Promotion of Science.

Bhaven N. Sampat is a doctoral candidate in the Department of Economics at Columbia University. His research interests include the economics of technical and institutional change, science and technology policy, and the history of economic thought. His dissertation focuses on the roles of research universities in the national innovation system.

Amit K. Sinha is a doctoral candidate in the Department of Systems and Control Engineering at Case Western Reserve University. His current research develops a fuzzy logic-based algorithm to analyze large data set patterns and a fuzzy measure theoretical basis for his model. A primary application has been the identification of patterns exhibited in the U.S. patent data. His other research areas include combinatorial optimization and network analysis.

Sheryl Winston Smith is a doctoral candidate in public policy at Harvard University. Her dissertation research focuses on international knowledge transmission in high-technology industries. From 1990 to 1992 she was a research analyst at the Congressional Office of Technology Assessment, after which she received a Rotary Scholarship to study science policy in the Czech Republic.

Yuan Sun is an associate professor at the Research Trend Division, National Center for Science and Information Systems. Her research interests include assessments of R&D activity, cross-national comparisons, cross-discipline comparisons, and measurements of the international flow of science and technology.

Katsuya Tamai is a professor at the Research Center for Advanced Science and Technology at the University of Tokyo, where he teaches patent and copyright law and conducts research on intellectual property rights and technology transfer.

Shinichi Yamamoto is a professor and director of the Research Center for University Studies, Tsukuba University. His main concern is the analysis of the various functions of the higher education system, including university research, administration, and management. He served at the Ministry of Education (Monbusho) for seventeen years, where he became familiar with issues in school education, university and research management, and international affairs. He was a research fellow at the National Science Foundation in 1988–1989. He is currently involved in the activities of the OECD, including research training, university funding, and similar issues, with the Committee on Science and Technology Policy (CSTP) in Paris.

Mariko Yoshihara is a doctoral candidate in political science, a Barbara Hillman Researcher at the Asia/Pacific Research Center, and an Asia/Pacific Scholar, at Stanford University. Her dissertation focuses on how institutional differences among bureaucratic organizations affect Japan's policymaking. In 1998 she was a visiting researcher at the Research Center for Advanced Science and Technology (RCAST) of the University of Tokyo.

Arvids A. Ziedonis is a research fellow at the Wharton School at the University of Pennsylvania. His research interests include the commercialization of new technologies, the valuation of intellectual property, and the technology strategies of firms.

I

Contrasting Two Systems of University-Industry Links

1

University Research as an Engine for Growth: How Realistic Is the Vision?

Fumio Kodama and Lewis M. Branscomb

There is intense public interest in the role of universities as the primary source of new skills, new knowledge, and new ideas for addressing almost every issue facing industrialized societies. Indeed, expectations of economic contributions from university research are so strong and widely held that many academics are increasingly concerned that these expectations may be unrealistic. They also worry about potential distortion of traditional academic values arising from links to industry. Political or commercial demands placed on the intellectual life of the university are far from uncommon, and universities have, as a result, sought insulation from these forces. At the same time, the dominant reality is that the best of the world's research universities are uniquely the sources of vitality, understanding, and skills in highly developed societies. Both as a matter of civic duty and in response to an opportunity to be acknowledged as vitally important institutions, universities in most industrialized nations are actively exploring just what their relationships with the commercial world should be.

Everywhere in this relationship there are questions and contradictions that are not easily resolved (Brooks and Randazzese 1998: 361–99). Universities are by tradition—one might say by intellectual necessity—open to participation by scholars from all over the world. Yet their sources of funding are almost entirely domestic, and in most countries (including the United States) primarily governmental. Politicians may be expected to ensure that the benefits of university research are effectively, if not primarily, captured by domestic workers and investors. When university research in the United States is carried out by guest researchers from foreign firms, or is conducted in collaboration with multinational industry, or when faculty inventions are licensed to foreign firms, politicians may ask hard questions about who captures the benefits from public funding of university research.[1] At the same time major science agencies of the federal government are providing strong incentives to the universities to expand their collaborations and business relationships with industry. The current wave of legislation intended to gain economic benefits from federally sponsored academic research began with the Bayh-Dole Patent Act of 1980, which allowed universities to own the patents arising from government-funded faculty inventions with the expectation that this would accelerate the diffusion of new technology to industry through licenses and new venture creation (see chapter 11).

In Japan, economic policy throughout the postwar period has rested on expanding exports based on rapid technical innovation, a strategy that has brought great success. The emphasis has been placed on industrial R&D investment, and the role of universities in the national research system has been much less visible internationally (Odagiri and Goto 1993: 76–114). However, despite very conservative budgetary management in the government, university research has received special and favorable treatment in both good times and bad. The Basic Plan for Science and Technology of 1996 declared that the government budget for Science and Technology should be doubled and reach to the level of ¥17 trillion in the year 2000. However, the financial and budgetary structure applied for Japanese national universities is such that the plan for doubling expenditure for university research may not actually be achieved (see Ohtawa chapter 6).[2]

Nevertheless, university faculties in the national universities are expected, with some recent exceptions, to restrict their collaborations with industry to time-honored informal relationships, even as the government gives high priority to expanding investments in university research despite severe budget stringency elsewhere. As civil servants, professors in national universities are, in general, not allowed to accept personal compensation for their research from outside sources such as commercial firms.[3]

There are also potential conflicts between individual faculty and their university administrations about who will capture the benefits from academic innovations. Universities may have a different view from that of the public agencies that fund their research over the balance between public and private interests in patents and copyrights. Private institutions may worry about the possible effect that business connections might have on philanthropy by businesses that might suffer competitively. The entire scene of university-industry collaboration is fraught with divided loyalties and accountability. But many of these debates are clouded by substantial uncertainty as to how much commercial interchange exists between universities and industry, and the extent to which these relationships actually produce economic payoffs. This book primarily addresses these underlying issues.

How important, from an economic perspective, is the growing linkage of academic research to industry? How extensive are these ties, which, if they become ubiquitous, could transform the priorities and even the mission of the universities? Are universities justified in looking to patents and industrial research sponsorship for material financial support? Are government's expectations reasonable, that economic benefits justify public policies that encourage university-industry collaboration? What might we learn from a study of these two countries, the United States and Japan, so similar in their economic and technological capabilities and so different in culture, tradition, and institutional structure of their universities?

Evolution of University-Industry Links in the Twentieth Century

University-industry links have an old and honorable history in both countries (Etzkowitz chapter 8; Geiger 1986, 1993; Hashimoto chapter 9; Odagiri chapter 10). In the era before the development of the modern corporate laboratory, inventions—mostly in chemistry—came directly from university faculty. Since technology was largely tacit and embedded, the researchers needed a critical assembly of experience and skill, so educational institutions tended to specialize in the needs of the local economy. Thus the University of Akron (Akron, Ohio) became a main source of expertise in polymers and elastomers, supporting the Akron tire industry (Love, Giffels, and Van Tassel 1998). Cornell pioneered the first American electrical engineering department; with Tesla as a faculty member they collaborated with George Westinghouse and built the first municipal electric power service for the mining town of Telluride, Colorado. Cornell students went to Telluride for a year to install and operate the system for Westinghouse (Spencer 1997).

As discussed in these papers, in the United States the flowering of Bell Labs, GE, and duPont as research centers led U.S. industrial growth, and from the university perspective they, with some financial help from a few foundations, were a primary source of demand for research outputs. These links became relatively weaker after the Second World War as the technical community turned its attention to science; government superseded industry as the primary customer for new technology from the universities. The direct link to industry was replaced, to a large extent, by an implicit "social contract" between university science and government, in which in return for a high degree of research autonomy and generous funding, universities accepted the premise that public benefits would, ultimately, far exceed public costs (Guston and Keniston 1994: 1–41).

In Japan something similar happened, but for quite different reasons. As documented by Hashimoto and by Odagiri in chapters 9 and 10, there were strong and effective links between the large firms in Japan and the imperial universities in the 1920s and 1930s,[4] linkages which were broken after the war with the breakup of the Zaibatsu and the rejection of a role for universities in support of the military-industrial complex. Instead, national policy focused on accelerating the market incentives for firms to create their own technical capability to "catch up" with the West, and then in selected areas to forge ahead of their international competitors.

Thus with a strong Japanese (and German) industrial capability challenging U.S. firms in the 1980s, U.S. government attention turned to the lack of "competitiveness" of American firms.[5] The U.S. trade balance in high-tech manufactures turned negative for the first time in 1986. This policy focus became even more intense with the fading of the cold war in 1989, so that by the time Clinton became President in 1993 the administration was committed to a major effort to shift the national focus from defense R&D to accelerated innovation in the commercial sector (Branscomb 1993: 15).

Similarly in Japan, as industry achieved parity with western firms in most technologies and forged ahead in several very important areas such as consumer electronics, autos, and ceramics, and U.S. firms responded to competition by reducing their manufacturing costs, shortening product cycles, and improving quality and responsiveness to customers, it became clear that further Japanese progress depended on creating an independent base for rapid innovation. Calls were heard for more intense investment in basic science and a more robust venture capital market.

Universities became a center of attention in both countries, again for different reasons. In Japan, research investments were primarily industrial. In government, the education ministry, the Monbusho, was best positioned to finance a broad investment in diversified "small" science. While the Science and Technology Agency (STA) was a vital source of research in the "big science" fields of nuclear energy, space research, and high-energy physics, STA never had the breadth of mission in basic research enjoyed by NSF, NIH, and the Department of Energy in the United States. In 1998 the government began the restructuring of its S&T organization, dismantling STA with parts expected to go to MITI and the main science emission going to the Monbusho.

In the United States the strategy for accelerating the competitiveness of U.S. industry also focused on universities, for two reasons. First, the higher education sector is not the responsibility of the national government, but rather of the states and of private institutions. Thus the primary source of federal support of universities came, not from an education ministry, but from the research agendas of a broad variety of government agencies, each of which had concerns about the economic impact of the research it funded. Second, because the government defense, space, and energy sectors had so successfully engaged the research universities in driving their innovation-based strategies in the 1950s and 1960s, it was natural for the Congress to seek to accelerate the diffusion of this knowledge to the commercial sector in the 1980s. Many U.S. universities responded by aggressively seeking to improve their connections to industry as competition for government research funding became more intense, institutional costs rose, and demographic changes meant a lowering of opportunities in academic careers for those holding S&E doctorates.

Forces Driving Interdependence of Research Universities and High-Tech Industry

These historical explanations for the current interest in university-industry collaboration are not the whole story. There are even more profound reasons why industry, both in the United States and in Japan, increasingly looks to the most creative technical people and institutions, and research universities recognize the most innovative firms as a major client for the services of their graduates, as needed sources of their research agenda, and as potential sources of financial support to supplement the increasingly constrained sources of public support for research.

Industry dependence on innovation has been accelerating dramatically since the Second World War, for a variety of reasons. The first is the creation of a scientific base for most engineering practice. There is now sufficient theoretical underpinning for engineering, and enough quantification of the behavior of matter and materials to permit engineering to become more predictable, less risky, and faster in its accomplishments. In many fields inventions can be systematically managed into being. With computer modeling and simulation of both product designs and production processes, the performance and cost of products can be predicted without the traditional "bread board" design verification and pilot line production engineering.

More important, the economic sectors with the most rapid growth are those closest to the science base: microelectronics, software, biotech, and new materials. These industries also have the most sought-after "social" qualities—high wages, good environmental characteristics, low barriers to market entry for small firms, independence from geographic constraints on where the firm (and its scientists and engineers) must be. National governments seek to drive progress in these areas; local governments compete for them, whether from direct foreign investment or relocation from a neighboring political jurisdiction. These growth areas are dependent on new, high-level skills and directly dependent on the latest research ideas. It is small wonder that the universities find themselves in a uniquely advantageous position to foster business growth in these very industries.

Why Examine Academic-Industrial Interactions in Japan and the United States?

Since Japan and the United States are the two leading economies in the world, especially from the perspective of their commitment to technology as a driving force for economic growth, it makes sense to look at the dynamics of university-industry interactions in these two countries. But there is another, more important reason for making this comparison.

The histories and cultures of the two nations are profoundly different, even as they meet one another on the same field of diplomatic and economic interchange. Research, diplomacy, and trade are intrinsically transnational activities. The forces of competition and the requirements for cooperation, whether intellectual, political, or commercial, cause the participants on each side to map their own assumptions and cultures onto the other. This can easily lead to misunderstanding when the mapping fails, as it often does (National Research Council 1999).

This is a particularly uncomfortable situation for Japanese participants, who feel that they have been expected to accept western assumptions, practices and values when engaging with their American counterparts. This became very apparent in a recent United States-Japanese Joint Task Force on Corporate Innovation, in which the two sides had great difficulty agreeing on how to examine the hypothesis that there had been a "convergence" of industrial policy in the two countries in the early 1990s. The difficulty was that the Japanese participants felt that the hypothesis implied that "convergence" necessarily meant greater acceptance by Japanese firms (or government) of American policies. The American participants insisted they did not intend this interpretation.

To avoid this difficulty in the present study, the Japanese and American scholars selected the issues for study and adopted criteria and metrics for their examination that were, in each country, most appropriate for an understanding of the system under study. This led to a result that could have a profound impact on United States-Japan relations in science and technology. What, in fact, do we know about the extent and modalities of interactions between university faculties and industry in Japan and in the United States? What metric should we adopt for assessing the extent and importance of economic activities engaging universities with firms? In the United States, where the leading research universities (both private and state) are free to own and exploit patents on government-sponsored research and are encouraged by government to solicit industry support for and participation in their research, the natural metric is to count the extent of these activities. How much industrial funding supports academic research? How many university-industry research centers on campus invite industrial experts to participate? How many patents and copyrights, owned by the university and arising from faculty research, are licensed to private firms, and how much revenue is returned to the university and its inventors?

If these American tests are applied to the leading (national) universities in Japan, it would appear that relative to American experience, little is happening in Japan. Indeed, as described by Niwa in chapter 5, when it is measured in terms of R&D funds paid by Japanese companies to outside institutions, the outsourcing of R&D to domestic universities has increased slowly, while R&D outsourcing to foreign institutions has grown rapidly. However, there are good reasons to believe that using American metrics might produce deceptive results. The national universities in Japan are severely limited in their ability to accept research funding from industry by their institutional structure as public institutions whose faculties are public employees. Only recently have these universities begun to establish patent portfolios which can be licensed to individual firms with the returns shared by the institution and its faculty inventors. However, there are other traditions of informal collaboration in Japan that cross institutional lines and might represent very significant flows of knowledge. Indeed, one of the most significant findings of this study (see chapter 4 by Pechter and Kakinuma) is that if coauthorship statistics rather than patent licenses and research funding are used as evidence of university faculty links to industry, faculty-firm collaboration appears to be at least as strong in Japan as it is in the United States.

This result will come as a major surprise to most Americans, and puts U.S. government policy toward U.S.-Japan relations in science and technology in quite a new light. Thus, it appears to be the case that different metrics should be applied in Japan and in the United States when attempting to quantify our understanding of the interactions of academics with commercial firms.

U. S. Assumptions Underlying Science and Technology Relationships with Japan

As the two leading economies in technologically intensive industries, the United States and Japan both cooperate and compete. Trade relations have been strained throughout the second half of the 20th century, partly because each side tends to project its own view of the interaction onto the behavior of the other side. In scientific and technological relations, the prevailing American view, at least since the 1970s, has been that there is a serious asymmetry between the American and Japanese "national systems" for research. The American view is that United States depends heavily for industrial innovations on a vast array of academic research laboratories which are very generously supported by government and are generally open to all with few proprietary constraints on scientific information generated by faculty and students. U.S. industry, at least in academic eyes, is seen as driven by quarterly profit reports and devoted to incremental technological improvements. The exception to this dreary image of American industry are the high-tech start-ups, funded by venture capital and often spun off from university research.

The Japanese economy, on the other hand, is seen by U.S. policy makers as driven by a set of very large, very aggressive firms led in many cases by executives with engineering backgrounds and devoted to fast and risk-prone commitments to new technology from which they seek a competitive advantage. These firms, like their American competitors, do not offer public access to their intellectual assets. Furthermore, this caricature goes, while the Japanese universities are open (at least to those who speak Japanese) they are isolated from industry, are bureaucratic, and are far behind the best of their western counterparts.

This oversimplified model has led American policy makers to create a demand for "symmetrical access" to the fruits of Japanese research. The asymmetry about which they complain arises from the assumption that flows of knowledge go from U.S. universities to Japanese industry, without a comparable flow in the other direction. This has been a theme in U.S.-Japan relations at least as far back as 1985, when it was extensively discussed in bilateral meetings of prominent scientists and engineers from both countries organized by the Office of Japan Affairs of the U.S. National Research Council and Committee 149 of the Japanese Society for the Promotion of Science. At the first meeting at Santa Barbara in 1985, Dr. Harold Brown was quoted as saying:

> We believe this concept of symmetrical rather than identical access to a broad range of high technology resources is what has been missing in previous discussions of US/Japan trade matters, which have concentrated heavily on markets. For example, the best Japanese scientific and technological research takes place in federally supported institutes and industrial cooperative ventures that have not, in the past, been readily accessible to American researchers. In contrast, much of our forefront high technology research takes place in association with open research universities and is published in widely read journals. The answer is not to limit access at US facilities, but to get symmetrical access to the best Japanese research results. (Brown 1986)

This point of view, which implies that Japanese universities are not seen as useful to industry, is questionable if the relationships of faculty members rather than of the university as an institution are taken as the example. If Japanese academics are, in fact, diligent collaborators with Japanese firms, an American might conclude that western firms have failed to understand the Japanese system and to participate in it in a wholehearted way. For this reason the situation may be even more "asymmetric" than Americans assumed because westerners find it difficult to participate in the informal networks that function so effectively in Japan.

A survey by Negishi and Sun in chapter 7 will show that Japan now ranks second in many fields of science and technology, below only the United States, in terms of papers registered in the science databases used worldwide. But the statistics are not as reassuring when the country size is taken into account, that is, when the number of papers is divided by the size of scientific communities. This is probably because the papers were written in the Japanese language for domestic readers and thus do not appear in all databases. There are also several structural factors involved specific to the Japanese university system, as discussed by Yamamoto in chapter 20 and Ogura in chapter 21. The accountability of the university in Japanese society and industry had been focused on the quantity of undergraduates; a shift toward quality and research at the graduate level is needed to activate university-industry linkages for industrializing knowledge. There are distributional issues in terms of supply and demand in engineering subdisciplines. Economic structural changes call for changes in distribution, but the government regulates admission quotas by discipline. In the national universities, new programs can only be established when the government allocates new admission quotas to them.

Patents, Venture Capital, and Industrial Transformation

The fact of extensive collaboration between faculty members and industry does not, of itself, indicate how economically productive such relationships might be. Evidence of extensive university-industry coauthorship simply refutes the assertion that the level of interchange is low. To test the economic impact, it would be helpful to know about new firm creation based on university research, and the subsequent success of these firms. In the United States there is a widely held assumption, at least in governmental circles, that the combination of government-sponsored basic research, university patents with rights shared by institution and inventors, and the creation of new firms to exploit the patents not only creates a "payback" on the government's investment but also provides a mechanism for regional economic rejuvenation.

Universities are struggling with the right balance between utilitarianism and independence. How close should the coupling be, between the academic and commercial worlds? Universities feel they should respond to the opportunity to benefit humanity through commercial realization of new ideas and discoveries (while bringing back to the university some needed unrestricted income). On the other hand, they realize that they are almost uniquely situated to view both the natural and social worlds from a distance, bringing perspective and perhaps vision that would be eroded by being too close to the "customer." One must start by recognizing that universities, especially in North America but in Japan as well, are quite diverse institutions. Princeton and Chicago, great private institutions with strong humanistic and scholarly traditions, may feel little need to encourage the faculty to cozy up to industry, while polytechnics like Rensselaer Polytechnic University, in New York, and Lehigh University, in Pennsylvania, may see partnerships with industry as a natural expression of their emphasis on engineering. The University of Georgia will provide a broad-gauge liberal arts program to the citizens of that state, but its engineering and agricultural extension programs will not neglect the needs of Georgia's peanut farmers. And within each university the links to industry tend to be highly concentrated in schools of engineering, in computer science and computer engineering departments, in molecular biology, and in clinical medicine.

But having noted that, what do we know about the effectiveness of government policies to encourage entrepreneurship in American research universities? What do we really know about the influence of the Bayh-Dole Patent Act? This question is investigated by Mowery, Nelson, Sampat, and Ziedonis in chapter 11 through a study of three major U.S. research universities. They find that indeed these universities and their faculties are benefiting from annual revenues of $40 million or more from licensed inventions, most of which resulted from federal research grants. But the rise of biotechnology as a promising new arena for firm creation happened at about the same time as the passage of the Bayh-Dole Act. At about the time that biotechnology firms began to mature and sell drugs as well as stock certificates, software applications and Internet service companies captured the attention of the venture capitalists.

The 1980 U.S. patent policy facilitated the licensing of university inventions, but it seems somewhat more likely that the real spur to entrepreneurship in university faculties was the appearance on the commercial scene of a business opportunity that seemed to arise straight out of basic science. Indeed Mowery et al. (chapter 11) describe the Bayh-Dole Patent Act as an "expression of faith in the linear model of innovation." They found in their study that although the universities they studied had each licensed a substantial number of patents, almost all of the revenue could be traced to a relatively few patents of a very basic nature, underlying some process or tool that everyone in the industry would need, and which were generally licensed non-exclusively. The larger number of exclusively licensed patents typically showed relatively little activity. Yet one of the arguments for passage of the Bayh-Dole Act at the time was the belief that only by licensing exclusively could the inventor shield the investor from competition long enough to provide the risk limitations required. In fact, experience from the most lucrative university patents in the United States seems to show that universities' interests in fundamental research, in the diffusion of knowledge, and in the need for resources are not necessarily incompatible with an entrepreneurial approach to faculty and staff inventions. However, experience with moderate rates of non-exclusive licensing of quite fundamental science is still limited, and hope springs eternal for winners among the inventions offered exclusively. Eventually the cost of patent filing, maintenance, and defense will provide the incentive for a more finely tuned intellectual property strategy.

In technology transfer dynamics at Japanese universities, as argued in this volume by Kneller, Yoshihara and Tamai, and Kano, the new Technology Licensing Office (TLO) system will not diminish transaction costs associated with having to obtain government permission to patent or license inventors, nor will it give universities control over the inventions made by their employees. Moreover, in order to realize technology transfer through creation of new firms in Japan, public policy must include university-industry coordination dynamics in its understanding of technology transfer dynamics. Thus it appears that public policy in Japan will have to reach more deeply into the structure of relationships among universities, their faculties, the firms, and their researchers if the desired economic benefits are to be enjoyed without the resurrection of historical concerns in the academic community about the implications of their association with corporate interests.

Finally, political interest in university contributions to economic progress is strong at the level of national governments, and it is also seen as a critical tool for the restructuring of regional economies mired in mature and declining industries, such as the "rust belt" in the midwestern United States. University-centered industrial development may also be pushed as way of achieving geographical dispersion of economic opportunity. For example, the new Tsukuba Science City was intended to build a center of research-based economic activity outside Tokyo that could attract industrial research and reduce the growth pressures in Tokyo. This "technopolis" idea, originating in Japan, has been picked up in Korea and other Asian countries, but its conceptual parents were the high-tech industrial successes around Route 128 in Boston, Silicon Valley in California, and Research Triangle Park in North Carolina. In every one of these cases one or more research-intensive universities were at the core of the project prior to its initiation.

Notwithstanding the success of these American and Asian examples, can this approach to university-based industrial restructuring be used as a deliberate policy tool in places where there is a large established industrial base which is in decline or stagnation? Michael Fogarty and Amit Sinha examine this question in chapter 18, and the contrast between their findings and those of Amy Candell and Adam Jaffe (chapter 19), who analyzed the economic leverage of publicly funded research on the economy of Massachusetts, is very informative.

These studies all suggest that disappointment awaits those who expect quick results from university-based high-technology strategies for industrial renewal. First-rank research universities can and most often do make a large and positive contribution to economic performance, regionally and nationally. But to understand the effects we should not focus on the style and content of the transactions with firms but rather look at the university as a pivotal part of a network of people and institutions who possess high skills, imagination, the incentive to take risks, the ability to form other networks to accomplish their dreams. For university-industry collaboration to work its magic without diminishing the power that is rooted in either culture, it must be a loose, flexible, continually rearranging network. In short, the quality of the social capital in a region will largely determine whether university-based entrepreneurial activity "sticks" to the region's economy or slides away to more fertile ground. The ecology of a tropical marine reef is a more appropriate metaphor for this process than is a spider's web or a telephone network. The challenge to sound public policy is evident, a point to which we shall return in the last chapter.

Finally, this book explores the question, What accounts for the apparent success of university-industry relationships in the United States, at least in terms of new firm creation and licensing of intellectual property? One important American advantage is a vigorous venture capital industry (Lerner chapter 15). But Feller (chapter 3) makes the point that it is flexibility, resulting from decentralization of control, diversity of institutions, and the spectrum of sources of support, that accounts in large measure for this situation. Lester (1998) finds a similar source of advantage in his explanation for the resurgence of industrial competitiveness of U.S. industry in the 1990s.

The book is organized into six parts. The first paints the scene and provides an overview of what we know about the interactions of universities and science-based industry in Japan and in the United States. Part II addresses the metrics through which the two systems can be observed and described. To place these contemporary assessments in appropriate context, Part III provides some historical and sociological perspective. Part IV looks at the mechanics of university industry interactions, the incentives and barriers to the kinds of interactions that take place. Part V asks, To what extent can policies for encouraging university activity with industry serve to address regional or national economic needs? Part VI concludes with some industrial perspectives and a final chapter that looks to the future and the challenge to public policy makers.

Notes

1. MIT was specifically singled out by Congressional critics for the extensive participation by Japanese scientists and engineers in U.S. government-sponsored research, criticism to which a faculty study group under the leadership of Eugene Skolnikoff responded (Skolnikoff 1994: 194–223).

2. A comparable situation prevails in the U.S. Congress, where bipartisan resolutions in the Senate calling for R&D budget doubling are not binding on appropriations committees in either house.

3. In 1998 a biology professor at the medical school of Nagoya University and the president of Otsuka Pharmaceutical Co. were each indicted on suspicion of accepting and giving, respectively, ¥60 million in "bribes for research results" [*The Japan Times*, Dec. 1, 1998, p. 4]. Prof. Hiroyoshi Hidaka had already been indicted on charges of "taking bribes" from two other firms. He was accused of receiving research students from the firm into his laboratory "as a reward ... for allowing the company to obtain the university's research results" and for receiving personal compensation from the firm through a "dummy research institute." This would probably not have been a violation under U.S. law, although it might well have violated the professor's contractual obligations to the university.

4. For example, Kageyoshi Noro, a professor in metallurgy at the University of Tokyo, fixed technical problems at the two major iron mills, Kamaishi and Yawata, by redesigning the imported furnaces and supervising the state of operation. Subsequently, his students became the chief engineers in these mills and helped improve the operation of the mills (see chapter 10).

5. In fact, there was a considerable history of government interest in promoting the contribution of scientific research to economic health, even outside the well-known case of agriculture (Hart 1998).

References

Branscomb, Lewis M., ed. 1993. *Empowering Technology: Implementing a US Strategy.* Cambridge MA: MIT Press.

Brooks, Harvey, and Lucien Randazzese. 1998. "University-Industry Relations: The Next Four Years and Beyond," in Lewis M. Branscomb and James Keller, eds., *Investing in Innovation: Creating a Research and Innovation Policy that Works.* Cambridge MA: MIT Press.

Brown, Harold. 1986. Quoted in press release entitled "Senior-Level Panel Calls for 'Symmetrical Access' to US/Japan High-Tech Resources." National Academy of Sciences/National Academy of Engineering, Washington DC, November 18, 1986.

Geiger, Roger. 1986. *To Advance Knowledge: The Growth of American Research Universities, 1900–1940.* New York: Oxford Press.

Geiger, Roger. 1993. *Research and Relevant Knowledge: American Research Universities Since World War II.* New York: Oxford University Press.

Guston, David and Kenneth Kenniston. 1994. "Introduction: The Social Contract for Science" in Guston and Kenniston, eds., *The Fragile Contract: University Science and the Federal Government.* Cambridge, MA: MIT Press.

Hart, David. 1998. *Forging the "Postwar Consensus": Science, Technology and Economic Policy in the United States, 1921–1953.* Princeton, N.J.: Princeton University Press.

Lester, Richard K. 1998. *The Productive Edge.* New York: W.W. Norton and Co.

Love, Steve, David Giffels, and Debbie Van Tassel. 1998. *Wheel of Fortune: The Story of Rubber in Akron.* Akron, OH: Univ. of Akron Press.

National Research Council. 1999. Report of the Joint Task Force on Corporate Innovation, Lewis M. Branscomb and Fumio Kodama, co-chairmen, Thomas Arrison, editor (forthcoming).

Odagiri, Hiroyuki, and Akira Goto. 1993. "The Japanese System of Innovation: Past Present and Future," in Richard Nelson, ed., *National Innovation Systems: A Comparative Analysis.* New York: Oxford University Press.

Skolnikoff, Eugene. 1994. "Research in US Universities in a Technologically Competitive World," in David H. Guston and Kenneth Keniston, eds. *The Fragile Contract: University Science and the Federal Government.* Cambridge, MA: MIT Press.

Spencer, Peter. 1997. "Power for the People." *Telluride Magazine* 15(1) Summer 1997, pp 32–35, 39. <www.shell.rmi.net/~pspencer/power.html>

2

Comparing University-Industry Linkages in the United States and Japan

Gerald Hane

Introduction

Trans-Pacific comparisons of innovation often draw attention to the different roles played by innovation partnerships in the United States and in Japan. Partnerships in Japan have developed the reputation in the United States for promoting decades of rapid industrial growth; partnerships in the United States are seen from Japan as having been key to the creation of new, high-technology industries. Over the past two decades, policies have been put in place in each country to gain the advantages seen in the other. The United States has given heightened priority to partnerships with industry that would enhance manufacturing and process advances, and Japan has given greater weight to policies targeted at promoting the creation of new industries.

With the globalization of markets and of innovation, competitive pressures on the two economies are increasingly similar, leading to similar demands on innovation policies. Thus, the question arises of whether the two nations are developing into mirror images of each other in innovation. Will the changes underway encourage similar approaches to mobilizing the nations' resources for innovation and perhaps increase the interchange between these two innovation systems?

This review of university-industry linkages in innovation argues that such a convergence is still distant, with substantial differences likely to remain in the forms of university-industry interaction and the nature of the innovations pursued. In spite of recent convergence in policies, key differences are reinforced by environments for innovation which are rooted in separate histories. Although the potential for greater mutual exchange appears to be growing, there are also substantial challenges ahead.

In the United States, Cohen (Cohen, Florida, and Goe 1994) has estimated that by 1990 there were 1,056 university-industry research and development (R&D) centers, which spent a total of $2.9 billion on R&D, more than double the National Science Foundation's $1.3 billion provided to academic research in that year. Cohen estimates that in 1990 as much as half of industry support for academic R&D went to university-industry R&D centers (Cohen, Florida, and Randazzese 1998).

In Japan, university-industry R&D centers have developed at a much more modest pace. In 1990, there were only a handful of such centers at national universities and by 1997 they numbered 49. A survey by MITI shows an even more modest level of activity in private universities, with 22 high-technology research centers identified in 1997 (MITI 1997). This difference in center activities reflects broader differences between the two countries in university-industry linkages. In 1994, industry support accounted for 7.0% of university research in the United States and 3.6% in Japan; university patent licenses are estimated to have drawn $266 million in the United States but only ¥29 million (less than $300,000) in Japan. What are the causes of these differences?

Cohen (Cohen et al. 1998) summarizes three factors to explain the recent rise of university-industry partnerships in the United States. The first was a decline in federal funding per full-time academic researcher, which dropped by 9.4% in real terms between 1979 and 1991. A second was the Economic Recovery Tax Act of 1980, which extended industrial R&D tax breaks to support research at universities. A third was the Patent and Trademark Laws Amendments Act of 1980, the Bayh-Dole Act. This Act permitted universities, small businesses, and nonprofit institutions to hold exclusive patent rights to the results of research sponsored by the federal government. This combination of events, it was argued, helped to strengthen linkages between these two sectors over the past two decades.

Yet similar factors were at play in Japan in the same period of time. Funding for university research was also decreasing, falling approximately 6.2% per researcher in real terms between 1979 and 1991. This took a particular toll on the number of research support staff at universities, which fell more than 50% between 1976 and 1996. Secondly, the Japanese government offered comparable tax breaks for research conducted with universities. However, rather than a break for 65% of the amount eligible for the 25% R&D tax credit in the United States, it was the full amount of a 20% credit in Japan. In addition, in 1996 the government allowed the deduction of 6% of the value of cooperative R&D with universities as a tax credit. Thirdly, as the result of a law passed in 1977, three years before Bayh-Dole, the Government of Japan offered patent rights back to university inventors for their exclusive use. So the question remains: Why the difference?

At least part of the difference appears to arise from the history of relations between universities and industry and the manner in which this has shaped linkages for innovation. Key factors include the cooperative practices that have developed over time and the forms of administration. Of particular importance are the treatment of appropriability and the management of transactions costs. In addition, environmental factors, such as human resource development and mobility and capital mobility, have acted to reinforce the historical patterns.

As a consequence of these differences, despite some similarities in policy trends, the two countries are likely to continue to evolve different strengths in university-industry linkages. The system in Japan will continue to exhibit a greater emphasis on informal networks nested in well-defined channels of resource flows, providing strengths in speeding the vertical movement of a technology to market. In the United States, the system will continue to exhibit a greater emphasis on the infrastructure that enables linkages to be made, facilitating the flow of resources to opportunities that arise from turns and discontinuities across a less predictable horizon of innovation.

Finally, both countries will continue to face challenges that are associated with the rise of closer university-industry cooperation due to the different incentives of the parties. These tensions arise in the dissemination of information, the nature of university research, and access to research.

The Evolution of the University-Industry Landscape in Japan

Recognizing the "win-win" possibilities of partnerships, programs to promote university-industry linkages have expanded substantially in both nations in recent years. Although there are similarities in the range of policies considered today, different histories have led to different practices and institutional relationships. In Japan, the history of these relations has created hesitancy on the part of universities to create close institutional linkages.[1] As a result, collaboration has been based on individual networks and the informal transfer of technology. The absence of this tension in the United States has enabled institutional involvement, which has played a large role in shaping the U.S. system.

In both the United States and Japan, the importance of universities to economic development was recognized by policymakers over 100 years ago. At the time when the Morrill Act of 1862 established Land Grant Colleges in the United States to contribute to regional agricultural development, Japan was building its system of national universities, beginning with the establishment of the University of Tokyo in 1877, to help that nation emerge economically from its centuries of isolation. Japanese professors and their universities played a central role in this economic rebuilding by absorbing and evolving knowledge from the Western nations.

Although the priority of that era in Japan was on industrial catch-up, there were already emerging arguments that policies in Japan needed to put the nation on firmer footing for innovative self-sufficiency by giving greater priority to creative science and technology. In the proposal to establish the Institute of Physical and Chemical Research in 1916, the government's first national physical science laboratory, this desire was reflected in the justification for the Institute offered by its proponents in the National Diet:

> Abandoning imitation of the advanced countries, this Institute will encourage creative research, produce inventions, contribute to world progress, promote the well-being of mankind and enhance national prestige. (Itakura and Yagi 1974)

The dual goals of economic growth and the promotion of creativity were thus evidenced in Japan soon after the turn of the century. Integrating universities into this challenge has, however, been a complex task.

During the first half of this century, several policy priorities developed that had an important effect on role of universities in the economy. Political, bureaucratic, and industrial elites collaborated closely on policies to draw on all sectors of the economy to support industrial growth. The power of the Ministry of International Trade and Industry (MITI) emerged from this period (Johnson 1982).

Elites at universities urged an active government role in promoting science and technology and achieved Diet approval for the Japan Society for the Promotion of Science (JSPS: *Nihon Gakujitsu Shinkokai*) in 1932 (Hiroshige 1974). In 1933, it launched its University-Industry Cooperative Research Committees, groups composed of the leading industrial interests of the time as well as prominent academics and militarists. Early subcommittees addressed themes such as metal processing, electrical storage, wireless communications, wear-resistant materials, and rice processing. By 1997, 166 committees had been established, of which 50 are active today (JSPS 1997).

Grant support in the JSPS soon reflected a priority on university-industry collaboration. In its first year, JSPS granted most of its awards to individual researchers: 78% of the approximately 500 grants were to individuals and 12% went to university-industry cooperative projects. However, this shifted quickly in subsequent years. In 1936, the ratio of cooperative projects had grown to over 40% of approximately 700 projects, and by 1939 almost 80% of approximately 1,400 projects were for university-industry cooperation. During the pre-war period, collaboration flourished (Inose, Nishikawa, and Uenohara 1982).

The end of the war brought a cessation of close formal programs to promote cooperative R&D between industry and universities. After the war there was an intellectual backlash among leaders in the academic community against working with industry because industry was considered to have been a central contributor to armaments and the war effort (Inose, Nishikawa, and Uenohara 1982).[2]

It was in this postwar environment that another influential body was formed, the Japan Science Council (JSC). In 1948, the JSC was formed as a large organization of scientists with a legislated mandate to provide advice to the government on matters involving science and technology. Influenced by the environment of that time, however, the JSC expressed broader views of society. The sentiment of the Council leadership was strongly anti-war. The Council emphasized academic autonomy and was wary of cooperation with industry. In 1960, during the highly charged public protests against the renewal of the U.S.-Japan security agreement, the Japan Science Council issued a declaration to "determine never to engage in scientific research which contributes to wars" (Anderson 1984). This anti-war, anti-industry atmosphere prevailed through the 1960s and has played a major role in the reemergence of university-industry cooperation.

In the postwar decades, university-cooperation did occur, but was largely informal and consultative, based on networks of human relations (Hicks 1993). In a labor market that was short of skilled labor, ties to university professors were very important to firms for success in recruiting. In exchange for access to students, industry contributed human and physical resources to the university laboratory. Professors also worked with firms as advisors, using as intermediaries industrial and professional associations, a tradition that developed during the Meiji Era (Sato 1996).

By the late 1970s, there was a growing consensus that university research should be strengthened and better connected to the economy. Japan was emerging from a period of postwar technological catch-up and was concluding that future competitiveness would require the creation of new technologies for new industries. The need for an enhanced role of the universities in the economy was highlighted in the 1977 report of the Council for Science and Technology, *The Basic Principles of an Overall Science and Technology Policy from a Long-term View.* As described by Okimoto and Kobayashi, Japan was placing a high priority on its transition from pursuer to pioneer (Okimoto et al. 1984; Kobayashi 1998a).

However, the history of institutional separation between these sectors has slowed the reemergence of formal university-industry relations in Japan. The desire not to confront lingering criticism of past ties has caused academics to stay away from high-visibility cooperation with industry. At the end of the 1970s, the principal formal means of industry linkage to university research were general grants and endowments, *shogaku kifukin*. This system was established in 1899, and as recently as 1993 it accounted for 80% of formal industrial support of university research (Suematsu 1996).

The level of such cooperation was also low at private universities, which were not constrained by regulations limiting cooperation with industry. Overall, private universities accounted for 38% of academic research in Japan in 1995 and they conducted a proportionate level of collaboration with industry. Much of the collaboration that occurred with industry tended to be smaller than at the national universities, emphasizing local industrial relationships and a heavy concentration in the areas of medicine and dentistry. In 1996, these universities and colleges reported 14,883 projects totaling ¥19.1 billion. Although this was approximately four times the number of projects reported at national universities, the average project was a fifth as large, leading to a 20% lower level of total funding.

When the first major postwar Program for Joint Research with the Private Sector began at national universities in 1982, the Ministry of Education (Monbusho) built upon the existing informal practices and networks of cooperation that had developed over the years. In this Joint Research Program, it is not the university or the department that is the central entity, but the individual professor. The focus was on research projects rather than institutional changes.[3] Between 1983 and 1996, the number of projects initiated under this program increased thirty-five-fold, from 56 to 2001.

In 1995, the Diet passed the Basic Law for Science and Technology, which called for the promotion of science and technology overall, and of university-industry linkages in particular. Restrictions on the ability of national university professors to consult with industry were substantially lightened and other Ministries were allowed to fund university research. In just the two years between 1994 and 1996, the number of joint projects increased over 30%, from 2,586 to 3,714, with the funding increasing almost four-fold, from ¥6.6 billion to ¥23.3 billion. Funding is expected to have increased at least another 50% in 1997. Commissioned research increased from 20% of the total industrial contribution to national universities in 1983, to approximately 40% in 1997.

Riding this wave, partnership programs have been developed across the government. The JSPS revived research in the JSPS University-Industry Research Cooperation Committees, with 26 committees in operation in 1997.[4] In 1995, MITI launched the Innovative Industrial Technology R&D Promotion Program for university-industry-government cooperation.[5] The Science and Technology Agency established a new program for Core Research for Evolutionary Science and Technology (CREST).[6] In addition, two programs that represent steps toward greater institutional involvement were created: the Cooperative Research Centers Program,[7] modeled after the NSF University-Industry Cooperative Research Centers, and the Venture Business Laboratory Program.[8]

All of the policies and programs described above emphasized the need to promote greater creativity and new industries: long-standing and well-accepted goals in Japan (Kobayashi 1998a). However, as explored in the remainder of this chapter, it is not clear that this goal is achievable through partnership policies. At the national level, the goal of creativity may have had greater value as a means of coalescing political support and of avoiding confrontation with the historical reluctance to engage universities in directly supporting industrial advances.

In addition to the changes occurring in the national universities, experiments are underway in other institutions in Japan to bring invention and commercialization closer together. Harvey Brooks has observed that government partnerships with society can be made more effective by distributing responsibility and taking advantage of local expertise and needs (Brooks 1986). Such experiments are occurring at new national graduate schools, regional research centers and high-technology incubators, and private universities. Some examples are provided in the text box. These will be interesting experiments to track as they test the limits of university-industry relations and entrepreneurial innovation in Japan.

Innovative Programs to Foster High-Tech Entrepreneurship in Japan

Two national graduate school programs that have been established to pursue interdisciplinary, advanced research and contribute to economic development are the Nara Institute of Advanced Science and Technology, established in 1991, and the Japan Advanced Institute of Science and Technology in Hokuriku, established in 1990. Both universities are in regions of Japan that have placed a priority on revitalization through high-tech industries. NAIST is located in Kansai Science City, which straddles the prefectures of Nara, Osaka, and Kyoto. With state-of-the-art facilities, an emphasis on interdisciplinary training and research, the nation's only fully computerized university library, and a philosophy that companies will more readily engage with the university if frontier research is performed, the Institute has emphasized a flexible research environment while seeking to integrate the surrounding industry into its research activities. In January 1998, the Institute had 115 grants and endowments from major corporations, a portfolio of commissioned research that exceeded research grants from the Ministry of Education, and several laboratories jointly run with industry, including companies such as NTT, Matsushita Electric, NEC , and Fujitsu in information systems; Sharp, Sanyo, Matsushita Electric, and Shimadzu in materials science; and Taisho Pharmaceuticals in biological sciences.

Prefectural governments are also taking an increasingly active role in promoting high-technology resources. Since Kanagawa Prefecture established an incubation facility in 1982, Kanagawa Science Park, and its own Kanagawa Academy of Science and Technology in 1989, the number of prefectures making similar commitments has increased year-to-year; by 1998 31 prefectures supported centers for advanced science and technology research. In Gifu Prefecture, for example, the prefectural government has made a major commitment to high-tech development through the opening in 1996 of Softopia, the country's first regional incubator dedicated to multimedia technologies. This incubator is targeted at filling a gap in support for entrepreneurs by providing space, greater visibility, and financing. Within two years Softopia was host to more than a dozen domestic and foreign universities, and 16 start-up firms used the incubation rooms, 25 firms used the technology development rooms, and 17 larger firms used laboratory space on-site. Softopia Japan is a pioneering effort in the country to mix a freer research atmosphere with the critical-mass benefits that arise when many research entities share interests and equipment needs. Companies that have successfully graduated from the incubation stage include Transoft, a company that makes Japanese software compatible with Windows, and Digital Wizard, a company that has successfully entered the digital movie animation business. In addition, Softopia is connected to a new International Academy of Media Arts and Sciences, which specializes in education in multimedia fields such as computer graphics and interactive art.

A third area in which new experiments are occurring is in the private universities. Private universities are not constrained by many of the regulations that guide the national universities and have been freer to engage in creative partnerships. Because their research budgets have typically been much smaller than those of the national universities, the focus of many private universities has been on regional industries. A particularly active university in this regard is Ristumeikan University, located on the shores of Lake Biwa outside of Kyoto. In February of 1995, the Board of Directors of the University decided to aggressively pursue opportunities for cooperative research with industry in light of the reforms underway in the Ministry of Education to promote these links (Sakamoto 1997; Tanaka 1997). A liaison office was established and the university pursued partnerships with local industry in areas that included robotics and factory automation, new materials processes, electronic devices, high performance computing, environment, construction management, and eco-technology. Beginning with the help of 300 companies in the local region that contributed ¥400 million in the first two years, the university constructed laboratory facilities to support the program. Between 1994 and 1996, the number of external research contracts and grants had more than doubled. from 110 cases to 230, with funding increasing five-fold in these two years. Eighty of these cases involved industrial contracts, of which 40 were from local small and medium enterprises, a priority of this university.

Characteristics of University-Industry Linkages in the United States

In the evolution of university-industry partnerships in the United States, there are several points of contrast with development in Japan. First, history did not involve a comparable breach between these two sectors. Military research has at times been a sensitive issue with universities, but the overall relationship has been continuous. This has allowed institutions to take a more active role in the partnerships, guiding the various forms of collaboration. Second, the cooperative research centers have focussed more on the goal of applied training and process advances than they have on creativity and breakthrough developments. Third, the university infrastructure developed for managing these interactions has led to different means of bringing inventions to the market that have encouraged high-tech entrepreneurship and spin-off ventures. Fourth, the different histories of these two countries have led to different incentives in the environment for innovation and to different paths of innovation that are most compatible with university-industry linkages.

University-industry linkages in the United States have been strengthened by the rise of U.S. research universities since the end of the Second World War. Although the relative extent of this relationship has been up and down over time, developments over the past two decades have promoted a continued expansion of ties between these sectors. According to NSF data, industry funding for

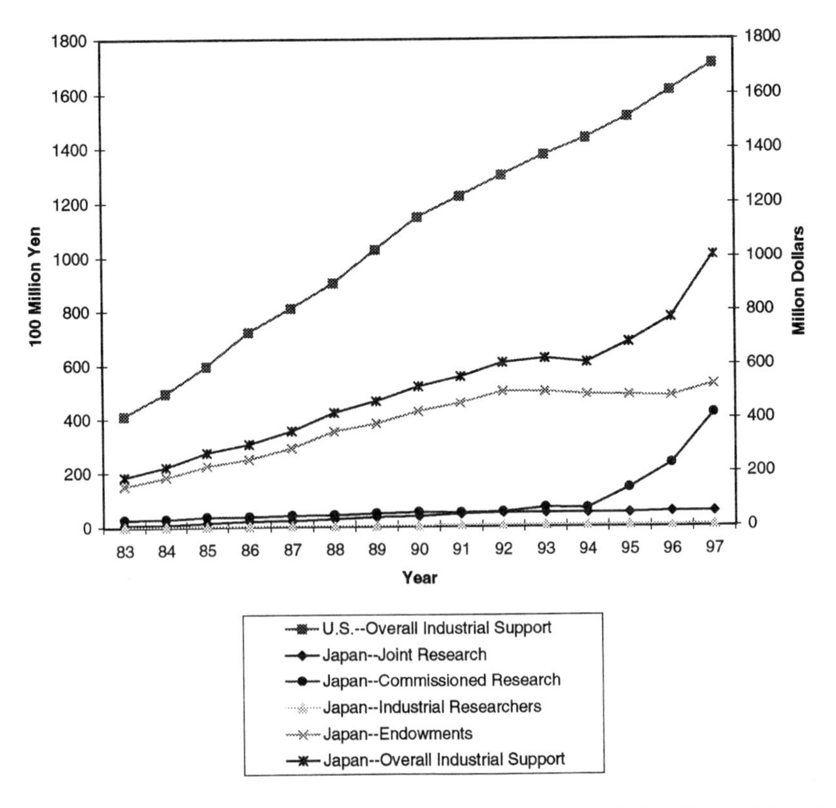

Figure 1 Comparison of Industrial Support of University R&D over Time in the United States and Japan (NSB 1998; MITI 1997; Sangaku Kaigi 1998). The left scale is in units of ¥100 million for funding in Japan. The right scale is in units of $1 million for funding in the United States.

university research was close to 8% through most of the 1950s, dipped to as low as 2.5% in the late 1960s, and has been on a steady rise through the 1980s and 1990s, reaching its current level of 7.1% in 1997. In Japan, both the government and industry have substantially increased contributions to university research since 1995, with increases of over 10% from the government each year and 15% to 30% from industry. A comparison of industrial contributions to university research in the United States and Japan is provided in Figure 1.

Through the 1970s and 1980s there was a growing sentiment that the capabilities that gave the United States some of the world's best science could be better connected with industry and the economy. There was a deepening sense that U.S. manufacturing was losing its competitive edge to international firms and that universities could better help to remedy the decline. During the administration of President Carter an Advisory Committee on Federal Policy on Industrial Innovation released a report that addressed this concern. The report noted that:

> There has been an ever-widening gap between the university and industrial communities and, as a result, this key national source of new technical knowledge is not being adequately tapped for its innovative potential by the private sector—the sector with a major responsibility for innovation in the United States. (Advisory Committee 1979)

In January 1985, the President's Commission on Industrial Competitiveness issued a widely circulated report that noted:

> Perhaps the most glaring deficiency in America's technological capabilities has been our failure to devote enough attention to manufacturing or "process" technology. It does us little good to design state-of-the-art products, if within a short time our foreign competitors can manufacture them more cheaply. (President's Commission 1985)

The Commission further recommended that:

> Universities, industry, and Government must work together to improve the quality and quantity of manufacturing-related education ... [and that the] Federal Government should expand support for manufacturing-related activities in universities.

In Japan, it was common for companies to invest several years of training in new hires, anticipating that these individuals would stay with the companies for the full term of their careers; in the United States greater job mobility has made such training investments by companies less attractive. Programs such as the National Science Foundation's University-Industry Cooperative Research Centers are an attempt to address that gap: Students gain a higher level of applied research experience, industry gains the expertise of faculty and students, and universities gain the resources contributed by industry for modernizing their facilities, equipment, and materials.

These centers further attempt to integrate research and education and emphasize a cross-disciplinary approach to problem solving, allowing students and faculty to move beyond the borders of departmental disciplines.

By the mid-1980s, university-industry centers were more widely seen as an effective means of promoting industrial partnerships and a more applied education. The number of such efforts grew from this period forward as NSF Director Eric Bloch made such partnerships with industry a much higher priority than before. By 1997, NSF contributed to 15 Engineering Research Centers, 24 Science and Technology Centers, 12 Centers of Research Excellence in Science and Technology, 25 Materials Research Science and Engineering Centers, and 53 University-Industry Cooperative Research Centers, with a combined federal contribution of $169 million. The goal was not invention, although invention occurred. Rather, NSF surveys of these programs indicate that center faculty assign greatest value to the contributions of these centers to a more applied education for their students (Gray and Walters 1997). In addition to the NSF centers, hundreds of other university-based cooperative R&D centers were established across the country, with 60% of these centers formed in the 1980s (Cohen, Florida, and Goe 1994).

Interestingly, it appears that a major impetus for the promotion of these policies and programs in the United States was the view that Japan was much more effective in bringing together institutions to commercialize innovations. Japan, which was struggling with a much more significant gap between universities and industry, was a major motivator for the renewed emphasis on this linkage in the United States.

Incentives Related to Administration: Appropriability and Transaction Costs

In the United States and Japan, the different histories of university-industry linkages have led to very different practices for administering these relationships.[9] Although recent policies to promote these linkages display growing similarities, major differences persist in the treatment of ownership (appropriability) and the mechanisms for bringing inventions to the market (transactions costs). These differences have implications for the ultimate impact of these linkages on society.

In discussions leading up to the passage of the Bayh-Dole Act in 1980, policy makers took note of the estimate that fewer than 5% of the 28,000 patents being held by federal agencies had been licensed, compared with 25% to 30% of the small number of federal patents for which the government had allowed companies to retain title to the invention (USGPO 1998). The Act provided universities and small businesses with that right. In 1980, universities were granted about 390 patents. By 1995, universities received approximately 2,373 patents and $270 million in royalties,[10] with $57 million at the University of California alone (Hashimoto 1998).

One consequence of the Bayh-Dole Act was an increase in university Technology Licensing Offices (TLOs). At most major universities in the United States, TLOs provide an institutional infrastructure to facilitate the commercialization of intellectual property. In 1980, there were only 25 technology licensing offices at major universities in the United States, but by 1996 there were more than 200 (AUTM 1997).

These offices reduce the costs of bringing technology to the market and ensure that both the universities and the researchers have an incentive to promote innovations by dividing the reward. At Stanford University, for example, the licensing office takes 15% of licensing revenue to cover its costs, and the remaining 85% is divided three ways between the inventor, his or her department, and the parent school at the university (Schrieberg 1998). At MIT, the Technology Licensing Office also takes 15% for administrative costs, one-third goes to the inventor, and the remaining two-thirds is divided between the department and MIT's general fund after deducting other processing costs; the University of California system in 1997 spent approximately 25% in legal and operational expenses, and of the remainder, about 40% went to the inventors, 15% to the State of California, and 30% to the university system. Distributions at many other major universities follow a similar pattern (USGPO 1998).

In Japan, the Ministry of Education issued regulations in 1978 that set up a review system that would either allow national university professors to take ownership of their patents or assign rights to the government.[11] In practice, most of the patent rights have been returned to the individual researchers. In 1996, approximately 90% of the 435 patent applications at national universities were assigned back to the inventing researcher, with only 10% going to government ownership (Tsusho Sangyo Sho 1998).

However, the university has no formal role. As a consequence, national universities have not been involved in the management or commercialization of patents, nor have they provided institutional support. In 1994, only 129 patent applications were submitted by the government from national universities, which compares with 326 patent applications that arose from the Joint Research Program between national universities and industry in the same year (Monbusho 1996). The government does return some of the royalties to the national universities and inventing researchers, but the amounts are small.

This relatively low level of patenting activity was also seen at the private universities, with 124 patent applications in 1994.[12] Tokai University led this group with 24. Private universities are less restricted by Ministry of Education regulations in general, but in most cases, with the exception of Tokai University, these universities also did not take ownership of the inventions of university researchers.

To promote the commercialization of patents held by professors, a law known as the *Gijutsu Iden Kikan* was passed in the spring of 1998 that authorized the formation of Technology Licensing Offices at the national universities. This law established a greater stake for universities in the successful use of their research results. In addition to authorizing the TLOs, the new law provides for cost-shared support for up to the first five years, and establishes grants to encourage university-industry cooperation at these centers.

An early entrepreneurial effort to take advantage of these developments was launched by a group of professors at the University of Tokyo, who established a Technology Incubation Company with ¥10 million of their own funds in start-up capital. Offices have also been developed at other national universities, including Hokkaido University (the Ambitious Fund), Nagoya University (Entrepreneur Supporting Investors' Association), and Tokyo Institute of Technology (Frontier Research Center) (Fujisue 1998). However, only the latter is formally a part of the university; the others are set up as related associations. A key point is that participation is entirely voluntary by the professors.

Although the establishment of TLOs is a positive policy development, other challenges lie ahead. Attaining a critical mass of research results and gaining long-term institutional support will be key issues. Overall, the experience in the United States has been that the costs of running the licensing operations can often absorb much of the return even when the intellectual property is pooled from the entire university. On average it is estimated that royalties at the more successful schools account for 0.5% to 2.0% of the university's research budget (Nelsen 1996), with most universities averaging less than 1% (MITI 1997).

Even the more successful universities rely on a few highly used patents for the bulk of their income. At Stanford, of the 1,271 licenses that it has granted to companies for products based on on-campus inventions, only four have generated more than $5 million each. In 1996, one patent, the Cohen/Boyer patent for recombinant DNA, brought in 72% of Stanford's licensing income. Another 14 patents have earned more than $1 million and most of the remainder brought in far smaller sums (Schrieberg 1998). At many universities, revenues often do not cover administrative and legal costs. Without a critical mass of research activities or some form of sustained support in Japan, long-term viability will be a major issue.

In addition, in Japan, the established informal networks of technology exchange between professors and industry may not be quick to change. Once assigned a patent, a researcher has traditionally received no financial, legal, or administrative support for processing. These costs, combined with the historical reluctance to formal cooperation with industry noted earlier, have led to a widespread practice of informal technology transfer. Patents are often transferred to industrial partners in return for laboratory support such as research equipment, materials, or visiting researchers from the firm. In some cases, professors do not even bother to submit for review or file the patents because of the costs involved (Tamai 1996).

To the extent that the current system of informal transfer works, it may be efficient and remain the preferred option. Professor Nagao, President of Kyoto University, observed that the current mechanisms of university-industry cooperation, although not easy to quantify, do seem to lead to appropriate interchanges between these two sectors.[13] In one industry that tends to rely more heavily on university research, biotechnology, the transfers appear to be very significant. Data from the Biotechnology Industry Association of Japan indicate that 60% of biotechnology patent applications in 1994 cited university research.[14] By comparison, in the United States it has been estimated that 70% of significant industrial patents cite university research (CHI Research 1997).

Thus, the direction of university-industry linkages in Japan remains an open question. From the perspective of society, the larger benefit is not in maximizing university licensing revenue, but in translating laboratory developments into products and services. For example, in the United States it is estimated that through the new businesses formed and induced investments—investments that are made to realize the commercial potential of a university innovation—university innovations directly contributed $24.8 billion to the economy in 1996 (AUTM 1997). The impact of both formal and informal transfers in Japan is unknown. Whether the new TLOs in Japan can survive the challenges of their balance sheets to provide this broader benefit to society is a question.

Incentives Related to the Environment for Innovation

Although policy measures to promote university-industry linkages show growing similarities in the United States and Japan, there are deeply rooted differences in the environment surrounding innovation that continue to reinforce traditional patterns. One of the factors is the mobility of human resources—the ability to turn the needed manpower to new opportunities; another factor is capital mobility—the ability to bring financing quickly to promising ventures; and a third factor is access—the ability to partner with institutions and integrate into the overall enterprise for innovation. These environmental factors both enable and limit the potential of university-industry linkages in both countries.

Mobility of Human Resources. A key element to the success of innovations is mobilizing human resources where they are needed, when they are needed. Different modes of innovation will benefit from different human resource patterns, with evolutionary change benefiting from continuity and discontinuous change benefiting from an ability for people to move quickly where opportunities arise. These differences are reflected in the patterns of personnel mobility in Japan and the United States—patterns that may not be quick to change.

Westney and Sakakibara have shown that continuity in personnel has played a key role in successfully managing innovations in Japanese companies. Successful companies that they studied in Japan advanced an innovation by moving personnel along with the innovation from research to commercialization (Westney and Sakakibara 1985a; 1985b). This system promotes interaction across the various stages of innovation and enables the carrying forward of the large amount of tacit information embodied in process development. This notion of integrating knowledge from the multiple dimensions of innovation was adopted to some extent in the United States and drove the emergence of product teams in the 1980s.

In Japan, however, this is not a recent phenomenon. Lower levels of mobility have been a central feature of business operation over the past five decades. As noted earlier, the tight postwar market for skilled labor led larger firms to adopt the promise of lifetime employment to attract recruits. Because many of these larger firms and industrial groupings were credited with propelling Japan's rapid economic growth, they became not only sources of stable employment but also of societal prestige. Changing companies was frowned upon and most major corporations had policies that discouraged job-hopping. Competing companies often recruited on the same days to prevent the need to bid for recruits. Success in business for an individual was commonly reflected by advancement in the same organization. These characteristics provide strength when stability is needed, but challenges when mobility is desired. A study by Japan's National Institute for Research Advancement that contrasted venture business cultures between the United States and Japan observed that the system in Japan is particularly challenged by a culture and climate of low mobility of human resources, lack of tradition in allowing the failed to rise again, and relative lack of societal prestige for entrepreneurs (NIRA 1997).

By contrast, in the United States the system of employment encourages far higher levels of horizontal mobility, as changing jobs is a common means of improving one's salary and position. The lower level of continuity that results can be a handicap if competitiveness relies on long-term process developments, embodied knowledge, or continuous changes. However, there are strengths in the flexibility that it offers in periods of discontinuous change—when the advance is not an incremental improvement over an existing technology but a new technology. For example, a survey conducted by MIT's Technology Licensing Office found that 77% of the investment in MIT technology and 70% of the jobs in the study were associated with start-up companies (Pressman et al. 1995). John Gee, the first head of a Silicon Valley Incubator set up by NASA in 1993, observed that "It used to be 5 percent of the Stanford masters in engineering wanted to start businesses. Then it was 10%, then 15%, then 20%" (Lewis 1998). Higher mobility allows attention to shift quickly to adapt to new opportunities.

Differences in mobility are borne out in broader studies of career patterns in the United States and Japan. Ishii has confirmed substantial differences in mobility when tracing the career paths of graduates from the University of Tokyo, the Tokyo Institute of Technology, and the Massachusetts Institute of Technology. Examining graduating classes of 1960, 1970, 1980, and 1985, he found much higher levels of mobility in the careers of MIT graduates than their counterparts in Japan (Ishii, Yokoo, and Hirano 1993). By the beginning of 1991, less than 20% of the MIT engineering class of 1960 were with their original employer, whereas approximately 75% of Tokyo graduates were with the same employer. Less than half of MIT graduates in 1985 were with their original employer, compared with over 95% of the graduates of the Tokyo schools.

This pattern is further reflected in more general data. For example, 1993 survey data from the Science and Technology Agency of 2,665 engineers indicated that only 20% changed jobs in their careers, with this mobility seen later in life rather than early in their careers (NSF 1995). One of three engineers over 50 years of age had experienced a job change, whereas far fewer changed jobs early in their careers. This late-career mobility is influenced by restructuring as well as the traditional practice of rotating senior staff to subsidiaries. For scientists and engineers as a whole in

Japan, a 1996 survey of the Science and Technology showed mobility to be a bit higher, with 31.6% responding that they had changed jobs. Those at universities had the highest positive response rate of 45%, followed by national laboratory researchers at 38%, and industry, still at 25% (STA 1997).

By contrast, in the United States movement between organizations is far more common. According to National Science Foundation data, it is common for scientists and engineers to change employers every four to five years.[15] Those in computers and software change the most frequently, with Ph.D.s averaging just over three years with an employer. Those in civil and chemical engineering move among the least frequently, with their scientists and engineers averaging tenures of seven to eight years.

Across all professions in the United States, a survey by the Bureau of Labor Statistics and Bureau of Census shows that men remain with one employer for a median period of 4 years; women have a median tenure of 3.5 years. Mobility tends to be greater at younger ages, with a median tenure of 3.0 years for men 25–34 years old, increasing to 6.1 years for men 35–44 years old and 10.5 years for men 55–64 years old. For women the tenures are slightly shorter, with median periods of 2.7 years for those 25–34 years old, 4.5 years for those 35–44 years old, and 9.9 years for those 55–64 years old (BLS 1997).

One consequence of the lower level of mobility in Japan is a concentration of researchers in the academic sector. Although the number of researchers per 10,000 labor population is roughly comparable in Japan and the United States—79.4 in Japan and 74.3 in the United States in 1993 (NSB 1998)—there are proportionately many more researchers in academia in Japan. In the United States academic researchers account for about 15% of all researchers, whereas in Japan the figure is approximately 25%. The number of academic researchers in Japan will most likely increase with the growth in fellowships made available by the government. However, the rigidity of the faculty chair, or *koza*, system limits the possibilities for advancement at a university, and a tradition of promoting from within a university limits possibilities of moving elsewhere in academia. This is a large pool of talent that can be tapped if the proper incentives are put in place.

Finally, there are the challenges that emanate from the overall incentives in the career system. The vertical paths to success in Japan exert a major influence on the goal of nurturing creativity at a young age. In general, incentives must be modified downstream to affect changes upstream. Firms look to obtain the best employees, which is generally reflected in heavy recruitment of students at the best universities. Thus, entrance into a prestigious university has been a critical, life-defining career step for students. The primary criteria to achieve admission to the best universities are scores on demanding college entrance exams. As a result, secondary and primary education focus on preparing individuals for these college entrance exams, emphasizing curricula that are the subject of the testing. Therefore, reforms at schools to promote more "creative" students are handicapped by clear downstream incentives that do not reward creativity. As long as the opportunities for success are embodied in a tight vertical system of career advancement, students do not have an incentive to stray from that model. This makes change difficult.

Mobility of Capital. Another factor that affects the shape and success of university-industry linkages is the mobility of capital for investment in commercializing university innovations. As in the case of human resource mobility, Japan's traditional strengths have been in committing funding for longer-term projects. In the United States, impatient capital has in the past been criticized for not enabling continuity for long-term investments through business cycles.

However, in an environment of rapidly emerging innovations, the kind of advances often hoped for in university research, finding the most efficient path for investment is a key step to market success. In the United States, start-up firms and venture capital have played a major role in bringing new innovations to the market. Venture capitalist John Doerr has calculated that between 1981 and 1990, the personal computer industry grew from virtually nothing to $100 billion, "the largest legal accumulation of wealth in history." More than 70% of these firms were backed by venture capitalists. In Japan, the venture capital sector has had less time to develop. However, its development, and the development of a supporting stock exchange, can have a major influence on the future of high-technology start-ups in the country.

In the United States, a growing route to the commercialization of university innovations is the formation of start-up, or venture, enterprises and adaptation by smaller, fast-moving firms. A survey by the Association of University Technology Managers (AUTM) estimates that between 1980 and 1996, academic licensing has helped to form 1,881 new companies. Of this total, 712, or 38%, were formed in the most recent three years, 1994 to 1996 (AUTM 1996; 1997). In 1996, AUTM further estimates that 3,261 patent applications were filed by U.S. and Canadian universities, with 2,741 new licenses and options granted in that year, and 10,178 invention disclosures submitted. Small firms and start-ups were major partners with the universities. In 1996, 64% of new university licenses and options executed were with newly formed or existing small companies. This percentage was as high as 80% at Stanford University.

In Japan, small and medium enterprises also play a significant role in the formal licensing of university innovations. In 1998, the Japan Society for the Promotion of Science had licenses for university patents with 58 companies. Just under half of those companies were small and medium firms, with only half a dozen of the very large corporations engaged in licensing with JSPS. However, virtually all of the small and medium firms with licenses were local manufacturers and subcontractors rather than venture start-ups. The lower presence of larger firms is not too surprising in this case, as these are the firms that can better afford the sustained relationships with professors that underlie the larger volume of informal transfers to industry.

Licensing to larger firms is not necessarily a handicap, provided that the advances in technology fit well with the portfolio and vision of the corporation. However, to the extent that these corporations do not have the flexibility to respond quickly to new technology-based market opportunities, these opportunities may not be fully realized. In a survey of 222 firms, 60% of small-sized companies surveyed viewed new business ventures as successful; this compared with 54% of the medium-sized companies and 35% of the large companies. The survey concluded that "for large companies, the primary objective of pursuing new business ventures is to reinvigorate the company organization and make effective use of human resources that may have become superfluous in the course of restructuring and re-engineering; the success of the ventures per se tends to be of secondary concern" (Kodama 1998).

In the United States, there was substantial growth in the venture capital sector through the 1990s. Figure 2 shows venture capital investment in both the United States and Japan. In the United States, venture capital investments have recovered from a dip at the beginning of the decade and rose from $2.65 billion in 1991 to $9.42 billion in 1996 (Rausch 1998). The cumulative amount under venture capital management was estimated at $37.2 billion in 1995. In Japan, there has been a slower growth in the venture capital market, with investments rising to ¥243 billion in 1996. Venture capital investments fell back about 20% in 1997, although investments in computers and electronics increased 40% over the previous year. Outstanding investments are estimated at ¥853 billion in 1995 and ¥826 billion in 1996 (Ono 1997; *Nihon Keizai Shimbun* 1998). The total number of firms in which Japanese venture capital firms invest is estimated at approximately 13,000 in 1996 (Kutsuna and Cowling 1998).

The growth of venture enterprises in the 1990s in the United States has been aided by a stock market that has helped to catalyze the further expansion of these high-tech start-ups, the NASDAQ. At the end of 1997, NASDAQ listed 5,466 companies, worth $1.9 trillion. The closest equivalent in Japan is the Over-the-Counter (OTC) market managed by JASDAQ. This market has also grown steadily in recent years, with the number of initial public offerings (IPOs) in the range of 100 per year, increasing eight-fold between 1982 and 1997. However, the trading volume and capitalization are far more modest, with only 1% the trading volume of the NASDAQ in 1997. In June 1998, there were about 855 companies listed for trading on the OTC, with a capitalization of ¥9.62 trillion, about 5% the capitalization of NASDAQ.

The JASDAQ continues to face challenges in fostering high technology growth because of the conservative character of the market and of investors. This challenge has been aggravated in recent years by a shift of investment away from these smaller firms, leaving overall valuations at very low levels. In 1997, the average price-to-earnings ratio of firms on the JASDAQ was under 15. In

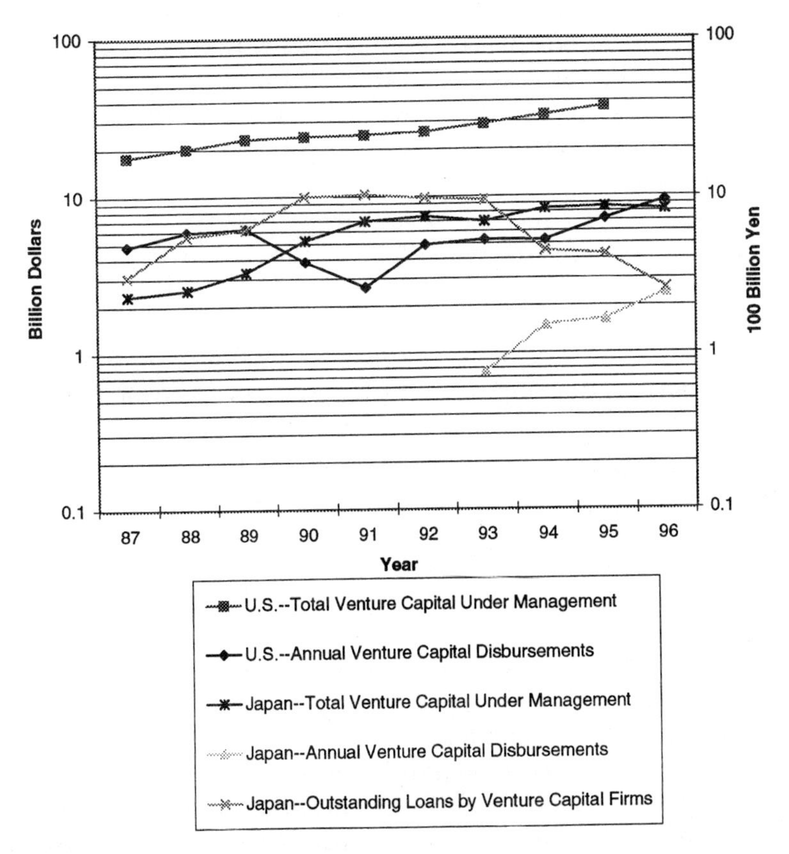

Figure 2 Comparison of Venture Capital Investment Trends in the United States and Japan (Ono 1997; Arakawa 1992; NSB 1998). The left scale is in units of $1 billion for funding in the United States. The left scale is in units of ¥100 billion for funding in Japan.

addition, the post-floatation performance of companies that register on the OTC has tended to decline, resulting in a declining stock price after the public offering.

In 1996, 110 companies registered with JASDAQ to trade their stock on the OTC. Although there are some recent cases of companies going public quickly, such as Yahoo Japan, which went public in a record 21 months, and DigiCube, which will issue an IPO 29 months after founding, these are still very much the exceptions. The average age of these IPO firms was 29 years, with a range of 5 years to going public to over 60 years. Regarding the investing venture capital firms, the *Nihon Keizai Shimbun* observed in 1996 that "Most are affiliates of financial institutions and are notorious for investing only in companies that are on the verge of going public" (*The Nikkei Weekly* 1996).

The distribution of funds by venture capital firms in Japan further reflects their conservative approach. In 1992, 46% of the annual income of venture capital firms derived from interest on loans, 22% from capital gains from publicly traded stocks, and 17% from stock dividends. This emphasis on loans has been reduced in recent years, as shown in Figure 2, but the tendency has nonetheless reinforced a conservative profile for these venture capital firms (Ono 1995).

One of the difficulties that smaller high-technology start-ups have with registration on the JASDAQ is meeting the strict registration requirements for listing on the OTC market: over ¥20 million in current profit and over ¥200 million in net assets. In practice, the norm is for firms to be significantly larger than this with a mean current profit of ¥1.8 billion and mean net assets of ¥6.9 billion. This has closed the OTC market to small start-up firms that might be losing money, as is the case with many of the Internet-related firms today in the United States.

In 1995, a second OTC was founded to stimulate IPOs among younger high-technology ventures by allowing the firms to value their research as part of their assets. This would allow firms still operating at a loss to register for public trading. However, only three firms had registered on this market in the first 18 months of its operation.

Post-IPO performance of firms in the 1990s has also been discouraging. Analysis of newly registered JASDAQ companies between 1990 and 1994 revealed that net profits peaked one year prior to flotation and subsequently decreased (Kutsuna and Cowling 1995). In addition, the average 5-year growth rates for firms registering in 1989/90 was lower than on the Tokyo Stock Exchange over this same period: –9.5% versus –8.3%. This trend is reflected in the movement of the stock prices of these firms. Kutsuna observes that "Broadly speaking a consistent pattern emerges which can be characterized as high initial return, decreasing but positive return, then ultimately negative returns" (Kutsuna and Cowling 1995).

Whereas initial returns are in the range of 15% to 20%, reflecting an over-subscription for shares at the time of offering, six months later prices have fallen to a range of 1% to –17%.[16] A March 1995 report of the Nippon Life Insurance company found that the volume of trading thinned out quickly as a consequence of this decline. The average trading volume of a company's share fell to 16% of the initial volume by the second month following the IPO and was only 8% of the initial volume by the sixth month. Because of the falling stock price after an IPO, it becomes difficult for a company subsequently to raise funds by issuing new shares at the market price (*Nikkan Kogyo Shimbun* 1996).

In addition, there has been a different type of relationship in Japan between venture capital firms and the firms in which they invest. In the United States, venture capital firms bring value not only in funds but in management expertise, with the latter being at least as important as the former. In Japan, the rise of venture capital firms since the late 1980s was due primarily to the financial institutions finding themselves flush with cash and in need of additional routes of investment. Although the investment targets changed, the investment philosophy and practices did not necessarily do so. In Japan, only 25% of venture capital firms have consulting groups to assist enterprises with management. Further, until August 1995, Japanese laws prohibited Board seats for venture capital investors, which impeded their ability to gain leverage in the management of the firms.

Despite the challenge, stimulating this market continues to be a priority. In June 1998, the Japan Securities Dealers Association announced that it would divide the OTC market and implement a market solely for blue-chip venture companies. The goal of this division is to enhance the appeal of this exchange for investors. The challenges of bringing capital to high-tech ventures will be a continuing and critical issue for innovation in Japan.

Getting There—the Trade-offs

One of the greater challenges to realizing the full potential of university-industry partnerships is in managing the different incentives of these two sectors. In both countries, the trend toward cooperation has begun to raise concern about its impact on the trade-offs between the public goals of education and knowledge generation and the private goal of commercial gain. Harvey Brooks has observed,

> The role of universities in the total research system of a country and of the world lies in the fact that they are uniquely the locus of both the generation and diffusion of publicly sharable knowledge, whereas industry naturally tends to be the locus of proprietary and competitive research. ... The Universities have few competitive advantages in the conduct of proprietary research, and if they did they would not be universities as we have known them. While cooperation between universities and industry is highly desirable and beneficial to both sides up to a point, it needs to be restrained by the fundamental notion of public knowledge which lies at the basis of science as a social system. (Brooks 1986)

Three tensions that arise due to the different roles and incentives of universities and industry are in the dissemination of information, the nature of research, and access to the research system by newcomers.

Altering the Dissemination of Information?

In basic science, research collaboration has been increasingly common, as it can widen the expertise applied to complex problems. Such collaboration is often a net benefit for all parties as well as for the public in general. However, it is not clear that the same is true for industry-university collaborations. On the one hand, there are data that show industry collaborations with academia can increase the diffusion of research results; on the other, there is evidence that some forms of publication are delayed or suppressed and that the nature of academic research may be affected.

Data in the United States show a rising level of collaborative publication among university and industrial researchers (NSB 1998). In 1981, 21.6% of industrial articles were co-authored with academia across all fields of science and engineering; by 1995 this had increased to 40.8%. A rise in collaborative publications occurred across a wide range of disciplines, with the largest percentage of increases in engineering and technology, 106% (15.6% to

32.2%), and physics, 101% (19.9% to 39.9%), and the highest level of collaboration in mathematics, 56.8% in 1995. Industry also increased its citations to academic research in scientific and technical articles, from 46% in 1990 to 50% in 1995.

A similar trend is seen in Japan. Shinichi Kobayashi (Kobayashi 1998a) has gathered publications data in the electrical equipment sector and the machinery sector which show substantial increases in the number of articles from industry in the principal society journals. Publications by industry in the electrical equipment sector increased from 24.9% of the total in 1980 to 44.5% in 1995; in the machinery sector they increased from 31.9% of the total in 1975 to 40.0% in 1995. In both sectors by far the largest of the increase was due to increased collaborations with universities. University-industry publications accounted for 81% of the growth in the electrical equipment sector and 67% in the machinery sector.

The effect of collaborative research on dissemination has been perhaps best studied in Europe, where Narin and Whitlow (1990) and Katz and Hicks (1997) have shown that collaborative papers in all categories are more highly cited than those that are not. An apparent reason for the high level of interest in industry-university collaborations is that they are often in "hot" areas of research or areas in which there is expanding potential for research to be applied.

However, even though publications stemming from university-industry research are more widely disseminated, there is concern in both nations regarding the impact of these growing collaborations on the openness of university research. A 1996 survey of life-science companies cooperating with universities revealed that 82% asked universities to keep information confidential until patents are filed; almost half (47%) asked that information be kept confidential longer than needed for patents to be filed. In another study in 1995, almost 80% of technology managers and over 50% of faculty with experience in university-industry cooperation reported occasions when firms asked for results to be delayed or kept from publication. In addition, a 1994 nationwide survey of University-Industry Research Centers showed that 53% of respondents reported that there are policies that might permit delayed publication, with 35% permitting the deletion of information prior to submission for publication (NSB 1998). Some universities have strict policies to guide these collaborations. MIT, for example, has guidelines that prohibit research results to be withheld from publication for more than 30 to 90 days. However the situation appears highly variable from university to university.

If the policies simply delay publication, then the result may be only a slowing of the advance of knowledge, as publication would be forthcoming. However, to the extent that these restrictions reduce the incentives to publish, either from proprietary concerns or lack of time, the penalty to the public is greater. Further study is needed of the trade-off between economic gain and the penalties that arise from altering the timeliness and form of the release of information.

Changing the Nature of Research?

A second area of concern is the effect of collaboration with industry on the nature of university research. As the interests of industry are typically applied to specific problems, there is the question for both nations about how greater pressures for applied research will affect fundamental inquiry. Also, there is a question of whether conflicts between industrial interests and the purer pursuit of knowledge might bias research results and negatively affect society.

One example of successful collaboration that is often considered a model of how this should occur in Japan is the case of high temperature superconductivity. In 1986, a major breakthrough in physics and materials science occurred when scientists in Japan and the United States independently confirmed superconducting properties in a new class of materials and at significantly higher temperatures than previously observed. The announcements of this phenomenon by Professors Shoji Tanaka, of the University of Tokyo, and Prof. Paul Chu, of the University of Alabama, at the Materials Research Society winter conference in 1986 ignited a flurry of research worldwide. Although the traditional superconductivity industry in Japan was among the world's most advanced, few firms were conducting any work in this area. Rather than corporate laboratories, it was the universities and national laboratories in Japan that were considered to be in the lead. Of the collaborative patent applications submitted in the two years following the discovery, university professors were cited as collaborators in 58% of the cases, with the government laboratories accounting for another 12% (Hane 1992).

In this case, although cooperation played a key role in diffusing knowledge to the broader economy, it was independent laboratory research that enabled the discovery. It is an example of "creativity to exploit cooperation" rather than "creativity through cooperation." It is the type of division in roles, however, that may be difficult to sustain if university research more directly serves the needs of industry.

In the United States, Henderson, Jaffe, and Trachtenberg (1995) found evidence of a change in the types of academic patents submitted in a shift toward the production of less widely cited patents over a 23-year period. Looking at university patenting behavior between 1965 and 1988, the research team found that university patents increased 15-fold while real university research spending almost tripled. They also found that up until the mid-1980s, university patents were more highly cited, and were cited by more technologically diverse patents, than a random sample of all patents. This was consistent with the notion that university patents are more widely used and more basic than the average patent. However the difference between the two groups disappeared in the middle part of the 1980s, with an increase in patents that received zero citations. Based on these data, the study concluded that the rate of increase of important patents from universities is now much less than the overall rate of increase of university patenting in the period covered.

Addressing the same concern, in 1993 the Industrial Research Institute (IRI), whose 260 member companies represent more than 80% of the nation's privately funded R&D in the United States, issued a position statement on industry-university interactions:

> Universities are more receptive to and interested in collaborating with industry at this time. However, academic research should focus its efforts on the long-term, fundamental needs of the United States in science and engineering, with input on those needs from private industry, government and other sectors. ... The value to be gained from interaction with

the academic research enterprise, namely insights, contacts, and early access to new information in science and technology, should be the driving force of industry's collaboration. (Roessner 1996)

In Japan, data are still emerging. Analyses of numerous industrial consortia of the 1980s have shown that in almost all cases the technical targets are incremental process improvements, not breakthrough advances (Hane 1992; Sigurdson 1986). Even though some programs were planned as 10-year R&D efforts, the technical targets were often known to be achievable at the outset. Even with a "long-term" effort, industrial needs press downward on the time horizons of collaborative projects.

In both the United States and Japan, guidance regarding these research relationships tends to vary from university to university. At times it is unclear whether there is a well-defined separation of university and industrial interests. For example, in February 1998, Keio University announced that it will participate in Japan Digital Contents, a multimedia venture to be set up by five private companies and Keio (*The Nikkei Weekly* 1998). The Chairman will be Keio Professor Hideo Aiso and the company will use graduate students from Keio. The financial pressures are significant, as this company has set a goal of creating ¥10 billion ($80 million) worth of intellectual property within five years. It will also manage a ¥1.5 billion Tokyo Multimedia Fund and serve as sole sales agent for products or services in return for its financial support. The impact of this type of venture on the traditional role of the university as a source of fundamental research and public knowledge is a growing question.

In the United States, there continues to be anecdotal evidence that industrial interests in university-industry partnerships lead to problems that include biased information and the withholding of information of value to society. For example, in 1996 two academic scientists criticized an article in *The New England Journal of Medicine* that noted a potentially severe side effect of the drug Fen-Phen. The critics failed to disclose that they were paid consultants to companies that distributed similar drugs. A study of the authors of 800 scientific papers found that in a third of the cases, at least one of the chief authors had a financial interest connected to the research (Blumenstyk 1998).

Reflecting on this issue, Hisashi Kobayashi, former Dean of Engineering at Princeton University, observed: "These subjects are by no means simple. Many U.S. universities are rather loose in allowing their faculty members to do consulting and/or run companies and face serious problems in maintaining a proper balance between their commitment to teaching and their involvement in business-oriented research" (Kobayashi 1996).

Limiting Access to University Research?

A related area of concern is the effect of collaboration on access to university research, particularly by newcomers. As collaborative projects often have their genesis in personal networks, one issue of equity is their openness to new partners. To the extent that cooperation is facilitated by procedures or offices that make access more transparent, access is likely to improve; but to the extent that collaboration is directed through project-specific efforts that draw on traditional networks or employ nontransparent processes, the effect of the growth of collaboration on access is unclear.

One measure of accessibility to newcomers is the level of international participation. Although data on this issue are sparse in both nations, this may be an area that warrants continued policy attention. Data from the NSF Engineering Research Center programs indicate a moderate level of international participation. Program directors estimate that approximately 50 of the 700 industry partners in the Engineering Research Centers are non-U.S. firms, approximately 7%. Non-U.S. firms are most strongly represented in the areas of biotechnology and bioengineering, information technologies and telecommunications, optoelectronics, and waste recycling.[17]

In Japan, project-oriented and group-based approaches to encourage collaboration appear to have a low level of international participation. For example, foreign participation has been very low in the Ministry of Education's Joint Research Program, which has been the principal vehicle for formally organizing cooperation between national universities and industry. In 1994, there were 5 foreign partners (7 of which were the same institute in Korea) in 1,488 projects; in 1995, 2 foreign partners in 1,704 projects; and in

1996, 3 foreign partners in 2,001 projects. A second example is the reemerging university-industry cooperation support of the JSPS University-Industry Research Cooperation Committee. These committees tend to organize around established networks of professors and industrialists, much as they did in their earlier years. In recent years, the committees have averaged 30 to 60 industrial partners. However, foreign firms typically account for only 0 to 3 of the representatives.

Over time, there has also been a trend toward more exclusive relationships within the Joint Research Program in Japan. Data gathered by Kobayashi show that whereas at the outset of this program, partnerships would involve multiple universities, over time this has shifted to a predominant reliance on one-on-one arrangements (Kobayashi 1998b). In 1983, more than half of the initial collaborations involved two or more universities, with two-thirds of those involving four or more universities in one partnership. This pattern continued through the mid-1980s, but the number of one-on-one cases continued to rise as the multiple partner arrangements decreased. By 1991, partnerships involving four or more universities had virtually disappeared, with one-on-one partnerships accounting for well over 90% of the total.

Another dimension that affects access is the manner in which the patents are assigned. The 1978 patent law in Japan established a system of Invention Review Committees, *hastumei iinkai*, to which patent applications are submitted to determine whether a patent should be assigned to the professor or to the government. However, these committees are often composed of professors from across the university, operate without staff funding, and typically do not have the services of legal advisors. The committees also do not provide for public participation and are not under a prescribed time-line for review. In the United States, guidelines are typically established in advance by the university.

This post-hoc review system in Japan makes it impossible for an industrial partner to know clearly, in advance, how rights may be allocated for an invention developed with its funding. For collaborators new to the system and without long-established relations, such as foreign firms, this leaves a substantial uncertainty and disincentive to cooperative work. One U.S. biotech firm seeking cooperative research with certain university professors in Japan noted that it found it too risky to fund research at a university in Japan with the ownership of intellectual property left undefined, subject to the decisions of a group of unrelated professors and administrators.

Elsewhere in this volume, Kneller argues that a common method used by firms in Japan is to provide a formally untied grant or endowment, *Shogaku Kifukin*, with an informal understanding that the benefits of research would accrue to the donor. This approach would, however, bear even greater uncertainty for newcomers.

Although both Japan and the United States have instituted policies to encourage an open public research infrastructure, there is concern in both countries about the impact on this openness of collaboration with industry. Overall, in Japan there continues to be an unusually low level of foreign participation in the industrial R&D enterprise.

Although the presence of foreign company R&D is rising in most advanced nations, reaching 17% of total industrial R&D in the United States in 1996, it continues to be on the order of 1% to 2% in Japan. Measures that would further limit access would penalize both Japan and the international community.

Conclusions

In light of recent policy initiatives, it might seem that the role of university-industry partnerships in innovation is converging in the United States and Japan. However, these relationships have evolved from different histories, are shaped by different norms and practices, and are set in environments for innovation with different incentives. The differences are particularly apparent when contrasting the relationship of universities with entrepreneurial, high-tech start-up companies.

An historic reluctance of universities in Japan to engage in institutional relationships with industry has shaped a system that relies more heavily on networks of individual professors and informal transfers of technology. The extent of the informal transfers is unclear, but the low transactions costs have made this a common practice. In the United States, the lack of such an historical reluctance combined with the convenience of universities as institutions to facilitate transactions have led to a more transparent infrastructure to support linkages with industry.

Important efforts are underway to open more avenues for university-industry linkages in Japan, but the shadow of institutional history remains and is reinforced by the environment for innovation. Incentives affecting the mobility of personnel favor a vertical flow of innovation in Japan and a more horizontal flow in the United States. In addition, a finance system that has grown up with large corporate leaders in Japan still bears this bias. The underavailability of smart capital in Japan may prove to be one of the more significant handicaps to catalyzing new industries. The administration of partnerships can be improved through policy measures in both countries, but it is the environment that constrains change, and reshaping the environment in which innovation occurs is a much taller order.

Considering the policies, incentives, and constraints surrounding university-industry linkages, it is expected that the two countries will continue to exhibit different strengths. The U.S. system will continue to show relative strength in facilitating the flow of resources to opportunities that arise from turns and discontinuities across a less predictable horizon of innovation. In Japan, relative strengths are in networks nested in well-defined channels of resource flows that contribute to innovation through the vertical movement of evolutionary research rather than as seedbeds for new industries.

The pursuit of creativity as a goal of cooperative policies is seen to be as elusive as it is desirable. The continued importance of this goal was captured in the recent statement of a Japanese computer company executive who commented that they had spent decades focussed on catching up to IBM, and when they did, they found that it was irrelevant. To better achieve creativity, it may be more important to distance university research from industry than to promote partnerships; to adopt a strategy of using cooperation to exploit creativity rather than cooperation to create creativity. To the extent that "creativity" is a political goal rather than a technical one, however, it may serve the purpose of coalescing support around a target that may be eternal. The pursuit of creativity may represent the great success of an unfulfilled promise.

Finally, both nations face the need to pay close attention to the trade-offs between the university's role in society to expand and disseminate knowledge versus the commercial desire for appropriation. In both countries, this balance is being struck case-by-case, and the lack of data makes it difficult to gauge the effect on innovation. Both nations also face the need to continue to assess the impact of these collaborations on the openness of their systems and on the nature of research. It is still unclear how the future will evolve, if it will advance to systems of greater openness and creativity, or systems in which university-industry consortia serve further to limit rather than expand the contribution to advanced research. For the innovation systems of both countries, this is a serious challenge.

Acknowledgments

The author expresses his appreciation to Professor Lewis Branscomb and Professor Fumio Kodama for their guidance and support, and to Mananobu Miyahara of the National Science Foundation's Tokyo Office for his detailed insights regarding changes occurring in Japan's policies on university-industry linkages.

Notes

1. This relationship is more thoroughly explored by Takehiko Hashimoto in chapter 9 in this volume.

2. The Ministry of Education supported a few postwar cooperative R&D projects such as the use of rice-bran oil, geothermal energy development, mass production of precision watches, and manufacture and clinical application of penicillin, but after three years support ceased.

3. Projects could nonetheless take on substantial scale, as in the Mesoscopic Electronics Project, which enlisted 10 major Japanese electronics firms, contributing $5 million over the five year term 1989–1994 (Arakawa 1997).

4. Between 1994 and 1997, the budget for the JSPS grew almost four-fold, from ¥11.3 billion to ¥39.5 billion, with most of that growth due to the Research for the Future Program. In this rapid growth environment, substantial expansion is also expected in R&D support related to the Research Cooperation Committees.

5. The program identified priority research themes that were clearly related to industrial interests: materials processing technology, biotechnology, electronics and information technology, medical and welfare equipment technology, hu-

man life engineering technology, and resources technology. In 1995, MITI received nearly 3,000 proposals, of which 109 were funded (NSF 1996a).

6. This STA program addresses somewhat longer-term research themes than the MITI program, such as brain functions, genetic programming, the immune mechanism, quantum effects, single molecule atomic reactions, and the phenomena of extreme conditions. Beginning in 1997, the CREST program also included industrial as well as university researchers. In 1995, from the 1,350 proposals received, 54 projects were funded (NSF 1996b).

7. In 1987, the Ministry of Education established the Cooperative Research Centers Program, modeled after the NSF University-Industry Cooperative Research Centers. By 1997, 49 such centers had been established. Until 1995, the centers were hampered by a lack of authority to accept funds directly from industry for commissioned research. Industrial support was limited to forms such as equipment, personnel, or endowing grants and contributions. With those rules now changed, there may be a rise in the prominence of these centers.

8. In 1996, the Ministry established a program to promote the formation of Venture Business Laboratories. Each of these laboratories has a primary theme, such as Advanced Electronic Materials, High Functionality Nanostructure Materials, Knowledge-based Multimedia, and Photonic Materials. By the end of 1997, Venture Business Laboratories had been established at 21 national universities. Although these are state-of-the-art laboratory facilities, there is typically little connection to venture business organizations.

9. An extensive discussion by Robert Kneller of intellectual property treatment in Japan can be found in chapter 12 in this volume.

10. No U.S. university approaches the level of royalty income seen by major U.S. firms. For example, IBM collected royalties of about $1 billion in 1996, nearly 20 times the income of the large University of California system.

11. Exceptions are work conducted with special government research equipment or on an applied, problem-oriented grant, in which cases the government would gain patent rights. This law emerged from recommendations of the Council for Science and Technology (June 1977) and was operationalized through a guidance published in March 1978.

12. Data from the NSF Tokyo Office.

13. Interview with Kyoto University President Nagao, January 22, 1998.

14. Data from the Ministry of International Trade and Industry, personal communication.

15. Unpublished data from the National Science Foundation, Division of Science Resource Studies. 1998.

16. One factor causing this fall is the use of bidding by the lead underwriter to set the stock price. To purchase stock at the time of offering, the lead underwriter solicits bids for the stock, which in 1995 were 20% higher than the stock value of similar companies, thereby creating pressure for a fall in price after its offering (*Asahi Shimbun,* January 14, 1996).

17. In addition, foreign student participation is surely high when one considers that approximately 50% of U.S. doctoral degrees in engineering are being granted to non-U.S. citizens.

References

Advisory Committee on Federal Policy on Industrial Innovation. 1979. *Report of Industrial Advisory Subcommittee on Direct Federal Support of Research and Development.* Washington, DC: U.S. Department of Commerce.

Anderson, Alun. 1984. *Science and Technology in Japan.* Essex, UK: Longman Publishers.

Arakawa, Hideo. 1992. "The Promotion of Venture Business and Venture Capital Industry." In James Borton, ed. *Venture Japan.* New York: Woodhead-Faulkner.

Arakawa, Yasuhiko. 1997. "Large-Scale University-Industry Cooperative Research with an International Outlook," *Gakugitsu Geppo,* 50 (7) (in Japanese).

Asahi Shimbun, January 14, 1996.

Association of University Technology Managers (AUTM). 1996. *Licensing Survey FY 1991–1995.* Norwalk, CT: AUTM.

Association of University Technology Managers (AUTM). 1997. *Licensing Survey, FY 1996.* Norwalk, CT: AUTM.

Blumenstyk, Goldie. 1998. "Conflict-of-Interest Fears Rise as Universities Chase Industry Support." *The Chronicle of Higher Education* (May 22).

Brooks, Harvey. 1986. "The Research University: Centerpiece of Science Policy." Working Paper Series of the College of Business, Ohio State University, WPS 86-120.

Bureau of Labor Statistics (BLS) and Bureau of the Census. 1997. "Employee Tenure in the Mid-1990s." USDL 97-25. Washington, DC: U.S. Census Bureau.

CHI Research. 1997. "Industry Technology Has Strong Roots in Public Science." *CHI Research Newsletter* V:1. Haddon Heights, NJ: CHI Research Incorporated.

Cohen, Wesley, Richard Florida, and W. Richard Goe. 1994. *University-Industry Research Centers in the United States.* Pittsburgh: Carnegie Mellon University.

Cohen, Wesley, Richard Florida, Lucien Randazzese, and John Walsh. 1998. "Industry and the Academy: Uneasy Partners in the Cause of Technological Advance." In *Challenges to Research Universities,* ed. Roger G. Noll. Washington, DC: Brookings Institution Press.

Fujisue, Kenzo. 1998. "Promotion of academia-industry cooperation in Japan—establishing the law of promoting technology transfer from university to industry in Japan." *Technovation.* 18(6–7) (June–July).

Gray, Denis and S. George Walters. 1997. *Managing the University/Industry Cooperative Research Center: A Handbook for Center Directors.* Arlington VA: National Science Foundation.

Hane, Gerald. 1992. *The Role of Research and Development Consortia in Innovation in Japan.* Ph.D. Dissertation. Cambridge, MA: Harvard University.

Hashimoto, Masahiro. 1998. "Desirable Form of Academia-Industry Cooperation." *Journal of Japanese Trade and Industry* No. 2.

Henderson, Rebecca, Adam Jaffe, Manuel Trajtenberg. 1995. "Universities as a Source of Commercial Technology: A Detailed Analysis of University Patenting 1965–1988." NBER Working Paper No. 5068.

Hicks, Diana. 1993. "University-industry research links in Japan." *Policy Sciences* No. 26.

Hiroshige, Tetsu. 1974. "Social Conditions for Prewar Japanese Research in Nuclear Physics." In *Science and Society in Modern Japan,* ed. Shigeru Nakayama et al. Cambridge, MA: MIT Press.

Inose, Hiroshi, Tetsuji Nishikawa, Michiyuki Uenohara. 1982. "Cooperation Between Universities and Industries in Basic and Applied Science." In *Science Policy Perspectives: USA-Japan,* ed. Arthur Gerstenfeld. New York, NY: Academic Press.

Ishii, Masamichi, T. Yokoo, Y. Hirano. 1993. *Comparative Study on Career Distribution of Engineering Graduates in Japan and the U.S.* Tokyo: National Institute of Science and Technology Policy, Report No. 25.

Itakura, Kyonobu and Yagi Eri. 1974. "The Japanese Research System and the Establishment of the Institute of Physical and Chemical Research." In *Science and Society in Modern Japan,* eds. Shigeru Nakayama et al. Cambridge, MA: MIT Press.

Japan Society for the Promotion of Science (JSPS). 1997. "*Nihon Gakujitsu Shinkokai no Sangaku Kyoryoku Jigyo* (University-Industry Cooperation and the JSPS)." *Gakujitsu Geppo* 50(7). (in Japanese)

Johnson, Chalmers. 1982. *MITI and the Japanese Miracle.* Stanford: Stanford University Press.

Katz, H.S. and Diana Hicks. 1997. "How Much is a Collaboration Worth? A Calibrated Bibliometric Model," *Scientometrics* 40:3.

Kobayashi, Hisashi. 1996. "Japanese and U.S. Universities: Their Issues and Challenges." Presented at Strategies for Academia-Government-Industry: Partnerships Towards the 21st Century-Science and Industry in the Network Society. Nara, Japan: Nara Institute of Science and Technology.

Kobayashi, Shinichi. 1998a. "Policy and Changes in the System of Knowledge Generation." In *Koutou Kyouiku Kenkyu Kiyou* (March). Tokyo: *Koutou Kyouiku Kenkyujo.* (in Japanese)

Kobayashi, Shinichi. 1998b. "A New Level in University-Industry Relations." In *Koutou Kyuouiku Kenkyu Kiyou* (March). Tokyo: *Koutou Kyouiku Kenkyujo.* (in Japanese)

Kodama, Toshihiro. 1998. "The Venturing Vitalities of Small and Large Companies." *Look Japan* (April).

Kutsuna, Kenji and Mark Cowling. 1995. "Price Formation on Japanese OTC Markets." *Securities Analyst Journal* (in Japanese).

Kutsuna, Kenji and Mark Cowling. 1998. "The Performance of JASDAQ Companies and Venture Capital Investment: Before and After Floatation." The Center for Small and Medium Sized Enterprises, Warwick Business School. Working Paper No. 55.

Lewis, Michael. 1998. "The Little Creepy Crawlers Who Will Eat You in the Night," *The New York Times Magazine,* March 4.

Ministry of International Trade and Industry (MITI). 1997. *Sangaku no renkei kyoryoku no arikata ni kansuru chosa kenkyu kyoryokusha kaigi (Chukan matome)* (Summary of a Conference on the Survey of the Promotion of University-Industry Cooperation—Mid-term Report). (December 12). (in Japanese)

Monbusho. 1996. *Research Cooperation Between Universities and Industry in Japan.* Tokyo: Ministry of Education.

Narin, Frances and E.S. Whitlow. 1990. *Measurement of Scientific Co-operation and Coauthorship in CED-related Areas of Science.* Luxembourg: Office for Official Publications in the European Communities.

National Institute for Research Advancement (NIRA). 1997. *A Study on How Venture Business Should be Supported.* NIRA Research Report 970100. Tokyo: NIRA.

National Science Board (NSB). 1998. *Science and Engineering Indicators.* Washington, DC: Government Printing Office.

National Science Foundation (NSF). 1995. "Japanese Engineers—Profile and Environment. Report Memorandum 95-6." Tokyo: NSF Japan Office.

National Science Foundation (NSF). 1996a. "Innovative Industrial Technology R&D Promotion Program. Report Memorandum 96-14." Tokyo: NSF Japan Office.

National Science Foundation (NSF). 1996b. "Core Research for Evolutionary Science and Technology (CREST)." Report Memorandum 96-15. Tokyo: NSF Japan Office.

Nelsen, Lita. 1996. "University/Industry Partnerships As a Source of New Technology for Industry." Conference on Strategies for Academia-Government-Industry Partnerships Towards the 21st Century-Science and Industry in the Network Society. Nara, Japan: Nara Institute of Science and Technology.

Nihon Keizai Shimbun. June 30, 1998. "*Benchya Kyapitaru Senbetsu Toshi Tsuyomaru* (Strengthening Investment in Venture Capital)." (in Japanese)

The Nikkei Weekly, May 27, 1996.

The Nikkei Weekly. 1998. "University joins companies in multimedia-content venture." (February 16).

Nikkan Kogyo Shimbun, January 19, 1996.

Okimoto, Daniel et.al. 1984. *Competitive Edge. The Semiconductor Industry in the U.S. and Japan.* Stanford: Stanford University Press.

Ono, Masato. 1995. "Venture Capital in Japan, Current Overview." Unpublished paper.

Ono, Masato. 1997. "*Benchya Kigyo to Toshi no Jissai Chishiki* (Venture Firms and Practical Investment Knowledge)." Tokyo: *Toyo Keizai Shimbunsha*, 1997. (in Japanese)

President's Commission on Industrial Competitiveness. 1985. *Global Competition, The New Reality*. Washington, DC: U.S. Government Printing Office. Vol. 1.

Pressman, Lori et al. 1995. "Pre-Production Investment and Jobs Induced by MIT Exclusive Patent Licenses: A Preliminary Model to Measure the Economic Impact of University Licensing." Cambridge, MA: MIT Technology Licensing Office.

Rausch, Lawrence. 1998. "Venture Capital Investment Trends in the United States and Europe." National Science Foundation Issue Brief, NSF 99-303. Arlington, VA: National Science Foundation.

Roessner, David. 1996. "Choose the Right Metric!." AAAS Forum on University-Industry Collaborations. Washington, DC: American Association for the Advancement of Science.

Sakamoto, Kazuichi. 1997. "*Ritsumeikan Daigaku ni okeru Sankangaku Koryu no Torikumi* (Methods of University-Industry Interchange at Ritsumeikan University)" *Gakujitsu Geppo* 50 (7) (July). (in Japanese).

Sangaku Kaigi. 1998. Sangaku no Renkei Kyoryoku no Suishin ni Kansuru Chosakenkyu Kyorokusha Kaigi (Sangaku Kaigi) (Council for the Promotion of Cooperative Relations between Universities and Industry (University-Industry Council). Summary Report (March).

Sato, Toshio. 1996. "Scientific Collaboration between University and Industry in Mechanical Manufacturing in Japan." Unpublished paper.

Schrieberg, David. 1998. "The Matchmakers," *Stanford Magazine* (Jan./Feb.).

Science and Technology Agency (STA). 1997. *White Paper on Science and Technology, 1997*. Tokyo: Science and Technology Agency.

Sigurdson, Jon. 1986. "Industry and State Partnership in Japan—The Very Large Integrated Circuits (VLSI) Project." Discussion Paper No. 168. Lund, Sweden: Research Policy Institute, University of Lund.

Suematsu, Yasuharu. 1996. "The Changing Role of Universities in Japan." Presented at Strategies for Academia-Government-Industry Partnerships Towards the 21st Century-Science and Industry in the Network Society. Nara, Japan: Nara Institute of Science and Technology.

Tamai, Katsuya. 1996. "The Problem of Patents at National Universities." Presented at Strategies for Academia-Government-Industry Partnerships Towards the 21st Century-Science and Industry in the Network Society. Nara Institute of Science and Technology. (in Japanese)

Tanaka, Tsunechichi. 1997. "Creation of New Industries and Businesses, and Collaboration between Industry, Government, and Academia," *Japan Close-Up* (March).

U.S. Government Accounting Office (USGPO). 1998. "Technology Transfer, Administration of the Bayh-Dole Act by Research Universities," GAO/RCED-98-126. Washington, DC: U.S. Government Accounting Office.

Westney, D. Eleanor and Kiyonori Sakakibara. 1985a. "Comparative Study of the Training, Careers, and Organization of Engineers in the Computer Industry in Japan and the United States." M.I.T.-Japan Science and Technology Program" (September).

Westney, D. Eleanor and Kiyonori Sakakibara. 1985b. "The Role of Japan-Based R&D in Global Technology Strategy," *Technology in Society*, Vol. 7.

II

Assessing the Two Systems

Here "assessing" does not necessarily mean performing direct statistical comparisons, which could obscure real historical and socioeconomic similarities and differences between the two systems. We have taken account of factors underlying the data in each system, recognizing that assessing them is as much an issue of determining the process of assessment for each system as it is of actually comparing the metrics resulting from the processes. The authors in this volume were given the freedom to choose their own research topics rather than asked to do a mechanical comparison of the two countries. The result is that the assessments are not necessarily symmetric, but instead highlight the critical features specific to each system. The Japanese system receives a bit more attention here, due to the ample information available on the American academic system and the need for additional information on the university-industry linkage in Japan in terms of the flows of research funds and knowledge.

While the authors range widely on a variety of topics, all have one thing in common: The underlying current throughout is a focus on assessment as the process of identifying and measuring key features of the systems, not simply as a rote comparison of metrics.

3

The American University System as a Performer of Basic and Applied Research

Irwin Feller

Introduction

Viewing the American university system's contributions to technological innovation through a comparative United States-Japan framework has a bipolar effect: stepping back offers a refreshed and in places new perspective on how the structure and conduct of American universities affect their performance in basic and applied research and technology transfer, and thus their contributions to technological innovation; stepping forward reveals the fine grains in this larger perspective.

Four features of the American system of higher education emerge as particularly important in this viewing: decentralization, competition, regionalism, and the coupling of research and graduate education. Singly and collectively, these features have been cited by several scholars of higher education as significantly differentiating American colleges and universities from their counterparts in Europe and Japan (Clark 1995; Gumport 1993; Trow 1991). Recent work by economists on the role of universities in the American innovation system likewise has begun to emphasize one or more of these features (Goldin and Katz 1998; Mowery and Rosenberg 1993; Noll 1998; Rosenberg and Nelson 1994).

Extending this latter line of inquiry, this essay presents both a long-distance and a close-up view of the relationships between the above four characteristics of the American university system and processes of technological innovation. Its thesis is that these four features combine to produce three further traits that provide the institutional base for the substantive contributions of U.S. universities to technological innovation: geographical ubiquity, functional comprehensiveness, and flexibility in interorganizational relationships. Multiple fracture lines in this base, however, threaten the viability and productivity of the American university system's contributions to scientific research and technological innovation.

First, this article briefly recounts and provides some fine-grained contextual data on the four features; next, it relates these features and related traits to contemporary frameworks for analyzing processes of technological innovation and national science and technology policy. Finally, it offers a brief account of the historical processes by which these features were formed, and exegesis on their stability and policy manipulability.

The analysis performed here has another effect: It points anew to the limited perspective on the contributions to technological innovation made by American colleges and universities offered by the basic science paradigm in *Science: The Endless Frontier* (Bush 1960). Seen from the Olympian heights of national innovation systems, the distinguishing characteristic of the role of U.S. universities in the U.S. national innovation system is indeed the degree to which publicly funded basic research is performed within public and private universities. This reliance is widely held to have led to the worldwide preeminence of the U.S. scientific enterprise and system of graduate education.[1]

In 1997, U.S. universities performed $16.1 billion, or approximately 50%, of the nation's estimated $31.2 billion total expenditures on basic research and development (R&D), and oversaw an additional $2.8 billion spent by Federally Funded Research and Development Centers (FFRDCs) (National Science Foundation 1998a, Appendix Table 4-7: 125). The role of universities as performers of publicly-funded basic research is even more evident in their performance of $10.1 billion, or 57%, of the $17.7 billion in basic research supported by the federal government in 1997 (National Science Foundation 1998a, Appendix

Table 4-9: 127). Academic research also is clustered in basic research as contrasted with either applied research or development. In 1997, an estimated 67% of academic research was directed at basic research, 25% at applied research, and 8% at development.[2]

Data on the role of universities as performers of basic research, however, are only part of the story. As Rosenberg and Nelson (1994) have recounted, American universities have a long history of performing both problem-focused and applied research, and of creating new modes of knowledge, such as in computer science and artificial intelligence, where research does not easily fit within the conventional R&D categories.

Moreover, an emphasis on the distinctive role of universities as performers of basic research in America does not mean that the U.S. national innovation system is necessarily more efficient or productive than those in other countries; the research functions performed by universities in the United States are in fact performed by other organizations in other countries. Indeed, an overemphasis on the basic research to which universities contribute and on break-through innovations relative to diffusion and adoption strategies has been cited as a flaw of postwar U.S. science and technology policy (Ergas 1987; Mowery and Rosenberg 1993). Japanese and European models of technology transfer and government–industry R&D partnerships were frequently championed (until quite recently at least) as worthy of U.S. emulation in its efforts to rekindle the fires of its international technological competitiveness (Kelley and Brooks 1989; U.S. Congress, Office of Technology Assessment 1990).

Four Distinctive Features

Decentralization

Decentralization refers to the multiple loci of institutional control and sources of funding for American colleges and universities. American colleges involve a mix of public and private institutions and, historically within the public sector, a mix of state and local government initiatives, control, and

Table 1 Revenue Sources, Colleges and Universities, 1995 ($ millions).

Revenue Sources	All Universities (n=3,483)	Public (n=1,538)	Private (n=1,945)	Percentage Distribution All Universities	Public	Private
Tuition	50,570	21,190	29,380	27.0	18.2	41.6
Government appropriations						
Federal	907	689	218	0.5	0.6	0.3
State	38,030	37,797	233	20.3	32.4	0.3
Local	3,804	3,800	4	2.0	3.3	0.0
Government grants and contracts						
Federal	21,929	14,296	7,633	11.7	12.3	10.8
State	4,500	3,239	1,261	2.4	2.8	1.8
Local	858	452	406	0.5	0.4	0.6
Private gifts, grants, and contracts	10,609	4,494	6,115	5.7	3.9	8.7
Endowment income	3,874	577	3,297	2.1	0.5	4.7
Sales and services of educational activities	5,535	3,557	1,978	3.0	3.1	2.8
Auxiliary enterprises	18,025	11,069	6,956	9.6	9.5	9.9
Hospitals	18,579	12,058	6,521	9.9	10.4	9.2
Other	6,383	3,068	3,315	3.4	2.6	4.7
Independent operations	3,492	206	3,286	1.9	0.2	4.7
Total	187,095	116,492	70,603	100.0	100.0	100.0

Source: U.S. Department of Education, Integrated Post-Secondary Education Data System (IPEDS) 1995.

financing.[3] In comparative terms, decentralization relates to the lesser degree of control exercised either by the national government or by any level of government over the activities of colleges and universities in the United States as compared with Japan or Europe (Clark 1995).

Decentralization is reflected in the multiple, although not necessarily independent, sources of revenue from which American colleges and universities derive their revenues. Table 1 presents data on revenue sources for all universities and colleges by type of control. In 1995, 3,483 colleges and universities reported total revenues of $187 billion. Tuition was the major single source, comprising 27% of revenues for all universities and 18.2% and 41.6% of revenues for public and private universities, respectively. Next in order of importance was state government appropriations, which constituted 20.3% of revenue for all universities—32.4% of revenue for public universities, but only 0.3% for private universities. Federal government grants for all educational purposes ranked third.

Similar patterns of decentralized control and multiple funding patterns are also evident for research-intensive universities. Tables 2 and 3 present data on absolute and percentage distributions of revenue sources for universities, organized by Carnegie Classification of Higher Education (Carnegie Foundation for the Advancement of Teaching 1994)—a measure of the degree of research and graduate degree intensiveness and locus of control.

Decentralized funding is also evident in the funding of academic research. Although the postwar system of support for academic researchers is described largely in terms of a federal grants economy, the sources of funding of academic research have become more distributed over time. The share of academic R&D funded by the federal government has declined in a slow but cumulative manner, from a high of 73.5% in 1965 to an estimated 59.6% in 1997 (a level above the low point of 58.2% in 1991) (National Science Foundation 1998a, Appendix Table 5-2: 198) even as total federal expenditures for academic R&D have risen steadily, if intermittently. A similar decline, from 9.8% to 7.5%, also has occurred in state and local governments' share of funds for academic R&D. Industry's role as a source of funds for academic R&D has increased from 3.3% to 6.8% over the full period, 1976–1996, but remained relatively unchanged between 1993 and 1995. Institutional funds have risen from 12.0% of the total to 18.4% (see Table 4).

Decentralization of research funding has had several, at times opposing, effects on the selection of individual faculty research agendas, institutional research portfolios, and the collective

Table 2 Sources of Revenue, Research-intensive Universities, 1995 ($ millions)

Revenue Sources	Total (n=236)	Research I (n=88)		Research II (n=37)		Doctoral I (n=51)		Doctoral II (n=60)	
		Public (n=59)	Private (n=29)	Public (n=26)	Private (n=11)	Public (n=28)	Private (n=23)	Public (n=38)	Private (n=22)
Tuition	22,139	6,628	6,179	1,640	1,557	1,347	1,970	1,309	1,509
Government appropriations									
Federal	645	314	164	123	0	0	0	44	0
State	18,876	12,006	82	2,686	4	1,967	11	2,113	7
Local	60	52	0	4	0	0	0	4	0
Government grants and contracts									
Federal	14,034	6,878	4,441	855	336	362	218	690	254
State	1,677	792	224	189	49	101	67	170	85
Local	474	170	230	17	1	27	6	21	2
Private gifts, grants, and contracts	6,793	2,633	2,580	374	412	150	187	232	225
Endowment income	2,390	379	1,479	52	209	28	78	14	151
Sales and services of educational activities	4,105	1,940	1,468	227	100	96	63	155	56
Auxiliary enterprises	9,828	4,549	1,792	969	557	704	319	632	306
Hospitals	13,357	7,678	4,273	0	613	20	419	350	4
Other	3,344	1,018	1,051	196	222	172	83	144	458
Independent operations	3,109	92	2,945	3	12	0	8	3	46
Total	101,067	45,129	26,908	7,335	4,072	4,974	3,429	5,881	3,103

Source: U.S Department of Education, Integrated Post-Secondary Education Data System (IPEDS) 1995.

Table 3 Percentage Distribution, Sources of Revenue, Research Universities, 1995

Revenue Sources	Total (n=236)	Research I (n=88) Public (n=59)	Research I (n=88) Private (n=29)	Research II (n=37) Public (n=26)	Research II (n=37) Private (n=11)	Doctoral I (n=51) Public (n=28)	Doctoral I (n=51) Private (n=23)	Doctoral II (n=60) Public (n=38)	Doctoral II (n=60) Private (n=22)
Tuition	22.0	14.7	23.0	22.4	38.2	27.1	57.5	22.3	48.6
Government appropriations									
Federal	0.6	0.7	0.6	1.7	0.0	0.0	0.0	0.7	0.0
State	18.7	26.6	0.3	36.6	0.1	39.5	0.3	35.9	0.2
Local	0.1	0.1	0.0	0.1	0.0	0.0	0.0	0.0	0.0
Government grants and contracts									
Federal	14.0	15.2	16.5	11.7	8.3	7.3	6.4	11.7	8.2
State	1.7	1.8	0.8	2.6	1.2	2.0	2.0	2.9	2.7
Local	0.5	0.4	0.9	0.2	0.0	0.5	0.2	0.4	0.1
Private gifts, grants, and contracts	6.7	5.8	9.6	5.1	10.1	3.0	5.5	3.9	7.3
Endowment income	2.4	0.8	5.5	0.7	5.1	0.6	2.3	0.2	4.9
Sales and services of educational activities	4.1	4.3	5.5	3.1	2.5	1.9	1.8	2.6	1.8
Auxiliary enterprises	9.7	10.1	6.7	13.2	13.7	14.2	9.3	10.7	9.9
Hospitals	13.2	17.0	15.9	0.0	15.1	0.4	12.2	6.0	0.1
Other	3.3	2.3	3.9	2.7	5.5	3.5	2.4	2.4	14.8
Independent operations	3.1	0.2	10.9	0.0	0.3	0.0	0.2	0.1	1.5
Total	100.0	100.0	100.0	100.0	100.0	100.0	100.0	100.0	100.0

Source: U.S Department of Education, Integrated Post-Secondary Education Data System (IPEDS) 1995.

Table 4 R&D Expenditures at Universities and Colleges, by Source of Funds, in Millions of Constant Dollars: Fiscal Years 1976, 1986, and 1996.

Source	FY1976	FY1986	FY1996
Total	3,729	10,928	22,995
Federal government	2,512 (67.4%)	6,712 (61.4%)	13,810 (60.1%)
State and local governments	364 (9.8%)	915 (8.4%)	1,725 (7.5%)
Industry	123 (3.3%)	700 (6.4%)	1,576 (6.8%)
Institutional funds	446 (12.0%)	1,869 (17.1%)	4,232 (18.4%)
All other sources	285 (7.6%)	732 (6.7%)	1,653 (7.2%)

Source: National Science Foundation/SRS, Survey of Research and Development Expenditures at Universities and Colleges, Fiscal Year 1996 Tables B-1 and B-2.

university system's capabilities to pursue research, development, and technology transfer functions. First, revenue sources—tuition, endowment, foundations—independent of government or industrial sponsors have permitted universities a degree of freedom to pursue research trajectories not beholden to the political or ideological mission of governmental or industrial sponsors.[4] The vaunted pluralism of funding streams favors multiple performers and thus multiple research approaches not bound to dominant paradigms.

Decentralized funding also provides a revenue base for the entry of new researchers and new institutions; it thus is inextricably linked to the competitive character of the U.S. higher education system. Whatever the current emphasis on selective focus now visible in university budgeting and planning, universities have felt free for the most part to enter any and all fields of research, either seeking to open up new frontiers of knowledge or taking on established institutions for preeminence in a field. While considerations of minimally efficient scales of operation and constraints on duplication of degrees and programs have existed within state borders, decentralized authority and funding across states have permitted interstate competition in academic research to flourish.

This financial and statutory independence from a central authority has a downside, however. Decentralization has meant a lack of assured, stable funding for the core of the university's research activities. Tuition and state appropriations to higher education are targeted primarily at an institution's undergraduate teaching mission. With notable exceptions, public universities historically have struggled to gain state support for their generic research activities. Where state funds for research have been forthcoming, this research has largely been tied to the needs of state-based industries. In general the university's research function is supported by revenues obtained in competitive processes, derived from multiple sponsors and increasingly from its own discretionary revenues.

Historically, this dependence on external sources of research funds is held to have produced a tighter coupling of industrial and academic research (at least in that performed by public universities) in the United States than in other European countries.[5] In addition to industry or state dollars for academic research in selected areas, this connection has entailed development by the university of a research program and curriculum relevant to the technological and labor force needs of influential state economic sectors.

Competition

Competition is a defining trait of the American university system. In seeking to explain why America's system of graduate research universities is the "world's leading magnet" for study, with the international quality of its graduate universities standing in sharp contrast to its standing in elementary and secondary education, Clark (1995: 117) has commented that "While full answers are inordinately complex, historically and organizationally, explanation centers on the initiative exercised by a plurality of institutions in a uniquely competitive arena. ... It was out of the competitive interaction of institutions that there developed a lasting combination of the graduate school married to specialized departments, a form that has produced elite results in a context of mass higher education." Similarly, Rosovsky, again offering a comparison with other countries, has observed that

> American universities exist in the real world, where leads are challenged and sometimes forced to make room for—even be replaced by—newcomers. For us, the comforts of Oxford, Cambridge, the University of Tokyo, and the University of Paris do not exist. At all times there is a group of universities clawing their way up the ladder and others attempting to protect their position at the top. That a large proportion of the world's leading universities are located in the United States I have already in part attributed to the effect of interinstitutional rivalry. (Rosovsky 1990: 226)

This competition extends across the spectrum of inputs—faculty, students, facilities—and outputs—research findings, placement of students, and the various metrics by which the reputation and standing of institutions are measured, whether by the National Research Council or *The U.S. News and World Report.*

Central to this competition is that for research funds. Public and private research institutions compete with one another for federal and foundation grants; competition also extends to state government funds (Zumeta 1997), especially for technology-based economic development programs where private universities (for example, Lehigh University, Pennsylvania, and Rensselaer Polytechnic Institute, New York) have been highly visible participants.

Competition is evident in data on the number of research-intensive universities and the distribution of academic R&D expenditures. According to the Carnegie Classification, the number of doctorate-granting institutions increased from 173 in 1970 to 236 in 1994, an increase of 36% (Table 5).[6] The number of Research University I's increased from 52 to 88, or by 69%; within this classification, the number of Public Research Universities increased from 30 to 59, or by 97%. Other sources, while also containing indistinct boundaries (Noll, 1998) among types of universities and colleges, point to a substantial increase

Table 5 Number of Institutions, by Carnegie Classification

	1970			1976			1987			1994		
	Total	Pub.	Priv.	Total	Pub.	Priv.	Total	Pub.	Priv.	Total	Pub.	Priv.
Doctorate-granting institutions	173	109	64	184	119	65	213	134	79	236	151	85
Research Universities I	52	30	22	51	29	22	70	45	25	88	59	29
Research Universities II	40	27	13	47	33	14	34	26	8	37	26	11
Doctorate-granting Universities I	53	34	19	56	38	18	51	30	21	51	28	23
Doctorate-granting Universities II	28	18	10	30	19	11	58	33	25	60	38	22
Comprehensive universities and colleges[a]	456	309	147	594	354	240	595	331	264	529	275	254

[a]Category redefined as master's colleges and universities in 1994.
Source: *A Classification of Institutions of Higher Education,* 1994 Edition (New York: The Carnegie Foundation for the Advancement of Teaching), Tables XI–XIV.

in the number of "research-intensive" universities. For example, immediately following the Second World War, national studies called for a doubling of the number of research institutions from the then current estimate of 20–30; more recent federal reports speak variously of between 150 and 170 research-intensive universities.

The distribution of academic R&D expenditures among institutions also shows evidence of an increased number of universities in which the faculty and the institutions are capable of successfully vying for merit review (as well as for congressionally earmarked federal research funds) and for industrial support. Historically, academic R&D expenditures, measured in terms of either total academic R&D expenditures or federal academic R&D expenditures, have been concentrated in a relatively small number of institutions (and states). In FY1996, the leading 20 research institutions accounted for 34% of federally sponsored R&D and 31% of total academic R&D, while the top 100 institutions received 81% of federally financed spending and 80% of total academic R&D expenditures (National Science Foundation 1998b).

These figures, however, mask several underlying trends that point to intensified competition among universities (and states) for research funds. First, the distribution of total academic R&D and federally funded academic R&D has become somewhat more dispersed over time (Geiger and Feller 1995). Table 6 shows that the share of the top 10 in total academic R&D expenditures fell from 20.2% to 16.5% between 1980 and1997. In terms of federal R&D funds, the share received by the top 10 institutions fell from 19% to 16.6%, while over this period the share received by institutions outside of the top 100 increased from 19% to 22% (Jankowski 1998). Second, the total number of institutions receiving federal support has increased noticeably over time, rising from 555 in 1975, to 648 in 1985, to 882 in 1995 (National Science Foundation 1998a: 5–13). Third, noticeable advances toward national research prominence and sizeable research programs have been achieved by universities that were established or reconfigured in the postwar period specifically to take on research programs, such as the State University of New York-Stony Brook, and the University of Alabama-Bir-

Table 6 Cumulative Percentage Shares, Total Academic Research and Development Expenditures, Top 100 Institutions

Institutions	1980	1990	1997
First 10	20.2	18.0	16.5
20	34.0	31.9	29.3
30	45.2	42.4	39.6
40	53.9	50.8	48.0
50	60.9	58.0	54.9
60	67.1	64.0	60.7
70	72.2	69.4	65.9
80	76.5	74.1	70.7
90	80.3	78.3	74.7
100	83.4	81.9	78.1
110	86.1	84.9	81.2
120	88.3	87.3	83.8
130	90.2	89.4	85.8
140	91.7	91.0	87.5
150	92.9	92.3	89.0
160	94.0	93.5	90.2
170	94.9	94.4	91.1
180	95.7	95.2	92.0
190	96.3	95.9	92.8
200	96.4	96.5	93.4

Source: National Science Foundation, Total R&D Expenditures for Fiscal Years 1980, 1990, and 1997, from Academic Research and Development Expenditures: Fiscal Year 1997, early release tables, B-32, WebCASPAR database (expenditures are excluded for The Johns Hopkins Applied Physics Laboratory). <http://caspar.nsf.gov/cgi-bin/web1c.exe>

mingham. Fourth, in selected major federal programs, such as that for National Science Foundation Engineering Research Centers, institutions ranked outside the top 100 in terms of academic R&D expenditures (for example, Mississippi State University, Montana State University, and Brigham Young University, which ranked 106, 149, and 196, respectively, in federal R&D in FY1996, or below the top ranks in National Research Council rankings of graduate programs) have successfully competed against higher-ranked institutions.

Competition enhances the American university system's ability to contribute to scientific research and technological innovation in several ways. It aids in increasing the range of activities and types of clients that will be served by each institution; it creates multiple sources of supply in terms of institutional scientific and technological research capabilities, as well as institutional cultures for the R&D tasks sponsors need to have performed; and in conjunction with resource dependence, induces institutions to search constantly for market niches that will simultaneously fulfill their own missions (and aspirations within the context of these missions) and the preferences of the purchasers of the (research) goods and services they produce.

One result of this competition is a rough approximation of the knowledge-generation and dissemination activities of institutions by Carnegie Classification type with the stages of a linear R&D model. Extending to all universities an earlier effort developed by SRI (1986) for the public university sector, it is possible to depict in a stylized manner Research I private universities and select public universities as specializing in fundamental research, Research I and II universities as working over a larger continuum encompassing basic and applied research and some developmental and dissemination work, and Doctorate-granting I and II institutions as working on firm-specific applied problems relevant to locally situated firms. Finally, to complete this model, two-year community colleges and specialized vocational schools may be seen as engaging in technical problem-solving and labor force training.

Another consequence of competition is the recurrent pressures on institutions, including the most prestigious, to adjust their R&D portfolios, policies, and costs to accommodate shifts in aggregate supply and demand relationships in the market for academic research. In recent years, in the context both of the relative decline in federal funding of academic research and the schizophrenic character of congressional support for basic research—enacting sizeable increases for NIH and NSF in FY1999 but preoccupied with tax cuts and reductions in domestic discretionary expenditures—universities have fervently pursued industrial R&D funding and income from intellectual property rights.[7] A not surprising consequence of this shift to a buyer's market for academic research has been a competitive relaxation in the "price" of academic R&D, whether this price be calculated in terms of strictures regarding disclosures of research findings (Blumenthal et al. 1996; Cohen et al. 1998) or the ability of federal agencies to shift an increased portion of the costs of academic research to universities (Feller 1998a: 345–354).

Regionalism

Institutions of higher education pervade the American landscape. As reported in Table 7, 40% of U.S. counties contain at least one college or university, and 94% of central counties with populations of one million or more contain at least one such institution (Adams 1998).[8] The geographic spread follows upon the importance of state and local government finance for bricks and mortar as well as operating expenses. It reflects the broad-based U.S. belief in education as a source of self-improvement, economic and social advance, geographic and social mobility—aspects of a democratic and capitalistic ethos—combined with distributive politics at the state level.

Colleges and universities have historically been sources of community boosterism and regional pride. Indeed, the contrast between a university being "of" rather than "in" a place is seen by scholars as another of the defining differences between U.S. and European systems of higher education. Nevins, for example, has observed that

> The names of the early state institutions—the University of Virginia, Georgia, North Carolina, Vermont, Wisconsin, Michigan—remind us of a significant fact in the history of higher education. Regionalism, or relevance to a special community, had never affected the pattern of German universities. Many of them were placed in small towns, where they cultivated a detached, unworldly spirit. Students migrated freely from seat to seat, without local or district ties. ... So, too, in Scotland and England the universities were national rather than regional. [In the United States, by way of contrast,] ... the idea that universities should have a regional function took firm root from the beginning. ... The country was so large that as higher education spread westward it had to find a state or regional pattern, just as the nation was so practical minded that the spirit of academic work became in increasingly degree utilitarian. America became the home of universities which combined a world outlook with adaptation to special environments. (Nevins 1962: 23)[9]

Table 7 U.S. Counties Containing at Least One College or University*

	Counties (total number in U.S.: 3,102)	
with 1 institution	703	(22.7%)
with 2 institutions	228	(7.4%)
with 3 institutions	98	(3.2%)
with 4 institutions	52	(1.7%)
with 5 institutions	23	(0.7%)
with 6 institutions	30	(1.0%)
with 7 institutions	17	(0.5%)
with 8 institutions	13	(0.4%)
with 9 institutions	13	(0.4%)
with 10 institutions	9	(0.3%)
with more than 10 institutions	53	(1.7%)
Total with institutions	1,239	(40.0%)

Source: Adams (1998)
*Central counties with populations of 1 million or more

A consequence of financial and social ties to state local communities was that many public universities developed research programs related directly to the resource bases of their states—Wisconsin, dairy products; Iowa, corn; North Carolina, tobacco; and Oklahoma and Texas, oil exploration and refining—while several states developed specialized institutes directed at state-specific industries (various schools of mines and Lowell Textile Institute, Massachusetts) (Goldin and Katz 1998; Rosenberg and Nelson 1994).[10] Similarly, in Pennsylvania, The Pennsylvania State University (Penn State) developed programs in railroad engineering and petroleum engineering in response to requests and support from firms (Bezilla 1996).

Another effect of the geographic spread of institutions, not unlike that attributed to the spatial location of nineteenth-century railroads with their divisional switching yards and repair facilities, was the spread of technical literacy and skills across the states. The pattern is most evident in engineering fields. As noted by Goldin and Katz (1998: 20), "Among the most striking differences between the curricula of public and private institutions in the formative period concerns engineering. In 1908, almost 30 percent of all students in the public sector were in engineering programs, although by 1929 the figure dropped to around 15 percent. The private sector percentage was about one-third that in the public sector."

The development of local institutional R&D capabilities and curricula and extension programs, especially among publicly funded universities, to support local industries is a familiar aspect of the U.S. setting. A less evident consequence of this pattern is that the development of a research capability in institutions to deal with industry-specific, often applied research programs also created the infrastructure and institutional ethos to pursue more basic research agendas. In other than truly capital-intensive fields, fundamental research may be the paradigmatic footloose industry. What researchers require is a modicum of resources, a supportive institutional environment, and an opportunity to pursue their own ideas (Teich and Gramp 1996). The opportunities for niche research abound, in part based on distinctive regional locations and in part on the richness in the number of research questions and research trajectories in science. Thus, for example, Montana State University has become a recognized national leader in research on biofilms, a new research area of economic value both to the public sector (where a primary use is to reduce corrosion on ships) and to the private sector (with multiple uses ranging from the cleaning of boilers to removal of dental plaque) (Portera 1996); Brigham Young University's (Utah) development of comprehensive computer-based combustion models likewise has contributed to changing practices in the boiler industry, nationally.

The ramifications of the increasing spatial dispersion of academic institutions in scientific research and technological innovation has been an evolving process whose importance was likely modest before the Second World War, but which has increased over time. To the extent that many public institutions provide teacher education and general liberal arts education or receive negligible support from state funds to foster institutional environments—salaries, laboratories, facilities, graduate assistantships—conducive to research programs that could compete on a national basis for federal government or industry funds, regionalism means little. But, as is evident in Table 5, the trend in American higher education has been toward a research- and graduate-degree orientation and upward movement (and aspirations) through the Carnegie ranks. Former second-tier regional institutions (examples are the University of Alabama-Birmingham and Arizona State University) have become major national performers of basic and applied research. Erstwhile normal institutions have become more comprehensive, slowly but steadily adding master's-level programs, internship programs, and a modicum of applied research; in the process they have thickened the functional and spatial network of institutions contributing to technological change.

A similar transition is evident for community colleges. First established in Joliet, Illinois, in 1901, "community colleges were largely liberal-arts-oriented institutions, providing many students with the first leg of their baccalaureate preparation and others with a terminal general education." Over the years, beginning slowly and "blooming only in the late 1960s and early 1970s," this orientation changed radically toward more vocationally oriented programs and to increased roles in state and federal technology-based economic development programs. As of 1994, vocational education is the dominant program in the community colleges, "enrolling between 40 and 60% (depending on the estimate) of community college students" (Dougherty 1994: 191, 192).

Research and Graduate Education

The coupling of research and graduate education in specialized departments is another widely cited distinctive feature of U.S. higher education. Observing, for example, that U.S. federal support for basic research has gone to universities rather than consortia of in-house, government-run research institutes, Kennedy (1986) has argued that the "most significant outcome" of this process was the "co-location of research and research training. Most of the basic science in America today is done by mixed groups of journeymen and apprentices: the result is that the nation's research trainees are being developed alongside the best scientists."

Dasgupta and David (1994) have contended that this arrangement also encourages risk-taking behavior on the part of researchers. Rather than competing in a winner-take-all priority race for scientific discoveries, the coupling of research and instruction in a single occupation provides a core level of income to the academic researcher, with additional income and prestige tied to achievements in research, and encourages risk-taking behavior in research.[11]

Coupling research and graduate education, as viewed here, is also important because of its effects on students as researchers and as sources of technology transfer. Industrial representatives have repeatedly stated that universities' primary contribution to technological innovation resides in the training of students; indeed, they have expressed concern that efforts by universities to become more actively involved in the downstream commercialization of academic research not come at the expense of training students (Government-University-Industry Research Roundtable 1991). Students also are a means by which new scientific findings and technologically relevant knowledge are transferred from the campus to the firm. Indeed, as new technologically relevant research findings become embedded in the tacit know-how of students regarding laboratory procedures and software, their importance as technology transfer agents is likely to increase.

Students make other contributions to the production of new knowledge. Research in the life sciences and engineering involves combinations of faculty, postdocs, Ph.D. students, M.S. students, and, increasingly, undergraduates, assisted by full-time technicians and other support staff, an arrangement which increases the total quantity of research directed by a faculty member. Students also provide a flexible supply of human capital to faculty as they seek to accommodate the different stages of the R&D continuum into which sponsor interests fall. A pattern seems to be emerging in some university laboratories, at least in engineering, in which a faculty member engaged in fundamental research conducts this work with a team consisting of him/herself, the postdoc, and the Ph.D., and then takes on more applied work by assigning it to master's level and undergraduate students who work under the supervision of Ph.D. students and postdocs.

Finally, students also contribute to the discovery process. For example, most of the early work beginning in the 1960s on the computer simulation of integrated circuits was done by the private sector, largely on an in-house, proprietary basis by vertically integrated chip users. A major advance toward an industry-wide approach—SPICE—occurred at the University of California, Berkeley, with students being credited for substantial contributions.[12] Introduced in the early 1970s, and subsequently released in several updated versions, SPICE became the dominant industry approach by the late 1970s. Coupled with the spread of personal computers in engineering, SPICE contributed to the development of computer-aided engineering and to the rise of an engineering software industry (Roessner et al. 1998).

Universities as Performers of Basic and Applied Research and More

As stated in the title of this paper, American universities perform several roles in contributing to technological innovation. The generic form and specific content of these contributions are described in several lists of major technological innovations that have originated in university research (Crow 1993); specific case histories (Etzkowitz 1993; Gelijns and Rosenberg 1995); estimates of the value of academic research to industry's technological innovations (Mansfield 1991; Mansfield and Lee 1996), the commercial import of academic patents (Henderson, Jaffe, and Tratjenberg 1998), and the contribution of academic publications to industrial patents (Narin, Hamilton, and Olivastro 1997). Publications by universities and university associations about their contributions to specific technologies and regional and/or national economic growth likewise have proliferated (BankBoston 1997; Columbia University 1996; National Association of State Universities and Land-Grant Colleges 1997).

These studies document the outputs and outcomes of academic research. The perspective here is process-oriented, focusing on how the four features described above lead to three further qualities—geographical ubiquitousness, functional comprehensiveness, and interorganizational flexibility—that intersect with processes of scientific research and technological innovation. The demonstration is in terms of two prevailing schematic models of linkages between scientific research and technological innovation; it leads to yet one more entry to the list of analogical metaphors for how to comprehend the conceptual and empirical complexities of these relationships.

Universities in Pasteur's Quadrant

The intersection between the two sets of structural characteristics of the American research university system may first be seen in terms of the Pasteur's Quadrant model developed by Stokes (1997). This model features increasingly in national policy discourse, as it offers redress from the Manichean debate about federal support of R&D, where choices have tended to be presented in dichotomous terms of basic or applied research alone (Ehlers 1997). Stokes' model involves a 2x2, yes/no matrix of research inspired by a "quest for fundamental understanding" and research inspired by "considerations of use." Three of the four quadrants in this matrix have captivating appellations: Bohr's Quadrant—pure basic research (quest for fundamental understanding but no considerations of use); Pasteur's Quadrant—use-inspired basic research (quest for fundamental understanding and considerations for use); and Edison's Quadrant—pure applied research (considerations of use but no quest for fundamental understanding). The fourth quadrant lacks a distinctive label, but is not empty.[13]

An important quality of the American higher education system's contribution to technological innovation is that it fills all the boxes in Pasteur's Quadrant: it does so functionally, attempting on its own or in collaboration with industrial, government, and other sponsors, to generate all the forms of knowledge contained within this framework; and it does so spatially, providing (to varying degrees) broadly distributed regional institutional capabilities. These spatial capabilities can be employed to capitalize on geographically-dependent research and technological opportunities; they also can be used in collaboration with world-class, R&D-intensive firms serving world markets and with local firms seeking specific technical solutions for mundane products. Moreover, the combination of decentralization, resource-dependence, and competition makes the American university system sufficiently plastic in organizational forms, institutional policies, and cultural norms to accommodate external sponsors' changing requirements for research performed in any and all of the Pasteur Quadrants (Geiger 1993; Lee 1996).

Some universities—usually private, elite research universities—may act in only one or two quadrants, concentrating and explicitly confining their activities to Bohr's and Pasteur's Quadrants; others, such as the elite public universities in the University of California system and the Big Ten, where institutional activities span world-class science, applied research, and extension programs by mission, mandate, and resource-dependence, may act in all quadrants. In other institutions—regional universities, community colleges, and specialized technical institutes—the center of gravity for research and teaching may be located either within or astride the Pasteur and Edison Quadrants, with outlier activities at times extending to Bohr's Quadrant.

The mix of activities for given institutions is shaped in part by history and future aspirations; in part by their internal assessments of the comparative advantage in the performance of the science and technology contained within any single quadrant relative to that of other performers within a state; and in part by external assessments by state government or industry within a state of the need for an institution to act within a quadrant. Indeed, much of the ferment in university–industry–government R&D collaboration since the 1980s has represented a sorting process for optimal combinations of partners and R&D and technology transfer capabilities (Feller 1990).

Two recent developments reinforce the functional comprehensiveness and geographical dispersion within and across these quadrants, albeit in a somewhat opposite manner. First, federal and state government initiatives in university–industry–government R&D programs have fostered improved cooperation among higher education institutions within a state, thus providing for a somewhat more coordinated or accessible set of R&D activities across quadrants to industrial firms. At the same time, competition for resources and market share among institutions for tuition and government and industrial support for R&D, technological advances related to distance education, and the growth of for-profit higher education institutions have led to a blurring of traditional boundaries among institutions in the delivery of both research and instructional services. Public universities, in particular, are extending the upstream and downstream reach of their R&D, technology transfer activities, and educational services, competing on the one hand for more basic research while extending their educational and outreach programs into the domains of community colleges on the other. Bypassing here any attempt to assess the import of these trends on the future structure of American higher education, their immediate impact is to increase the system's aggregate capacity (if not necessarily its efficiency or effectiveness) to perform the knowledge-generating and knowledge-disseminating functions called for in a national innovation system.

Universities in a Chain-linked R&D Model

Matching institutional capabilities to R&D quadrants is more than a static, classificatory exercise, however; it has dynamic qualities that go to the heart of processes of technological innovation. Consider, for example, the Kline-Rosenberg chain-linked model of innovation (1986). This model contains a complex flow of interactions between research and knowledge generation, including the support of scientific research by instruments, machines, tools, and procedures of technology. It also involves interactions between scientific and technological developments, on the one hand, and recognition of potential markets and specific market-oriented activities such as design, production, distribution, and marketing, on the other.

General correspondence exists between the research and knowledge components of this model and those found in Stokes' model; the analytical contributions of the Kline-Rosenberg model, however, are its depiction of the complex interactions and feedback loops that exist among the R&D and transfer activities that link scientific research to technological innovation. The model also emphasizes that processes of technological innovation are not linear or unidirectional, but can begin from any of a set of initial conditions—including "disinterested" scientific discovery that leads to technological opportunities, or, conversely, efforts to solve specific production bottlenecks that press against existing knowledge to stimulate more fundamental investigations—and from many other intermediate R&D settings.[14]

American universities, as described above, if not generating most of the R&D tasks contained within the chain-linked model, have the ability to perform many of them. They are particularly active in the research- and knowledge-generating phases of technological innovation, braking possibly at the stages of process technology, industrial design and production technology, and stopping (for the most part) at the stage of marketing and distribution. More important than precise demarcations of boundaries, however, in keeping with the dynamic interactive emphasis to the chain-linked model, is that the contributions of universities to technological innovation are not constrained to any functional or temporal order. Universities can set the process moving, as in the recent case of biotechnology where academic basic research has been the launching pad for subsequent basic, applied, and developmental work by others; alternatively, they may keep the process moving, as in rapid injection molding where fundamental research in universities helped overcome technical bottlenecks to products and processes developed by industry.[15]

The benefits of the flexible and changing character of leader-follower roles also is seen in the recent flourishing of university–industry–government R&D collaborations (Cohen, Florida, and Goe 1994). Universities have been active promoters of these collaborations, but in some cases their involvement in a technological field lags well behind industrial practice.[16] Moreover, although the paradigmatic case for public-sector support of university–industry–government R&D programs is that universities will conduct work on pre-competitive generic research problems that relate to the effectiveness of an industry's dominant technologies, a challenge encountered by universities in these collaborative situations is to convince firms that the research that they, the universities, conduct indeed has commercial relevance. To overcome this perception, in fact, at times they have initially focused on well-defined questions related to an industry's core technology. As they make progress on this initial agenda, they may then move on to a more basic set of research questions that opens up new technological vistas for firms. In turn, the receptivity of firms to these new technological trajectories may be facilitated by their confidence in the university's earlier demonstrated capacity to understand the technical and economic parameters within which commercial products are introduced.

A Soccer Model of the Contributions of Universities to Technological Innovation

Technological innovation, as sketched above, is the outcome of the interaction of two dynamic processes: (a) between and among linked sequences in which bits of knowledge are generated, refined, adapted, disseminated, and targeted at specific market demands; and (b) between and among institutions—firms, universities, others—that perform the activities contained in these stages. Multiple metaphors, analogies, and schematics—pipelines, quadrants, chain-links, helixes—have been employed in efforts to describe these interactions. Unquestionably influenced by much watching of the 1998 World Cup series, I see the interactions as resembling a soccer match: as in soccer, goals can result from long passes involving few players, from unexpected, long, breathtaking runs, starting from anywhere on the field, from a single player who moves around and about defenders, or from deft passing and maneuvering of the ball back and forth among players; goals can be made by any player who touches the ball.

Historic Antecedents and Future Prospects

This interpretation of how the ubiquitous, comprehensive, and plastic traits of the American higher education system contribute to technological innovation is not an argument by design. Nor is the description of the system's current congruity with mainstream models of science and technology processes a prediction of its future.

As frequently described, the American higher education system has evolved in a "largely unplanned fashion" (Clark 1995: 116). Or, as Graham and Diamond (1997: 15) have observed: "What is most striking about the pre-1994 American system of higher education is its inadvertent, unplanned quality, in sharp departures from the centralized and state-dominated European pattern, and its uncanny instinct for survival and growth in a shifting market environment."

The interpretative force and historical accuracy of these assessments again relate to the bipolar perspective noted at the opening of this essay. They are most compelling when seen in the context of international comparisons, less so when examined up close. The assessments hold in comparative terms—again, say, between the United States and Japan. In the United States, for example, unlike Japan, the federal government has no constitutional authority to operate the nation's higher education system. In the United States, federal support of academic research has flowed largely from policy decisions to fund the basic research activities of mission agencies through universities rather than through government or other research laboratories.

Still, even in the absence of direct constitutional provisions or formal planning, federal policies have significantly affected the system's characteristics. The Morrill Act, Hatch Act, and the Smith-Lever Act helped shape the fundamental features of the public research university. The research grants model adopted by NIH, NSF, and, to varying degrees, other mission agencies provided federal support for research and graduate programs on essentially a competitive, all-comers basis—public and private university, established institutions, and new-comers alike. Congressional interest in the distribution of federal academic R&D awards by institution and by state was present at the creation of NSF, and continues apace today.

For purposes of this paper, though, the most salient of the historic changes that have affected the current U.S. higher education system is the dynamically evolving balance between basic and applied research. Viewing the current U.S. system in terms of the university's dominant role as a performer of basic research in effect masks the long history of tensions about the contents of the academic research agenda.

A few brief examples highlight these tensions. Academic agricultural scientists supported by Hatch funds struggled for the time and resources to conduct research rather than being assigned instructional duties, and for legitimacy as scientists rather than as assayers of the qualities of soil fertilizer (Marcus 1988).[17] Intra-institutional conflicts between William Walker and Arthur Noyes over the balance between industrially-oriented research and more basic research at MIT during the early part of the twentieth century is an often-told story (Geiger 1986; Servos 1980). What emerges more forcibly in recent accounts is less the conflict between individuals but the haphazard process that involved coalitions of competing interests, producing "fragile alliances between groups within academia and between academia and industry" (Lecuyer 1995: 39).

Battles over the proper balance between "fundamental" and "applied" research have occurred repeatedly in the history of academic engineering.[18] Efforts were made throughout much of the twentieth century to instill an engineering science orientation to a "hands-on" craft, while since the 1980s, reforms toward more "systems-oriented," industrially relevant R&D have dominated (National Academy of Engineering 1985). Not surprisingly, given the decentralized character of U.S. higher education, these tensions reflect broader ones found in the history of public support of research in the United States by the federal government (duPree 1957), by state governments (Nash 1963), and by industry of its own research agendas and relationships with universities (Swann 1988).

The present congruence between the characteristics of the U.S. higher education system and processes of technological innovation described above is thus the product of multiple adjustments made in response to many influences. Many of the adjustments made by universities are those of de facto market-oriented organizations to changes in the supply and demand for selected types of research. Some are attributable to public policy, notably the establishment of the land-grant university system and the postwar pattern of channeling federal support for basic science into the university sector. More recent policies have made for a yet better fit. The establishment since the mid-1980s of government–industry–university collaborative R&D programs, coupled with university initiatives to find industrial support for their research programs and with industrial firms seeking contract performers of work formerly conducted by in-house laboratories, has legitimized and supported the development of closer coordination between the research and technology development efforts of firms and universities. As described by Mowery et al. in this volume, the Bayh-Dole Act has helped spawn university activities in technology patenting and licensing, although the net effects of the legislation on patent trends and technological innovation remain uncertain (Feller 1997).

Another Bipolar View

Fitting this analysis into a comparative, policy-oriented U.S.-Japan framework serves to create a new bipolar set of questions, one relevant to Japan, the other to the United States. The first set pertains to the applicability of the type of U.S. policies noted above relating to the role of universities in technological innovation across national borders. The second pertains to the stability of the present "high-level" equilibrium between the characteristics of the U.S. higher education system and processes of technological innovation. Guarded answers must be offered to each question.

To the first question, although it is tempting from a policy perspective to focus on highly visible features of the American landscape (such as placement of public funds for R&D within university laboratories) as a source of U.S. scientific leadership, the above analysis suggests that the contributions of American colleges and universities to technological innovation emerged slowly, and for varying periods were both too close to and too far from commercial ends to make substantial contributions to technological innovation. The often-cited salutary impacts on technological innovation of the postwar public policy decision to place much of the nation's investments in basic research and large amounts of its mission-oriented research in universities, depended on the larger set of characteristics of the U.S. higher education system—decentralization, competition, regionalism, and the coupling of research and graduate education—that are deeply woven into the political, economic, and cultural features of American society, and may not be relevant to conditions in other countries.

Similar caution must be used in answering the second question. Multiple tensions and flaws exist within the U.S. academic research system, in terms of both its overall features and its contributions to technological innovation. The U.S. system is not necessarily efficient; decentralization and regionalism subject it to considerable duplication of effort and reduced levels of output, while political constraints dampen the effects of merit-based competitions for public funds. The system is prone to overreaching by institutions as they seek to take on new roles as performers of research or technology transfer. Regionalism contributes to recurrent (and possibly growing) political pressures to produce broad-based geographical distributions of federal R&D funding, and to institution-specific incentives to obtain funds via political patronage rather than merit review procedures. The R&D capacities of the system may be overextended, with the number of institutions now committed to a research orientation (supplemented by those on the cusp of such a commitment) exceeding realistic projections of external support likely to come from government and industry.

The system exists in a state of continuing flux, subject to pressures and criticisms relating to priorities and performance. A concerned and often pessimistic outlook on the health of the American research university pervades contemporary discourse and exegesis (Cole, Barber, and Graubard 1993; Ehrenberg 1997; Guston and Keniston 1994). Economists have likewise raised cautions about the combined effects of university efforts to commercialize technology as a revenue-enhancing strategy, its increased dependence on industrial funding, the shifting research agenda from the Bohr and Pasteur to the Edison Quadrants (Nelson and Romer 1996), its apparent willingness to accept restrictions on the disclosure of research findings technology (Cohen et al. 1998), and the shift from public to private science (Dasgupta and David 1994) of both research agendas and the diffusion of knowledge.

How the academic research industry will respond to any sort of shakeout caused by the structural imbalances between supply and demand for resources is uncertain. Different institutions hold quite different perspectives on what a new equilibrium will (or should) look like (Feller 1998b). Even as universities promote their contributions from research, education, and outreach activities to technological innovation, their individual moves may not necessarily produce a dynamically stable system.

Acknowledgments

In a paper that highlights the coupling of research and graduate education, it is only appropriate that I acknowledge the contributions of two Penn State graduate students, Jennifer Adams, who generously shared with me her work on the spatial distribution of U.S. universities, and Darryl Farber, for his assistance in creating statistical tables.

Notes

1. "After World War II, the United States made a rather remarkable national policy decision: the federal government should invest heavily in scientific research, and most federally sponsored research should be done in the nation's universities. This policy was unique in world history, and it has helped to elevate scientific research in the United States to the highest standards on the globe. Other nations (most notably Russia, but also Japan) separate teaching institutions (the universities) from research institutions (government or industry laboratories), and lose the opportunity for synergy that characterizes our system" (Likens and Teich 1994: 179).

2. The relative distribution of academic R&D has remained roughly unchanged since about 1978. This pattern flows from the National Institutes of Health (NIH)'s dominant role as the agency source of federal funds for academic R&D (57% in 1995), and the related concentration of this research in the life sciences (55% in 1995). The share of basic research, however, has declined from the 1960–1975 period, in which it represented about 75 percent of total academic R&D.

3. Illustrative of this decentralization is Geiger's depiction of 15 (of the fewer than 25) universities seriously committed in 1920 to research as an institutional goal;

five were state universities (four in the midwest and one in California), five began as colonial colleges, and five were private universities founded in the late nineteenth century (Geiger 1986: 3).

4. In comparing research output between private and public universities across fields of knowledge, Graham and Diamond (1997) report, "In humanities scholarship, even more decisively than R&D funding and scientific journal publications, the elite private universities are a world apart. For generations and in some cases for centuries, tradition has nourished the arts and humanities at the newly endowed private universities" (Graham and Diamond 1997: 67).

5 "The politics of state funding meant that both the curriculum and research of U.S. public universities were more closely geared to commercial opportunities than was true in many European systems of higher education. Especially within emerging fields of engineering and, to a lesser extent, within mining and metallurgy, state university systems often introduced new programs as soon as the requirements of the local economy became clear" (Mowery and Rosenberg 1993: 35).

6. Research Universities I are defined as those institutions that are committed to graduate education through the doctorate level, and give a high priority to research. Quantitatively, in the 1994 survey, each of these institutions awarded 50 or more doctoral degrees each year and received $40 million or more in federal support (Carnegie Foundation for the Advancement of Teaching 1994: xix). Definitions used in the Carnegie Classifications changed across surveys, so Table 5 represents an approximation.

7. Academic R&D expenditures grew in real terms by 5.7% per year (1992 dollars) between 1984–1994; the rate of increase fell to 1.6% annually between 1994 and 1997 (National Science Foundation 1998a: 5–8). Between FY1995 and FY1996, federal agency obligations for academic science and engineering fell by two-tenths of 1% in current dollars and by 2 % in constant dollars (National Science Foundation 1998c).

8. The geographical spread, according to Adams, is the result of two quite different historical thrusts. In the nineteenth century, when about 1,100 colleges and universities were formed, the combination of the search for an Arcadian environment for educating young adults and an anti-urban bias, reinforced by the Morrill Act's emphasis on agriculture, led to interior locations being the favored sites. In the twentieth century, the burgeoning growth, especially in the period 1964–1994, of two-year colleges, which were predominantly located in cities of over 100,000 people, gave a more urban cast to the location of institutions (and even more so to geographic patterns of enrollment).

9. Similarly, Ratcliff has observed that "Each state established one or more state colleges to advance its reputation for an educated citizenry. And many towns established colleges to provide evidence, along with the museum, library, opera house, and symphony band shell of its cultural stature" (Ratcliff 1994: 6).

10. This pattern continues today. The universities most dependent on industrial

funding for their academic R&D expenditure "tend to be smaller, specialized institutions—frequently ones with a single R&D specialty that is closely linked with a local industry" (National Science Foundation 1993: 137).

11. Noll has expressed a similar view: "The uniqueness of U.S. research universities derives from their integration of teaching and research. Were education and research completely separate, employment as a faculty member in a research university would resemble having two part-time jobs, one as a teacher and the other as a researcher. This circumstance is not far from the case in much of Europe" (Noll 1998: 19).

12. "SPICE is the result of the work of a number of talented graduate students in the Department of Electrical Engineering and Computer Science at the University of California at Berkeley, who had a mandate to produce the best computer program for the simulation of practical integrated circuits, ICs, under the guidance of D. Pederson and R. Rohrer" (Vladimirescu 1994: 1).

13. According to Stokes (1997: 74), this quadrant (no quest for fundamental knowledge, no quest for considerations of use) represents "research that systematically explores particular phenomena without having in view either general explanatory objectives or any applied use to which the results can be put. ..." As the type of activity fitting into this quadrant, Stokes cites *Peterson's Guide to the Birds of North America* as highly systematic research for which bird watchers are grateful.

14. "... [M]odern innovation is often impossible without the accumulated knowledge of science and ... explicit development work often points up the need for research, that is, new science. Thus, the linkage from science to innovation is not solely or even preponderantly at the beginning of typical innovations, but rather extends all through the process—science can be visualized as lying alongside development processes, to be used when needed" (Kline and Rosenberg 1986).

15. "As in many other areas, large scale commercial use of RIM (rapid injection molding) preceded a full understanding of the technology from a theoretical point of view" (Alberino, as quoted in Roessner et al. [1998: 35]). See also Branscomb (1986).

16. In data storage technology, for example, in the period between 1957 and the early 1980s, "almost all research on magnetic and optical recording technologies was performed by industry (mostly IBM), in a proprietary atmosphere. As a result, there was almost no U.S. university, either in research or education, in magnetic or optical recording technologies. ..." (Carnegie Mellon University 1998: 8).

17. In the early history of Penn State's experiment station, the university's president, George Atherton, expected station staff to concentrate their efforts on "applying and disseminating scientific knowledge rather than uncovering new knowledge through basic research" and to "do a significant amount of teaching at the undergraduate level" (Bezilla 1987: 48).

18. In 1933, Penn State President Ralph Hetzel, pleased with the levels of external research received by the Schools of Engineering, Chemistry and Physics, and Mineral Industries, yet concerned about their dependence on private firms and

industrial groups, wrote to the College of Engineer's Council on Research, "There is a danger that applied research in cooperation with industries may overshadow and stifle institutional research of a more fundamental nature. The question is not whether such investigations should be eliminated, but how far can we go in this line of work without prejudicing the institutional and research programs of an institutional that is supported by public funds and unrestricted by contract" (quoted in Bezilla 1987: 105). Hetzel's efforts to subsequently implement a policy that gave the central administration review power over College of Engineering cooperative research projects to insure that they yielded results of public value as contrasted with value primarily to the donor were opposed by the College and effectively blocked.

References

Adams, J. 1998. *Educating Places: Urban Universities and Colleges in the U.S.* University Park, PA: The Pennsylvania State University, Department of Geography.

BankBoston. 1997. *MIT: The Impact of Innovation.* Boston, MA: BankBoston.

Bezilla, M. 1987. *The College of Agriculture at Penn State.* University Park, PA: The Pennsylvania State University Press.

Bezilla, M. 1996. *The College of Engineering at Penn State.* University Park, PA: The Pennsylvania State University Press.

Blumenthal, D., D. Causino, E. Campbell, and K. Seashore Louis. 1996. "Relationships Between Academic Institutions and Industry in the Life Sciences—An Industry Survey." *New England Journal of Medicine* 334: 368–373.

Branscomb, L. M. 1986. "IBM and U.S. Universities—An Evolving Partnership." *IEEE Transactions on Education* E-29(2): 69–77.

Bush, V. 1960. *Science: The Endless Frontier.* Washington, DC: National Science Foundation.

Carnegie Foundation for the Advancement of Teaching. 1994. *A Classification of Institutions of Higher Education.* Princeton, NJ.

Carnegie Mellon University, Data Storage Systems Center. 1998. Eighth Annual Report, Volume 1 (January 31).

Clark, B. 1995. *Places of Inquiry.* Berkeley, CA: University of California Press.

Cohen, W., R. Florida, and R. Goe. 1994. *University-Industry Research Centers in the United States.* Pittsburgh, PA: Carnegie Mellon University.

Cohen, W., R. Florida, L. Randazzese, and J. Walsh. 1998. "Industry and the Academy: Uneasy Partners in the Cause of Technological Advance." In *Challenges to the Research University,* ed. R. Noll. Washington, DC: Brookings Institution, pp. 171–199.

Cole, J., E. Barber, and S. Graubard. 1993. *The Research University in a Time of Discontent.* Baltimore, MD: The Johns Hopkins University Press.

Columbia University. 1996. *Columbia University's Contribution to the New York City Economy.* Executive Summary. Report prepared for Columbia University by Appleseed.

Crow, M. 1993. "The University as a Catalyst for Scientific and Industrial Development." In *Industrial Policy for Agriculture in the Global Economy,* eds. S. Johnson and S. Martin. Ames, IA: Iowa State University Press, pp. 109–127.

Dasgupta, P., and P. David. 1994. "Toward a New Economics of Science." *Research Policy* 23: 487–552.

Dougherty, K. 1994. *The Contradictory College.* Albany, NY: State University of New York Press.

duPree, A. H. 1957. *Science in the Federal Government.* New York: Harper Torchbooks.

Ehlers, V. 1997. "A Scientist in Congress Looks at Science Policy." In *AAAS Science and Technology Policy Yearbook 1996/97,* eds. A. Teich, S. Nelson, and C. McEnaney. Washington, DC: American Association for the Advancement of Science, pp. 67–75.

Ehrenberg, R. (ed.). 1997. *The American University: National Treasure or Endangered Species.* Ithaca, NY: Cornell University Press.

Ergas, H. 1987. "The Importance of Technology Policy." In *Economic Policy and Technological Performance,* eds. P. Dasgputa and P. Stoneman. Cambridge, UK: Cambridge University Press.

Etzkowitz, H. 1993. "Enterprises from Science: The Origins of Science-based Regional Economic Development." *Minerva* 31: 326–360.

Feller, I. 1990. "University–Industry R&D Relationships." In *Growth Policy in the Age of High Technology,* eds. J. Schmandt and R. Wilson. Boston, MA: Unwin Hyman, pp. 313–343.

Feller, I. 1997. "Technology Transfer from Universities." In *Higher Education: Handbook of Theory and Research,* ed. J. Smart. New York, NY: Agathon Press.

Feller, I. 1998a. "Matching Fund and Cost-sharing Requirements in Federal Support of Academic Research." In *AAAS Science and Technology Policy Yearbook,* eds. A. Teich, S. Nelson, and C. McEnaney. Washington, DC: American Association for the Advancement of Science, pp. 345–354.

Feller, I. 1998b. "A Preliminary Assessment of the Strategies Used by Research Universities to Maintain and Enhance Their Research and Graduate Degree Competitiveness in a Period of Challenged Priorities and Austere Funding: II." Discussion paper prepared for the Social Science Research Council Workshop, Boundaries of U.S. Higher Education Institutions and the U.S. Higher Education System.

Geiger, R. 1986. *To Advance Knowledge.* New York: Oxford University Press.

Geiger, R. 1993. *Research and Relevant Knowledge.* New York: Oxford University Press.

Geiger, R., and I. Feller. 1995. "The Dispersion of Academic Research in the 1980s." *Journal of Higher Education* 66: 336–360.

Gelijns, A., and N. Rosenberg. 1995. "From the Scapel to the Scope: Endoscopic Innovations in Gastroenterology, Gynecology, and Surgery." In *Sources of Medical Technology: Universities and Industry*, eds. N. Rosenberg, A. Gelijns, and H. Dawkins. Washington, DC: National Academy Press, pp. 67–96.

Goldin, C., and L. Katz. 1998. *The Shaping of Higher Education: The Formative Years in the United States, 1890–1940.* National Bureau of Economic Research Working Paper 6537. Cambridge, MA: NBER.

Government-University-Industry Research Roundtable. 1991. *Industrial Perspectives on Innovation and Interaction with Universities.* Washington, DC: National Academy of Sciences.

Graham, H., and N. Diamond. 1997. *The Rise of American Research Universities.* Baltimore, MD: The Johns Hopkins University Press.

Gumport, P. 1993. "Graduate Education and Organized Research in the United States." In *The Research Foundation of Graduate Education,* ed. B. Clark. Berkeley, CA: University of California Press, pp. 225–260.

Guston, D., and K. Keniston (eds.). 1994. *The Fragile Connection.* Cambridge, MA: The MIT Press.

Henderson, R., A. Jaffe, and M. Trajtenberg. 1998. "Universities as a Source of Commercial Technology: A Detailed Analysis of University Patenting, 1965–1988." *Review of Economics and Statistics* 81: 119–127.

Jankowski, J. 1998. "Statistical Data and Their Impact on University Research Efforts: Trends and Patterns of Academic R&D Expenditures." Presentation to the International Quality and Productivity Center's Conference on Performance Measurements for Research and Development in Universities and Colleges, Washington, DC, May 18–19.

Kelley, M., and H. Brooks. 1989. "From Breakthrough to Follow-Through." *Issues in Science and Technology* 5: 42–47.

Kennedy, D. 1986. "Basic Research in the Universities: How Much Utility." In *The Positive Sum Strategy*, eds. R. Landau and N. Rosenberg. Washington, DC: National Academy Press, pp. 263–274.

Kline, S., and N. Rosenberg. 1986. "An Overview of Innovation." In *The Positive Sum Strategy*, eds. R. Landau and N. Rosenberg. Washington, DC: National Academy Press, pp. 275–305.

Lecuyer, C. 1995. "MIT, Progressive Reform, and 'Industrial Service,' 1890–1920." *Historical Studies in the Physical and Biological Sciences* 26(1): 35–88.

Lee, Y. S. 1996. "Technology Transfer and the Research University: A Search for the Boundaries of University–Industry Collaboration." *Research Policy* 25: 843–864.

Likens, P., and A. Teich. 1994. "Indirect Costs and the Government–University Partnership." In *The Fragile Connection*, eds. D. Guston and K. Keniston. Cambridge, MA: MIT Press, pp. 177–193.

Mansfield, E. 1991. "Academic Research and Industrial Innovation." *Research Policy* 20: 1–12.

Mansfield, E., and J. Y. Lee. 1996. "The Modern University: Contributor to Industrial Innovation and Recipient of Industrial R&D Support." *Research Policy* 25: 1047–1058.

Marcus, A. 1988. *Agricultural Science and the Quest for Legitimacy*. Ames, IA: Iowa State University Press.

Mowery, D., and N. Rosenberg. 1993. "The U.S. National Innovation System." In *National Innovation Systems*, ed. R. Nelson. New York: Oxford University Press.

Narin, F., K. S. Hamilton, and D. Olivastro. 1997. "The Increasing Linkage Between U.S. Technology and Public Science." *Research Policy* 26: 317–330.

Nash, G. 1963. "The Conflict Between Pure and Applied Science in Nineteenth-Century Public Policy: The California State Geological Survey." *Isis* 54: 217–228.

National Academy of Engineering. 1985. *New Directions for Engineering in the National Science Foundation: A Report for the National Science Foundation by the National Academy of Engineering*. Washington, DC: National Academy Press.

National Association of State Universities and Land-Grant Colleges. 1997. *Value Added: The Economic Impact of Public Universities*. Washington, DC.

National Science Foundation. 1993. *Science and Engineering Indicators–1993*. Arlington, VA: U.S. Government Printing Office.

National Science Foundation/SRS. *Survey of Research and Development Expenditures at Universities and Colleges, Fiscal Year 1996*. Arlington, VA: U.S. Government Printing Office.

National Science Foundation. 1998a. *Science and Engineering Indicators–1998*. Arlington, VA: U.S. Government Printing Office.

National Science Foundation. 1998b. "Academic R&D Expenditures Maintain Steady Growth in FY1996." Data Brief, NSF 98-303, March 3, 1998.

National Science Foundation. 1998c. "Federal Academic Science and Engineering Obligations Decreased Slightly in FY1996." Data Brief, NSF 98-308, April 27, 1998.

Nelson, R., and P. Romer. 1996. "Science, Economic Growth, and Public Policy." In *Technology, R&D, and the Economy*, eds. B. Smith and C. Barfield. Washington, DC: The Brookings Institution/American Enterprise Institute, pp. 49–74.

Nevins, A. 1962. *The Origins of the Land-Grant Colleges and State Universities*. Washington, DC: Civil War Centennial Commission.

Noll, R. 1998. "The American Research University: An Introduction." In *Challenges to the Research University*, ed. R. Noll. Washington, DC: Brookings Institution.

Portera, C. 1996. "Biofilms Invade Microbiology." *Science* 273(27): 1795–1799.

Ratcliff, J. 1994. "Seven Streams in the Historical Development of the Modern American Community College." In *A Handbook of the Community College in America*, ed. G. Baker III. Westport, CT: Greenwood Press, pp. 3–16.

Roessner, D., R. Carr, I. Feller, M. McGeary, and N. Newman. 1998. *The Role of NSF's Support in Enabling Technological Innovation, Phase II.* Arlington, VA: SRI International. Report to the National Science Foundation.

Rosenberg, N., and R. Nelson. 1994. "American Universities and Technical Advance in Industry." *Research Policy* 3: 323–348.

Rosovsky, H. 1990. *The University: An Owner's Manual.* Cambridge, MA: Harvard University Press.

Servos, J. 1980. "The Industrial Relations of Science: Chemical Engineering at MIT, 1900–1939." *Isis* 7: 531–549.

SRI International. 1986. *The Higher Education–Economic Development Connection.* Washington, DC: American Association of State Colleges and Universities.

Stokes, D. 1997. *Pasteur's Quadrant.* Washington, DC: Brookings Institution.

Swann, J. 1988. *Academic Scientists and the Pharmaceutical Industry.* Baltimore, MD: The Johns Hopkins University Press.

Teich, A., and K. Gramp. 1996. "Competitiveness in Research: Perceptions of Practitioners." In *Competitiveness in Academic Research,* ed. A Teich. Washington, DC: American Association for the Advancement of Science, pp. 73–112.

Trow, M. 1991. "American Higher Education: Exceptional or Just Different." In *Is America Different? A New Look at American Exceptionalism,* ed. B. Shafer. Oxford: Clarendon Press, pp. 138-186.

U.S. Congress, Office of Technology Assessment. 1990. *Making Things Better: Competing in Manufacturing.* Washington, DC: U.S. Government Printing Office.

U.S. Department of Education. 1995. Integrated Post-Secondary Education Data System (IPEDS). Washington, DC: U.S. Government Printing Office.

Vladimirescu, A. 1994. *The SPICE Book.* New York: John Wiley & Sons, Inc.

Zumeta, W. 1997. "State Policy and Private Higher Education: Past, Present, and Future." In *Higher Education: Handbook of Theory and Research,* Vol. 12, ed. J. Smart. New York: Agathon Press, pp. 43–106.

4

Coauthorship Linkages between University Research and Japanese Industry

Kenneth Pechter and Sumio Kakinuma

Introduction

It is commonly accepted that weak links between university research and industrial innovation are a distinctive characteristic of the Japanese innovation system. Recently, this long-held view has been strengthened by the expanding role of university resources in industrial innovation in the United States, highly visible through university-held patents, faculty-based entrepreneurial activity, and academia-oriented venture capital, and by the paucity of similar visible signs in Japan. The assessment of the university-industry linkage in Japan described in this chapter, however, paints quite a different picture: Analysis of the coauthorship of scientific and technical papers by university and industry researchers indicates that the linkage is as strong as, if not stronger than, that in the United States.

It is not surprising that Japanese industry is commonly assumed to be disinterested in Japanese academia. As described by Hashimoto in this volume, a "hesitant relationship" between academia and industry resulted from the policy frameworks that emerged after the Second World War. Furthermore, it is well known that much of the innovation that drove Japanese postwar growth took place in industry rather than academia, and often involved foreign sources of technology. However, these experiences, so different from those of the United States, have led to quite different institutional arrangements in the two countries. For precisely this reason, we must exercise caution in using the metrics suitable for one arrangement to measure activity in the other.

When the university-industry linkage in the United States displays such visible signs as university patenting and Japan's does not, it is tempting to say that the Japanese systems lags the American system and must catch up. But such logic ignores the fundamental differences between the institutional arrangements in Japan and the United States. For instance, because the leading universities in Japan are national, their faculties civil servants of the central government, there are severe restrictions on individual profit-seeking and use of corporate funds, a situation immensely different from that in the United States, which has no national university system. A specific effect of such differences is that the movement in the United States to transfer ownership of intellectual property resulting from government-sponsored research to universities and researchers is irrelevant to Japan, because in the Japanese system university inventors in most cases already possess the rights to their inventions. The low university patent counts in Japan are not simply a result of Japan's failure to follow the United States' lead, but instead result from institutional factors such as the civil service status of faculty members, which inhibits patenting activity in the universities and creates incentives for inventions to be passed along to industry without the faculty member filing for a patent. The subtlety of this point is underscored by the fact that it is often overlooked not just by American policy makers, but by Japanese policy makers as well.

The normative question of whether the Japanese system should adopt aspects of the emerging American system is addressed elsewhere in this volume; the chief lesson of this chapter is that due to institutional differences, patents and licensures make poor yardsticks for empirically measuring the university-industry linkage in Japan. In order to address the many important normative questions confronting policy makers, it is crucial that metrics be found that allow meaningful comparisons to be made. For instance, asymmetric access, a key topic in the United States–Japan science and technology policy dialogue of the past two decades, arises from the claim that because Japanese researchers have access to U.S. university laboratories while Japanese research is conducted behind the closed doors of industry, Japan has a strong advantage. The university-industry coauthorship analysis we present in this chapter, which indicates a significant role for Japanese universities in Japanese industrial innovation, erodes the foundation of the asymmetric access claim and forces a reconsideration of the issue.

Assessing the University-Industry Linkage

We have chosen to look at university-industry research collaboration in Japan through the window of coauthored papers, a useful approach because coauthorship is an indicator of a broad range of collaborative activity, and may result from a collaboration involving the flow of financial, capital, human, or intellectual property resources. Furthermore, unlike in mass data such as R&D funding flows, the parties to the collaboration can be identified. Of course, coauthorship is not a perfect indicator: Publication is not normally a top priority for industry; much meaningful interaction may not result in any published paper; methodological issues like dual affiliations by a single author and variations on institutional names create errors; such biases as the tendency to publish more in one field than another, in fundamental as opposed to applied research fields, in one language rather than another, may be misleading. However, as one piece in the complex puzzle of university-industry research interaction, analysis of coauthorship data provides a rich view of the interaction between industry and university research.

Although coauthorship data may be thought of as an output indicator, with the input being collaboration-enhancing variables like inter-organizational funding flow, transfer of human resources, or change in institutional routines, we think of coauthorship data as a simple indicator of the incidence of collaborative activity. For example, we are not claiming a straightforward causality between coauthorship and R&D funding flows from industry to university, but rather view coauthorship as a byproduct of collaboration between university and industry, regardless of the causes of the collaboration. The significance of this distinction is the fact that since publication is not the ultimate goal of firms it makes a poor output indicator of firms' activities, but as a gauge of the incidence of collaborative activity it provides a reasonable surrogate measure.

In order to assess the level of university-industry research interaction through bibliometric data on coauthorship, we analyzed coauthorship between university and industry researchers from the perspective of industry. We chose a 16-year period, 1981–1996, for our study. Searching a comprehensive database from the Institute of Scientific Information of publications in which at least one author is affiliated with an organization located in Japan, we created a subset containing all papers in the database published with at least one author from a firm located in Japan. This subset contains 110,588 papers. We then performed various processes in order to determine the coauthorship patterns of these papers, by assigning each author affiliation to an institutional sector based on information in the author's address. The sectors we used were academia, industry, and other (including national laboratories, public corporations and non-university hospitals). For details on the methodology used throughout this chapter, see Pechter and Kakinuma (1999) and Kakinuma and Pechter (1999).

Macro View of University-Industry Coauthorship in Japan

Figure 1 shows a breakdown of papers produced by Japanese industry from 1981 to 1996: those authored intramurally (within a single firm), by multiple firms, by a firm and a university, and by a firm and an organization in the "other" category. These data contain both domestic and international collaborators of the Japan-based firms. Total publication by researchers in Japanese industry, the sum of the four lines in the figure, increased steadily, from 3,433 papers in 1981 to three times that figure at 10,450 papers in 1994. From 1994 to 1996, however, growth has faltered. Of these papers, in 1981 70.3% were authored intramurally (this includes single authors as well as multiple authors in the same firm), but by 1996 this had dropped to 43.3%. On the other hand, papers authored jointly with a university researcher went from 23.1% in 1981 to 46.4% in 1996, overtaking intramurally authored papers. The portion of papers authored jointly with another firm was only 2.8% in 1981 and grew much less dramatically to 3.5% in 1996. Coauthored papers with the "other" category rose from 3.8% to 6.8% over the period.

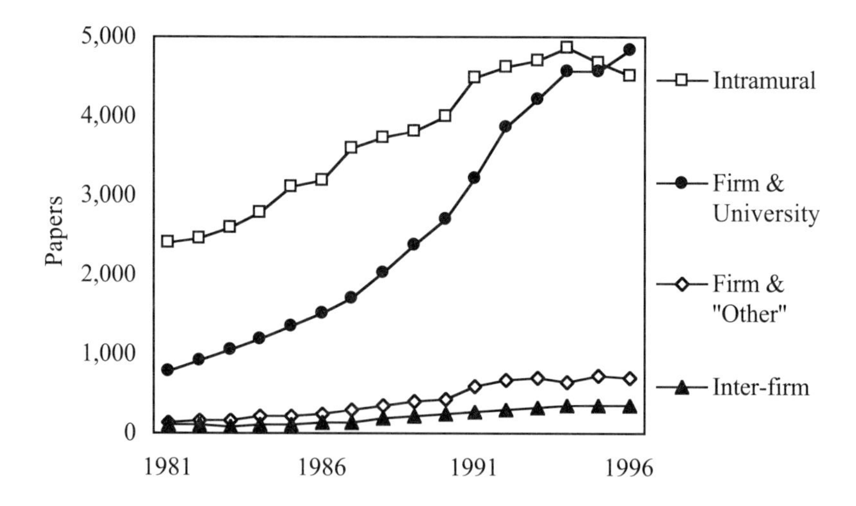

Figure 1 Japanese Industry Papers by Type of Collaboration.

This analysis reveals that university researchers have been significant collaborators with industry researchers, and that this significance has intensified over the past two decades, while inter-firm collaboration, the often-claimed characteristic of Japanese industry, is of only minor importance when it comes to publication. Low inter-firm coauthorship is not necessarily surprising, considering that inter-firm collaboration is likely to involve proprietary information not suitable for publication; but it does undercut the anecdotal evidence that university research labs are of little interest to industrial researchers in Japan.

These data are truly significant in the context of debates over asymmetric access and other issues premised on the weak university-industry linkage view of Japan. This is because these figures are remarkably close to corresponding figures in the United States. According to *Science & Engineering Indicators—1998* (National Science Board 1998), the portion of U.S. industry papers coauthored with academia grew from 21.6% in 1981 to 40.8% in 1995. Thus, in spite of anecdotal evidence to the contrary, the empirical evidence suggests that industry and academia in Japan interact in the research process at least as much as, if not more than, in the United States. Although methodological differences must be considered before making strict comparisons, the fact that this study and the

National Science Board data are based on the same data source and the same methodology lends credibility to this comparison. At the very least, the evidence suggests that there may indeed be more to the university-linkage in Japan than is commonly assumed, and warrants further investigation.

Another piece of conventional wisdom which this empirical evidence forces us to reconsider concerns Japanese dependence on foreign, particularly American, fundamental research, a view that would lead us to believe that U.S. universities are a large and growing component of our data base. Figure 2 shows the "Firm & University" component of the previous figure with Japanese and foreign universities treated separately, and indicates that coauthorship with foreign universities has indeed increased over 1981–1996, with the rise accelerating in 1990. As a percentage of all university-industry papers, foreign university papers have doubled from 8% to 16% during this period.

As we will show later, however, the portion of coauthorship in which a U.S. university researcher is the collaborator has been declining steadily over the period of this study, with non-G7 countries growing the most. While it can only be speculated that some of this rise is due to the move of Japanese firms abroad due to the rising yen and trade friction as well as the growth of innovative competitiveness in many regions throughout the world, what is certain is that the empirical evidence does not support the conventional view of Japan's growing dependence on the United States as a source of research.

Coauthorship Data by Firm

In order to add more detail to this analysis, we looked at coauthorship at the firm and industrial sector levels, first identifying each firm and university to which an author of a paper in the data set is affiliated, then creating a running tally for each firm and university. We then determined each firm's corresponding industrial sector and each university's class and location (national, public, private, or foreign). Since bibliometric analysis is most effective on firms that actively publish, and since we expect these firms to be the leading innovators, we chose the top 100 paper-producing firms—

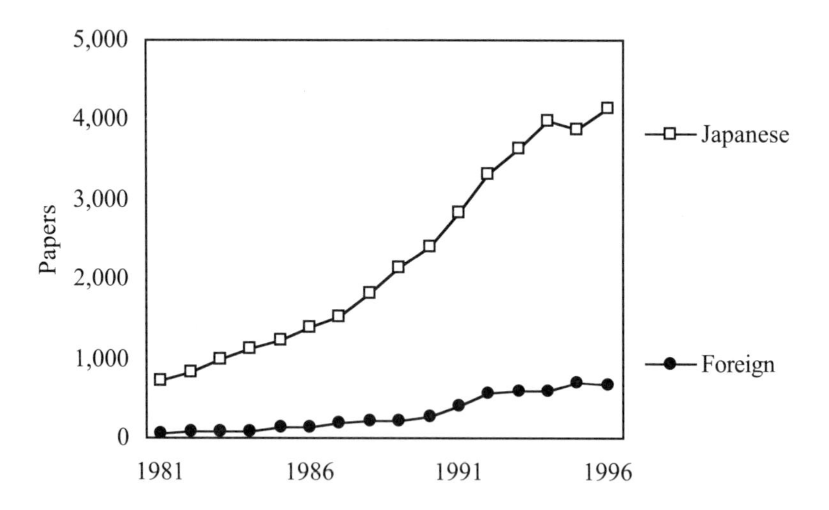

Figure 2 University-Industry Coauthored Papers by Location of University.

all papers, not only university-industry coauthored papers—from the study set over the entire 1981–1996 period.

Table 1 shows the 100-firm set. For each firm the table displays the corresponding industrial sector, and for the period 1981–1996 the total number of papers the firm has published, the total number of coauthored papers, the coauthorship ratio (defined as the ratio of coauthored to total papers) over the period, and the average yearly growth and R^2 of the coauthorship ratio time trend. Together, these 100 firms produced 82,013 papers, which comprise 74% of the 110,588 papers produced by Japanese industry in this period.

Average yearly growth is defined here as the slope of the coauthorship ratio time trend, and is thus the average yearly rise in percentage points of the ratio. R^2, or the coefficient of determination, measures the degree of linearity in the time trend. An R^2 close to one indicates a steady rise or fall at the average yearly growth rate, while one close to zero indicates extreme year-to-year volatility in the coauthorship ratio. In this table and the analyses to follow, unless otherwise specified, a coauthored paper refers to a paper that an author affiliated with a Japanese industrial sector organization published jointly with an author from academia, whether the academic author is located in Japan or abroad.

Table 1 Top 100 Paper-Producing Firms in Japan, 1981–1996

	Firm	Sector	Papers	Coauthorship			
				Papers	Ratio	Yearly Growth	R^2
1	Nippon Telegraph & Telephone	SVC	11,127	1,051	9.4%	0.9%	0.9
2	Hitachi	ELC	6,997	1,295	18.5%	1.0%	0.7
3	NEC Corp	ELC	4,112	552	13.4%	0.8%	0.8
4	Toshiba Corp	ELC	3,889	725	18.6%	0.6%	0.5
5	Fujitsu	ELC	2,689	385	14.3%	1.3%	0.7
6	Mitsubishi Electric Corp	ELC	2,431	501	20.6%	0.5%	0.4
7	Matsushita Electric Industries	ELC	2,117	424	20.0%	1.0%	0.5
8	Shionogi & Co	MED	2,075	770	37.1%	1.3%	0.6
9	Nippon Steel Corp	STL	2,055	536	26.1%	2.1%	0.9
10	Takeda Chemical Industries	MED	1,702	514	30.2%	0.9%	0.5
11	Mitsubishi Kasei Inst Life Sci	LAB	1,318	659	50.0%	1.7%	0.6
12	Kyowa Hakko Kogyo Co	CHM	1,178	356	30.2%	1.4%	0.6
13	Sankyo Co	MED	1,108	386	34.8%	0.5%	0.1
14	Sumitomo Chemical Co	CHM	1,107	323	29.2%	2.4%	0.6
15	Sumitomo Metal Industries	STL	1,085	240	22.1%	1.8%	0.5
16	Sony Corp	ELC	1,080	187	17.3%	1.5%	0.6
17	Fujisawa Pharmaceutical Co	MED	988	270	27.3%	0.2%	0.0
18	Toyota Motor Co	TRS	946	269	28.4%	–0.5%	0.1
19	Tanabe Seiyaku Co	MED	942	202	21.4%	1.3%	0.5
20	Kobe Steel	STL	937	310	33.1%	1.7%	0.7
21	Ajinomoto Co Inc	FOD	937	484	51.7%	1.9%	0.5
22	Daiichi Pharmaceutical Co	MED	870	304	34.9%	1.4%	0.3
23	Asahi Chemical Industries Co	CHM	836	391	46.8%	3.0%	0.8
24	NKK Corp	STL	830	251	30.2%	0.8%	0.1
25	Otsuka Pharmaceutical Co	MED	791	541	68.4%	–0.3%	0.0
26	Mitsubishi Heavy Industries	TRS	784	320	40.8%	–0.7%	0.1
27	Japan Tobacco Inc	FOD	779	285	36.6%	3.0%	0.7
28	Eisai & Co	MED	758	347	45.8%	0.5%	0.1
29	Chugai Pharmaceutical Co	MED	687	360	52.4%	0.5%	0.1
30	Taisho Pharmaceutical Co	MED	679	333	49.0%	1.9%	0.4
31	Oki Electric Industries Co	ELC	627	98	15.6%	0.5%	0.0
32	Suntory	FOD	622	408	65.6%	2.7%	0.6
33	JEOL	ELC	607	399	65.7%	–0.3%	0.0
34	Yamanouchi Pharmaceutical	MED	602	244	40.5%	1.8%	0.3
35	Sanyo Electric Co	ELC	581	131	22.5%	–1.1%	0.1
36	Sumitomo Electric Industries	NFM	572	186	32.5%	1.0%	0.2
37	Toray Industries	FIB	566	304	53.7%	3.0%	0.7
38	Kirin Brewery Co	FOD	561	261	46.5%	3.5%	0.6
39	Teijin	FIB	530	276	52.1%	1.2%	0.3
40	Tosoh Corp	CHM	515	303	58.8%	2.8%	0.6
41	Kao Corp	CHM	506	270	53.4%	3.2%	0.7
42	Mitsubishi Kasei Corp	CHM	486	249	51.2%	3.5%	0.2
43	Dainippon Pharmaceutical Co	MED	478	109	22.8%	0.7%	0.1
44	Meiji Seika Kaisha	FOD	472	234	49.6%	1.9%	0.5
45	Nissan Motor Co	TRS	465	170	36.6%	–0.3%	0.0
46	Nippon Kayaku Co	CHM	446	249	55.8%	2.5%	0.5
47	Kanebo	FIB	431	181	42.0%	2.3%	0.6
48	Ishikawajima Harima Heavy Ind	TRS	428	209	48.8%	0.8%	0.1
49	Kokusai Denshin Denwa Co	SVC	408	20	4.9%	–0.1%	0.0
50	Sharp Corp	ELC	397	88	22.2%	0.7%	0.1

Table 1 (continued)

Firm	Sector	Papers	Coauthorship			
			Papers	Ratio	Yearly Growth	R^2
51 Tokyo Electric Power Co Inc	SVC	395	83	21.0%	1.4%	0.4
52 Mitsubishi Chemical Industries	CHM	391	198	50.6%	2.5%	0.2
53 Snow Brand Milk Products	FOD	379	155	40.9%	3.3%	0.7
54 Furukawa Electric Corp	NFM	372	119	32.0%	1.0%	0.1
55 Fuji Photo Film Co	CHM	371	123	33.2%	2.2%	0.4
56 Shimadzu Corp	PMC	364	227	62.4%	2.5%	0.4
57 Asahi Glass Co	CER	353	130	36.8%	2.5%	0.5
58 Nippon Roche	MED	352	107	30.4%	3.3%	0.6
59 Banyu Pharmaceutical Co	MED	346	141	40.8%	–0.1%	0.0
60 Kawasaki Steel Chemical Ind	CHM	344	72	20.9%	1.3%	0.2
61 Ube Industries	CHM	335	183	54.6%	2.1%	0.3
62 Protein Engineering Res Inst	LAB	333	200	60.1%	5.4%	0.7
63 Kawasaki Steel Corp	STL	333	64	19.2%	1.1%	0.0
64 Nippon Oil & Fats Co	CHM	328	218	66.5%	2.0%	0.2
65 Takara Shuzo Co	FOD	316	176	55.7%	0.5%	0.0
66 Canon Inc	PMC	306	118	38.6%	1.8%	0.5
67 Idemitsu Kosan Co	PET	305	151	49.5%	1.6%	0.2
68 Taiho Pharmaceutical Co	MED	304	189	62.2%	2.1%	0.3
69 Hamamatsu Photonics	ELC	299	153	51.2%	5.1%	0.9
70 Fuji Electric Co	ELC	298	116	38.9%	1.2%	0.2
71 Yakult Honsha Co	FOD	288	119	41.3%	0.9%	0.2
72 Meiji Milk Products Co	FOD	288	149	51.7%	1.2%	0.1
73 Hoechst Japan	CHM	286	167	58.4%	2.8%	0.3
74 Tsumura & Co	MED	284	167	58.8%	1.0%	0.0
75 Kikkoman Foods Inc	FOD	271	51	18.8%	0.7%	0.1
76 Mitsui Toatsu Chemical Inc	CHM	267	139	52.1%	–2.5%	0.4
77 IBM Japan	ELC	267	120	44.9%	0.6%	0.0
78 Toyobo Co	FIB	264	136	51.5%	2.4%	0.3
79 Sumitomo Pharmaceutical Co	MED	259	92	35.5%	2.5%	0.3
80 Kawasaki Heavy Industries	TRS	257	113	44.0%	–0.4%	0.0
81 Kanegafuchi Chemical Industries	CHM	244	121	49.6%	2.8%	0.6
82 Mitsubishi Materials Corp	NFM	236	108	45.8%	4.4%	0.6
83 Kuraray Co	FIB	236	137	58.1%	1.7%	0.2
84 Yoshitomi Pharmaceutical Ind	MED	235	86	36.6%	0.3%	0.0
85 Toray Res Center Inc	LAB	234	109	46.6%	0.4%	0.0
86 Nippon Shinyaku Co	MED	231	112	48.5%	–0.9%	0.0
87 Denso Corp	ELC	225	161	71.6%	–1.0%	0.1
88 Kansai Electric Power Co Inc	SVC	216	70	32.4%	2.2%	0.3
89 Osaka Gas Co	SVC	215	136	63.3%	0.0%	0.0
90 Mitsui Petrochemical Industries	CHM	212	94	44.3%	2.5%	0.2
91 ATR	LAB	212	68	32.1%	2.7%	0.1
92 Olympus Optical Co	PMC	209	136	65.1%	–1.6%	0.2
93 Nisshin Steel Co	STL	207	87	42.0%	4.3%	0.6
94 Nikon Inc	PMC	207	90	43.5%	3.9%	0.6
95 Green Cross Co	MED	205	125	61.0%	–3.1%	0.4
96 Matsushita Res Inst Tokyo Inc	LAB	204	75	36.8%	2.2%	0.2
97 Mitsubishi Petrochemical Co	CHM	203	105	51.7%	0.5%	0.0
98 Unitika	FIB	199	126	63.3%	1.0%	0.1
99 Sumitomo Metal Mining Co	NFM	199	81	40.7%	2.1%	0.2
100 Hitachi Cable	NFM	195	46	23.6%	0.6%	0.0

What can we expect of a sample biased toward large paper-producing firms? Obviously, it is not representative of all firms. As Diana Hicks said of firms that publish papers in her seminal study of coauthorship in 28 Japanese firms in the years 1980, 1984, and 1989, "A company that performs science must have extra money, must use one of the more science-based technologies, must believe long-range research can result in profit, and must be open enough to allow publication of its discoveries. Few companies fit this description" (Hicks 1993). The companies that do fit this description, though, are those most likely to interact with universities, and are the main drivers of industrial innovation. Although this sample is certainly biased toward larger firms, there is still a wide variation in the size of firms as measured by sales revenue and R&D expenditure.

In order to get an idea of how firms performed in terms of coauthored papers over the entire time period, we have presented the same data in a somewhat different form in Table 2. In this table, the top 20 firms are listed for both the number of coauthored papers and the coauthorship ratio. Data are shown for each of the three subperiods, 1981–1986, 1987–1991, and 1992–1996, as well as for the overall 1981–1996 period. It is interesting that in general the top producers of coauthored papers are not the firms with the largest coauthorship ratio. We will consider this relationship later. Three firms, however, are on both lists for the overall period: Otsuka Pharmaceuticals, Suntory, and JEOL.

Other characteristics visible in this table are the rise in coauthored papers by firms from the iron and steel sector as a result of their diversification efforts, and the sudden disappearance of the Green Cross pharmaceuticals firm from the coauthorship ratio list as a result of its involvement in the AIDS-infected unheated blood products scandal.

Abbreviations used in Tables 1–3: CER=ceramics. CHM=chemicals. ELC=electrical machinery. FIB=textiles. FOD=food. LAB=laboratories. MED=drugs & medicines. NFM=non-ferrous metals. PET=petroleum & coal. PMC=precision instruments. STL=iron & steel. SVC=transport communications & public utilities. TRS=transport equipment.

Coauthorship Data by Industrial Sector

Table 3 shows the results of aggregating the firms of the 100-firm set into industrial sectors over the entire period of the study. We must be careful in generalizing this table to all of Japanese industry, as the sector data here consist of extremely small numbers of firms. In attempting to draw conclusions regarding a sector, we may instead be seeing only the characteristics of the particular firms in our set. For instance, the transportation, communications, and public utilities sector is dominated by a single firm, NTT. These 100 firms do, however, account for three-fourths of all publications by Japanese industry. Since the top six sectors are all represented in the 100-firm set by over five firms, we focus on these firms in our analyses. Figure 3 shows the total papers and university-industry coauthored papers time trends for these six sectors.

The electrical machinery sector has by far the largest number of papers, nearly tripling in our time period, from 1,710 papers to 4,592 papers; its nearest performer is the drugs and medicines sector which produced 1,140 papers in 1981 and 2,442 papers in 1996. All sectors showed fairly steady growth from 1981 to 1993, with growth faltering in 1994–96.

If we look at the plot of coauthored papers, we see that the electrical machinery and drugs and medicines sectors produce many more than the others, and have increased the gap further since 1993. Following these two sectors, the next actively coauthoring sectors are chemicals and food.

From the table and figure we see that the key attribute of the industrial sectors is that in all sectors other than transportation, the number of coauthored papers has grown while the total number of papers has leveled off and even declined since the early 1990s.

Foreign University Collaborations with Japanese Industry

Where are the universities located, with which Japanese industry collaborates? We saw earlier that over the period covered in this study, coauthored papers involving foreign universities grew from 8% to 16% of all coauthored papers; we next plotted foreign collaboration by region, based on the 100-firm set. Figure 4 shows

Table 2 Top 20 Firms by Coauthored Papers and Coauthorship Ratios

Papers 1981–1986

Firm (Sector)	Papers
1 Hitachi (ELC)	232
2 Shionogi (MED)	198
3 M Kasei Inst (LAB)	196
4 Toshiba (ELC)	188
5 Takeda Chem Ind (MED)	148
6 Otsuka Pharm (MED)	125
7 Sankyo (MED)	112
8 NTT (SVC)	110
9 JEOL (ELC)	109
10 Mitsubishi (ELC)	102
11 Chugai Pharma (MED)	85
12 NEC (ELC)	79
13 Eisai & Co (MED)	76
14 Ajinomoto (FOD)	74
15 Matsushita (ELC)	69
16 Mitsub. Heavy Ind (TRS)	69
17 Teijin (FIB)	69
18 Toyota Motor Co (TRS)	67
19 Japan Tobacco (FOD)	66
20 Meiji Seika Kaisha (FOD)	60
20 Nippon Steel (STL)	60

Papers 1987–1991

Firm	Papers
1 Hitachi (ELC)	376
2 NTT (SVC)	347
3 Shionogi (MED)	279
4 Toshiba (ELC)	204
5 M Kasei Inst (LAB)	196
6 Ajinomoto (FOD)	192
7 Suntory (FOD)	167
8 Takeda Chem Ind (MED)	164
9 Mitsubishi (ELC)	160
10 NEC (ELC)	152
11 Otsuka Pharma (MED)	143
12 JEOL (ELC)	142
13 Nippon Steel (STL)	142
14 Eisai & Co (MED)	118
15 Sankyo Co (MED)	116
16 Matsushita (ELC)	113
17 Taisho Pharma (MED)	111
18 Asahi Chem (CHM)	110
19 Kyowa Hakko (CHM)	110
20 Chugai Pharma (MED)	107

Papers 1992–1996

Firm	Papers
1 Hitachi (ELC)	687
2 NTT (SVC)	594
3 Nippon Steel (STL)	334
4 Toshiba (ELC)	333
5 NEC (ELC)	321
6 Shionogi (MED)	293
7 Otsuka Pharm (MED)	273
8 M Kasei Inst (LAB)	267
9 Matsushita (ELC)	242
10 Mitsubishi (ELC)	239
11 Asahi Chem (CHM)	235
12 Fujitsu (ELC)	230
13 Ajinomoto (FOD)	218
14 Takeda (MED)	202
15 Kyowa Hakko (CHM)	193
16 Suntory (FOD)	192
17 Kirin (FOD)	183
18 Tosoh (CHM)	180
19 Sumitomo Met (STL)	173
20 Taisho Pharma (MED)	173

Ratio 1981–1986

Firm (Sector)	Ratio
1 Mitsubishi Kasei (CHM)	100.0%
2 Osaka Gas (SVC)	88.9%
3 Olympus Opt (PMC)	81.3%
4 Green Cross (MED)	77.5%
5 Otsuka Pharma (MED)	70.2%
6 Tsumura (MED)	67.4%
7 JEOL (ELC)	67.3%
8 Denso (ELC)	66.7%
9 Mitsui Toatsu Chem (CHM)	66.1%
10 Nippon Oil & Fats (CHM)	56.8%
11 Meiji Milk Prod (FOD)	55.6%
12 Unitika (FIB)	54.5%
13 IBM Japan (ELC)	54.1%
14 Taiho Pharma (MED)	53.3%
15 Chugai Pharma (MED)	51.8%
16 Kuraray (FIB)	50.8%
17 Banyu Pharma (MED)	50.0%
18 Kawasaki Steel (CHM)	50.0%
19 Ishikawajima (TRS)	48.7%
20 Nissan (TRS)	48.6%

Ratio 1987–1991

Firm (Sector)	Ratio
1 Denso (ELC)	78.3%
2 Green Cross (MED)	69.7%
3 Suntory (FOD)	66.5%
4 Osaka Gas (SVC)	65.3%
5 Kuraray (FIB)	64.4%
6 Otsuka Pharma (MED)	63.6%
7 JEOL (ELC)	62.8%
8 Unitika (FIB)	62.3%
9 Mitsub. Petrochem (CHM)	62.2%
10 Olympus Opt (PMC)	61.4%
11 Taiho Pharma (MED)	61.1%
12 Mitsub. Chem Ind (CHM)	60.8%
13 Prot Engn Res Inst (LAB)	59.7%
14 Hoechst Japan (CHM)	58.6%
15 Kao (CHM)	58.4%
16 Mitsub. Mat Corp (NFM)	56.8%
17 Nikon (PMC)	55.8%
18 Kawasaki Heavy Ind (TRS)	54.9%
19 Nippon O&F (CHM)	54.3%
20 Teijin (FIB)	52.0%
20 Nippon Kayaku (CHM)	52.0%

Ratio 1992–1996

Firm (Sector)	Ratio
1 Shimadzu (PMC)	77.7%
2 Nippon O&F (CHM)	76.0%
3 Suntory (FOD)	71.1%
4 Tosoh (CHM)	70.6%
5 Otsuka Pharma (MED)	70.4%
6 Denso (ELC)	68.9%
7 Nippon Kayaku (CHM)	68.5%
8 JEOL (ELC)	67.6%
9 Toray Ind (FIB)	67.2%
10 Unitika (FIB)	66.7%
11 Taiho Pharma (MED)	66.2%
12 Ube Ind (CHM)	66.0%
13 Olympus Opt (PMC)	64.4%
14 Hamamatsu (ELC)	63.5%
15 Takara Shuzo (FOD)	63.5%
16 Kanegafuchi (CHM)	62.9%
17 Meiji Milk (FOD)	61.9%
18 Osaka Gas (SVC)	61.1%
19 Kao (CHM)	61.1%
20 Tsumura (MED)	61.1%

Table 2 (continued)

Overall Papers 1981–1996		Overall Ratio 1981–1996	
Firm (Sector)	Papers	Firm (Sector)	Ratio
1 Hitachi (ELC)	1,295	1 Denso (ELC)	71.6%
2 NTT (SVC)	1,051	2 Otsuka Pharma (MED)	68.4%
3 Shionogi (MED)	770	3 Nippon O&F(CHM)	66.5%
4 Toshiba (ELC)	725	4 JEOL (ELC)	65.7%
5 M Kasei Inst (LAB)	659	5 Suntory (FOD)	65.6%
6 NEC (ELC)	552	6 Olympus Opt (PMC)	65.1%
7 Otsuka Pharma (MED)	541	7 Unitika (FIB)	63.3%
8 Nippon Steel (STL)	536	8 Osaka Gas (SVC)	63.3%
9 Takeda Chem Ind (MED)	514	9 Shimadzu (PMC)	62.4%
10 Mitsubishi Electr (ELC)	501	10 Taiho Pharma (MED)	62.2%
11 Ajinomoto (FOD)	484	11 Green Cross (MED)	61.0%
12 Matsushita (ELC)	424	12 Prot Engn Res Inst (LAB)	60.1%
13 Suntory (FOD)	408	13 Tosoh (CHM)	58.8%
14 JEOL (ELC)	399	14 Tsumura (MED)	58.8%
15 Asahi Chem (CHM)	391	15 Hoechst Japan (CHM)	58.4%
16 Sankyo (MED)	386	16 Kuraray (FIB)	58.1%
17 Fujitsu (ELC)	385	17 Nippon Kayaku (CHM)	55.8%
18 Chugai Pharma (MED)	360	18 Takara Shuzo (FOD)	55.7%
19 Kyowa Hakko (CHM)	356	19 Ube Ind (CHM)	54.6%
20 Eisai (MED)	347	20 Toray Ind (FIB)	53.7%

Table 3 Industrial Sector Aggregation of 100-Firm Set, 1981–1996

				Coauthorship			
Industrial Sector		Firms	Papers	Papers	Ratio	Yearly Growth	R^2
1	MED	20	13,896	5,399	38.8%	1.1%	0.88
2	CHM	17	8,055	3,561	44.7%	2.1%	0.96
3	ELC	15	26,616	5,335	19.8%	0.9%	0.84
4	FOD	10	4,913	2,322	47.3%	2.3%	0.96
5	STL	6	5,447	1,488	27.0%	1.8%	0.84
6	FIB	6	2,226	1,160	52.1%	2.1%	0.81
7	SVC	5	12,361	1,360	11.0%	1.1%	0.91
8	TRS	5	2,880	1,081	37.4%	–0.3%	0.03
9	LAB	5	2,301	1,111	48.4%	1.2%	0.50
10	NFM	5	1,574	540	35.5%	1.0%	0.31
11	PMC	4	1,086	571	52.4%	1.3%	0.38
12	CER	1	353	130	36.8%	2.5%	0.52
13	PET	1	305	151	49.5%	1.6%	0.18
Aggregate		100	82,013	24,209	29.5%	1.4%	0.98

that from 1981–1996, 61.4% of foreign coauthors have been in the United States, followed by 9.5% in the United Kingdom, 5.6% in Germany, 5.0% in Canada, 1.6% in France and 1.2% in Italy. The total for non-G7 countries is 23.1% (the percentages add up to over 100% due to papers coauthored with researchers in more than one country).

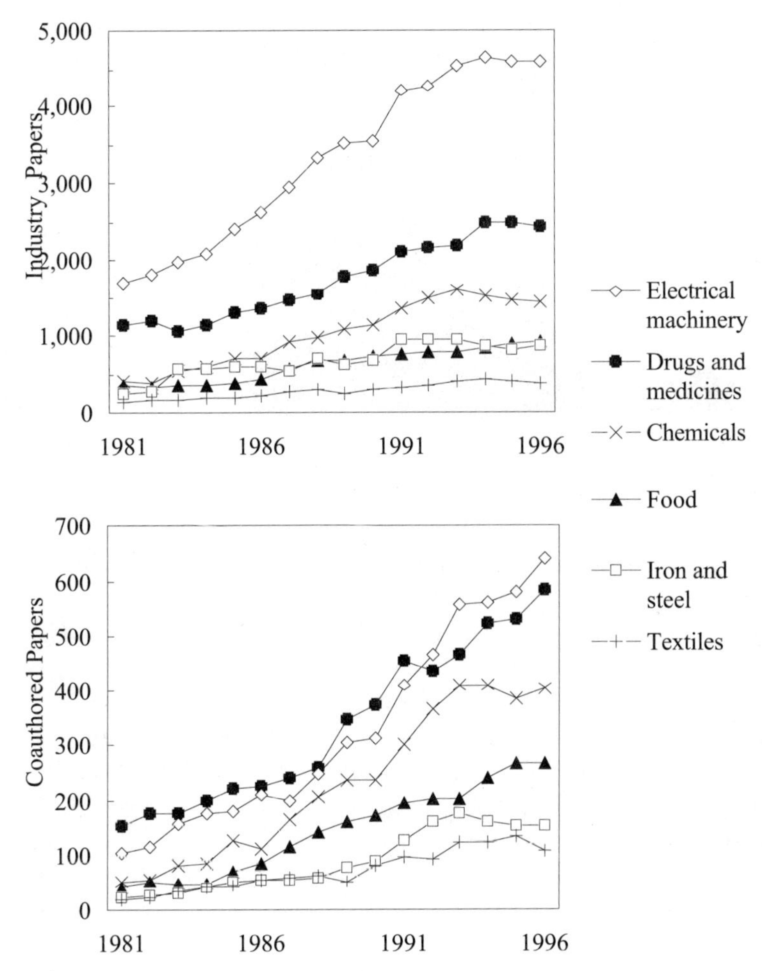

Figure 3 Time Trend of Papers and Coauthored Papers by Industrial Sectors.

At first glance, this appears to support the common assertion that Japanese industry is highly dependent on U.S. academic research. However, as mentioned earlier, as a percentage of all foreign collaboration, coauthorship with the United States has dropped, from a high of 80% in 1983 to 51% in 1996. Over the same period, coauthorship with university researchers in the non-G7 countries went from 9% to 25% of all internationally coauthored papers. Even though the United States is still by far the most frequent collaborator with Japanese firms, the empirical evidence does not

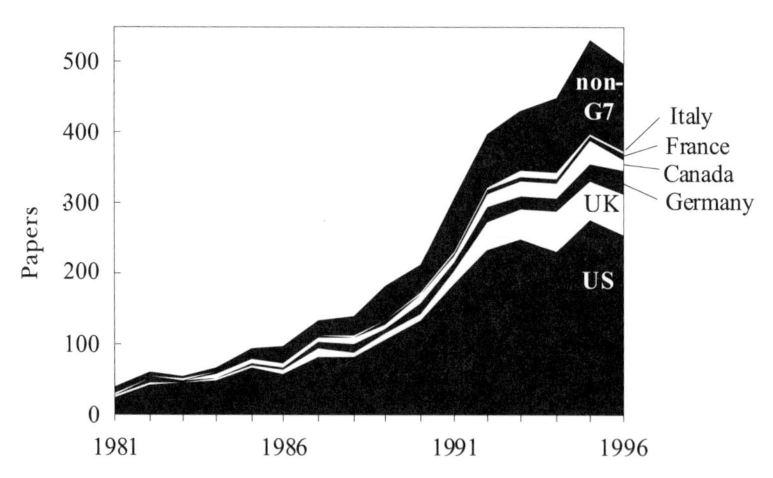

Figure 4 Foreign University Collaborations with Japanese Industry.

support the view of Japan's growing dependence on U.S. academic research.

Table 4 shows the top 50 universities (domestic and foreign) that collaborate with Japanese industry. It is not surprising that the top ten Japanese universities are the major national universities, seven of which are former imperial universities. These are the best-funded of Japan's universities, and thus have the strongest research infrastructures. Slots 11 and 18 are the leading private universities, Keio and Waseda. MIT and Stanford University rank 27 and 47 among all universities in terms of numbers of papers coauthored.

Table 5 shows the top 50 foreign universities that collaborate with Japanese industry, led by the United States, with England, the People's Republic of China, Canada, Belgium, Australia, Sweden, Germany, and Switzerland following in that order.

Analysis of Coauthorship in Relation to Firm and Sector Structural Characteristics

The data thus far have clearly shown that coauthorship with academia is prevalent and rising in Japanese industry. In this section we examine the issue of whether certain structural characteristics of firms and industrial sectors are related to the degree of coauthorship. In particular, we focus on three commonly accepted

Table 4 Top University Collaborators, 1981–1996

University	Class	CP	University	Class	CP
1 U Tokyo	national	2,847	27 MIT	US	187
2 Osaka U	national	1,957	28 Yamaguchi U	national	174
3 Kyoto U	national	1,670	29 Yokohama Natl U	national	173
4 Tohoku U	national	1,302	30 Teikyo U	private	164
5 Tokyo Inst Tech	national	993	30 Shizuoka U	national	164
6 Nagoya U	national	872	32 Kanazawa U	national	162
7 Kyushu U	national	812	33 Tokyo Med & Dent U	national	161
8 Hokkaido U	national	683	34 Nagoya Inst Tech	national	155
9 U Tsukuba	national	555	35 Nihon U	private	152
10 Hiroshima U	national	494	36 Niigata U	national	146
11 Keio U	private	343	37 Nagaoka U Tech	national	144
12 Chiba U	national	301	38 Toho U	private	143
13 Okayama U	national	287	39 Tokyo Metropolitan U	public	142
14 Gunma U	national	272	40 Kyoto Inst Tech	national	141
15 U Tokushima	national	270	41 Tokyo U Pharmacy	private	140
16 NLHEP	national	269	41 Kagoshima U	national	140
17 Kobe U	national	259	43 Showa U	private	136
18 Waseda U	private	254	43 Shinshu U	national	136
19 Kitasato U	private	245	45 Kyushu Inst Tech	national	134
20 Science U Tokyo	private	237	46 Mie U	national	133
21 Tokyo U Agr & Tech	national	231	47 Stanford U	US	128
22 Kumamoto U	national	224	48 Kinki U	private	125
23 U Osaka Prefecture	public	218	49 Nagasaki U	national	122
24 Tokai U	private	205	50 Meijo U	private	121
25 Gifu U	national	195	50 Kyoto Pharmaceutic U	private	121
26 Osaka City U	public	188			

CP=number of coauthored papers. NLHEP=National Laboratory for High Energy Physics.

views concerning industry motives for collaboration with academic researchers. The first view is that coauthorship is more likely to occur in smaller firms, since larger firms possess sufficient financial resources and research infrastructure to carry out research without dependence on university research. Applied to industrial sectors, this view says that sectors with larger resource bases would exhibit less coauthorship. We call this the *size hypothesis*, and state it this way:

Size Hypothesis: Firms or industrial sectors with larger resource bases exhibit lower degrees of coauthorship.

To operationalize this hypothesis, we calculated the coefficients of correlation between the 16-year coauthorship ratios and both annual sales revenues and annual R&D expenditures. We obtained mutually comparable 1997 sales figures for nearly all of the firms in the 100-firm set (Toyo Keizai Shinposha Publishers 1998). Since

Table 5 Top Foreign University Collaborators, 1981–1996

University	Country	CP	University	Country	CP
1 MIT	US	187	27 U Colorado	US	28
2 Stanford U	US	128	28 U Toronto	Canada	26
3 U Calif Berkeley	US	103	28 Catholic U Leuven	Belgium	26
4 U Illinois	US	80	28 Carnegie Mellon U	US	26
5 Harvard U	US	63	31 Yale U	US	25
6 U Texas	US	54	31 Princeton U	US	25
6 U Cambridge	England	54	31 Columbia U	US	25
8 Vanderbilt U	US	53	34 U Calif Riverside	US	24
9 U Washington	US	51	35 U Oxford	England	23
9 U Pennsylvania	US	51	35 Arizona State U	US	23
11 U London ICS	England	41	35 Academy Sinica	China	23
11 U Hawaii	US	41	38 U Calif Davis	US	22
13 U Arizona	US	40	38 Purdue U	US	22
14 U Calif San Diego	US	39	40 U Melbourne	Australia	21
14 Ohio State U	US	39	40 Tulane U	US	21
16 CWRU	US	38	40 Royal Inst Tech	Sweden	21
17 Penn State U	US	36	43 U Pittsburgh	US	20
17 Northwestern U	US	36	43 U Birmingham	England	20
17 Cornell U	US	36	43 Johns Hopkins U	US	20
20 U Wisconsin	US	34	46 U Massachusetts	US	19
21 U Calif SB	US	33	46 U Heidelberg	Germany	19
21 UCLA	US	33	46 U Alabama	US	19
21 CAMS	China	33	46 Calif Inst Tech	US	19
24 U Michigan	US	32	50 U Calif San Francisco	US	18
25 Washington U	US	31	50 Swiss Fed Inst Tech	Switzerland	18
26 UBC	Canada	30	50 McGill U	Canada	18

CP=number of coauthored papers. CAMS=Chinese Academy of Medical Science. CWRU=Case Western Reserve. ICS=Imperial College of Science. SB=Santa Barbara. UBC=University of British Columbia.

mutually comparable R&D figures, however, are more difficult to obtain, we decided that we would use the 1998 R&D figures reported by the Nikkei Shimbun Newspaper (Nikkei Shimbun Company 1998). While these figures are more uniform in definition than those available elsewhere, the published data only include figures for 55 firms of the 100-firm set, requiring that we narrow our sample to these 55 firms.

The results of this correlation analysis are shown in Table 6 under the heading 55-firm set. Although the coefficients are in the rather weak range of –0.5 to –0.4, their negative signs do indeed support the size hypothesis.

Because correlation analysis reduces a wealth of data to a single number, however, it is important to observe the underlying data when interpreting what that single number means. We therefore consulted scatter plots of all the data on which we performed correlation analysis. Since all the plots in general displayed the same characteristics, to save space we have shown just one of them in Figure 5, sales vs. coauthorship ratio for the 100-firm set.

By observing the plot of sales revenue against coauthorship ratio, we can see a decline in coauthorship with rising sales; this relationship is most profound at the tail containing the small number of firms with sales over 1 trillion yen; within the majority of firms, the relationship is less obvious. The small number of firms with extremely large sales revenues and R&D expenditures tend to dominate the calculation of the correlation coefficient. In order to test the hypothesis on the remaining firms, we removed the outliers to construct a set of 43 firms, defining an outlier as a firm having at least one parameter of papers, sales, or R&D expenditure which differs from the median value of all 55 firms by more than 1.5 times the difference between the 75th and 25th percentile values. The outlier firms thus identified were Toyota, NEC, Hitachi, Fujitsu, Toshiba, Sony, NTT, Nissan, Canon and Mitsubishi Electric, Denso and Shionogi. As shown in Table 6, the correlation coefficients for the 43-firm set are much closer to zero, indicating that the size hypothesis is not well-founded for the set. From this we conclude that the size hypothesis holds weakly when contrasting firms with tremendous differences in scale, but does not hold in general across all firms.

For the industrial sector level analysis, we narrowed the sample to the six sectors well-represented in the 100-firm set: chemicals, drugs and medicines, fibers, food, electrical machinery, and iron and steel. While the firms included in the 100-firm set are undoubtedly the key innovators of each sector in Japan, the data of the set come from only a small portion of all firms in each sector. In order to act as a check on our sector aggregates for R&D, therefore, we have also calculated the coefficients of correlation analysis using the sector-wide R&D expenditure statistics published by the government of Japan's Management and Coordination Agency (MCA), the most comprehensive statistics on R&D activity in Japanese industry. As Table 6 shows, the size hypothesis is fairly well supported among industrial sectors. This conclusion is reinforced by the fact that the two entirely different R&D measures give such similar results.

Table 6 Correlation Coefficients for Hypothesis Testing

Hypothesis	Factor 1	Factor 2	55-Firm Set	43-Firm Set	Sector Data
Size	Coauthorship Ratio	Sales Revenue	–0.46	–0.32	–0.78
	Coauthorship Ratio	R&D Expenditure (firm data)	–0.42	–0.21	–0.75
	Coauthorship Ratio	R&D Expenditure (MCA statistics)	NA	NA	–0.72
Intensity	Coauthorship Ratio	R&D Intensity Ratio (firm data)	0.12	0.16	–0.11
Funding Flow	Coauthorship Ratio	Industry to Universities Flow (MCA, %)	NA	NA	0.29
	Coauthorship Ratio	Industry to Foreign Flow (MCA, %)	NA	NA	0.06

NA = not applicable

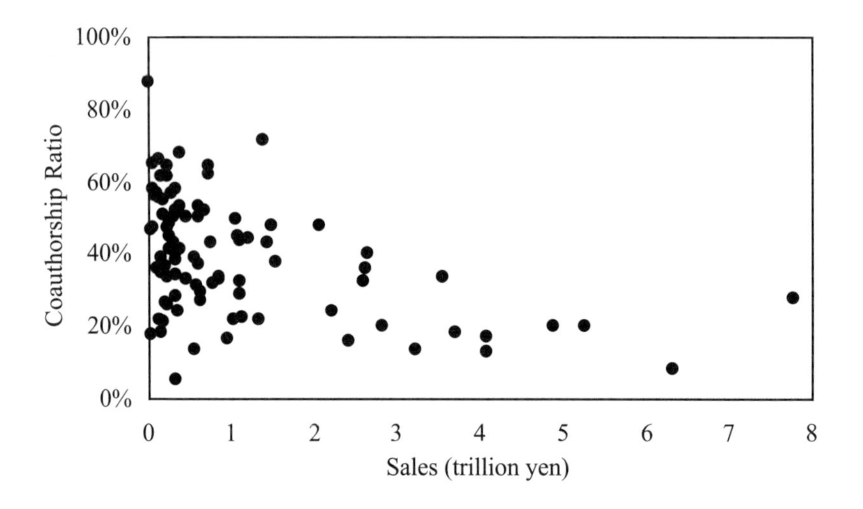

Figure 5 Relationship between Sales and Coauthorship Ratio (100-Firm Set).

The second view concerning coauthorship holds that more research-intensive firms will be more likely to benefit from the basic research carried out in universities, and thus are more likely to coauthor papers with university researchers. We call this the *intensity hypothesis,* and state it this way:

Intensity Hypothesis: Firms or industrial sectors with higher research intensities exhibit higher degrees of coauthorship.

To operationalize this hypothesis, we calculated the coefficients of correlation between the R&D-to-sales ratios and the coauthorship ratios. Again, we performed this for the 55-firm set, the 43-firm set and at the sector level. The results are again shown in Table 6. The near zero correlation coefficients for all three cases indicate that the intensity hypothesis has no validity at all.

The third commonly accepted view concerns the linkage between cross-organizational funding and collaborative publication. If coauthorship were often the result or by-product of outsourced research, we would expect a correspondence between the flow of R&D funds from industry to academia and the degree of university-industry coauthorship. We call this the *funding flow hypothesis,* and state it this way:

Funding Flow Hypothesis: Firms or industrial sectors with larger funding flows to universities exhibit higher degrees of coauthorship.

Since the funding flows of individual firms are not widely available, we limited this analysis to the sector level at which we have access to the Management and Coordination Agency statistics on funding flow. We calculated the correlation coefficients between the 16-year coauthorship ratios and the 16-year industry-university R&D funding flows (expressed as percentages of total R&D expenditures of the sectors). Since a small but growing portion of university collaborators are not in Japan, we also analyze R&D funds being paid abroad. The data on foreign expenditures, however, do not distinguish between money paid to foreign universities and to foreign firms. The results are shown in Table 6.

There is no correlation at all between foreign funding flow and the coauthorship ratio, indicating that the funding flow hypothesis does not apply to the R&D funds paid abroad. For the flow to universities in Japan, however, the correlation of 0.29 suggests there may be a weak relationship. We therefore refer in Figure 6 to the plot of coauthorship ratio against funding flow in order to interpret this value.

According to the plot, there indeed may be a relationship between funding flow and coauthorship if we ignore the food and the drugs and medicines sectors as outliers. In the case of the food sector, there may be a justification for doing so, since a large portion of the funding flow of the sector over the entire period of this study occurred in a single year, probably the result of a specific program (see Niwa in this volume). There may also be a justification for treating the drugs and medicines sector as a special case, since the nature of research in this field is such that it is very close to the research of universities, and the sector is also dependent on university hospitals for drug trials (see Kneller's chapter 16, in this volume). If we thus treat these two sectors as outliers, then we can indeed see a direct relationship between funding flow and coauthorship.

On the other hand, however, there is ample reason to accept the relatively low correlation coefficient at face value. To begin with, comparison of funding flows and coauthorship ratios over time indicates that the ongoing rise in coauthorship ratios is not a direct result of a rise in funding flows: The rise in R&D funding flows from industry to national universities does not correlate with the rise in coauthorship. Furthermore, the MCA statistics themselves may obscure the relationship even if one did in fact exist. The reason is that, as Niwa, Ohtawa, and Odagiri discuss in this volume, assumptions must be made in order to treat the MCA statistics as industry-university funding flows; and even if these assumptions are valid, the coverage and collection methodology of the statistics introduce additional sources of error. As a result, the data may include funds that did not actually support research or fail to include funds that did. In this less-than-transparent environment, we do not expect that coauthorship performance will show up as a result of specific transfers that appear in official statistics. All things considered, the validity of the funding flow hypothesis remains in question, and will

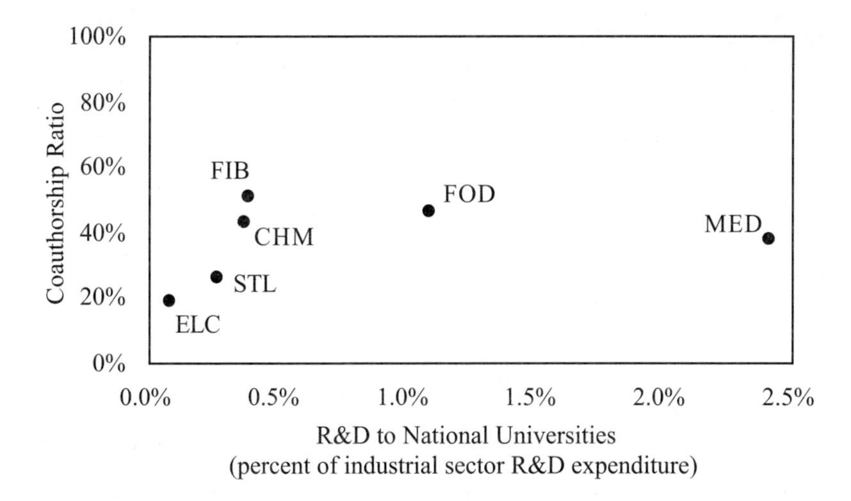

Figure 6 Relationship between R&D Funding Flow and Coauthorship, 1981–1996.

until more work is carried out on the collaborative mechanisms of the university-industry linkage.

As an example of the need to further elucidate collaborative mechanisms, it is interesting to note that although the coauthorship ratio correlates highly with neither funding flow nor R&D intensity, funding flow and R&D intensity do correlate quite well with each other. This is shown in Table 7.

The implication is that while it is the high-sales sectors that conduct much R&D and publish many papers, it is the R&D-intensive sectors that support R&D in universities and abroad. Lack of a significant correlation between R&D intensity and the coauthorship ratio suggests that this increased investment in university and foreign R&D does not translate directly into coauthored publications. We hope that further work on these critical mechanisms will lead to a better understanding of the relationship between external funding and coauthorship.

Conclusions

Despite widely held views that the linkage between industry and university research in Japan does not play a strong role in Japan's

Table 7 Correlation Coefficients for R&D Intensity

Hypothesis	Factor 1	Factor 2	55-Firm Set	43-Firm Set	Sector Data
Intensity	R&D Intensity Ratio	Papers	0.00	0.00	0.51
	R&D Intensity Ratio	Coauthorship Ratio	0.12	0.16	–0.11
	R&D Intensity Ratio	R&D Expenditure (MCA)	NA	NA	0.31
	R&D Intensity Ratio	Industry to Universities Flow (MCA)	NA	NA	0.83
	R&D Intensity Ratio	Industry to Universities Flow (MCA, %)	NA	NA	0.62
	R&D Intensity Ratio	Industry to Foreign Flow (MCA)	NA	NA	0.90
	R&D Intensity Ratio	Industry to Foreign Flow (MCA, %)	NA	NA	0.81

NA = not applicable

industrial development, the empirical evidence of bibliometric analysis draws a very different picture. Our analysis shows that the coauthoring of scientific and technical papers between Japanese industry and academia both in Japan and abroad is prevalent and has been rising over the 1981–1996 period covered in this study. In spite of anecdotal evidence to the contrary, we found that university-industry coauthorship in Japan is slightly more common than in the United States.

We also found empirical evidence that disputes another widely held view, that when it comes to utilization of university resources, it is foreign universities—primarily those in the United States—from which Japanese industry draws; our study shows that from 1981 to 1996 only about one-tenth of the university collaborators were outside of Japan. While the rise in the foreign component of coauthorship over this period, from 8% to 16%, does lend credence to the common view, American universities have, on the contrary, continually declined relative to the universities of non-G7 nations as collaborators of Japanese industry.

While there are many ways to assess the university-industry linkage and we thus do not claim ours to be the only and final one, our assessment at the very least requires that the common wisdom be reconsidered. Since effective policy formulation depends on the accuracy of such assessments, this reconsideration must extend to the policy frameworks which emerged based on the common wisdom. The debate over asymmetric access, the adoption of

American-style intellectual property rights policies in Japan, even the innovation strategies of companies, are all issues that require rethinking based on this analysis. For instance, in a recent study of the linkage between public science and industrial technology, Narin, Hamilton, and Olivastro (1997) showed that the average number of references to scientific papers on the U.S. patents of Japanese firms has nearly tripled over the period 1985–1995. The conventional wisdom would be apt to attribute this rapid increase to a growing dependence of Japanese industry on academic research. There is little doubt that this is indeed at least one cause of the increase, as growing dependence on academic research is a trend common to all innovations systems. According to our analysis, however, we first must acknowledge that since at least 1981, the first year covered by our study, Japanese industry had already been very closely involved with university researchers. The fact that this did not translate into more citations on Japanese industry patents is probably a result of the specific institutional arrangement that evolved in Japan, in which university researchers' contributions were often not officially documented. With this in mind, an additional explanation for the rapid rise in science citations on Japanese industry patents is that the institutional arrangement is changing and existing dependence on public science is being acknowledged on patents for the first time, rather than the dependence itself just beginning. While the two views are complementary to each other, they have different implications for policy formulation. In this way, the assessment based on coauthorship enables more informed policy formulation than would be the case when relying on the conventional view alone.

Given the importance of measured assessments in the policy making process, why have these inaccurate assessments taken hold in the first place? The cause is most likely the interpretation of the features of the Japanese innovation system in an American context. Scarcity in Japan of the familiar landmarks of the university's role in industrial innovation that exist in the United States—university-held patents, faculty-founded firms and so on—leads many observers to assume that the university's role in Japan is insufficient. However, the policy frameworks shaping university-industry interaction in Japan evolved out of very different circumstances from those in the United States; in light of the funding regulations of Japanese universities, the civil servant status of national university faculty, and other pieces of the institutional arrangement in Japan, we should not expect to see these landmarks in great numbers. Therefore, their paucity is not necessarily a sign of weakness, but may be instead just a sign of difference.

Our point is *not* to say that the university-industry linkage in Japan should not learn from the ongoing changes of the linkage in the United States. The linkage is considered by many to be a growth engine of American economic development, and Japanese policy makers, facing the longest economic slump of the postwar era, are naturally looking to the United States for solutions. But before adopting components of the American approach, the Japanese system and the American approach must be critically assessed. As the coauthorship analysis we described in this chapter demonstrates, this requires the development of specific measurement techniques suitable to the respective contexts of each country, or, even better, general techniques suitable to the contexts of both.

References

Hicks, Diana. 1993. "University-industry research links in Japan." *Policy Sciences* 26:377.

Kakinuma, Sumio and Kenneth Pechter. 1999. "University-industry research collaboration in Japan: Industry dependence on academic research" ("Nihon kigyo to daigaku no kyodo kenkyu—Daigaku kenkyu e no izon"). *Research Bulletin of the National Center for Science Information Systems (Gakujutsu Joho Senta-Kiyo)* 10 (March): 197–205.

Management and Coordination Agency. 1980–1997. *Report on the Survey of Research and Development (Kagaku Gijutsu Kenkyu Chosa Hokokusho)*. Tokyo: Government of Japan.

Narin, Francis, Kimberly S. Hamilton, and Dominic Olivastro. 1997. "The increasing linkage between US technology and public science." *Research Policy* 26(3): 317–330.

National Science Board. 1998. *Science & Engineering Indicators—1998*. Arlington, VA: National Science Foundation.

Nikkei Shimbun Company. 1998. "R&D investments" ("Kenkyu kaihatsu toshi"). *Nikkei Industry Newspaper (Nikkei Sangyo Shimbun)*. July 20, page 1.

Pechter, Kenneth and Sumio Kakinuma. 1999. "Coauthorship as a measure of university-industry research interaction in Japan." *Proceedings of the Portland*

International Conference on Management of Engineering and Technology, IEEE Engineering Management Society, July.

Toyo Keizai Shinposha. 1998. *Company Quarterly Handbook* (*Kaisha Shikiho*). Tokyo: Toyo Keizai Shinposha Publishers.

5

An Inter-industrial Comparative Study of R&D Outsourcing

Fujio Niwa

Introduction

During the period 1986–1996 Japanese industry, constrained by various obstacles within Japan, increased the amounts spent on R&D performed abroad by about six times, while increasing its funding of Japanese R&D by only about 1.5 times. Recently the obstacles have been identified and various measures taken to remove them, with the result that R&D funds paid by industry to Japanese universities have increased. However, even this modest improvement shows signs of slowing.

This suggests that there must be some substantial problems in Japanese universities rather than institutional obstacles acting against a smooth flow of R&D funds from companies to universities, such as intricate account management. Under the current global competition, Japanese companies have relied much more on R&D sources in foreign countries; the position of Japanese universities as sources of creativity and knowledge has been continuously declining. Many Japanese have not recognized the seriousness of the situation.

The main purposes for which companies pay R&D funds to universities and foreign countries are to acquire technology and its results, to collect scientific and technological information, and to hire top-level scientists and engineers, functions that are thought of as a kind of outsourcing of R&D. This paper presents a quantitative analysis of the trends in outsourcing by Japanese companies and interprets them based on interview surveys. Next, the real R&D outsourcing activities of each Japanese industry are arranged according to a conceptual framework. Finally, from these analyses some policy implications are presented.

Trends in R&D Outsourcing by Japanese Companies

The use of external resources opposes the traditional methods of some large Japanese companies, which carry out all activities "in-house," preparing a full set of goods "from cradle to tombstone," and disseminate little information outside the company.

The recent concept of outsourcing attaches much more importance to strategic value than to the activity of using external resources itself. According to the definition of The Outsourcing Institute, outsourcing is a very effective strategic tool (Ozeki 1997:16). The function of outsourcing has changed over time, from "cost reduction "aimed at reduction of indirect and personnel expenses at an early stage, to "pursuit of added value," aimed at improvement of efficiency in the work flow and establishment of a new organization. From the viewpoint of organizational relation theory, this shift corresponds to a conversion from the classic resource approach to the more recent learning approach.

The recently used definition of outsourcing in Japan is "to externalize all works from planning to execution of business which have been carried out or will be started by organizations of a company with clear defined strategic purposes such as (1) concentration of resources on its core businesses, (2) attainment of specialization, and (3) reduction of costs." Under this definition R&D itself and its component parts must be objectives of outsourcing (Ozeki 1997: 18–19).

Dr. Mitsuyo Hanada, a professor at Keio University, drew up a useful table which broke down outsourcing into two dimensions: (1) planning and design of business, and (2) execution of business Ozeki (1997: 17). He classified only one model as outsourcing, a case where both these areas are carried out with the close cooperation of external partners. However, R&D activity is in itself highly strategic, and does not always fit this model. A different classification model is used in this paper. Only the case in which a company

Table 1 Types of R&D Outsourcing

		R&D Execution	
		Self-implementing	Carried out with partners
R&D Planning and Design	Self-implementing	Information collection, patent purchase, cross-license	Contract R&D
	Carried out with partners	Consultant	Concurrent R&D, collaboration

From information kindly supplied by Dr. Mitsuyo Hanada.

Table 2 Partners of R&D Outsourcing

	Sector	Outsourcing Partner
Domestic	Private sector	Companies Private R&D institutions Private Survey and Consultant institutions Private universities
	Central and local governments	National and public universities National and public research institutions
	Special corporations (*Tokushu Hojin*)	Government-affiliated agencies and R&D institutions Public corporations and enterprises
Overseas	Private sector	Companies Private R&D institutions (including ventures) Private survey and consultant companies Private universities
	Governmental organizations	Public universities Governmental R&D institutions

Data compiled by author.

carries out all its own R&D planning, design, and execution is not considered strategic outsourcing. The partners of R&D outsourcing, classified on the basis of their strategic characteristics, are indicated in Table 2.

There are three statistical sources concerning R&D outsourcing that are continuously surveyed: The Management and Coordination Agency (MCA) annual reports entitled "Report on the Survey of Research and Development"; the Ministry of International Trade and Industry (MITI) annual reports entitled "Overseas Business Activities of the Japanese Enterprises"; and the Ministry of Education, Sports and Culture (Monbusho) "Report by the Committee on the Investigation of Promoting Cooperation between Industry and Academia" (Monbusho 1997). None of these three surveys measures all kinds of R&D outsourcing. Therefore in this paper the statistics surveyed by the MCA are used for the main analysis with the other statistics introduced to fill gaps in the MCA data.

Trends of External R&D Funding by Companies

First, total R&D expenditures paid by companies are examined. As Figure 1 indicates, they increased from ¥6.120 to ¥10.584 trillion in the eleven years from 1986 to 1996. The average increase ratio was 5.3%.

Second, external R&D funds paid by Japanese companies as surveyed by the MCA are analyzed. "External R&D funds" includes those for R&D collaboration as well as contract research, but excludes technology acquisition costs such as patent purchases and construction costs for R&D institutes in foreign countries.

Figure 2 shows the trend in external R&D funds broken down by industry. The total increased from ¥497 to ¥1,008 billion in the same period. The average increase ratio was 7.5%. That is, external R&D funds increased by over two percentage points more than total R&D expenditures; the ratio of external R&D funds to total R&D expenditures increased from 8.13% in 1986 to 10.02% in 1996. This indicates that the ratio increased from 1986 to 1991, with less fluctuation to 1994. It increased rapidly from 1994, despite the severe Japanese economic situation.

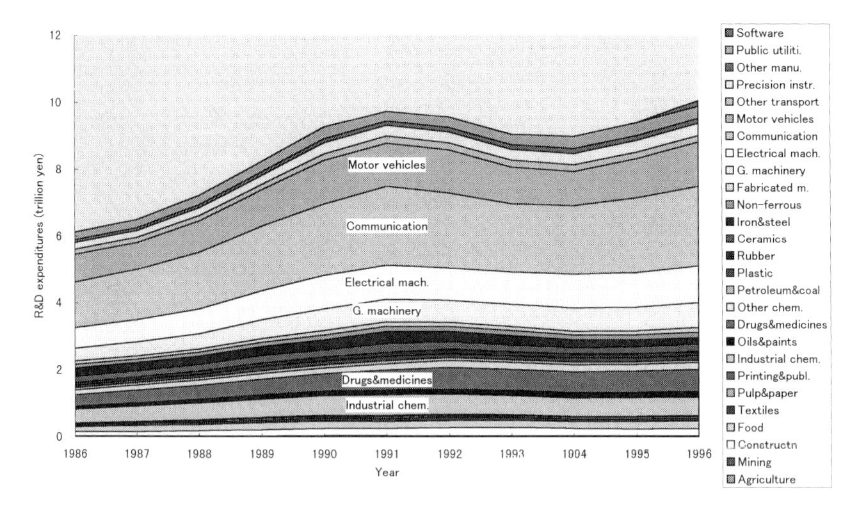

Figure 1 R&D Expenditures of Companies Broken Down by Industry.

Trends by Industry

Figure 2 shows that external R&D funding differs greatly across industries. The motor vehicle industry spent the largest external R&D funds, ¥341 billion (the value in 1996, a share of 33.8%), followed by ¥180 billion (17.9%) for the transport, communication, and public utilities industry; ¥150 billion (14.9%) for the communications and electronics equipment industry; and ¥119 billion (11.8%) for the drugs and medicines industry. The electrical machinery, equipment, and supplies industry (¥33 billion) followed them, but the gap was very big. All these industries except the transport, communication, and public utilities industry made large R&D expenditures.

External R&D funds are next compared with total R&D expenditures to calculate their ratio. The average ratio for all industries was 7.24%, a slight increase over the period, particularly since 1990. The motor vehicle industry had the highest ratio of external R&D funds to total R&D expenditures, an average of 23.7%. This decreased slightly in the 1980s and increased slightly in the 1990s. The drugs and medicines industry followed, with an average of 15.5% which did not fluctuate. The petroleum and coal industry had a similar ratio to that of the drugs and medicines industry, but

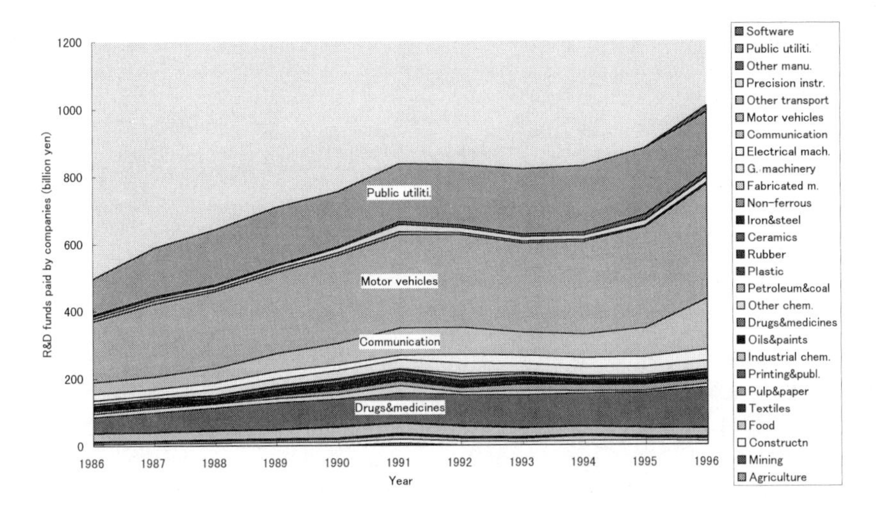

Figure 2 External R&D Funds by Companies Broken Down by Industry.

different from other industries in that its ratio increased rapidly over the period.

The "other" industries had lower ratios than the 7.2% average for all industries, with "other manufacturing industries" largest at 6.0%. The main player in this sector was said to be manufacturers of musical instruments and phonograph records. The ratio of this segment fluctuated greatly but on the whole increased, a trend shared by most other industries. The reasons for these less steady trends seem to be (1) that the ratio was small and in consequence a small change of external R&D funds effected a large change in the ratio, and (2) that external R&D funds per research project were very large.

Outsourcing Partners

In Table 3, the trend in external R&D funds by companies is broken down by outsourcing partner (the total value is different from that shown above because of the different classification). Private R&D institutions received the largest external R&D funds, ¥416 billion, a share of 49.1%, nearly half the total. This was followed by companies, ¥252 billion, 26.9%; foreign countries, ¥117 billion, 8.8%; and national and public universities, ¥60 billion, 6.1%.

Table 3 External R&D Funds by Companies, Broken Down by Partners (1996)

Sector	External R&D funds by companies (billion¥)	Outsourcing Partner	External R&D funds by companies (billion¥)
Domestic			
Private sector	739.39 (78.4%)	Companies	21.91 (26.7%)
		Private R&D institutions	463.47 (49.1%)
		Private universities	24.01 (2.5%)
Central and local governments	64.85 (6.9%)	National and public universities	62.41 (6.6%)
		National and public research institutions	2.44 (0.3%)
Special corporations (*Tokushu Hojin*)	21.90 (2.3%)		
Overseas	116.96 (12.4%)		

Data compiled by author from MCA reports.

Application of regression analysis to these trends shows that the average increase of total external R&D funds by Japanese companies amounted to about ¥37.3 billion. The biggest contributing outsourcing partner was private R&D institutions, ¥18.3 billion. This was followed by companies, ¥10.7 billion, and by foreign countries, ¥9.1 billion. These three outsourcing partners showed an increasing trend. In the case of the growth ratio, foreign countries had the highest average, 19.5%.

R&D Funds Paid by Japanese Companies to Japanese Universities

Figure 3 shows the trend in R&D funds paid by Japanese companies to Japanese universities. The MCA Report does not break down statistics by industry, therefore we estimated the R&D funds paid by companies to national and public universities broken down by industry, while those to private universities were not broken down by industry, because of a lack of reliable statistics. Our estimation

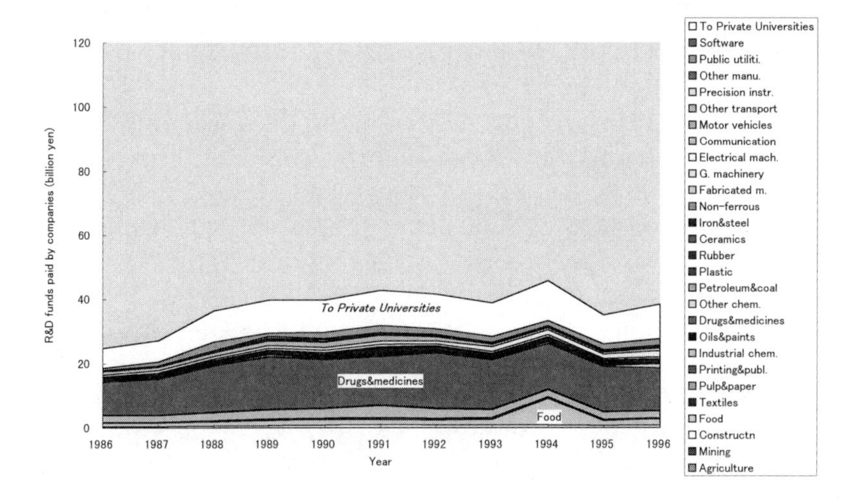

Figure 3 R&D Funds Paid by Companies to Universities Broken Down by Industry.

regarding the industrial value broken down to national and public universities seems reliable.

R&D funds paid by companies to universities increased from ¥24.8 billion in 1986 to ¥38.4 billion in 1996, a growth ratio of 5.5%. The growth ratio for national and public university funding was 5.0%, while that for private universities was 7.2%, the latter exceeding the former. Furthermore, the ratio of private universities to national and public universities was 35.3% on average. However, this increased from 33.2% in 1986 to 38.5% in 1996, while the R&D funding ratio of public universities to national universities also increased. Both are clear evidence of a slight but steady decline of funding to national universities for R&D.

The drugs and medicine industry spent the largest R&D amounts in universities, about ¥14.5 billion on average, 51.0% of the total. This was followed by the industrial chemicals and chemical fibers industry, with ¥1.6 billion and 9.0 %, and the food industry, ¥2.2 billion and 7.8%. While the food industry spent about five times the average in 1994, this seems to have been due to a single extraordinarily large contract. This means that R&D funds are not supplied steadily, as has frequently been claimed.

R&D Funds Paid by Japanese Companies to Foreign Countries

The MCA statistics report does not break down the "foreign countries" category into sub-categories such as companies, national universities, or national research institutes. This is a very serious limitation to further detailed analysis. According to the MCA definition, R&D funds cover expenditures for contract research, collaboration, and information collection, but not expenditures on tangible fixed assets of affiliated research institutes of Japanese companies. R&D funds paid by Japanese companies to foreign countries (as measured by the MCA) increased from ¥22.4 billion in 1986 to 117 in 1996. This can be divided into three periods. In the first, from 1986–1991, R&D funds increased rapidly; in the second, 1991–1994, they decreased very slightly; and in the third, 1994–1996, they increased rapidly again. The average growth ratio was 19.5%.

Total R&D expenditures showed a similar trend to external R&D funds paid by companies to foreign countries, that is, an increase in the first period, a slight decrease in the second, and a rapid increase again in the third, although the growth ratio was smaller. This means that the trend of R&D funds paid to foreign countries was proportionate to total R&D expenditures of companies. Furthermore, the amount of R&D funds paid to Japanese universities in 1986 was similar to that of R&D funds paid to foreign countries. The former increased gradually, with an average growth ratio of 5.5%, while the latter increased rapidly, with an average growth ratio of 20.0%.

Figure 4 reveals great differences among industries. The drugs and medicines industry showed a major share and a rapid increase. Its amount increased from ¥4.2 billion in 1986 to ¥50.9 in 1996 with a growth ratio of 31.2%. In consequence, its share amounted to 42.5 % (the average of 1995 and 1996). This industry had different characteristics from others, in that the amount has increased steadily and rapidly in recent years. Next came the communications and electronics industry, whose average share in 1995 and 1996 amounted 17.4%. R&D funds paid by this industry increased from ¥1.3 to ¥19.1 billion in the same period, with an average growth ratio of 125%. This had increased rapidly by 1994. The third

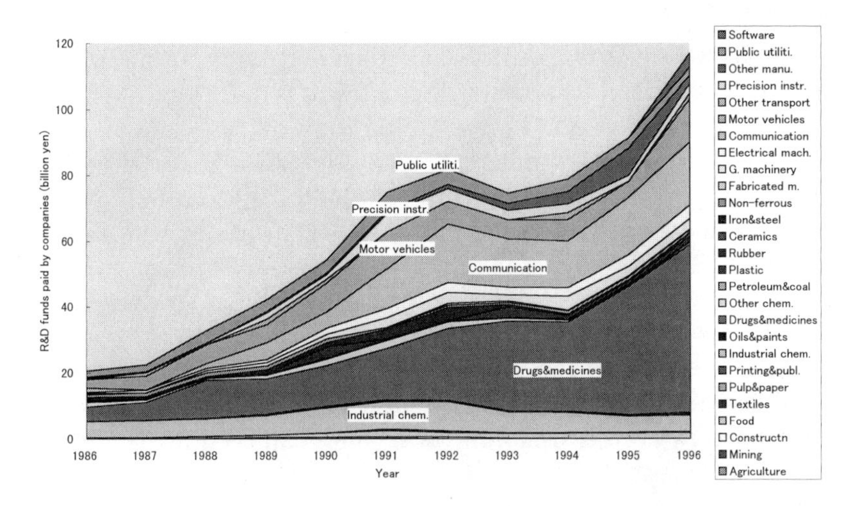

Figure 4 R&D Funds Paid by Companies to Foreign Countries Broken Down by Industry.

major player was the motor vehicles industry. Its R&D funds increased from ¥2.4 to ¥13.1 billion in the same period with a 29% average growth ratio. The most recent share amounted to 8.2 %. The increasing trend was similar to that of the communications and electronics industry, that is, a rapid increase by 1991 and less fluctuation after 1992.

R&D Outsourcing Indicators

Table 4 shows some indicators which were developed in order to comprehend different characteristics of R&D outsourcing among industries.

The first indicator, the ratio of external R&D funds to total R&D expenditures, indicates the extent to which each company attempts to use external resources based on its business and technology strategy. Needless to say, this depends on the characteristics of the industry to which the company belongs. Therefore, the first indicator shows the degree of openness to R&D outsourcing in each industry. When the characteristics of R&D outsourcing are open, for instance, the flow of R&D can be clearly divided into modules, which can be consigned externally.

The table shows that the indicator for the drugs and medicines industry was the third highest, which indicates that this industry is much more open in R&D outsourcing than others. Actually, the industry has used external resources aggressively, for instance seeking out technological seeds such as biotechnology to develop new products, and entrusting external organizations with safety checks on newly developed medicines. Its outsourcing partners are mainly Japanese universities and foreign countries (universities, research institutes, and ventures). It is believed that in the case of Japanese universities the major function of R&D outsourcing is safety examination, while in the case of foreign organizations developing new areas is given more importance. Furthermore, because "international harmonization" of standards will be introduced in 2003, R&D outsourcing in safety examination is expected to increase.

The motor vehicles industry had the highest indicator (Table 4, a) of 26.19%, suggesting that the industry has very open R&D characteristics. However, the indicator does not always reveal actual characteristics. There are close and strong relationships (a *Keiretsu* relation) between motor vehicles manufactures and their affiliated R&D institutes. These relationships have been gradually loosened, but they still exist. Therefore, the motor vehicles industry is categorized as having open R&D or as a less R&D-outsourcing industry, although its indicator was the highest. When we compare actual R&D activities of this industry with those of the drugs and medicines industry, this classification is accepted as reasonable.

The second-largest indicator was in the petroleum and coal products industry. This is also a misunderstanding. First, its total R&D expenditures were very small; second, there is no broad science and technology frontier in the Japanese model of the industry; third, the industry distributes small amounts of R&D expenditures to develop new products in close cooperation with downstream industries; and finally, cooperation partners of the downstream industries are frequently within an affiliated company and enjoy *Keiretsu* relations. Therefore, this industry is also classified as a less open R&D outsourcing industry.

Table 4 R&D Outsourcing Indicators

Industry	Ratio to total R&D expenditures[a]	Growth ratio in recent 3 years (%)[b]	Partners (Receivers)[c] (%) National universities	Private sector	Foreign countries
Food	4.54	10.49	36.61	41.70	19.07
Textiles	4.06	10.76	11.15	78.18	7.62
Pulp & paper products	3.42	–4.56	6.48	90.19	2.20
Printing & publishing	0.63	25.18	—	—	—
Industrial chemicals & chemical fibers	4.29	–1.48	9.84	64.48	22.52
Oils & paints	1.32	1.79	10.65	61.07	26.03
Drugs & medicines	16.45	7.53	13.64	49.42	35.97
Other chemicals	3.69	11.13	8.70	74.93	8.05
Petroleum & coal products	23.09	–9.16	0.69	95.78	1.27
Plastic products	1.81	6.11	5.21	91.58	1.79
Rubber products	2.93	–16.90	4.54	58.36	35.29
Ceramics	1.64	17.66	9.89	75.60	10.38
Iron & steel	3.43	4.44	9.19	85.23	1.83
Non-ferrous metals & products	4.82	–8.87	3.26	80.04	14.31
Fabricated metal products	3.12	0.85	3.12	91.05	4.33
General machinery	3.72	1.20	2.12	84.16	12.90
Electrical machinery, equipment & supplies	2.85	10.45	5.52	82.80	11.42
Communication & electronics equipment	4.53	33.60	0.81	81.03	17.98
Motor vehicles	26.19	8.85	0.06	96.98	2.60
Other transport equipment	2.35	–10.11	3.94	67.44	24.65
Precision instruments	4.50	13.66	1.51	81.52	16.10
Other manufacturing	8.54	21.37	1.36	60.26	37.89
Manufacturing (average)[d]	7.90	9.00	4.06	81.99	13.11

(a) The ratio of external R&D funds by companies to total R&D external R&D funds fluctuated greatly in some industries.
(b) Growth ratio of external R&D funds in the most recent three years: This indicator shows the recent trend in R&D funds paid.
(c) Partners' (receivers') share of R&D funds paid by companies: The indicators show shares of three R&D fund receivers. The total does not amount to 100% because the share of special corporations was omitted.
(d) Average values of all manufacturing industries, which are useful to compare with each industry's values.
Source: Data compiled by the author from MCA reports.

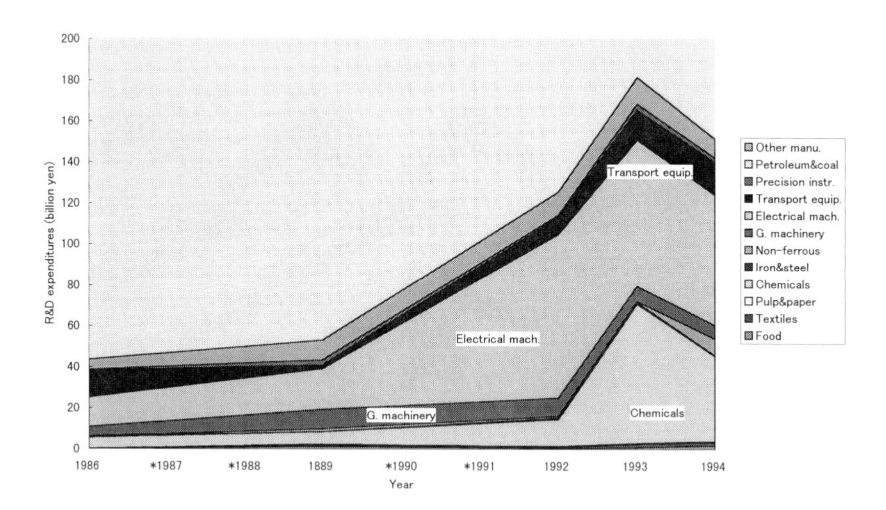

Figure 5 R&D Expenditures of Japanese Affiliates in Foreign Countries.

R&D Expenditures of Japanese Affiliates in Foreign Countries

MITI has been investigating R&D expenditures of Japanese affiliates located in foreign countries, published annually in its "Overseas Business Activities of Japanese Enterprises." R&D expenditures here are those invested by Japanese affiliates for their research and development activities, and comprise labor costs, materials, expenditures on tangible fixed assets, and other expenses. R&D funds (such as contract research and collaboration) paid by Japanese companies and received by Japanese affiliates in foreign countries are included in both the MCA reports and the MITI reports. There are no separate statistics on these funds, nor do other R&D funds seem to be included in both sets of statistics.

R&D expenditures of Japanese affiliates in foreign countries were investigated from 1986 to 1994. However, there are no statistics for 1987, 1988, 1990 and 1991, when statistics were compiled only every three years. The total R&D expenditures of Japanese affiliates in the manufacturing industry increased from ¥43.6 billion in 1986 to ¥151.6 billion in 1994. In particular we find a rapid increase in the three years from 1991 to 1993. The largest annual amount was ¥181.4 billion in 1993.

The electrical machinery industry (in the MCA classification, this industry is broken into two sub-categories: electrical machinery, equipment, and supplies, and communications and electronics equipment, which are not always classified the same way) expended the largest amount, ¥71.8 billion, a share of 47.0%, nearly half that of the total manufacturing industry, using the average of the most recent three years. This was followed by the chemicals industry (industrial chemicals and chemical fibers, oils and paints, drugs and medicines, and other chemicals, in the MCA sub-categories, with the main player the drugs and medicines industry), ¥40.9 billion and 26.7%. The transport equipment industry (the motor vehicles and other transport equipment industries are included, but the main player is the motor vehicles industry) expended ¥12.9 billion and 8.5%, and the general machinery industry ¥7.7 billion and 5.0%. These are the Japanese industries most active in R&D globalization.

The statistics on which Figure 5 is based break down expenditures by regions, North America (mainly the United States), Europe (mainly European Union countries), and Asia. The average R&D expenditure distribution of the most recent three years was 47.8% in America, 45.0% in Europe, and 7.2% in Asia. The trend for North America was similar to that of the total, that is, a gradual increase until 1992 and a rapid increase after that. R&D in Europe increased rapidly from 1989 (either 1987 or 1988 could have been the start of the rapid increase because of the three-year publication schedule). This means that the rapid increase began in Europe at least one year earlier than in North America. The main player in the rapid increase was the electrical machinery industry. Asia showed a U-pattern of decrease, leveling off, and increase. The recent increase was due to the electrical machinery and transport equipment industries.

In North America the chemicals industry had a share of 41.3% (the most recent three-year average), and the electrical machinery industry 33.5%. In Europe the electrical machinery industry had a share of 65.9%, followed by the chemicals industry with a 14.9% share. In Asia the electrical machinery industry had 32.6%, the transport equipment industry 23.4%, the other manufacturing industries 17.3%, and the chemicals industry 10.2%. In North America the major players were the chemicals and electrical machinery industries, in Europe the electrical machinery industry, while in Asia there were more players.

The purposes of the R&D of these affiliates can be divided into two categories: (1) conducting basic research: hiring high quality research staff and/or closely cooperating with top-level research institutions to conduct frontier research and to acquire technological seeds; and (2) conducting development, in particular product development: developing more appropriate products for regional markets. In North America and Europe Japanese affiliates emphasize both purposes roughly equally, while in Asia the major target is product development.

In the first case, research staff are expected to activate Japanese researchers by their different approaches, as well as to produce creative research themselves. This form of R&D outsourcing can be termed brain outsourcing. Some internationalized Japanese companies, such as the Sony Corporation, commission their foreign affiliates to conduct basic research, and treat them the same as other research institutions, selecting the best from among all research institutions. Based on these real situations, in this paper R&D activities in basic research are classified as R&D outsourcing, although this classification may be controversial. According to this definition of R&D outsourcing, some parts of R&D expenditures by Japanese affiliates in North America and Europe exhibit the characteristics of R&D outsourcing.

The first indicator in Table 5 is the ratio of R&D expenditures of Japanese foreign affiliates to total (domestic) R&D expenditures. Only three industries, chemicals (2.59), electrical machinery (2.31), and non-ferrous metals (2.19), had larger indicators than the average for the manufacturing industry (1.77). In the chemicals industry, the North American share was 72.5%, while the Asian share was 2.7%. As mentioned above, the main player is the drugs and medicines industry, and this indicator shows its R&D activities fairly precisely. The second largest industry, electrical machinery, had more than a 60% share in Europe, while its Asian share was small. This is very different from what the previous table (Table 4) showed. The electrical machinery industry, mainly the communications and electronics equipment industry, seems to rely much more on R&D outsourcing; the share of foreign countries was more than the previous table shows.

The non-ferrous metals industry seems to be active in R&D outsourcing, as seen in its high ratio. However, total R&D expendi-

Table 5 R&D Expenditures of Japanese Affiliates in Foreign Countries

Industry	Ratio to total R&D expenditures	Share of regions (%) North America	Europe	Asia
Food	0.49	74.80	5.68	19.50
Textiles	1.32	48.62	22.20	29.16
Pulp & paper products	0.87	45.48	33.12	21.39
Chemicals	2.59	72.53	24.75	2.70
Iron & steel	0.20	56.91	32.77	10.30
Non-ferrous metals	2.19	88.12	4.76	7.10
General machinery	1.14	57.65	34.61	7.73
Electrical machinery	2.31	33.31	61.80	4.88
Transport equipment	0.96	47.40	29.31	23.28
Precision instruments	0.59	58.32	27.30	14.36
Petroleum & coal	0.09	14.35	0.00	85.64
Other manufacturing	0.92	29.64	53.65	16.69
Manufacturing (average)	1.77	47.78	45.01	7.19

Data compiled by author based on MITI survey reports.

tures of its Japanese affiliates were small, and according to the previous table R&D funds paid to foreign countries were relatively small when compared with total R&D expenditures. Therefore, as Table 5 shows, this industry is relatively closed to outsourcing.

On the other hand, petroleum and coal products, textiles, transport equipment, pulp and paper products, and the food industries had small ratios and large Asian shares. This indicates that their main R&D purposes were product development, for which outsourcing would not be helpful. In the case of the petroleum and coal products industry our clarification in Table 4 is supported by Table 5. Finally, Table 5 indicates the R&D outsourcing characteristics of the transport equipment industry, where the motor vehicles industry is the major player, as Figure 1 shows, and where the ratio was small and the Asian share very large.

Payment for Technology Imports

In this paper patent purchase is considered a form of R&D outsourcing, although it is not strategic. Because there are no statistics concerning patent purchases and cross-licensing in Japan, I have used statistics on technology trades from the annual reports of the Management and Coordination Agency. International patent purchasing appears to be more strategic than domestic purchas-

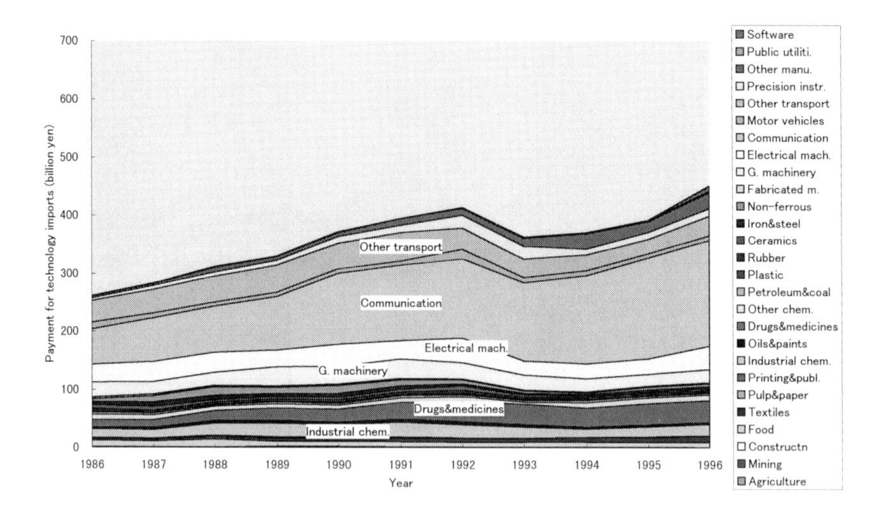

Figure 6 Payment for Technology Imports.

ing; therefore patent imports were used for an analysis of R&D outsourcing (see Figure 6). Some average values for the most recent three years are shown in Table 6.

We start with the indicator in the first column of the table: the ratio of R&D expenditures of Japanese affiliations in foreign countries to total R&D expenditures, which can indicate an aspect of companies' efforts to use external R&D resources—international R&D outsourcing. The "other manufacturing" sector (manufacturers of musical instruments and phonograph records especially) had the largest indicator, 16.7%, and the "other transport equipment" industry (where the major player was aircraft and parts) had the second largest indicator, 14.7%. Yet both sectors had less than a 0.3 ratio of receipts of technology exports to payment of technology imports, the indicator in the second column. In fact, the technological levels of these industries are very low, and they are not strong international competitors. They must rely on the technology of foreign countries and are among the "technology catch-up" industries at present.

The textiles industry had the third largest first indicator, 11.4%, while its receipts of technology exports slightly exceeded its payment for technology imports, as the second indicator shows. Because the Japanese textiles industry's survival strategy has been

Table 6 Payment for Technology Imports

Industry	Ratio of payment to R&D expenditures	Ratio of receipts to payment	Share of regions (%)	
			United States	Europe
Food	3.92	1.14	29.7	70.3
Textiles	11.37	1.15	32.6	67.4
Pulp & paper products	0.94	4.37	—	—
Printing & publishing	2.28	0.22	60.8	39.2
Industrial chemicals & chemical fibers	3.01	1.76	56.0	44.0
Oils & paints	2.10	1.42	78.4	21.6
Drugs & medicines	5.48	1.12	56.4	43.6
Other chemicals	4.64	0.35	72.5	27.5
Petroleum & coal products	4.24	0.27	29.2	70.8
Plastic products	1.44	1.89	78.7	21.3
Rubber products	2.88	2.02	42.8	57.2
Ceramics	1.27	4.48	70.0	30.0
Iron & steel	1.46	5.31	53.9	46.1
Non-ferrous metals & products	2.56	1.21	56.5	43.5
Fabricated metal products	1.41	5.06	52.5	47.5
General machinery	3.17	0.96	60.8	39.2
Electrical machinery, equipment & supplies	2.94	2.03	81.2	18.8
Communication & electronics equipment	7.57	0.79	82.5	17.5
Motor vehicles	0.71	20.87	32.8	67.2
Other transport equipment	14.66	0.26	71.2	28.8
Precision instruments	3.35	0.69	87.5	12.5
Other manufacturing	16.69	0.30	83.5	16.5
Manufacturing (average)	4.53	1.42	72.2	27.8

Data compiled by author from MCA reports.

continuous diversification, these indicators show that the industry has been obtaining technological seeds for new product development by purchasing foreign patents. This interpretation is also strongly supported by the evidence shown in Table 5: The major domestic partner of its R&D outsourcing is universities. Universities and foreign patents are good tools for textiles companies to use in venturing into new enterprises.

These three industries were followed by the communications and electronics equipment industry with 7.6%, and the drugs and medicines industry with 5.5%, which exceeded the average of the manufacturing industry. Because the technology levels of both industries were fairly high, as the ratio of receipts to payment showed, their patent purchases seem to be strongly strategic; furthermore, their R&D characteristics are open.

Industry Classification and R&D Outsourcing

The manufacturing industries are classified in Table 7 by their R&D outsourcing characteristics based on the three analyses above. Here, "closed R&D" means R&D is conducted intensively within the company. This may be due to one or more factors: (1) collaboration may be unnecessary because the industry's own technologies are competitive; (2) collaboration partners are not easily found, because the technology is frontier in nature; or (3) There are strong non-technological limitations on R&D outsourcing, such as *Keiretsu* relationships.

When a university is the outsourcing partner the industry's purposes are to seek new technological seeds, to acquire core technologies, to foster potential technology for expanding into new markets, to train high-quality researchers, and so on. When a member of the private sector is the R&D outsourcing partner, the purposes will be different according to its position in the industry, whether upstream, equal, or downstream. The purposes are mainly materials and parts development with upstream partners, and product and parts development with downstream partners. In the case of equals in the same industry, the purpose is mainly precompetitive technology development, reduction of development risks, sharing of development costs, and so on. In the case of foreign countries, all the purposes above mentioned are included, with the addition of a global leader in internationally competitive markets to the list of possible partners.

We have classified all manufacturing industries (Table 6) based on their R&D outsourcing characteristics as revealed by the statistics and our survey interviews. First, R&D in the drugs and medicines industry and the communications and electronics equipment industry were classified as "open." This classification is supported by reality: Both industries must keep their R&D open to survive in global competition, and both have sufficient but incomplete technological bases. Although their major partners are different, foreign countries for the communication and electronics industry and foreign countries and universities for the drugs and medicines industry, both indicate that Japanese universities are less appropriate partners for development of high-technology seeds.

Table 7 Industry Classification and R&D Outsourcing

	Major receiver of R&D funds paid (R&D outsourcing partner)			
R&D characteristics	Private sector	Universities	Universities and foreign countries	Foreign countries
Open			Drugs & medicines	Communication & electronics equipment
Closed	Motor vehicles Petroleum & coal Pulp & paper Plastic products Fabricated metal	Food Ceramics Iron & steel	Textiles Industrial chemicals Oils & paints	Rubber products Other transport Other manufacturing

Non-characteristic: Other chemicals, Non-ferrous metals, General machinery, Electrical machinery, Precision instruments

Source: Data compiled by author.

All other industries are in the "closed R&D" category. The main R&D partner of the motor vehicles, petroleum and coal products, pulp and paper products, plastic products, and fabricated metal products industries is the private sector. This is because their technology level is as high as the international level, or because the market is not internationalized. The food, ceramics, and iron and steel industries select Japanese universities as R&D partners. While market orientations are different among the industries, mainly the domestic market for the first two industries and the global market for the iron and steel industry, the main purposes of R&D outsourcing must be to seek high-technology seeds in order to expand into new businesses. The textiles, industrial chemicals and chemical fibers, and oils and paints industries rely more or less equally on both Japanese universities and foreign countries. The background for this is serious global competition as well as the business strategy of expanding into new markets under these situations. The rubber products industry, as well as the "other transport equipment" and "other manufacturing" industries, selected foreign countries as their main R&D outsourcing partners. At the bottom of Table 7 is a group labeled "non-characteristic": R&D funds paid by these industries had a distribution similar to the average of the manufacturing industry. Roughly 78% of their external R&D funds were paid to the domestic private sector, roughly 12% to foreign countries, and roughly 10% to Japanese universities and others. The other chemicals, non-ferrous metals and products, general machinery, electrical machinery, and precision instruments are assigned to this category.

Strategies of R&D Outsourcing

Because of the limitations in the statistics in the previous section, I have described the trends in industrial R&D outsourcing. However, R&D outsourcing is carried out by each company based on its own business strategy, therefore differences among companies in the same industry might be greater than that among industries. Therefore, I would like to discuss R&D outsourcing at the company level based on the results of the previous quantitative analyses and my survey interviews.

Business Strategy

At present it is generally recognized that Japan is at a significant turning point. While there have been various turning points in the past, this one, it is agreed, is extremely serious. Until the mid 1990s Japan was proud of its technological achievements, but this confidence has since waned. There are many reasons: only a small number of companies knew the meaning of the mega-competition era and how to deal with it successfully; many companies believed the management methods that had succeeded before were the best and therefore did not change them; and they misunderstood the qualitative aspect of the information revolution, which is more significant than its quantitative aspect. All these factors are closely related to the substance of the mega-competition era.

Mega-competition can be characterized by market (output) and resource acquisition (input). In the output aspect, a company can sell its products everywhere in the world, but finds competition everywhere in the world, regardless of the region or industrial sector, and it is very difficult to predict the emergence of rivals. To survive these situations it is necessary for the company to develop varied and differentiated new products quickly. On the other hand, in the mega-competition era a company can procure all the resources used to develop new products, such as capital, technology, talent, materials, parts, and facilities: It is possible to procure optimal goods with minimum cost and optimal methods. For this it is necessary for a company to obtain a broad range of quality information, to establish a situation under which such information is automatically absorbed, and to understand that information correctly. The conditions for acquisition of good information are that the company's technology must be top-quality and that the company can use it effectively.

Under such situations, it is necessary for a company to produce the best global products by using the best global resources, that is, to aim at global productivity. To use the best resources it is necessary to consider the best combination of products made by its own subsidiaries, by companies of the same country, and by foreign companies. For producing the best products it is no longer sufficient to pursue only efficiency; a company must pursue global productivity by aiming at new value development, based upon a clear core competence (Prahalad and Hamel 1990).

Technological Strategy

Continuous product development is needed to realize such a management strategy and long-term R&D, that is, a clear technology strategy is needed, which must be fused with the management strategy. Its pre-condition is the existence of a clearly defined management strategy, which must be broadly understood and equally shared by all in the company. This will make it relatively easy to fuse the two strategies. Technology development has shifted from being curiosity-driven, to market-driven, to vision-driven in the mega-competition era (Matsui 1998: 8–14). In the vision-driven technology strategy, there must first be a fusion of management and technology strategies, creating new values based on a clearly articulated core competence. That achieved, it is necessary to identify not only the current core technologies but also those of the future. Outsourcing is recognized as necessary in R&D as in other business areas. Technology areas that are not recognized as core may be outsourced, and future core technology must be identified for consideration for possible outsourcing.

In the catch-up era, Japanese companies' technology strategy was a simple matter of studying existing technologies in advanced countries, and management strategy and technology strategy went in tandem. However, this is not possible if a company is to be a front runner. With no existing technologies to aim at, it becomes hugely difficult to establish and fuse management and technology strategies. Although many Japanese companies have loudly insisted since the beginning of the 1990s that they had to be front runners in various areas, they seem not to comprehend the full substance of mega-competition and to have not yet completely discarded the catch-up mind-set. Many companies seem unable to identify their own core competence, and to establish clear management and technology strategies, particularly a global technology strategy.

Core competence is literally the core of a management strategy, as core technology is the core of a technology strategy. The core technology concept and its management, initiated by NEC Corporation, has been adopted by the most technology-intensive manufacturing companies. It is displayed by matrix relations between technology and product. Because technologies which correspond to each product can be identified, the importance of each technology can be recognized. The portfolio method is useful to design a future strategy of product development. The combination of the technology-product matrix and the portfolio is one of the most frequently used methods for identifying the core technology of a company.

Technologies that do not belong to the future core technology of the company will be candidates for R&D outsourcing. A key factor of the technology strategy is whether to develop such technologies itself or to acquire it by outsourcing. Technology development, in particular development of so-called high-technology, is highly uncertain, even higher in this mega-competition era. Therefore it is vitally important to be well informed of seed technologies and where they are being developed, and to evaluate and foresee their future development. The high-quality manpower required for this is extremely important, and must be acquired, either through staff development or recruiting

Until recently there was a belief in Japan that bigger is better, that a company should develop and maintain its own technologies as much as possible. This is the so-called full-set principle. In the situation of non-global competition, this technology strategy was very effective. However, in mega-competition, it is disadvantageous because the company holds redundant and inefficient divisions.

The company must shift to becoming a strategic organization by major restructuring of inefficient and disadvantageous divisions based on the new principle of "selection and concentration."

However, it is not appropriate to throw away the entire full-set principle, which has its advantages: it can establish and maintain several or many complex technology platforms to develop new products continuously; and it can activate synergistic effects among a variety of technology platforms to develop new, niche or differentiated products. We must not forget these synergistic effects when judging R&D outsourcing of "selection and concentration." Sony Corporation, one of the most widely globalized Japanese companies, attaches great importance to the synergy effects among technology platforms, in particular between related platforms.

In this paper, purchasing patents is assumed to be one form of R&D outsourcing, though less strategic than others. The reasons to purchase a patent may be that (1) it is identified as a seed of part of the future core technology, for which related technologies will be developed; in this case it is vital to decide whether the technology will be acquired externally or developed by the company itself. Or it may be (2) that the patented technology, while not identified as a part of the future core technology, is necessary or useful in developing new products. It is a strategic decision as to whether to produce the product or not. The first case is closely involved in technology strategy, therefore strategic in R&D outsourcing, while the second, less strategic, is the more common in Japan.

Decisions on R&D outsourcing are extremely important. The main factors to consider are: management strategy and core competence; technology strategy and core technology; fusion of the above strategies; rules or principles of R&D outsourcing; and high-quality human resources.

Strategies of R&D Outsourcing

In order to classify R&D outsourcing measures, two dimensions are used: (1) R&D execution and (2) R&D planning and design (see Table 1). First, we will discuss the dynamism of R&D outsourcing, which in a given technology area proceeds in four stages: (1) the activities are conducted independently; (2) R&D execution is conducted in cooperation with an outsourcing partner; (3) R&D

planning and design are conducted in cooperation with the outsourcing partner; and finally, (4) both activities are conducted in cooperation with the partner. The stage number reflects the extent of the strategy. Needless to say, companies move back and forth between these stages depending on management and technological strategies. In extreme cases a company will transcend its present outsourcing level and undertake mergers and acquisitions and/or restructuring.

The purposes of R&D outsourcing have been classified (Misawa, Ito, and Shimizu 1996: 55–60) into short-term or long-term. Short-term purposes include:

(1) Reduction and effective control of R&D costs. In the short term significant effects can be expected.

(2) Effective use of R&D assets, which can be concentrated in the company's own core technology, and the long-term investment of assets can be promoted.

(3) Acquisition of R&D resources that cannot be found in the company, or specialties that could not be constructed in the company.

(4) Improvement of management ability in divisions whose management is difficult or impossible, through acquiring experts who can grasp completely and continuously the state of the outsourced areas.

Short-term purposes are defensive, whereas long-term purposes are aggressive. A key word among long-term purposes is "objectivity," the acquisition of more objective judgement; outsourcing tends to make company decisions more objective.

Long-term purposes include:

(1) To secure the flexibility of company management. By R&D outsourcing the company can first disperse the risks and gain objective, flexible judgments. These added resources can make its core technology more productive and profitable by selection and concentration.

(2) To establish a core business, reinforcing its competence by outsourcing non-core technology.

(3) To reinforce globalization. By using specialized R&D outsourcing partners, the company can acquire globally competitive technology and business.

While there are great advantages in R&D outsourcing as described above, it is necessary to understand its disadvantages and the conditions for success and to be prepared to deal with them. There are three main issues concerning R&D outsourcing (Misawa 1996: 55–60):

(1) Failure to acquire the specialty. The problem most frequently pointed out is that neither the specialty nor its know-how are accumulated in the company. Because there is the worst-case possibility of destroying the organization or its core competence, the company must not leave everything to the outsourcing partners. Information gathered by outsourcing partners must be accurately evaluated, digested in the company, and new values produced. This requires that the company treat its own human resources well, since it is they who can accurately evaluate technologies in the concerned areas. With a few exceptions, this has not been achieved in Japanese companies.

(2) Secret leakage: R&D outsourcing inevitably involves some risks of secret leakage. Therefore, it is necessary to make rules by which information is classified.

(3) Loss of job security for scientists and engineers: When an R&D project is assigned to outsourcing partners, scientists and engineers presently engaged in that project are made extremely anxious by the possibility that they will have to change their jobs, or may even be dismissed—a very serious event in the Japanese life-long employment system. Unclear management and technology strategies add to their fears. If large-scale R&D outsourcing is contemplated, an off-the-job training system is needed to retrain scientists and engineers , such as the "Metamorphosis University" of Matsushita Electrical Machinery Industry.

Table 8 summarizes various R&D outsourcing used by the three kinds of partners.

Table 8 Characteristic R&D Outsourcing Strategies, by Partner of Outsourcing

Private sector	Universities	Foreign countries
1. Joint development of materials and parts with upstream industries 2. Risk-sharing and cost reduction 3. Joint development of parts and products with downstream industries	1. Information collection, consulting, seeking seed projects 2. Acquisition of advanced technology seed projects	1. Collection of foreign information 2. Introduction of foreign advanced technology 3. Use of social institutions of foreign countries 4. Hiring foreign workers for their creativity and different approaches

Data compiled by author.

Conclusions

Based on the above analyses, several policy implications seem clear. Fundamental to each is the expectation that R&D outsourcing of Japanese companies will increase in the future. As "mega-competition" in a global market becomes widely recognized, business competition there, as well as in the domestic market, will become more severe. When the world becomes a single market for each product, and any company can get the best resources from all over the world, a company must continuously offer new value by "selection and concentration." As this situation is acknowledged by more and more industrial sectors, R&D outsourcing by Japanese companies will progress at a rate similar to those shown in the statistics. In order to survive such a situation, it is necessary for companies to establish clear and well-fused management and technology strategies. Moreover, an appropriate public policy is needed, which promotes smooth R&D outsourcing of companies with universities and national research institutes. This will involve deregulation.

R&D outsourcing of Japanese companies to foreign countries will also increase, with a higher growth ratio than that for domestic partners. In the present environment of mega-competition, deregulation, the progress of the information infrastructure, and so on, some industrial sectors have yet to adopt international R&D outsourcing, therefore R&D outsourcing to foreign countries must increase, not at its recent rapid rate, but at a steady pace, while an appropriate public policy for the promotion of domestic R&D outsourcing is developed.

According to some evaluations, many R&D outsourcing projects undertaken with foreign partners were unsuccessful, and in extreme cases a huge amount of capital was paid in return for poor results. As R&D outsourcing projects with foreign partners increase, success and failure will be experienced and know-how accumulated; companies must share their experiences to assure successful R&D outsourcing with foreign partners in the future.

There is an urgent need for talented experts who can master the skills of R&D outsourcing: decision-makers who comprehend management and technology strategies and understand the concerned technologies in depth, including their future trends; and managers, the technology-bridging personnel who bring the company and its partners together and fuse the technologies of both sides. Japan has developed neither a supply of talented human resources, nor the means to nurture their talents and develop their skills. Corporate policies must be developed that will give first priority to solving these problems.

Our study shows that R&D outsourcing by Japanese companies to domestic universities has increased much more slowly than to foreign countries. It is expected that once industry has assimilated the experience gained from foreign outsourcing, and has learned to evaluate it, R&D outsourcing to domestic universities will grow, its main purpose changing from hiring the best students, to R&D outsourcing based on high-quality evaluation of projects. Actually, the situation of the universities as partners of R&D outsourcing has widely improved in recent years, although this has not been recognized by companies. As university facilities improve, they will increasingly be used as outsourcing partners, encouraging competition among researchers and the advancement of research activities in the universities. While in general a major change of direction within the universities is unlikely or even impossible, the demands of industry are bound to have a significant influence.

Finally, as R&D outsourcing becomes more widespread it will grow more strategic and more challenging. Its techniques will also differ greatly among industrial sectors. Therefore public and corporate policies must be prepared to deal successfully with this future scenario.

References

Management and Coordination Agency (MCA), eds. (annual reports) "Report on the Survey of Research and Development." Tokyo: Printing Bureau, Ministry of Finance.

Matsui, Konomu. 1998. "Innovation of RD&E Strategy and Technology Management," In Proceedings of the Annual Meeting on Development and Technology, Management Innovation, JMA.

Ministry of International Trade and Industry (MITI), eds. (annual report). "Overseas Business Activities of the Japanese Enterprises." Tokyo: Printing Bureau, Ministry of Finance.

Misawa, Kazufumi, Hiroshi Ito and Hiroshi Shimizu. 1996. "Future R&D Strategy based on Outsourcing." In *Practice and Organizational Evolution of Outsourcing*. Tokyo: Diamond Press.

Ministry of Education, Sports and Culture (Monbusho). 1997. "Report by the Committee on the Investigation of Promoting Cooperation between Industry and Academia." Tokyo: Ministry of Education, Sports and Culture.

Ozeki, Tomoyasu. 1997. "Outsourcing Revives The American Companies" (in Japanese). In, *Introduction to Outsourcing*. Tokyo: Japan Management Association.

Prahalad, C. K. and Gary Hamel. 1990. "The Core Competence of the Corporation," *Harvard Business Review* 90: 3.

6

Public Financing of University Research in Japan

Yoshiyuki Ohtawa

Introduction

Many Japanese mistakenly believe research and development (R&D) in Japanese universities is well-funded; it is, in fact, much less so than in the United States. Most Japanese university researchers and faculty would like to expand their R&D funding, but cannot because of the limited funds available through the government.

Present statistics cannot reliably show the real expenditures on research in Japan, due to structural problems that make the standard data sources inappropriate to this purpose. This chapter discusses some of the pitfalls of using locally-gathered data for international comparisons. Next it examines the policy and linguistic difficulties of interpretation in the two government agencies responsible for scientific research funding in Japan, and finally it offers some suggestions for the future.

Problems in Statistics for University R&D Expenditures

The OECD Database

Existing international statistics, such as those of the OECD, consist of data gathered from countries with various R&D and funding systems, which makes international comparisons very difficult. For example, the Japanese data source for the OECD Main Science and Technology Indicators (OECD 1998) was the "Survey of Research and Development" (R&D Survey) conducted by the Statistics Bureau of the Management and Coordination Agency of the Japanese government (Management and Coordination Agency 1997), which in turn collects information based on the Frascati Manual (OECD 1994), the standard for OECD science and technology statistics. Under this protocol the R&D Survey data are gathered not only from universities but also from junior colleges and colleges of technology, and the research fields include the humanities, social sciences, and natural sciences. Additionally, expenditure data cover all current expenses, from labor cost to capital expenditures, and labor costs include the salaries of all regular teachers, from assistants to professors. In Japan every regular teacher is counted as one regular researcher,[1] nor are the full time equivalent (FTE) data on the basis of a time-budget survey. A further complication is that the R&D Survey includes all other educational expenditures.

On the other hand, the NSF survey of the U.S. system (NSF 1996) counts only expenditures for organized research, called "separately budgeted R&D," and does not include expenditures for research activities performed as a part of educational activities; it also omits expenditures for the humanities and some of the social sciences,[2] as well as capital expenditures.

The problems outlined above show why an international comparison of R&D expenditures cannot be made on the basis of the OECD database information, particularly between Japan and the United States.

The R&D Survey and the Basic School Survey

The R&D Survey presented by the Japanese government and described above, had earlier been compared with the Basic School Survey conducted by the Ministry of Education, Science, Sports, and Culture (the Monbusho) (Ohtawa 1995).

The Basic School Survey collected expenditure data from all Japanese universities. From the national universities it collected information on the settlement of the Special Account for National Educational Institutions (the Monbusho Special Account). In the expenditure data of the Basic School Survey, data directly connected with education and research activities yielded almost the same values as the expenditures data of the R&D Survey; this anomaly called for further study.

Table 1 Japan and U.S. Higher Education R&D Expenditures (1996)

	Japanese R&D Expenditures original data	estimated data	U.S. R&D Expenditures (original OECD data)
National currency	¥3,013,120 mill.	¥987,748 mill.	$27,800 mill.
Exchanged at IMF rate	$27,694 mill.	$10,497 mill.	$27,800 mill.
Exchanged at at OECD PPP	$17,022 mill.	$5,612 mill.	$27,800 mill.

Source: Main Science and Technology Indicators 1997/2 (OECD 1998); 1997 Report on the Survey of Research and Development (Management and Coordination Agency 1998).
Notes: U.S. data included expenditures of FFRDC. Japanese data include expenditures of universities, junior colleges, colleges of technology, and inter-university research institutes. PPP = purchase power parity. $US = ¥108.8 (IMF rate). $US = ¥175.19 (PPP rate).

To resolve the question, the actual respondents in the R&D Survey were surveyed by questionnaire (Ohtawa 1996). It became evident during a study of their responses that most of them could not understand the concept of Full Time Equivalent (FTE): Most of them had answered the part of the Basic School Survey directly connected with education and research activities as R&D data, which made the figure for education and research expenditures in the Basic School Survey nearly equal to the figure for R&D expenditures in the R&D Survey. This example goes far to explain why there are no reliable statistics on R&D expenditures in Japanese universities.

Up to now, there have been some studies on the FTE coefficient, but every study has had its drawbacks, and it is difficult at this writing to cite an adequate FTE coefficient for Japanese universities. But if the Japanese data can be adjusted in order to compare with U.S. data on the same expense ranges and scientific fields, assuming that the FTE coefficient is 50%, the figure used in French statistics, and that the coefficient is applied not only to labor cost but to all other costs, the result becomes as shown in Table 1, where it can be seen that U.S. university R&D expenditures are from 2.6 to 5.0 times greater than Japanese university expenditures.

The Effects of Misleading Information

Otherwise generally knowledgeable people in Japan are often misled by the information in official publications such as the "White Paper on Science-Technology" (ST White Paper) (ST Agency 1997), where data for international comparison are sometimes used, even though they are inappropriate to this purpose. R&D specialists can obtain university R&D expenditure data from the OECD Science and Technology Indicators database; people who are only interested in R&D read the ST White Paper. This leads to unfortunate misunderstandings.

The ST White Paper (1997) says, concerning national R&D expenditures, "If we compare every R&D statistic of the main countries, the U.S. spends the most, followed by Japan and Germany" (ST Agency 1997: 108). In a figure, the White Paper indicates that at the IMF rate of exchange (in parentheses as purchase power parity, PPP) U.S. total R&D expenditures are ¥16.8 (31.4) trillion, Japanese total R&D expenditures ¥14.4 (14.4) trillion. R&D expenditures of Germany, France, and the United Kingdom are shown as ¥5.2 (6.7) trillion, ¥3.4 (4.8) trillion, and ¥2.1 (3.8) trillion, respectively, in FY 1995 (ST Agency 1997: 109). Consequently Japanese R&D expenditures appeared high from an international standpoint.

In a figure in the same report that indicates the changing real R&D expenditures in leading countries as expressed by indexes, FY1990 is 100 and Japan has the highest score, 106.9, in FY1995. In a figure showing the changing R&D expenditure rates per GDP in the leading countries, Japan also has the highest score, 2.95% in FY1995; regarding university R&D expenditure rates compared to national expenditures, Japan again has the highest score, 20.7%. However, the Japanese government's share of expenditures scored lowest, at 22.9% (ST Agency 1997: 110–113).

Regrettably, all these scores are based on factual errors. As mentioned above, Japanese R&D expenditure data, especially for university R&D, are overestimated because of a failure to account for FTE data; on the contrary, the other countries' data are underestimated in the context of the Frascati Manual standard.

In the section related to expenditure data, the White Paper states: "When we compare R&D expenditures internationally, as the contents and the survey methods etc. are different from country to country, it is difficult to make a simple comparison, but the indicators show the rough trends of every country. ..."(ST Agency 1997: 108). The data on which this statement is based may be adequate to see R&D expenditure trends within one country, but they are not useful for international comparisons. Especially concerning university R&D expenditures, it is impossible to compare among several countries, because their survey methods may be very different; it would be possible to double or even triple some of the resulting figures, depending on their interpretation.

Because of these grave structural defects, the readers of the White Paper, most Japanese academics, and important government figures are misinformed about Japan's R&D expenditures, especially the funding of university R&D. They believe Japan's total R&D expenditures are close to those of the United States, and that the universities' share compared to the national share is the largest in the most advanced countries; therefore they are convinced that Japan's university research efforts are adequately funded.

A New Approach: The Research Environment Survey and the U.S. NCES Survey

In 1996, in order to clarify the actual situation of university R&D expenditures in Japan, our research group circulated a questionnaire in a project entitled The Research Environment Survey (Ohtawa, Kakinuma et al. 1997).[3] We obtained data on individual R&D expenditures of university researchers for FY1995, expressly excluding those expenses managed by the university, faculty, research institute, or other institutions: facilities maintenance, heating, lighting, and so on. Each research expenditure was distinguished as much as possible from non-research expenditures such as education and administration. If a researcher used the fund together with other researchers, the expenditure was divided proportionally. We ignored capital expenditures. The national current R&D university expenditures calculated from the survey is expressed in Table 2.

Data from the U.S. Department of Education (NCES 1997) were used for comparison to our results. This is possible because the NCES definitions and coverage are similar to ours. Whereas the

Table 2 Breakdown of Current R&D University Expenditures in Japan (¥ million)

	Total	1	2	3	4	5	6	7	8	9
Total	262,479	88,376	73,624	16,277	15,196	2,549	39,758	11,690	363	14,646
By control										
National university	162,052	44,973	55,603	14,802	10,602	1,203	22,331	7,411	262	4,864
Local public university	14,694	4,399	3,988	613	1,109	456	3,349	506	3	272
Private university	86,980	39,798	14,536	1,258	3,806	1,009	14,416	3,623	94	8,440
By field										
Humanities/ Social Sciences	23,629	14,108	5,547	11	329	167	1,224	841	54	1,348
Natural Sciences	205,075	59,407	59,641	14,595	12,424	1,960	35,403	9,139	216	12,290
Multidisciplinary and others	33,775	14,861	8,437	1,671	2,443	421	3,131	1,710	93	1,008

Columns: 1: Research funds from university authorities. 2: Grants-in-aid for scientific research from the Monbusho. 3: Special research funds of investment projects from public corporations. 4: Research funds from ministries other than the Monbusho and from public corporations except those under 3. 5: Research funds from local governments. 6: Research funds from private companies. 7: Research funds from foundations. 8: Research funds from abroad. 9: Others.

Source: *Research Environment Survey* (Ohtawa, Kakinuma et al. 1997).

fields of the NSF statistics are limited to Science and Engineering, the NCES statistics are not; nor do they include overhead costs that are not always related to research. Moreover, the NCES statistics cover all universities and colleges, including junior colleges, a wider coverage than the NSF's.

Table 3, based on the Research Environment Survey and NCES statistics, is the result of a comparison of current university R&D expenditures between Japan and the United States, not including labor costs, using the exchange rates of both the IMF and the OECD (PPP). The results reveal U.S. universities' current R&D expenditures to be three times greater than Japan's on the IMF scale, and six times greater on the OECD PPP scale, although real Japanese expenditures on the IMF scale may be lower because of IMF rate fluctuations during these years (Ohtawa 1998a).

University Researchers and the Shortage of Research Funds

The Research Environment Survey indicated that many university researchers suffer from a shortage of research funds. Most of them express a desire for an improvement in their research environ-

Table 3 Current University R&D Expenditures in Japan and the United States (1995)

	R&D expenditures (IMF exchange rate)	R&D expenditures (PPP exchange rate)	Rate against GDP
Japan	$2,789 million	$1,491 million	0.054%
United States	$8,762 million	$8,762 million	0.121%

Source: Ohtawa 1998a.
Note: $US = ¥94.1 (IMF rate). $US = ¥176.76 (PPP rate).

ment, expansion of their research space, and an increase in government funding.

Of the respondents to the questionnaire, 72% said government should do everything possible to expand research funds; particularly higher in the national and public local universities (about 80%). When asked what sort of funds they wanted to expand, almost half wanted to expand general university funds and the Monbusho Grants-in-aid for scientific research. (See Table 4.) These two funds appear to be essential to university researchers.

However, probably because of the general misinformation described above, under the Fiscal Structural Reform Law the Monbusho could not help instituting deep cuts in the 1998 university budget, affecting fundamental expenses such as facility and equipment costs. The president of the Science Council of Japan, which represents Japan's scientists and encourages the spread of science, protested these cuts, demanding in May 1998 that the Monbusho reinstate the funds. He said the cuts had caused great confusion and difficulty for the Japanese university researchers, who as well as being demoralized were afraid there would be a worldwide lack of confidence in Japanese science.

On the other hand, the Japanese Science-Technology Basic Plan of 1996–2000 was promulgated by the government to enlarge the selectivity and discretionary powers of researchers over their research funds, and to expand largely competitive funds, which contribute to the formation of a competitive research environment. Since the establishment of the Basic Plan, the Monbusho Grants-in-aid for scientific research and new research funds offered by public corporations have greatly expanded; as a result, a few capable researchers have more research funds than ever. The

Table 4 Rate of Demand for Expansion of Research Funds (%)

	1	2	3	4	5	6	7	8	9
Total	49.3	45.0	5.3	8.9	7.2	14.6	17.7	0.6	0.5
By control									
National universities	56.7	54.1	6.3	9.7	6.7	14.4	19.0	0.7	0.3
Local public universities	59.4	50.2	6.4	7.8	13.4	10.3	20.7	0.3	0.2
Private universities	41.7	36.5	4.2	8.0	6.9	15.1	16.4	0.6	0.6
By field									
Humanities/social sciences	49.7	32.1	1.5	3.3	8.0	6.1	16.8	1.2	0.9
Natural sciences	48.3	52.3	6.9	11.7	6.6	19.6	18.0	0.3	0.3
Multidisciplinary and others	52.2	43.2	7.5	9.3	7.8	12.5	18.3	1.0	0.3

Columns: 1: Research funds from university authorities. 2: Grants-in-aid for scientific research from the Monbusho. 3: Special research funds of investment projects from public corporations. 4: Research funds from ministries other than the Monbusho and from public corporations except those under 3. 5: Research funds from local governments. 6: Research funds from private companies. 7: Research funds from foundations. 8: Research funds from abroad. 9: Others.
Source: *Research Environment Survey* (Ohtawa, Kakinuma et al. 1997).

mass media ridicule this situation, calling it the "research funds bubble." But this is caused by the existing funding system, and only a few researchers have benefited. Some of them say their research funds are enough for the present, under the existing research system.

Causes of Inadequacy in Public Financing of University Research

While the Japanese government claims to operate on a small scale, and is underfunded in many areas, it is deeply in debt. In March 1998 its public debt was over ¥250 trillion, 50% of the GDP and five times the national budget. Not only are there budgetary shortfalls, there are also many other issues to be resolved. The causes that affect the subjects of our study, however, arise from the system of cost sharing and research funding.

The "Founder Pays" Rule

At their founding during the Meiji era, Japanese universities adopted the German system. The School Education Law provides that "The university, as a center of scientific research, shall aim at teaching and studying in depth professional arts and sciences and developing intellectual, moral, and practical abilities as well as giving a broad knowledge" (Article 52). The Japanese university is thus a research institution as well as an educational institution.

The law also provides that "The establisher of a school shall manage the school which he established and defray the expenses of the school except in the cases specifically stipulated by laws or ordinances" (Article 5). This means that the founder of a Japanese university must pay all its educational and research costs.

Since this is the rule, the expenses of a national university, a local public university, or a private university are defrayed by the national government (the Monbusho), the local government, or the legally incorporated educational institution respectively. Accordingly, each founder's contribution to its research and educational expenses is constrained by its income. As there are many private universities in Japan, among total university research expenditures it follows that the share of the private universities' self-financed R&D funds is relatively large, while that of publicly financed research funds is shrinking.

Whatever the funding share of either sector, in general local public and private university R&D funds are insufficient. When the founder cannot afford the R&D funds, the Monbusho takes several steps to promote university research, such as providing research grants or post-doctoral fellowships. At present, Monbusho Grants-in-aid for scientific research are the essential funds for university researchers, of course including those in national universities. But the budget for the grants—¥117.9 billion (about $842 million, in FY 1998 at $1=¥140)—is equivalent to only about 30% of the NSF budget; there is a strong demand that the grants program be greatly expanded.

The National Administration and the Research Funding System

The national authority with the responsibility for promoting university research is the Monbusho, also the founder of the national universities. In general, university R&D expenses are paid through the Monbusho budget. This is very different from the U.S. system, in which there are no national universities and state governments bear the R&D expenses of local public universities and private universities.

On the other hand, Japanese national and local governments have founded many Test and Research Institutions (TRIs). Consequently university R&D activities have been mainly driven by the curiosity of individual researchers rather than governmental or social demand or a national need for research, such as generic technology.

Under the Japanese system, no national government agency except the Monbusho pays R&D funds directly to university researchers, although there has been some de facto support from other agencies to private companies or national TRIs, which in turn may provide R&D funding to the university researchers. National universities receive such funds as commissioned research expenses. However, they totaled only ¥5 billion in FY1996.

In FY1995 a new funding system was established. The resources of the new funds are investments made in government bonds for the establishment of public corporations such as the Japan Science Promotion Society (JSPS), the Research Development Corporation of Japan (JRDC), and the New Energy and Industrial Technology Development Organization (NEDO). On average the allocated amount is higher than other funds (about ¥100 million in the case of one project); however, the total number of newly adopted projects in FY1997 was only 240, of which 85% were university R&D projects.[4] This is many fewer than the 23,000 projects supported by the Monbusho Grants-in-aid program. It should be noted that the funds are for applied and developmental research that promises quick results in the form of intellectual property. Therefore they cannot make up for the general shortage of university R&D funds.

University Research, the Monbusho, and the ST Agency

The 1871 law that established the Monbusho gives it the responsibility to promote *Gakujutsu*: the humanities, social sciences, natural sciences, and applied research pertaining to them. Another agency, the *Kagakugijutsu* Agency, was established in 1956. There is no English-language equivalent or legal definition of the word *Kagakugijutsu*; generally it means "science and technology," but sometimes, especially in law, it is used in the narrower senses of "scientific technology," or "technology to which science is applied." Therefore, in this discussion we will use Science-Technology (ST) to indicate the new agency's sphere of responsibility, and we will refer to the *Kagakugijutsu* Agency as the ST Agency. The jurisdictions of the Monbusho and the ST Agency do not overlap.

The ST Agency's mandate is to contribute to the development of the national economy through comprehensive ST administration, but excluding ST related to university research and ST related only to the humanities and social sciences, which are the responsibility of the Monbusho. Complicating the picture is the fact that ST research is also being carried out in universities as well as other venues, while the national administration system is divided between promoting science (university research) and promoting only Science-Technology (ST).

With university research expanding, and because of the drawbacks of this complicated administrative system, a Science-Technology Council has been established in the office of the Prime Minister, who serves as its chairman. Its task is to draft basic and comprehensive policies regarding ST in general, including ST-related university research. Although the secretariat of the Council is dealt with by both the ST Agency and the Monbusho, the ST Agency takes the lead in managing the Council.

In reality, it is sometimes very difficult to discuss science separately from technology: Recent high technology is closely related to modern science. As a result, both administratively and in reality, *Kagakugijutsu*'s complex shades of meaning have resulted in the problems discussed here. While Japanese ST policy is led by the ST Agency, and the ST budget and ST Basic Plan (to be discussed later) are in its domain, national science policy and university R&D policies belong to the Monbusho. And the main problem is that most people mistake ST (Science-Technology), as above, for Science *and* Technology.

The National Financing System

Currently the Japanese national budget operates under the so-called "Ceiling System." During the period of economic growth, through the 1970s, each Ministry and Agency could expand its

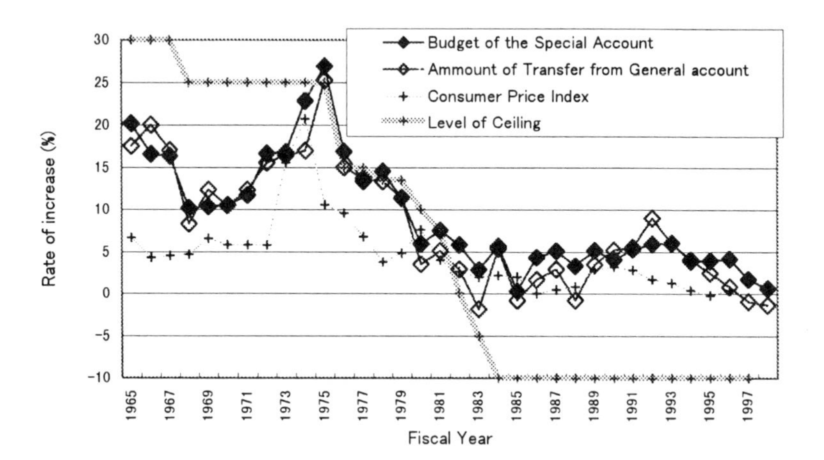

Figure 1 Increase rate of the Monbusho Special Account and Transfers (1965–1998). Source: The Monbusho; the Ministry of Finance.

budget within certain limits, justifying increased funding for some projects by pointing to their growing size. However, beginning in the 1980s fiscal restraints were imposed under which each Ministry or Agency had to cut its funding requests equally. For every fiscal year, each prepares a draft of its planned projects and a rough cost estimate, which are submitted to the Ministry of Finance. The proposed budget must remain under a "ceiling"—its upper limit as a percentage of the current year's budget totals for all Ministries, with certain exceptions such as labor costs. The Ministry of Finance then negotiates with each Ministry and Agency to arrive at a total government budget bill.

From FY 1983 until FY1997, the ceiling was set at minus (the "Minus Ceiling" system), and each Ministry and Agency had to operate on less funding than in the previous year. There is no "General Fund," as in the United States, from which needed funds can be shifted to projects in some needy Ministry; each must manage alone.

In the Monbusho General Account, labor costs accounted for 78% in FY1997, by far the largest portion of expenses. As labor costs increase every year because of wage increases, it has been difficult for the Ministry to maintain its mandated program and remain within its budget. During the two decades the Ceiling System has

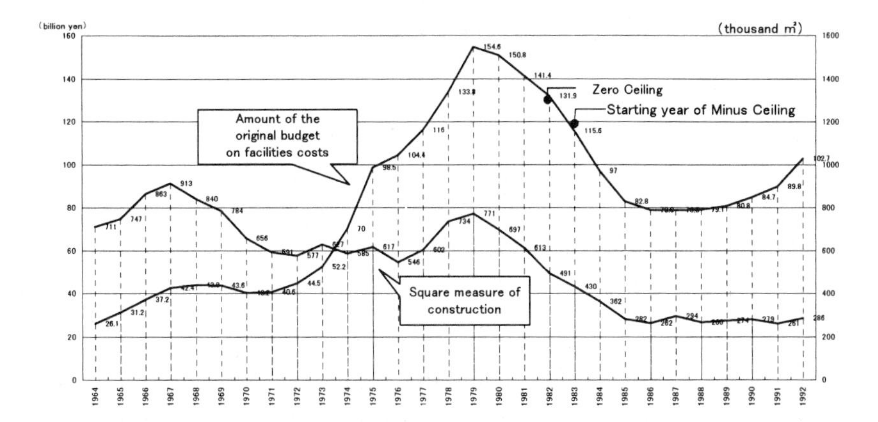

Figure 2 Facilities Costs of National Universities and Square Measure of Construction, (1964–1992). Source: The Monbusho. Note: In the 1970s the amount of the budget was expanding, while the square measure of construction was not. The rise in prices caused this situation.

existed, labor costs have increased annually for a total increase of 15%, which has forced a reduction in the funds available to carry out its mandate.

Within the Monbusho financial structure there is a Special Account for National Educational Institutions, which receives disbursements through transfers from the Monbusho General Account. Figure 1 shows that while the Ceiling dropped precipitously, transfers from the General Account also dropped, remaining between zero and about a 5% increase between 1980 and 1998. Figure 2 shows that in the same time period facilities costs and construction of new facilities were dropped in the national universities. Thus the burden of the Ceiling System has fallen heaviest on facilities, equipment, and operations costs, leaving less and less for the performance of actual research and nothing for improving the research environment.

Then in 1993, 1995, and 1998, in order to revive the economy, supplemental funds were made available through a public investment scheme. Although this had the effect of bolstering the research effort, it was not enough to update the old research university facilities due to their lack of space, nor could these old facilities be easily replaced by modern ones.

The University Budget under Fiscal Structural Reform

In November 1997 the "Special Measure Law for the Promotion of Fiscal Structural Reform" (the Fiscal Structural Reform Law) was passed. It assigned targets for fiscal reform through FY 2003, quantified targets for expense reduction in an initial period of FY 1998–2001, and set forth fundamental principles under which all government expenditures were thoroughly reconsidered in the FY1998 budget.

In the field of universities, one of its requirements was that transfers to a Special Account from a General Account, and subsidies for current expenditures of private universities, be smaller from year to year. The law also required that in FY1998 the amount of ST Promotion Expenses be increased up to 5% from FY1997, and that in FY1999 and FY2000 the amount of annual increase be severely restrained.

The Monbusho is also responsible for ST Promotion Expenses related to university R&D within the General Account. Its largest budget items are: Monbusho Grants-in-aid for scientific research; JSPS research fellowships and new JSPS funds for projects such as the "Research for the Future." Total funds have amounted to about ¥200 billion.

But in the national universities, R&D funding is in the Special Account, transferred from the Monbusho General Account, not a part of the ST Promotion Expenses. So far this has been possible because the Monbusho has reduced allocations to elementary and secondary education; but because of the Fiscal Structural Reform Law's requirement that such transfers be reduced each year, the Special Account must further reduce its funds for equipment and facilities upkeep in the national universities.

The ST-related Budget and University Research

The portion of the national budget that funds the promotion of ST but is strictly limited to the natural sciences, is labeled by the government the "ST-related Budget." The trend toward increasing this budget year to year is shown in Figure 3. In the late 1970s the ST-related Budget increased at a lower rate than the national

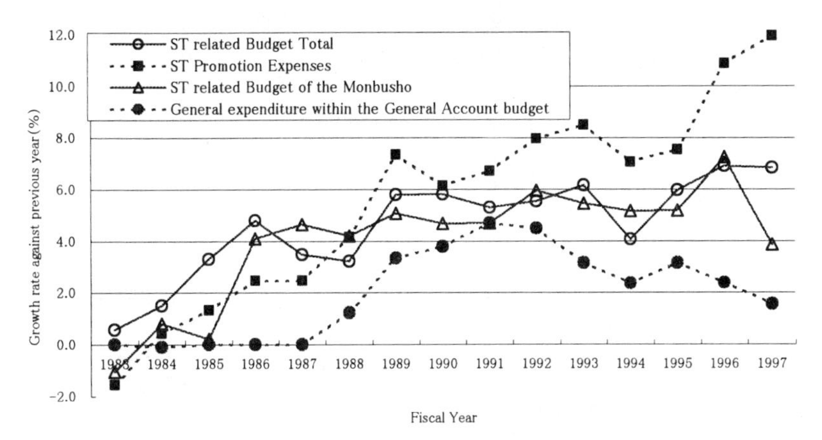

Figure 3 Growth Rate of ST-related Budget (1983–1997). Source: ST Agency, the Monbusho, and the Ministry of Finance.

general expenditure budget; during the following 15 years, the average growth rate of the ST-related Budget was 4.6%, much higher than the rate of the national general budget, which averaged 2.0%.

The ST-related Budget, shown in Figure 4, is divided between the ST-related Budget within the General Account (ST Promotion Expenses and other R&D expenses) and the ST-related Budget within the Special Account. At present, as shown in Figure 3, priority has been given to ST Promotion Expenses, the average growth rate of which was 9.2% for 1993–1997. In the ST-related Budget, the Monbusho budget,[5] especially the Monbusho ST-related Budget within the Special Account, grew more slowly than other expenses, while the share of the Monbusho ST-related Budget decreased year by year, from 49% in 1982 to 43% in 1997.

Thus, university R&D expenses, while considered one of the "other" R&D-related expenses, are not among the highest in the ST-related Budget. At present, the Japanese government prioritizes ST Promotion Expenses in the General Account, while (despite the provisions of the ST Basic Law and the ST Basic Plan, discussed below) university R&D expenses in the Special Account receive no special consideration. One of the reasons for this is the organization of the Japanese ST administration system, as described earlier.

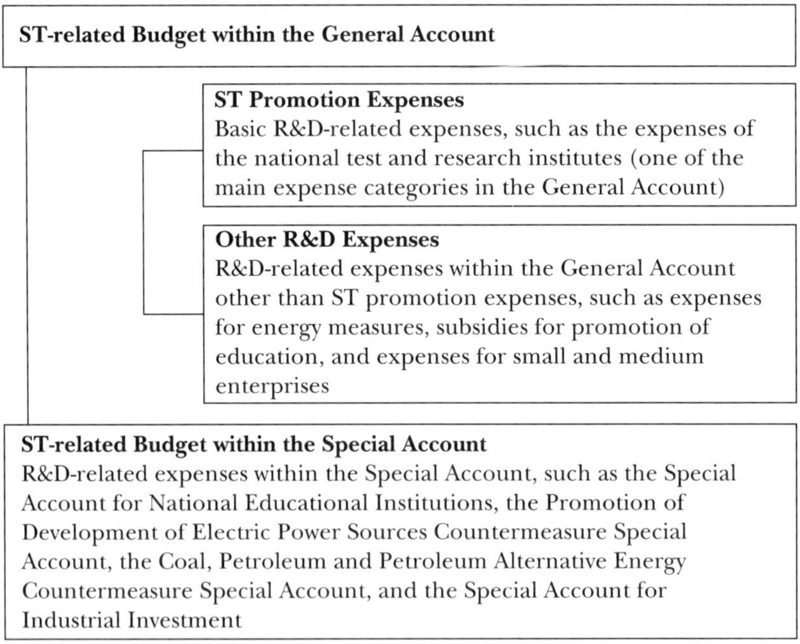

ST-related Budget within the General Account

ST Promotion Expenses
Basic R&D-related expenses, such as the expenses of the national test and research institutes (one of the main expense categories in the General Account)

Other R&D Expenses
R&D-related expenses within the General Account other than ST promotion expenses, such as expenses for energy measures, subsidies for promotion of education, and expenses for small and medium enterprises

ST-related Budget within the Special Account
R&D-related expenses within the Special Account, such as the Special Account for National Educational Institutions, the Promotion of Development of Electric Power Sources Countermeasure Special Account, the Coal, Petroleum and Petroleum Alternative Energy Countermeasure Special Account, and the Special Account for Industrial Investment

Figure 4 The Structure of the Budget for ST-related Expenses. Source: Ministry of Finance (1998).

The Nature of the Monbusho Special Account

The Japanese national account is divided into General and Special accounts. The Financial Law requires that the national government establish a Special Account only when it is (1) carrying on special business; (2) needs special funds for that purpose; and (3) needs to separate these expenses from the General Account revenue and expenditures (Article 13).

In 1964, the Special Account for National Educational Institutions was established to provide for the operation of national educational institutions, especially universities. This Monbusho Special Account is managed separately from the Monbusho General Account, in order to simplify accounting in national universities while contributing to their enhancement.

In FY1998, the Special Account budget was ¥2,701 billion. Its largest source of funding was a transfer from the Monbusho

Table 5 Breakdown of ST-related Budget (1998, ¥ 100 million)

	ST-related Budget in General Account		ST-related Budget in Special Account	ST-related Budget, Total
Ministries	ST Promotion	Other R&D		
Monbusho	1,761 (5.9%)	1,667 (5.5%)	9,683 (0.4%)	13,111 (1.7%)
ST Agency	3,996 (4.6%)	1,855 (–1.9%)	1,550 (–4.9%)	7,401 (0.8%)
MITI	821 (5.2%)	451 (4.7%)	3,656 (4.1%)	4,928 (4.4%)
Other ministries	2,329 (4.5%)	2,121 (12.2%)	430 (0.9%)	4,879 (–3.7%)
Totals	8,907 (4.9%)	6,094 (–3.6%)	15,319 (0.7%)	30,319 (1.0%)

Source: ST Agency
Note: Percentages show rate of increase against previous Fiscal Year.

General Account of ¥1,534 billion: 57% of its total funding. Revenue from university hospitals contributed ¥507 billion (19%), and tuition and other student fees ¥335 billion (12%). The transfer from the General Account made up 82% of the Special Account budget when the system was first established, and peaked at 84% in FY1971. The percentage declined gradually in subsequent years, due to the severe budgetary constraints facing the Japanese government, and by FY1998 it had dropped to 57%. On the other hand, there has been a gradual increase in contributions from self-earned revenues, including not only tuition and other student fees, but also external revenue—donations and income from joint research and from that commissioned by the private sector.

In this way, the Special Account seems to focus on raising profitability. Transfers have been reduced, and it has come to be treated as a self-supporting system, especially in the FY1998 budget, since the Fiscal Structural Reform Law clearly required that transfers from the General Account be smaller than in the previous year.

Meanwhile, by FY1998 labor costs had gradually increased to 54% of expenditures, and educational and research costs had been reduced to 18%, too small to carry out researchers' full activities. In the FY1998 budget, the Monbusho was forced to cut fundamental R&D allocations; those for research equipment were cut by 30%, and those for facilities and their operation were also sharply curtailed.

The ST Basic Law and the ST Basic Plan

In 1995 the Japanese government moved to double its ST-related Budget in five years, establishing the ST Basic Law, followed in 1996 by the ST Basic Plan (Johnson 1997). But because of severe financial restraints, these have not at this writing been fully implemented, nor has it been possible to expand the ST-related Budget.

The ST Basic Plan mandates that as costs increase, government will increase the various R&D funds: competitive funds; priority funds to diversity R&D; fundamental funds; and funds to promote R&D in the private sector in accordance with its aims. It gives priority to increasing funds necessary to secure and train researchers; those necessary to promote their movement within the research community; and those necessary to create the infrastructure for a new R&D system. Elsewhere the Plan requires competitive funds to be "substantially expanded," priority funds to be "expanded," and fundamental funds to be "increased."

In spite of these clear requirements, only competitive funds have been significantly increased, while fundamental funds, such as unit R&D funds and those for equipment and facilities costs and maintenance, have been cut off.

The Plan also requires that the physical facilities of national universities be systematically improved and expanded to deal with new research needs, and that the necessary funds for their laboratories be provided, to replace outmoded equipment and create a frontier research environment, while subsidies will also be increased to fund the facilities and equipment of private and public local universities.

Sadly, and despite these forward-looking plans, the university R&D budget is still insufficient to meet global standards. As an example, the regular budget allocations for facilities and equipment of national universities remained inadequate until. in a move to stimulate the economy. the government allocated additional funds under a supplementary budget. This stopgap measure is not enough to put the Plan into practice or to meet increasing R&D expenses. It is hoped that the ST Basic Plan will be realized as soon as possible.

Administrative Reforms

In neither the ST Basic Law nor especially the ST Basic Plan, which sets out systematized concrete measures for the promotion of R&D, is the role of universities as research institutions correctly evaluated. While the establishment of national TRIs is an encouraging sign, their actual role is minor, especially when compared to the R&D potential of Japanese universities. In FY1997 there were 142,000 university researchers—a count that includes only regular teachers, not students or post-doctoral researchers—while the national TRIs had 11,000; in FY1998, the Monbusho budget for ST was ¥1,311 billion as compared to the budget for national TRIs of ¥442 billion.

To examine Japan's R&D potential we conducted a study of the papers published as a result of research activities (Ohtawa 1998b). Figure 5 shows that in the engineering field, where national TRIs have a slight advantage over universities, the number of papers published by university researchers was 200,107, or 60% of the national total, while those published by national TRI researchers totaled 19,144, 6% of the national total. (It is interesting that scientific papers published by researchers in private companies accounted for 26% of the total.) In terms of simple quantities and comparative costs, universities would appear to be better performers as R&D institutions than national TRIs.[6]

Few of the people concerned are aware of these differences in publication data between the two domains; indeed, in the ST Basic Plan, they are treated almost equally, nor have university R&D allocations from the Special Account been increased.

A possible reason is the compartmentalized nature of the Japanese government, separating university R&D administration through the Monbusho from ST administration through the ST Agency, with the result that the ST Agency has lost touch with university administration as performed by the Monbusho. We suggest that the ST administration be reorganized to reflect the meaning of "Science and Technology"; not Science-Technology.

A hopeful sign is that among administrative reforms planned for the near future is the union of the Monbusho and the ST Agency as a new Ministry, a very significant and well-timed action, which

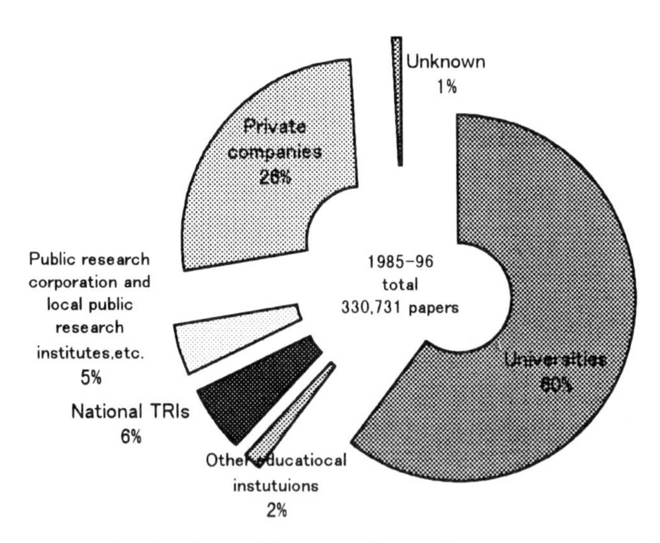

Figure 5 Numbers of Japanese Papers in COMPENDEX PLUS. Note: "Universities" means university and inter-university research institutes. "Total number of papers" is the integrated number of Japanese papers included in COMPENDEX PLUS from 1985 to 1996.

should help Japanese science take its rightful place in the 21st century.

Reforms in the National Budgeting System

Under the "Minus Ceiling System" adopted in the 1980s, the Monbusho has been forced to raise funds for universities through manipulations within its own budget, changing some projects, reorganizing others, scrapping still others. But, as labor costs account for most of the Monbusho budget, its ability to reassign the balance is limited, and university funding has stagnated even as expenses rose.

In order to resolve these issues, we propose that Japan legislate for comprehensive policy adjustments that would establish an overall budget allocation system. Because administration needs are different among the Ministries, it is unfeasible to cut (or, in better times, to increase) their allocations equally.

Although the promotion of ST is supposedly of great importance in Japanese policy, recently the government has favored ST Promotion Expenses. The Fiscal Structural Reform Law provides that ST

Promotion Expenses from the national General Account are to be used for promotion of ST projects in national TRIs as well as universities and other institutions. Thus the Japanese people mistakenly believe that if ST Promotion Expenses are expanded, all university R&D funding will also expand. On the contrary: The main R&D funds of national universities are in the Monbusho Special Account; being outside the ST Promotion Expenses, they are not given special consideration. Government should pay special attention not only to the ST Promotion Expenses, but also to the university R&D funds in the Monbusho Special Account.

Transfers from General Accounts to Monbusho Special Accounts should also be expanded and their oversight clarified. As the Monbusho Special Account is established only to provide a separate accounting from that of its parent General Account, it should not be considered self-supporting, while the Fiscal Structural Reform Law said that transfers to a Special Account from a General Account must be smaller than those of the previous year.

The strenuous efforts at productive management within the Monbusho budget have come to nothing. If the government attaches importance to the research function of universities, it should allow for appropriate transfers to the Monbusho Special Account from its General Account. In December 1998, in a government attempt to revive the economy, the Fiscal Structural Reform Law was temporarily suspended, but the problems with these transfers remain.

R&D expenses cannot be covered by tuition and other fees paid by students; these should be borne by the government agency responsible for awarding the research funds. As the present system can be looked upon as designed to transfer labor costs, there is nothing in the transfer from the General Account to support R&D expenditures. Therefore the new special transfers of R&D funding should be established in addition to the existing ones, and new special transfers expanded and included in ST Promotion Expenses. The new transfer funds should moreover be viewed as an investment in the creation of intellectual property.

These actions, taken timely and firmly, may help restore Japanese university R&D to its rightful place in the estimation of the world's scientific community.

Notes

1. According to the OECD Main Science and Technology Indicators, Annex 2, National Specifications, "In Japan data for R&D personnel is expressed as the number of physical persons rather than in terms of full-time equivalent. In consequence R&D personnel and labor cost data are overestimated by international standard. ... OECD has calculated an 'adjusted' Japanese series of data for the Higher Education sector, based on proportions observed in other Member countries (less 50% of researchers and hence about 35% of HERD)" (OECD 1998).

2. Omitted fields are jurisprudence, accounting, and so on. In Japan there are many university teachers in these fields.

3. The subjects of the survey were 8,400 university researchers who were registered in the NACSIS Information Retrieval Service "Directory of Researchers." The response rate was about 60%.

4. As these funds were open to all who wished to apply, some university researchers received direct aid from them. Their total amount was ¥57 billion in FY1997. The amount transferred to university researchers from public corporations, not including the JSPS, was about ¥27 billion.

5. The ST-related Budget of the Monbusho consists of R&D expenses within its General Account, such as Grants-in-aid for scientific research, JSPS post-doctoral fellowships, and R&D expenses within the Special Account. The budget is fundamentally limited to R&D expenses for the natural sciences. Labor costs in the Special Account count only 50% of the cost; thus, the FTE coefficient is regarded as 50%.

6. Very little work has been done in the area of comparative numbers of papers published, although a related study (Negishi and Sun, this volume) describes some methodological difficulties in doing so.

References

Johnson, J. M. 1997. "Japan Hopes to Double its Government Spending on R&D." NSF Issue Brief 97-310.

Management and Coordination Agency. 1997. 1996 Report on the Survey of Research and Development. Tokyo: Management and Coordination Agency.

Ministry of Finance. 1998. "Comment on Terminology 'ST Promotion Expenses and ST-related Budget'" ["Yogo Kaisetsu 'Kagakugijutsu Shinkohi to Kagakugijutsu Kankei Keihi'"]. *Finance* 33(12): 29.

NCES. 1997. *Current Funds Revenues and Expenditures of Institutions of Higher Education, Fiscal Years 1987 through 1995*. Washington D.C.: U.S. Department of Education.

NSF. 1996. *Academic Science and Engineering R&D Expenditures: Fiscal Year 1994.* Arlington, VA: National Science Foundation.

OECD. 1994. *The Measurement of Scientific and Technological Activities: Frascati Manual 1993.* Paris: OECD.

OECD. 1998. *Main Science and Technology Indicators 1997*: No. 2. Paris: OECD.

Ohtawa, Y. 1995. "Study on the Japanese Higher Education Sector's R&D Statistics." National Center for Science Information Systems *Research Bulletin* 7:181–192.

Ohtawa, Y. 1996. "A Review of the Management and Coordination Agency *Survey of Research and Development*" ["Soumucho Kagakugijutsu Kenkyu Chosa no Saiekntou"]. In *A Study on Time of Life Concerning University Researchers [Diagakutou ni okeru Kenkyusha no Seikatsujikan ni kansuru Chosakenkyu]: A Report of Grants-in-Aid for Scientific Research Funded by Monbusho,* ed. H. Takuma. N.p. pp 111–136.

Ohtawa, Y. 1998a. "Comparative Study on Current R&D Expenditure of Universities and Colleges between Japan and the United States." In *Comparative Study on Academic Research and Funding Systems: A Report of Grants-in-Aid for Scientific Research Funded by Monbusho,* ed. S. Kakinuma. N.p., pp. 91–110.

Ohtawa, Y. 1998b. "A Study on the Number of Japanese Engineering Research Field Papers through Analyzing 'COMPENDEX PLUS.'" *Journal of Information Processing and Management* 41(7): 509–516.

Ohtawa, Y., S. Kakinuma, et al. 1997. "A Survey Report for Research Environment of the Japanese Academic Researchers." In *Fundamental and Substantial Study on the Science Base: A Report of Grants-in-Aid for Scientific Research Funded by Monbusho,* ed. Y. Ohtawa. N.p.

ST Agency. 1997. *White Paper on Science-Technology 1997 [Heisei 9 Nenban Kagakugijutsu Hakusho].* Tokyo: Ministry of Finance Printing Office.

7

Trends in Scientific Publications in Japan and the United States

Masamitsu Negishi and Yuan Sun

Introduction

It is difficult to compare levels of scientific achievement among nations. The number of Nobel prizes awarded is often considered a significant indicator; however, research eligible for the Nobel prize is necessarily outstanding and statistically special. For purposes of making national science policy, it is more useful to look at general or collective indexes of scientific activity, such as the number of papers published and the number of citations to these papers (as an indicator of quality). We might expect a statistical analysis of papers and of citations to give us an objective evaluation of scientific levels from the aspects of both quantity and quality of research (Negishi and Adachi1991; Ichikawa, Uenohara, and Negishi 1995). Unfortunately, there are serious problems in the methods used to obtain these statistics, for which there are limited practical alternatives.

We have done a systematic search of the major abstracting databases to collect statistics on numbers of papers by countries, years, and fields. This paper summarizes the results, shows some characteristics observed in the data, and then describes some interesting results of a Correspondence Analysis.

Survey Method

We searched four databases for the numbers that we then summed and compiled into statistical tables:

(1) INSPEC, compiled by the Institution of Electrical Engineers, UK, covering 260,000 papers a year from 4,200 journals in physics, electricity and electronics, and computer-related areas.

(2) CA (Chemical Abstracts), compiled by the Chemical Abstracts Service, a division of the American Chemical Society, covering 450,000 chemical papers a year from 8,000 journals.

(3) COMPENDEX (EI Compendex Plus), compiled by Engineering Information, U.S., covering 180,000 papers a year in engineering from 4,500 journals.

(4) EMBASE (*Excerpta Medica*), complied by Elsevier Science Publishers, The Netherlands, covering 360,00 medical papers a year from 3,500 journals.

The survey covered the 18-year period 1976 through 1993. As we made the survey on the databases in October 1995, the statistics for 1994 and 1995 had also become available, however since it would take some two years to stabilize the counts because of the database compilation process, we did not analyze those two years.

The seven countries covered in the survey are Japan, the United States, the United Kingdom, Germany, France, Canada, and Russia. These are the countries identified as producing the most papers in our 1987 survey, as we will describe later. The databases show a field for country identification of each paper, derived from the affiliation address of the first author. In our survey, the country field was used as the search term for country identification.

As for pre-unification Germany, counts for the West and the East were searched and summed up; for Russia, searches were made for both the USSR and Russia, and aggregated. Russia now has about half the population of the former USSR, however the statistics gain continuity when the following is taken into account: Scientific activity had gone on mainly in the Russian area, which is still true. As a trial we searched Volume 119 of CA (second half of 1993) and found 13,763 papers from Russia and 4,439 papers from the other former USSR countries, Ukraine leading with 2,301, suggesting that adding 30% to the Russian numbers would give an estimation for the total counts for the former USSR countries.

COMPENDEX before 1981 does not record country names for all the papers from the United States, or for a part of the United

Kingdom, or for Canadian papers, and only includes state and province names in abbreviated forms, such as PA (Pennsylvania) and ONT (Ontario). In the survey, searches were made for the 50 U.S. state abbreviations, and the counts were summed up to represent the United States.

Each database adds searchable classification codes for each paper according to its content. In the survey, we designed classification tables for each database based on the original classes of databases as shown below. As the databases other than CA adopt multiple classification systems (with several classes assigned to a single paper), the total number of papers counted by fields far exceeds the number of papers actually registered in the database.

(1) INSPEC: Section A, B, and C of INSPEC correspond to Physics, Electricity and Electronics, and Computers and Control, respectively. These original sections were used as they were. As for Physics, we can see different research environments for Condensed Matter-related research and Nuclear/Elementary Particle-related research. To clarify the difference between the two groups, we collected their statistics separately based on the detailed classification within Section A.

(2) CA: CA classifies papers into 80 Sections. As this classification was too detailed for our analysis, we designed the following 11 classes by merging the original classes: Pharmaceutical Chemistry, Biochemistry, Agricultural Chemistry, Organic Chemistry, Polymer Chemistry, Chemical Engineering, Industrial Chemicals, Chemical Metallurgy, Physical/Analytical Chemistry, Energy and Nuclear Chemistry, and Physical Properties. CA includes patents as an important portion and the Japanese share of patents is known to be large. In our survey, only journal articles and conference papers were searched and counted, patents excluded.

(3) COMPENDEX: COMPENDEX has a classification system consisting of six major classes: Civil Engineering, Mining Engineering, Mechanical Engineering, Electrical Engineering, Chemical Engineering, and Industrial Engineering. These classes were used in the survey.

(4) EMBASE: In EMBASE, the classification system, EMCLAS, comprises 43 major classes. We merged them into the following

seven classes: Basic Medicine, Biochemistry and Genetics, Clinical Medicine, Cancer, Cardiology, Neurology, and Pharmacology. As EMCLAS was revised several times during our survey period, we checked the details of revisions and designed the search specifications to give consistent results.

The 1987 Survey

The same type of survey was conducted at NACSIS in 1987 (Negishi 1988a, b). In this earlier survey, the same four databases were surveyed for the period from 1976 to 1985. The methodological differences from the present survey are as follows.

For EMBASE, in 1987 only five selected fields, genetics, physiology, and others were surveyed, while the present survey covers the entire EMBASE. As is described above, before 1981 COMPENDEX did not include country codes for the United States, and we abandoned an attempt to collect the counts due to the cost. The detailed statistics in INSPEC for condensed matter research and nuclear research in physics were not collected in the previous survey. Thus the present survey, compared to that of 1987, not only gives updated long-range statistics but also provides more comprehensive data for international comparison.

Outline of the Results

Table 1 summarizes the results of the survey, where the counts for the entire database are included along with the counts by fields. The numbers of papers for 1976, 1986, and 1993 are shown by fields and countries, which are converted into the country shares (%) in the world totals in the right-hand column. The annual rankings of Japan are also included to directly indicate the trend of Japanese papers.

As is shown in Figure 1, which counts Japanese rankings by fields, Japan's ranking has steadily risen in many fields. The figure includes counts of Japanese rankings in 27 fields, excluding those for the breakdowns for condensed matter and nuclear research in physics. The counts indicate that Japan did not rank second in any field in 1976, while in 1986 Japan held second place in many fields

Table 1 Number and Percentage of Papers by Country and Field

DB	Field	Year	Number of papers									Country share (%)						
			Japanese Rank	JPN	USA	UK	FRA	GER	RUS	CAN	World Total	JPN	USA	UK	FRA	GER	RUS	CAN
INSPEC	All fields	1976	5	8046	41159	9430	5442	9399	11095	4258	142147	5.7	29.0	6.6	3.8	6.6	7.8	3.0
		1986	2	18151	63219	13796	8903	13730	14376	6609	220712	8.2	28.6	6.3	4.0	6.2	6.5	3.0
		1993	2	25534	78975	16193	12250	16746	10303	8783	259756	9.8	30.4	6.2	4.7	6.4	4.0	3.4
	Physics	1976	5	5868	27811	6546	4487	6110	10127	2972	92956	6.3	29.9	7.0	4.8	6.6	10.9	3.2
		1986	3	11660	37318	7803	6499	9075	13262	4151	131105	8.9	28.5	6.0	5.0	6.9	10.1	3.2
		1993	2	15644	43149	8799	7968	11275	9155	4416	150855	10.4	28.6	5.8	5.3	7.5	6.1	2.9
	Physics (Condensed Matter, etc)	1977	3	5638	21041	5493	3761	4940	8350	2255	75085	7.5	28.0	7.3	5.0	6.6	11.1	3.0
		1986	3	9734	27716	6289	5129	6878	10911	2994	101628	9.6	27.3	6.2	5.0	6.8	10.7	2.9
		1993	2	13890	32902	7276	6597	9229	7537	3433	121739	11.4	27.0	6.0	5.4	7.6	6.2	2.8
	Physics (Nuclear / Elementary Particle, etc)	1977	5	1341	11202	1992	1152	1911	1906	1122	30626	4.4	36.6	6.5	3.8	6.2	6.2	3.7
		1986	4	2581	12567	2196	1700	2818	2848	1504	38594	6.7	32.6	5.7	4.4	7.3	7.4	3.9
		1993	2	2801	13788	2104	1861	2727	2345	1305	40355	6.9	34.2	5.2	4.6	6.8	5.8	3.2
	Electrical and Electronic Engineering	1976	4	3275	15825	3578	1505	3862	2950	1392	56855	5.8	27.8	6.3	2.6	6.8	5.2	2.4
		1986	2	7421	19567	4663	2097	4107	2388	1836	70270	10.6	27.8	6.6	3.0	5.8	3.4	2.6
		1993	2	10411	28458	5546	3308	4685	2160	3093	87599	11.9	32.5	6.3	3.8	5.3	2.5	3.5
	Computers and Control	1976	4	1337	10344	1894	746	1948	574	1052	31961	4.2	32.4	5.9	2.3	6.1	1.8	3.3
		1986	3	3709	19794	4234	1796	3238	663	1861	60651	6.1	32.6	7.0	3.0	5.3	1.1	3.1
		1993	2	6469	27552	5551	3426	4145	830	3499	79307	8.2	34.7	7.0	4.3	5.2	1.0	4.4
CA	All fields	1976	3	27026	72158	13268	8212	22538	70647	20731	308822	8.8	23.4	4.3	2.7	7.3	22.9	6.7
		1986	3	42914	98642	16433	11529	27646	50562	21671	377388	11.4	26.1	4.4	3.1	7.3	13.4	5.7
		1993	2	62994	124957	22130	14548	34978	25058	26746	467155	13.5	26.7	4.7	3.1	7.5	5.4	5.7
	Pharmaceutical Chemistry	1976	4	2124	7578	2453	988	1947	2308	743	25235	8.4	30.0	9.7	3.9	7.7	9.1	2.9
		1986	2	5951	13399	3036	2233	2506	1810	1435	43856	13.6	30.6	6.9	5.1	5.7	4.1	3.3
		1993	2	7519	15733	3172	2329	2572	667	1772	49193	15.3	32.0	6.4	4.7	5.2	1.4	3.6
	Biochemistry	1976	3	6507	24435	6497	3428	5386	8304	2424	75253	8.6	32.5	8.6	4.6	7.2	11.0	3.2
		1986	2	10817	32776	6907	4372	6033	5782	3404	94653	11.4	34.6	7.3	4.6	6.4	6.1	3.6
		1993	2	17135	42450	8718	6084	7641	2736	4480	123231	13.9	34.4	7.1	4.9	6.2	2.2	3.6
	Agricultural Chemistry	1976	3	1716	3985	1136	503	1275	3938	538	20982	8.2	19.0	5.4	2.4	6.1	18.8	2.6
		1986	2	2090	4456	1170	619	1216	1683	784	19877	10.5	22.4	5.9	3.1	6.1	8.5	3.9
		1993	2	2912	4509	935	936	1368	481	1237	22384	13.0	20.1	4.2	4.2	6.1	2.1	5.5
	Organic Chemistry	1976	3	3134	4856	2011	1562	2433	6758	818	28542	11.0	17.0	7.0	5.5	8.5	23.7	2.9
		1986	3	3612	4858	1412	1220	2572	4119	747	27346	13.2	17.8	5.2	4.5	9.4	15.1	2.7
		1993	2	4127	6150	1712	1596	2637	1765	781	29862	13.8	20.6	5.7	5.3	8.8	5.9	2.6
	Polymer Chemistry	1976	3	1686	2170	807	394	983	4636	157	13501	12.5	16.1	6.0	2.9	7.3	34.3	1.2
		1986	3	1790	3034	627	456	1292	3346	280	14861	12.0	20.4	4.2	3.1	8.7	22.5	1.9
		1993	2	3404	5073	1201	912	1729	1303	497	21983	15.5	23.1	5.5	4.1	7.9	5.9	2.3
	Chemical Engineering	1976	3	444	714	248	114	338	1569	105	4539	9.8	15.7	5.5	2.5	7.4	34.6	2.3
		1986	2	959	1638	299	147	448	864	146	6122	15.7	26.8	4.9	2.4	7.3	14.1	2.4
		1993	2	1219	2945	684	479	846	756	311	10308	11.8	28.6	6.6	4.6	8.2	7.3	3.0
	Industrial Chemicals	1976	3	2503	4642	1018	592	1673	6687	476	22744	11.0	20.4	4.5	2.6	7.4	29.4	2.1
		1986	3	3257	5491	1187	815	2474	4523	795	26841	12.1	20.5	4.4	3.0	9.2	16.9	3.0
		1993	2	5095	9006	1807	1507	3994	2348	1240	39510	12.9	22.8	4.6	3.8	10.1	5.9	3.1
	Chemical Metallurgy	1976	3	2274	3864	1236	953	1703	11484	702	27686	8.2	14.0	4.5	3.4	6.2	41.5	2.5
		1986	3	2804	3990	1087	850	1678	7468	691	25292	11.1	15.8	4.3	3.4	6.6	29.5	2.7
		1993	2	3904	5889	1377	1343	2366	3491	1158	31412	12.4	18.7	4.4	4.3	7.5	11.1	3.7
	Physical / Analytical Chemistry	1976	3	1635	4210	1450	1167	1504	8980	637	26473	6.2	15.9	5.5	4.4	5.7	33.9	2.4
		1986	3	2737	7261	1776	1402	2708	6536	899	34636	7.9	21.0	5.1	4.0	7.8	18.9	2.6
		1993	2	3991	8192	1870	1805	3281	3211	932	39279	10.2	20.9	4.8	4.6	8.4	8.2	2.4
	Energy and Nuclear Chemistry	1976	4	1792	6088	1076	1020	2024	6346	521	23902	7.5	25.5	4.5	4.3	8.5	26.6	2.2
		1986	3	3052	8478	1366	1240	2860	5293	1077	32647	9.3	26.0	4.2	3.8	8.8	16.2	3.3
		1993	2	3863	8747	1463	1424	2840	2826	946	33818	11.4	25.9	4.3	4.2	8.4	8.4	2.8
	Physical Properties	1976	3	3129	8871	2548	2500	3041	9355	1028	37834	8.3	23.4	6.7	6.6	8.0	24.7	2.7
		1986	3	5773	12633	2580	3011	3698	9061	1234	49543	11.7	25.5	5.2	6.1	7.5	18.3	2.5
		1993	2	9677	15543	3322	3624	5426	5385	1449	63963	15.1	24.3	5.2	5.7	8.5	8.4	2.3

after the United States, and third place after the United States and the USSR in others. In 1993, it was in the second rank in 24 fields, partly due to the decline of the former Soviet Union and Russia. It should be noted that the three fields in which Japan ranks third are all medical areas (basic, clinical, and cardiology), while the United Kingdom holds second place in these fields.

Figure 2, industrial chemistry in CA, and Figure 3, cancer in EMBASE, are included to exemplify the pattern of chronological

Table 1 (continued)

DB	Field	Year	Japanese Rank	Number of papers: JPN	USA	UK	FRA	GER	RUS	CAN	World Total	Country share(%): JPN	USA	UK	FRA	GER	RUS	CAN
COMPENDEX	All fields	1976	5	5043	31379	6117	2576	5967	5814	2856	92405	5.5	34.0	6.6	2.8	6.5	6.3	3.1
		1986	2	13534	42468	9216	4990	7913	7454	5065	151246	8.9	28.1	6.1	3.3	5.2	4.9	3.3
		1993	2	17113	57759	9339	6631	8065	6498	6830	181380	9.4	31.8	5.1	3.7	4.4	3.6	3.8
	Civil Engineering	1976	4	1118	10924	1862	604	1428	1030	1039	27041	4.1	40.4	6.9	2.2	5.3	3.8	3.8
		1986	2	3129	13356	2997	1251	2079	1734	1993	45132	6.9	29.6	6.6	2.8	4.6	3.8	4.4
		1993	2	3669	18739	2932	1482	1825	1403	2401	52049	7.0	36.0	5.6	2.8	3.5	2.7	4.6
	Mining Engineering	1976	5	1457	6481	1578	695	1804	2268	731	24368	6.0	26.6	6.5	2.9	7.4	9.3	3.0
		1986	2	4267	8445	2037	1544	2469	2974	1152	39241	10.9	21.5	5.2	3.9	6.3	7.6	2.9
		1993	2	3621	8220	1622	1262	1698	1381	1035	33853	10.7	24.3	4.8	3.7	5.0	4.1	3.1
	Mechanical Engineering	1976	5	1069	8660	1531	571	1734	1310	661	24493	4.4	35.4	6.3	2.3	7.1	5.3	2.7
		1986	2	3060	10775	2401	1175	2318	1699	1122	38975	7.9	27.6	6.2	3.0	5.9	4.4	2.9
		1993	2	2957	13673	2016	1129	1577	1278	1472	39677	7.5	34.5	5.1	2.8	4.0	3.2	3.7
	Electrical Engineering	1976	5	1972	11040	2196	1084	2231	2218	918	32486	6.1	34.0	6.8	3.3	6.9	6.8	2.8
		1986	2	6799	21321	4585	2397	3586	2496	2147	70081	9.7	30.4	6.5	3.4	5.1	3.6	3.1
		1993	2	11141	35903	5478	4218	4996	2500	3886	103828	10.7	34.6	5.3	4.1	4.8	2.4	3.7
	Chemical Engineering	1976	5	1168	7882	1334	536	1352	1290	659	21217	5.5	37.1	6.3	2.5	6.4	6.1	3.1
		1986	2	4606	12986	2661	2252	2698	3446	1735	48803	9.4	26.6	5.5	4.6	5.5	7.1	3.6
		1993	2	7638	16042	3219	2774	3307	2296	1988	58821	13.0	27.3	5.5	4.7	5.6	3.9	3.4
	Industrial Engineering	1976	5	1322	12713	2322	837	1683	1737	1093	32745	4.0	38.8	7.1	2.6	5.1	5.3	3.3
		1986	2	5579	21260	4402	2530	3697	3404	2457	71193	7.8	29.9	6.2	3.6	5.2	4.8	3.5
		1993	2	11060	39519	6415	4736	5503	2998	4511	116406	9.5	33.9	5.5	4.1	4.7	2.6	3.9
EMBASE	All fields	1976	5	12585	74901	18725	14173	21101	11800	6820	225285	5.6	33.2	8.3	6.3	9.4	5.2	3.0
		1986	3	18589	90505	20869	12290	15943	3326	8979	237522	7.8	38.1	8.8	5.2	6.7	1.4	3.8
		1993	3	31115	123984	33515	20487	23960	2706	13213	362686	8.6	34.2	9.2	5.6	6.6	0.7	3.6
	Basic Medicine	1976	6	5682	38665	10206	6397	10028	5752	3740	111383	5.1	34.7	9.2	5.7	9.0	5.2	3.4
		1986	3	7790	46969	10856	5594	7188	907	4875	114393	6.8	41.1	9.5	4.9	6.3	0.8	4.3
		1993	3	14112	63913	18006	9828	11118	1224	7173	181622	7.8	35.2	9.9	5.4	6.1	0.7	3.9
	Biochemistry and Genetics	1976	5	2650	16139	4072	2283	3292	2993	1548	43795	6.1	36.9	9.3	5.2	7.5	6.8	3.5
		1986	3	3841	20972	4326	2610	2531	528	2005	49011	7.8	42.8	8.8	5.3	5.2	1.1	4.1
		1993	2	8924	33779	7902	5069	5693	886	3413	90281	9.9	37.4	8.8	5.6	6.3	1.0	3.8
	Clinical Medicine	1976	5	5987	38318	9755	9307	12484	5292	3394	122671	4.9	31.2	8.0	7.6	10.2	4.3	2.8
		1986	4	8164	50473	12217	6621	8321	542	4804	125297	6.5	40.3	9.8	5.3	6.6	0.4	3.8
		1993	3	16388	60015	18374	11039	13378	1062	6136	186985	8.8	32.1	9.8	5.9	7.2	0.6	3.3
	Cancer	1976	3	1306	6747	1284	1245	1610	821	392	18241	7.2	37.0	7.0	6.8	8.8	4.5	2.1
		1986	2	2487	10192	2001	1182	1434	61	727	23991	10.4	42.5	8.3	4.9	6.0	0.3	3.0
		1993	2	4631	11649	2948	2303	2544	155	972	36483	12.7	31.9	8.1	6.3	7.0	0.4	2.7
	Cardiology	1976	6	860	4818	990	1163	1667	883	363	15330	5.6	31.4	6.5	7.6	10.9	5.8	2.4
		1986	2	1545	8322	1437	1076	1482	118	722	19861	7.8	41.9	7.2	5.4	7.5	0.6	3.6
		1993	3	2603	9345	2613	1708	2417	285	891	29594	8.8	31.6	8.8	5.8	8.2	1.0	3.0
	Neurology	1976	5	1395	6126	1554	1447	1585	1131	691	19782	7.1	31.0	7.9	7.3	8.0	5.7	3.5
		1986	3	2321	11073	2333	1476	1587	114	1253	26695	8.7	41.5	8.7	5.5	5.9	0.4	4.7
		1993	2	3586	11760	3095	1944	2458	255	1450	34619	10.4	34.0	8.9	5.6	7.1	0.7	4.2
	Pharmacology	1976	4	2295	12639	3350	2073	3005	1605	1211	36311	6.3	34.8	9.2	5.7	8.3	4.4	3.3
		1986	3	5018	21314	5118	3159	3809	419	2177	57621	8.7	37.0	8.9	5.5	6.6	0.7	3.8
		1993	2	7111	21581	5845	3893	4648	612	2361	71406	10.0	30.2	8.2	5.5	6.5	0.9	3.3

* Russia includes counts for the former Soviet Union.

change over the period. In EMBASE a temporary decline in the number of registrations in the database, with the lowest in 1987, is shown. According to the EMBASE publisher, they were at that time working on a revision of the thesaurus, which caused a decline in the number of registrations. The same type of decline is observed in COMPENDEX in that the number of registrations dropped from 1985 through 1992 but made a remarkable recovery in 1993, which was also caused by structural reforms of database compilation at the publisher. It is evident that the number of registered records in databases is not only affected by the general situation of paper publication but also by the particular conditions at publishers, and

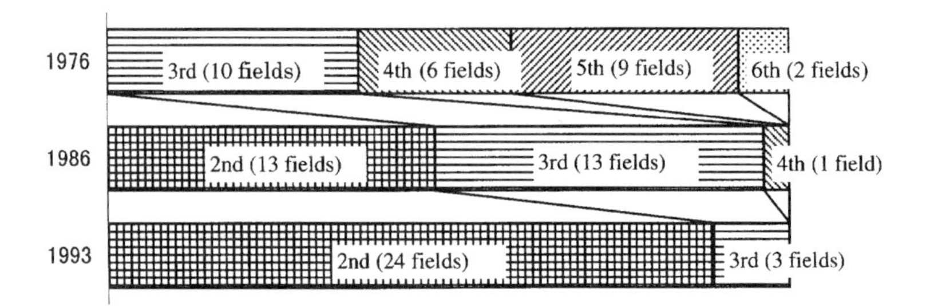

Figure 1 Japanese Rankings in 27 Fields, 1976, 1986, 1993.

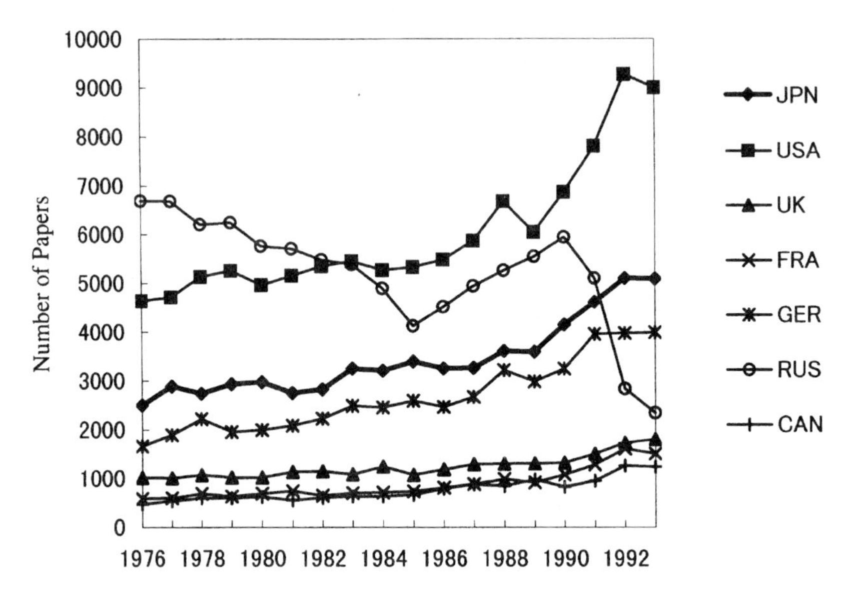

Figure 2 Number of Papers on Industrial Chemistry, by Country, 1976–1993. Source: CA database.

this points up how carefully we must evaluate statistics involving absolute numbers.

Trends in the Numbers of Papers

Regression indexes were calculated to get a bird's-eye view of the number of papers by fields and countries, shown in Table 2. In order to obtain a simplified index to depict the increase of Japa-

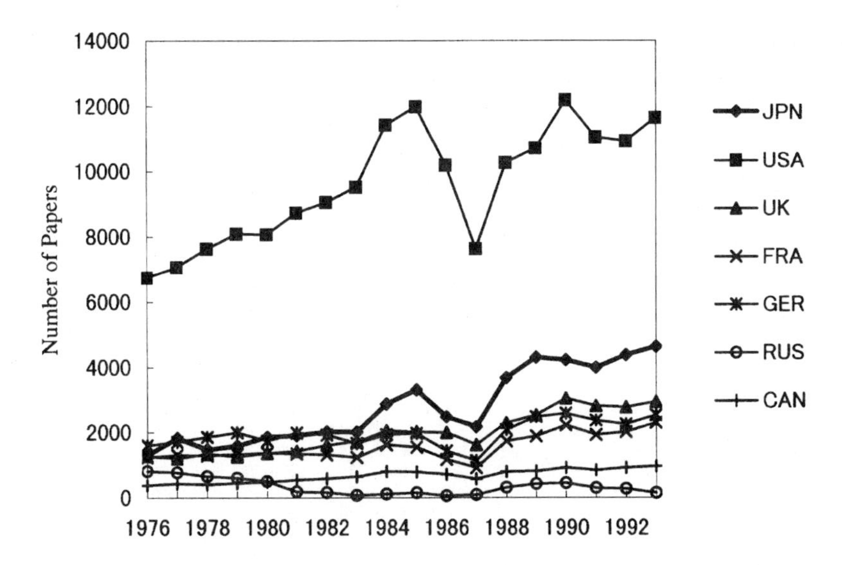

Figure 3 Number of Papers on Cancer, by Country, 1976–1993. Source: EMBASE database.

nese papers, the growth rates were calculated for Japan and the whole database by fitting compound interest curves ($T = a*b^t$, where a: initial value, b: growth rates, t: year), as shown on the left. As in COMPENDEX, external fluctuations caused a poor fit, while providing some useful yardsticks.

Here we see, for example, that in INSPEC a 3.8% growth was realized throughout the 18 years for papers in the entire database, while Japanese papers grew at a rate of 6.9%, causing a parallel growth in world shares. This trend can be seen in all the databases and fields excluding organic chemistry, polymer chemistry, and chemical metallurgy in CA.

The change of paper shares is shown at the right of the table by calculating a linear regression for the statistics by countries and fields. Here the change rates are shown as the index adjusted to 100 in 1976. This shows that the Japanese share for INSPEC in 1993 was 1.64 times as large as that in 1976; the numbers are 1.42 times for CA, 1.61 for COMPENDEX, and 1.47 for EMBASE. Growth rates of double or more were realized in the fields of chemical engineering and industrial engineering in COMPENDEX, electricity and electronics in INSPEC, and pharmaceutical chemistry in CA. In medi-

Table 2 Percentage Change in Publication Number by Country and Field, 1976–1993. (Annual growth rates are estimated by fitting compound interest curves.)

DB	Field	Growth rates of papers (%)		Growth trend of papers (Estimates of 1993 by linear regression, with 1976=100)							
		Japan	Whole DB	Japanese Rank	JPN	USA	UK	FRA	GER	RUS	CAN
I	All fields	6.9	3.8	1	164	106	91	116	96	71	111
N	Physics	5.6	2.8	1	159	96	76	103	107	80	95
S	(Condensed Matter, etc)	5.9	3.1	1	158	101	76	99	106	73	96
P	(Nuclear / Elementary Particle, etc)	4.9	2.3	1	154	88	78	112	106	128	98
E	Electrical and Electronic Eng.	7.4	3.4	1	198	114	104	157	86	55	149
C	Computers and Control	9.6	6.5	2	171	111	123	213	86	77	126
	All fields	4.2	2.0	1	142	114	82	107	101	36	117
	Pharmaceutical Chemistry	7.6	3.9	1	181	101	67	118	67	26	120
	Biochemistry	4.6	2.5	1	152	107	81	108	89	30	118
C	Agricultural Chemistry	1.9	-0.1	3	152	109	103	153	114	16	161
	Organic Chemistry	1.2	1.4	2	115	121	81	96	95	42	86
A	Polymer Chemistry	3.3	5.3	4	108	152	72	134	99	31	224
	Chemical Engineering	5.9	8.9	4	111	181	89	161	106	31	142
	Industrial Chemicals	3.7	3.6	4	110	107	98	147	133	34	138
	Chemical Metallurgy	1.9	2.0	4	125	129	81	99	128	46	146
	Physical / Analytical Chemistry	4.3	4.3	1	139	137	87	110	133	32	98
	Energy and Nuclear Chemistry	2.4	1.4	2	117	100	93	98	102	45	142
	Physical Properties	6.2	3.4	1	174	110	73	83	97	48	78
C	All fields	4.7	1.7	1	161	74	73	110	76	113	102
O	Civil Engineering	4.8	2.0	2	153	69	77	126	81	188	104
M	Mining Engineering	5.1	1.3	1	177	71	67	116	87	101	88
P	Mechanical Engineering	3.8	1.1	2	152	66	67	104	66	444	121
ED	Electrical Engineering	7.1	3.8	1	164	79	78	102	75	67	95
NE	Chemical Engineering	10.7	5.4	1	234	60	78	143	91	148	100
X	Industrial Engineering	9.5	5.5	1	181	73	73	114	92	133	95
	All fields	5.2	2.7	1	147	102	115	83	66	12	126
E	Basic Medicine	5.5	3.0	1	151	100	111	85	64	5	123
M	Biochemistry and Genetics	7.3	4.7	1	152	101	100	101	81	9	112
B	Clinical Medicine	5.9	2.5	1	172	101	131	74	63	5	121
A	Cancer	7.6	4.1	1	177	84	133	86	69	-2	123
S	Cardiology	5.8	3.5	1	141	103	135	70	71	6	130
E	Neurology	5.1	2.9	1	142	106	117	70	83	6	118
	Pharmacology	8.5	5.5	1	159	86	87	92	74	25	95

cal fields Japan was not outstanding except for cancer research, with an increase of 177%. Thc table also shows the Japanese rankings in share growth rates, where Japan is in the first rank in many fields, while polymer chemistry, industrial chemicals, and chemical engineering in CA show lower rates, although the former two fields already had large shares in 1976. As to industrial chemistry, recent U.S. growth in the field may affect the apparently low Japanese rate.

The United States has long maintained large shares in INSPEC and EMBASE. It is still increasing its shares in CA, perhaps partly because of the recent Russian decline. Although the United States appears to be losing shares in COMPENDEX, it is because

COMPENDEX had been registering more papers than ever from countries other than the United States and the United Kingdom. Shares of the latter seem to be stable or slightly declining in INSPEC, CA, and COMPENDEX, while its shares in EMBESE are notably increasing. In contrast, Germany is rapidly losing its share in EMBASE.

Statistics Revised by National Population

We have thus far examined relative scientific advancement through numbers of papers and world shares by country. However, it would also be useful to introduce some general factors representing the size of the research activity in each country, a comparison based on the numbers of papers discounted by the relative population size of each country. The indexes of the size of a country's research activities would be expected to be the number of researchers, national expenditures for research, gross domestic product, and so on. But these indexes involve problems of statistical definitions of researchers and research expenditures, and the fluctuations of foreign exchange rates, which introduces many complexities. Thus we employ the simplest index, population, as our first approach.

Table 3 shows paper shares in 1993 divided by population shares in the world for each country, with the values thus obtained converted into indexes. The United States represents 100. This indicates the ratio of paper share to the population share, relative to the United States. Actual ratios are shown only for entire databases. For example, the 1993 Japanese paper share in INSPEC was 4.4 times as large as its population share. As scientific research is convergently conducted in developed countries, the indexes for all countries in the table naturally far exceed 1. It should be noted that some of the small countries like Israel and North European countries, which were not surveyed here, should rise up to high ranks in this type of comparison. Thus, the present discussion compares only the countries big enough for our study in absolute numbers of papers.

Using statistics discounted by country size, Japanese rankings amazingly go down to sixth place in INSPEC and EMBASE, and fourth in CONPENDEX, with the U.S.-based indexes of 67, 52, and

Table 3 Country Share of Publication Numbers Compared to Population Share, 1993

DB	Field	Japanese Rank	JPN	USA	UK	FRA	GER	RUS	CAN
			Index to USA (USA=100)						
INSPEC	All fields	6	67	100	91	69	67	23	99
	(actual ratio)		*4.4*	*6.5*	*6.0*	*4.5*	*4.4*	*1.5*	*6.5*
	Physics	6	75	100	91	83	83	37	91
	(Condensed Matter, etc)	6	87	100	99	90	89	40	93
	(Nuclear / Elementary Particle, etc)	6	42	100	68	60	63	30	84
	Electrical and Electronic Engineering	4	76	100	87	52	52	13	97
	Computers and Control	5	49	100	90	56	48	5	113
CA	All fields	2	104	100	79	52	89	35	191
	(actual ratio)		*6.0*	*5.7*	*4.5*	*3.0*	*5.1*	*2.0*	*11.0*
	Pharmaceutical Chemistry	3	99	100	90	66	52	7	100
	Biochemistry	4	84	100	92	64	57	11	94
	Agricultural Chemistry	2	134	100	92	93	96	19	245
	Organic Chemistry	1	139	100	124	116	136	50	113
	Polymer Chemistry	1	139	100	106	80	108	45	87
	Chemical Engineering	5	86	100	104	73	91	45	94
	Industrial Chemicals	3	117	100	89	75	141	46	123
	Chemical Metallurgy	2	137	100	104	102	128	104	175
	Physical / Analytical Chemistry	4	101	100	102	99	127	68	101
	Energy and Nuclear Chemistry	4	92	100	75	73	103	56	96
	Physical Properties	1	129	100	95	104	111	61	83
COMPENDEX	All fields	4	61	100	72	51	44	20	105
	(actual ratio)		*4.2*	*6.8*	*4.9*	*3.5*	*3.0*	*1.3*	*7.2*
	Civil Engineering	4	41	100	70	35	31	13	114
	Mining Engineering	3	91	100	88	69	66	29	112
	Mechanical Engineering	4	45	100	66	37	37	16	96
	Electrical Engineering	4	64	100	68	53	44	12	97
	Chemical Engineering	3	99	100	89	77	66	25	111
	Industrial Engineering	4	58	100	72	54	44	13	102
EMBASE	All fields	6	52	100	120	74	61	4	95
	(actual ratio)		*3.8*	*7.3*	*8.8*	*5.4*	*4.5*	*0.3*	*7.0*
	Basic Medicine	6	46	100	126	69	55	3	100
	Biochemistry and Genetics	5	55	100	104	67	54	5	90
	Clinical Medicine	6	57	100	136	82	71	3	91
	Cancer	4	82	100	113	89	69	2	74
	Cardiology	6	58	100	125	82	82	5	85
	Neurology	6	63	100	117	74	66	4	110
	Pharmacology	6	68	100	121	81	68	5	98

67 respectively. By contrast, Japan ranks second in CA with Canada being the first and the United States third, a noteworthy difference. In breakdowns by fields in CA, Japan is first in organic chemistry, polymer chemistry, and physical properties.

In physics, under INSPEC, a great difference is observed between condensed matter research and nuclear research, so that Japan is close to the United States in condensed matter, whereas in nuclear research, an area in which the United States and Canada hold

Table 4 Distribution of Publication Number by Field for Japan and the United States in Each Database, 1993 (%)

INSPEC	JPN	USA	World	Japan's Index to USA
Physics	50	45	49	109
(Condensed Matter, etc)	41	32	37	129
(Nuclear/Elementary Particle, etc)	8	13	12	62
Electrical and Electronic Eng.	31	28	27	112
Computers and Control	19	27	24	72
	100	*100*	*100*	

C A	JPN	USA	World	Japan's Index to USA
Pharmaceutical Chemistry	12	13	11	95
Biochemistry	27	34	26	80
Agricultural Chemistry	5	4	5	128
Organic Chemistry	7	5	6	133
Polymer Chemistry	5	4	5	133
Chemical Engineering	2	2	2	82
Industrial Chemicals	8	7	8	112
Chemical Metallurgy	6	5	7	132
Physical / Analytical Chemistry	6	7	8	97
Energy and Nuclear Chemistry	6	7	7	88
Physical Properties	15	12	14	124
	100	*100*	*100*	

COMPENDEX	JPN	USA	World	Japan's Index to USA
Civil Engineering	9	14	13	65
Mining Engineering	9	6	8	145
Mechanical Engineering	7	10	10	71
Electrical Engineering	28	27	26	102
Chemical Engineering	19	12	15	157
Industrial Engineering	28	30	29	92
	100	*100*	*100*	

EMBASE	JPN	USA	World	Japan's Index to USA
Basic Medicine	25	30	29	82
Biochemistry and Genetics	16	16	14	98
Clinical Medicine	29	28	30	101
Cancer	8	5	6	147
Cardiology	5	4	5	103
Neurology	6	6	5	113
Pharmacology	12	10	11	122
	100	*100*	*100*	

overwhelmingly dominant positions, Japan's ratio is less than half that of the United States. Japan's rank is about half that of the United States in computers and control, a field characterized by great differences between the English-speaking countries and a group consisting of Japan, Germany, and France. The author has already made an analysis of the leadership of the English-speaking world in information-related fields (Yasuda et al. 1995).

In COMPENDEX, Japan's figures in mining engineering and chemical engineering are comparable to U.S. ratios, whereas it lags far behind in the other fields. In EMBASE the Japanese index for cancer research is 82, whereas its index is only 46 in basic medicine. As is mentioned in our comparison of the absolute number of papers in Table 1, the United Kingdom leads Japan, ranking second in the three medical fields. Here its strength in EMBASE is confirmed. In normalizing for the population size of the seven countries, the United Kingdom leads all countries in all the medical fields.

Patterns of Distribution Ratios across Fields

We compared the distribution ratios of number of papers by fields between Japan, the United States, and the world average, an analysis from the other axis than that for our analysis based on country shares. Table 4 shows distribution ratios of papers by fields in 1993 for Japan, the United States, and the world. The table also includes Japan's index relative to the U.S. index. As all the databases except CA use multiple classification systems, we used aggregations of numbers by fields, instead of the world total (the actual number of papers), as the population parameter in calculating the ratios.

In physics, in INSPEC, there are three times as many papers for condensed matter as for nuclear research, a difference that widens to five times for Japan. In CA, biochemistry papers occupy 26% of the world total, rising to 34% for the United States, indicating its activities in this field. In COMPENDEX, papers in industrial engineering and electrical engineering occupy a large portion of the output for both Japan and the United States. In EMBASE, we found that medical research is conducted at a similar scale in both basic and clinical areas, papers in biochemistry and genetics accounting for about half of them. The United States is close to this worldwide pattern, but Japan has somewhat fewer papers in basic medicine.

As Japan's indexes relative to the United States in Table 4 give the same values as calculated by multiplying Japanese number of papers by fields by the figures for whole database in Table 3, they give the same patterns of distribution over fields. However, in Table 4 the difference in the number of papers between Japan and the United States is removed, and here we can clearly see the characteristics of the distribution patterns among fields for Japan. Although there is no great difference between the Japanese and U.S. figures in CA, some strong Japanese biases toward material properties can be seen in INSPEC, mining and chemical engineering in COMPENDEX, and cancer research in EMBASE.

Further Findings through Correspondence Analysis

Correspondence analysis is a useful tool to analyze categorical data like our publication statistics, where categories of countries and fields are attached to numerical data, the number of publications. We have applied this method to these statistics to make some points clearer that were mentioned earlier, as this method assigns scores to categories, or converts categories into numerics, and lays them on an elegant but simple graphical display. The relative positions

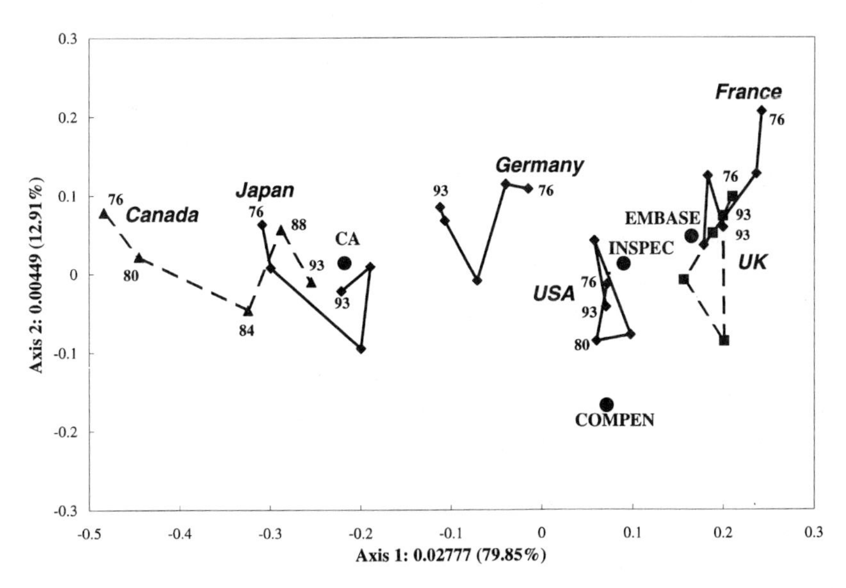

Figure 4 Trajectories of Six Countries Relative to Four Databases, 1976–1993 (selected years).

of two categories in the display indicates how they are highly associated in the data with respect to each other, therefore we can obtain a global view of the information; correspondence analysis thus makes possible a more rapid interpretation and understanding of the data (Sun and Negishi 1998).

Differences between Databases and Relation of Countries to Databases

Figure 4 plots the position of each database together with country trajectories over time, showing the relative position of databases and countries. Nearly 80% of the total variance is represented by the first principal axis. Here we can see that CA is at the leftmost position, then in the middle, INSPEC and COMPENDEX are at identical positions on the first axis, with EMBASE in the rightmost place. The distance between CA and INSPEC is about 2.5 times that between INSPEC and EMBASE. The figure shows CA's unique nature, characterized by Japanese strength in these fields, as has been shown by Japan's high ranking in Table 3.

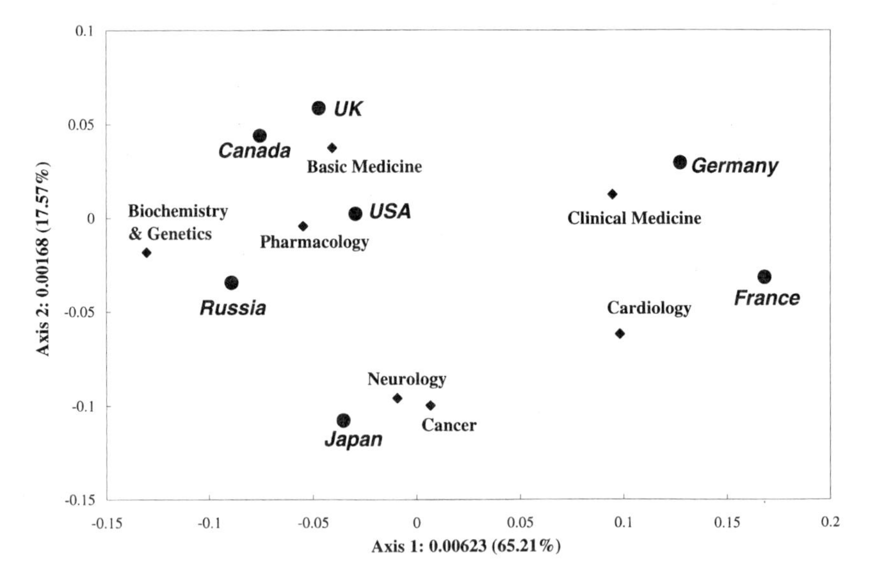

Figure 5 Relative Positions of Seven Countries to Seven Medical-Related Fields, 1976.

Looking at the positional relation between databases and countries in Figure 4, Japan and Canada are found close to CA. This means both countries have similar distribution patterns over the four databases, with relative large number of publications in CA compared to other countries. The leadership of the United Kingdom in EMBASE is confirmed here by the closeness of their positions, while France appears to be close to EMBASE because of its strength in medicine in relation to other areas. The yearly transit of Japan and Canada to the right or to the zero point indicates that their data patterns are becoming balanced, or that Japan in particular is getting more share in databases other than CA.

Characteristics of Japan and the United States in Medical Research

Correspondence analysis of the EMBASE data gives the following findings on medical research in Japan and the United States. Figures 5 and 6 show the disposition of countries and fields in 1976 and 1993. In both figures the same country groups are found, the first being composed of the United States, the United Kingdom,

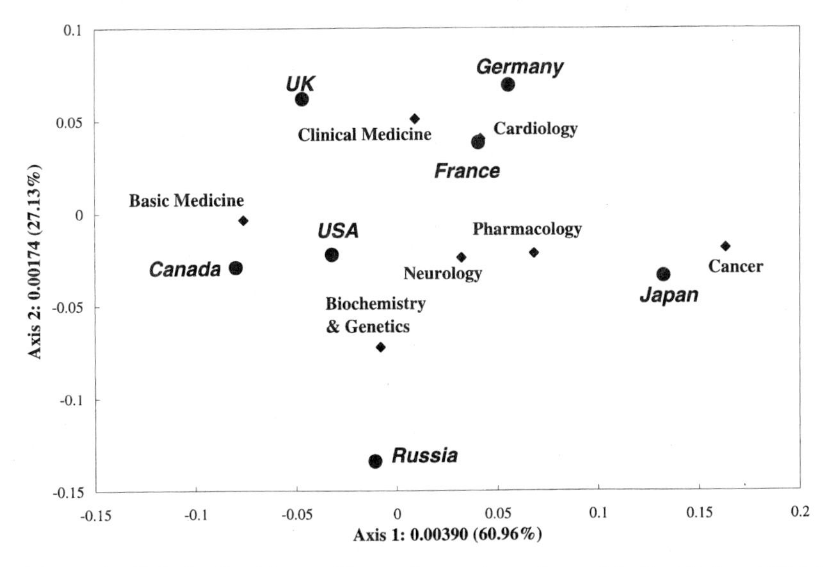

Figure 6 Relative Positions of Seven Countries to Seven Medical-Related Fields, 1993.

and Canada, and the other of Germany and France. In 1976 the former group were quite near to each other, and apparently specialized in basic medicine, pharmacology, and biochemistry/ genetics. In 1993 the connection among the three countries became weaker with the U.S. shift toward biochemistry/genetics, while the United Kingdom remained in the same position in basic and clinical medicine and in biochemistry/genetics. Japan emphasized both cancer and neurology in 1976, but seems to have shifted to cancer research in 1993.

The chronological progression of Japan and the United States is shown in detail in Figures 7 and 8. Figure 7 shows that Japan stressed cardiology and neurology in the early 1980s, but later in the decade became strongly specialized in cancer research. Recently Japan seems to attach more importance to biochemistry/ genetics, but it still stresses cancer research. The U.S. path appears simple, as it emphasized neurology, cardiology, and cancer research in 1980s and since around 1990 has rapidly shifted to biochemistry/genetics. This development is impressive when compared with Japanese movement in these fields.

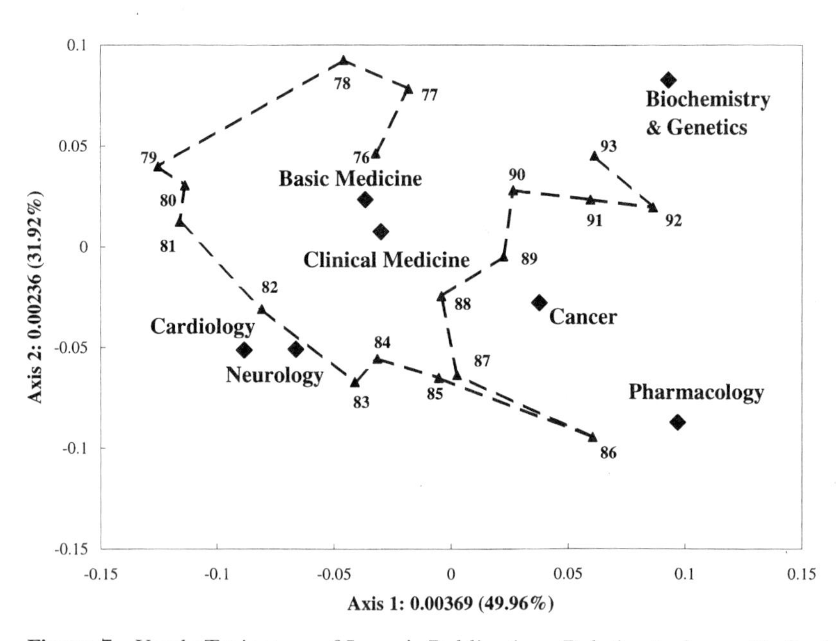

Figure 7 Yearly Trajectory of Japan's Publications Relative to Seven Medical-Related Fields.

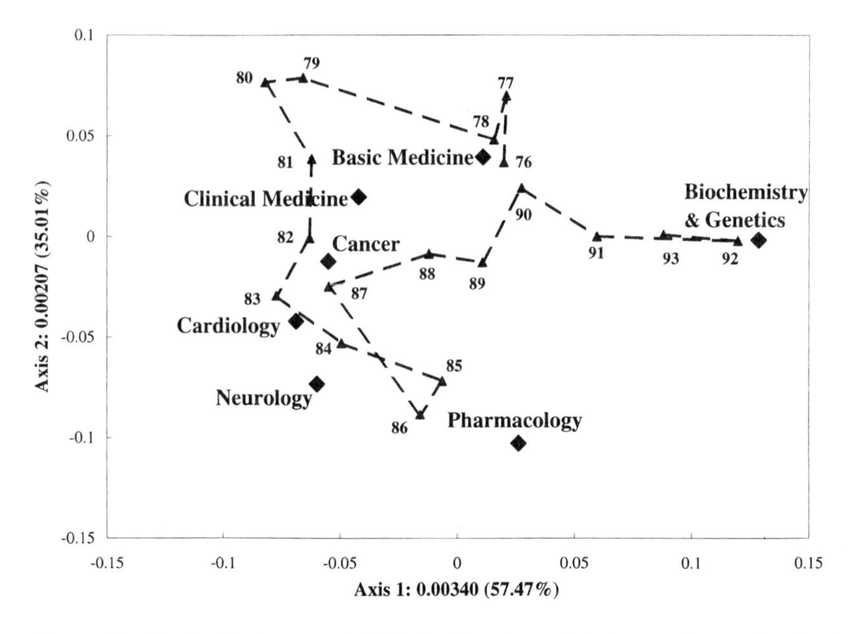

Figure 8 Yearly Trajectory of U.S. Publications Relative to Seven Medical-Related Fields.

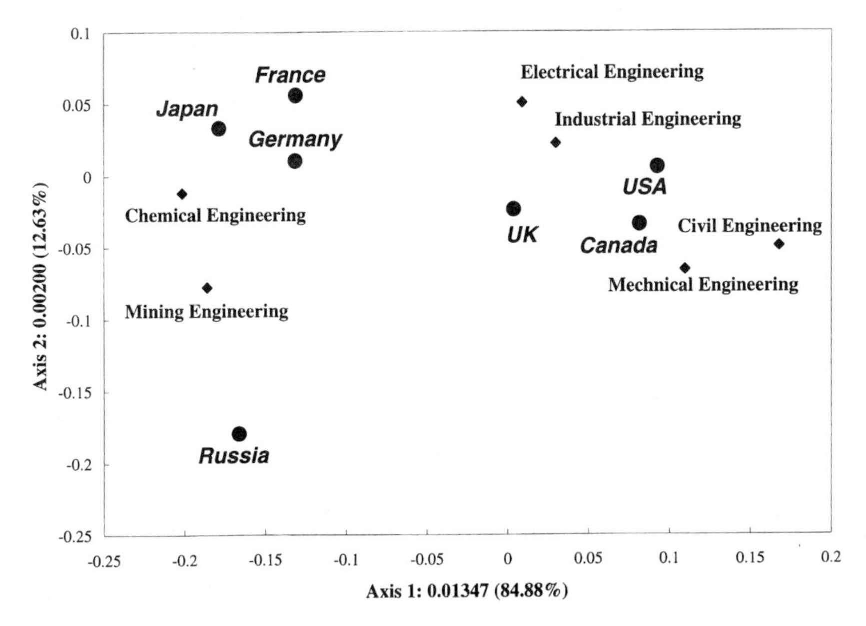

Figure 9 Relative Positions of Seven Countires to Six Fields from COMPENDEX, 1993.

The Language Factor

As we have seen in Figures 5 and 6, the English-speaking countries tend to form a group in our analyses; a clearer picture emerges from our analysis of COMPENDEX data (Figure 9), where the three Anglophone countries are placed far to the right while the others are on the far left with a sizeable gap between. The English-speaking group has relatively more publications in electrical, industrial, mechanical and civil engineering, and the non-English-speaking group, consisting of Japan, Germany, France, and Russia, has more in chemical and mining engineering.

Conclusion

When the number of Japanese scientific publications is viewed as a ratio of papers to population share (Table 3, Figure 4), a significant difference between CA and the other three databases appears. While Japan's index in 1993 was 104 in CA (using U.S. paper numbers as 100), it is 67 in INSPEC, 61 in COMPENDEX, and 52 in EMBASE, a situation that also appeared in the 1987 survey. Table 2 shows that Japan's share grew remarkably in the three databases, yet it does not reach the CA level. Japan is in a position comparable to Germany and France, and surpasses them in COMPENDEX, however in EMBASE it does not reach their levels (Negishi 1990).

One interpretation of this situation is that Japan has achieved the status of a big power in chemical research, surpassing the United States, but is far behind in physics, engineering, and medicine; another possibility is that many of the Japanese papers were submitted to journals published in Japan and inadequately covered by the three Western-made databases. When we realize that CA is compiled with a well-organized system of comprehensive worldwide coverage, the latter view has greater significance. Japan maintains one of the world's largest scientific communities, and thus discussions among people inside the country should be effective and functional. This is particularly applicable to the applied sciences, where the needs of the Japanese market establish the direction of research and development. It should also be noted that the major Japanese academic societies publish two kinds of journals, one in Japanese and the other in English, with different topics according to their readership.

Our survey shows that Japan now ranks second, below only the United States, in many fields in terms of the number of papers registered in the science databases used worldwide, which should confirm that Japan has contributed significantly to the promotion of scientific activities worldwide. But the statistics are not as reassuring when the country size is taken into account. This is probably because the papers were written in the Japanese language for domestic readers and thus do not appear in all databases. If globalization of markets and the consequent globalization of research and development make further progress in the years ahead, we can expect to see more Japanese papers in the international languages.

References

Ichikawa, A., M. Uenohara, and M. Negishi. 1995. "Evaluating university research and education capability: Correlation analysis of evaluation indices in electric and electronic engineering." *The Journal of the Institute of Electronics, Information and Communication Engineers,* 78(6): 552–559. (in Japanese)

Negishi, M., and J. Adachi. 1991. "Overseas acceptance of Japanese scientific papers as seen in a citation analysis." In Preprints of the 3rd International Conference on Japanese Information in Science, Technology and Commerce. Nancy, France: INIST-CNRS, pp. 243–259.

Negishi, M. 1990. "Research activities in Japan and Japanese articles registered in western databases." In *Japanese Information in Science, Technology and Commerce: Proceedings of the 2nd International Conference.* ed. D. Monch et al. Amsterdam: IOS Press, pp. 184–197.

Negishi, M. 1988a. "National activities as seen in publication data: A survey and analysis." *Japanese Scientific Monthly* (Japan Society for the Promotion of Science) 41(7): 40–47. (in Japanese)

Negishi, M. 1988b. "National activities as seen in publication data (II): A comparison among social sectors: Industry, universities and government." *Japanese Scientific Monthly* (Japan Society for the Promotion of Science), 41(11): 42–49. (in Japanese)

Sun, Y., and M. Negishi. 1998. "Cross-national comparison of publication output in Engineering based on correspondence analysis." In Preprints of the Sixth Conference of the International Federation of Classification Societies on Data Science Classification and Related Methods. Rome: IFCS-98, pp. 293–296.

Yasuda, Y., M. Negishi, M. Ishizuka, and J. Adachi. 1995. "A summary of the report: A study on the level of Japanese information engineering research and policies for the promotion." *EAJ Information* (Japan Engineering Academy), 47: 2–17. (in Japanese)

III

Sociological and Historical Perspectives

The roots of university-industry relationships in the histories and cultures of the two nations are described in the next three chapters, in an attempt to explain the difficulties facing those who push to break down old barriers and test new missions for higher education in an advanced, industrialized global economy. These disparate backgrounds are explored in depth, to provide an understanding of why American universities tend to be entrepreneurial about exploring commercial ties, while Japanese universities are more cautious and more protective of their independence from inappropriate influences of powerful industrial or political forces.

8

Bridging the Gap: The Evolution of Industry-University Links in the United States

Henry Etzkowitz

Introduction: The Endless Transition

A variegated system of university-industry relationships has spread from a specialized academic sector throughout U.S. academia in recent years. In the early 20th century, Vannevar Bush and his MIT colleagues created models for academic consultation (the one-fifth rule), patenting, and firm formation (Etzkowitz 1999). Compromises between proponents and opponents of linkage resulted in acceptable formats; in response to objections that industrial involvement diverted academic attention from students, it was argued that real-world experience invigorated teaching. The Research Corporation, founded by Frederick Cottrell, a Professor of Chemistry at the University of California at Berkeley, introduced the principle of utilizing income generated by patents to seed-fund new research. A potentially self-generating system of academic research funding from the sale of intellectual property was initiated which was extended by the Patent and Trademark Act of 1980 to all universities that were recipients of federal research funds. More familiarly known by the names of its authors as the Bayh-Dole Act, the law renewed the original intent of the patent clause of the U.S. Constitution by mandating some portion of the financial rewards to the inventor as well as his or her employer.

Bayh-Dole assigned the intangible property of scientific and technological knowledge emanating from federally funded research to the university in which the discovery was made, regulating technology transfer by clarifying ownership rights and responsibilities. The Morill Act of 1862 had donated federal land to support the development of higher education for the improvement of agricultural and industrial practices, systematizing and expanding an agricultural research system that had arisen since the founding of the first experiment station in Connecticut in 1816; in a knowledge society, the Bayh-Dole Act's transfer of intellectual property rights is the virtual equivalent of a land grant to the entire research university system. As economic potential in academic research spreads from the biomedical and computer fields to other scientific and technical areas, as well as to the social sciences and the humanities, the effects of the law will become even more significant (Rosenberg and Nelson 1994). "Bayh-Dole" is also a shorthand for an academic mission of economic development that extends well beyond the sale of intellectual property to existing companies, to the capitalization of knowledge in new firms that generate equity for their academic progenitors and owners.

The MIT/Stanford model, combining basic research and teaching with industrial innovation, is displacing Harvard as the academic exemplar. Until quite recently, pursuing the "endless frontier" of basic research was the primary ideological justification of elite United States academic institutions, with Harvard University the model and numerous schools identifying themselves as the "Harvard" of their respective regions. With an entrepreneurial mode increasingly followed at Harvard, and at academic institutions that model themselves upon it, the prediction that MIT would eventually conform to the traditional United States research university mode has been disconfirmed (Geiger 1986). Instead, the reverse process has occurred as liberal arts research universities adopt a mission closer to the "land grant" tradition of regional economic development, MIT's founding purpose and historic forte. In this paper I discuss the proximate and long-term causes of the growth of industry-university links in the United States as a co-evolution of academia and industry, encouraged by government initiatives.

Academic Revolutions: The Transformation of the University

The university, originating as a medieval institution for the conservation and transmission of knowledge, has twice been radically transformed in the United States. The discovery of new knowledge became an academic goal during the late nineteenth century—the so-called "academic revolution" (Jencks and Riesman 1968); the translation of knowledge into economic activity as an academic function alongside research and teaching is a second academic revolution. During the past two decades a broad range of universities, both private and public, have established one or more of the following mechanisms for academic-industrial relations: offices to manage patenting and licensing of technology, interdisciplinary research centers with industrial participation, research parks, and incubator facilities. In addition, many schools have established procedures to manage potential conflicts of interest and commitment as faculty members play dual roles on both sides of the academic and industrial divide.

The first revolution created the necessary conditions for the second: a research tradition and a cadre of research teams. The sufficient conditions for the second revolution appeared as technological and business opportunities arose closer to the research frontier. Change was spurred as it became more difficult to obtain research funds from traditional sources. On the one hand, the competition for federal research funds intensified as more academic institutions attempted to become research universities, while on the other, fields such as molecular biology produced technological and commercial possibilities even as technologies such as computers were recognized as academic disciplines. Moreover, universities became more salient to industry as R&D providers in response to increased international competition. Even though industry provides a relatively small proportion of academic research funds, it is significant because it represents an area of growth and its presence is now the prerequisite for attracting many kinds of state and federal funds offered in support of technological innovation and local economic development.

The formation of firms out of research activities occurred in the late nineteenth century at Harvard, as well as MIT, in the fields of industrial consulting and scientific instrumentation (Shimshoni 1970), however these commercial entities were viewed as anomalies rather than as a normal outcome of academic research. During the past two decades, an increasing number of academic scientists have taken some or all of the steps necessary to start a scientific firm by writing business plans, raising funds, leasing space, and recruiting staff (Blumenthal et al. 1986a; Krimsky, Ennis, and Weissman 1991). These studies likely underestimate the extent of faculty involvement, especially in molecular biology. For example, in the biology department at MIT where surveys identified half the faculty as industrially involved in the late 1980s, an informant could identify only one department member as uninvolved at the time. In succeeding years, a complex web of relationships has grown up among university originated start-ups in emerging industries and older and larger firms in traditional industries. Often the same academic scientists are involved with both types of firms, managing a diversified portfolio of industrial interactions (Powell, Koput, and Smith-Doerr 1996).

Since 1980, an increasing number of academic scientists have broadened their professional interests from a single-minded desire to contribute to the literature to making their research the basis of a firm. A faculty member at Stanford University, in the mid-1980s, reviewed his colleagues' activities: "In psychiatry there are a lot of people interested in the chemistry of the nervous system and two of them have gone off to form their own company." Another Stanford professor, during the same period, estimated that "In electrical engineering about every third student starts his own company. In our department [computer science] it's starting as well. That's a change in student behavior and faculty acceptance because the faculty are involved in companies and interacting a lot with companies and the attitude is: we talk to them, we teach them. 'Why not try it: this is my experience.'" Formerly largely confined to a specialized academic sector, firm-formation has spread to a broad range of universities, public and private, elite and non-elite. In a recursion from science studies research to practice, a University of New Orleans professor requested a copy of a study of entrepreneurial activities at State University of New York at Stony Brook to encourage his colleagues in the marine research center to found a firm.

If the disjuncture between theory and invention is accepted, the emergence of entrepreneurial science is an anomaly. Entrepreneurial scientists' research is typically at the frontiers of science and leads to theoretical and methodological advance as well as the invention of devices. Their activities involve sectors of the university, such as basic science departments, that heretofore, in principle, limited their involvement with industry. One explanation for the emergence of entrepreneurial science is that academic scientists, such as the founders of biotechnology firms in the late 1970s and early 1980s, suddenly awakened to the financial opportunities emanating from their research. Implicit in this explanation is the notion that there were recent scientific advances in molecular biology, polymers, and materials science that could be quickly developed as sources of profit. However, opportunities for commercial utilization of scientific research had often been available to scientists in the past, such as Marie and Pierre Curie and Pasteur, who did not believe in crossing the boundary between science and business themselves, even though they evinced a strong interest in the practical implications of their findings (Etzkowitz 1983).

Recognition of a congruence between basic research and invention vitiates the ideological separation of these spheres of activity. Until quite recently most academic scientists assumed that the advancement of knowledge was synonymous with theoretical innovation, however recent examples of research in which theoretical advances have occurred in tandem with the invention of devices or innovation in methodology in transistors/semiconductors, superconductivity, and genetic engineering have called into question the assumption of a one-way flow of from basic to applied research to industrial innovation (Gibbons et al. 1994). The acceptance of dualisms such as patents vs. publication and basic vs. applied research goals were the surface expressions of a theory of knowledge based on an underlying dichotomy that placed scientific advance, that is, the development of theory, in opposition to technological advance.

In recent years, a non-linear recursive interaction between theory and practice, academia and industry, individual and group research has become an alternative academic mode. A significant number of faculty members have adopted multiple objectives, "to not only run a successful company and start a center here at the university which would become internationally recognized but to retain their traditional role as individual investigator," directing a research group. An ideal-typical entrepreneurial scientist held that the "interaction of constantly going back and forth from the field, to the university lab, to the industrial lab, has to happen all the time." These relationships involve different levels of commitment (financial and otherwise) by industrial sponsors, including the involvement of industrial sponsors in problem selection and research collaboration. Conversely, the level of commitment required of a university and its faculty in the commercialization of research varies in intensity according to the mechanism selected (Matkin 1990).

What is new in the present situation is that many academic scientists no longer believe in the necessity of an isolated "ivory tower" to the working out of the logic of scientific discovery. Heretofore, in the hiatus between scientific discovery and application, industry was expected to have its scientists and engineers pursue applied research and product development. The model of separate spheres and technology transfer across strongly defined boundaries is still commonplace. However, academic scientists are often eager and willing to marry the two activities, nominally carrying out one in their academic laboratory and the other in a firm to which they maintain a close relationship. Thus, technology transfer is a two-way flow from university to industry and vice versa, with different degrees and forms of academic involvement: (1) the product originates in the university but its development is undertaken by an existing firm; (2) the commercial product originates outside of the university, with academic knowledge utilized to improve the product; or (3) the university is the source of the commercial product and the academic inventor becomes directly involved in its commercialization through establishment of a new company.

Entrepreneurial Scientists and Entrepreneurial Universities

The original reason the U.S. system makes faculty compete for research grants (instead of receiving assured institutional funding as from Monbusho or the old United Kingdom University Grants Commission) resides in the constitutional reservation of higher education to the states. When federal initiatives are taken, they must be in partnership with the states, such as the jointly funded agricultural research system, or at least formally open to all to compete for resources. Indeed, a significant portion of decision-making authority has devolved outside of the federal granting agency, even to academics themselves, as in the NSF peer review system. As research has become more germane both to academic advance and economic competitiveness, government at various levels has revised its view of universities as a potential source of economic growth well beyond the traditional contribution of student spending in a local economy (Weber 1922).

The university has become a marketer of intellectual property and a venture capitalist to spin-off firms. Although most universities still conservatively confine investments of their endowment to stocks and bonds, a small but increasing number of schools invest in start-up firms, often founded by their own faculty and students, in exchange for equity. This trend is in line with the overall shift of university investment strategies, once confined to preferred stocks and bonds, then including common stocks, to a diversified strategy of riskier, higher reward investments. The transformation of the university into a risk investor was underway even before MIT took the lead in inventing the format for a venture capital firm as a vehicle for sharing risk and rewards among a group of universities and financial institutions, in funding the establishment of high-tech firms from academic research.

The founding of a venture capital firm as an intermediary organization was an initial step to creating a series of mechanisms within and between academia and industry for the purpose of capitalizing intellectual property. Foundations acting on behalf of the university, such as the University of Wisconsin's Alumni Research Foundation, generated income from a series of patents, sufficient to make the university a leader in the biological sciences during the 1930s. Technology transfer offices were established at some schools as a direct arm of the university administration. More recently, incubator facilities provide subsidized space and a support structure to assist new firms, both those generated from campus research and those that wish to develop links to academia. As a result, we now have a system in which the university plays a significant role in generating, capturing, enhancing, and marketing intellectual capital. Rules have also been worked out for an equitable distribution of revenues: typically one third to the inventor, one third to the department and one third to the university. While people on the firm and the university sides of this equation are often spoken about as if they were very different entities, quite often we are examining situations in which the university professor is in a relationship with a firm in which he or she is a major participant, as founder, as a member of the scientific advisory board, as a consultant, or all of the above.

Some conflicts of interest between disinterested research and pecuniary interest have been subsumed into confluences of interest through innovations in industry-university links. The university as well as a professor may be involved with a company as an owner of the intellectual property that has been turned over to the firm without a licensing fee in exchange for an equity share in the firm. Under these circumstances the so-called "Chinese walls" that were constructed between a researcher's firm and his or her academic research group, with each entity working in different topic areas to preclude conflicts of interest, have been torn down. Indeed, some involved academics hold that potential conflicts are resolved by integrating these activities and interests. With the university as the holder of the intellectual property on which the firm is based, information can flow freely between the firm and the academic research group.

A university surrounded by a community of firms which emanated from academic research is displacing the "ivory tower" exemplar. Such conurbations around MIT and Stanford were earlier misidentified as unique instances which others could hardly hope to emulate. A patient strategy can be formulated taking advantage of local strengths and making up for weaknesses; often taking decades to realize, it can work in a variety of academic and regional circumstances. In the late 1950s, North Carolina used its congressional influence to induce federal agencies to locate branch laboratories at Research Triangle Park, established on a tract of land anchored by three universities. Building upon this base, the state's economic and political leaders were able to persuade major corporations such as IBM to follow suit and establish branch laboratories at the site. Decades later, as some R&D facilities were closed, even as others opened, scientists and engineers who did not wish to leave the region founded their own firms.

Utilizing a university as a base for local economic development has become a more targeted and explicit process (Feller 1986). In the 1980s, by attracting the research laboratories of the semiconductor and computer industries and building up its own research in these fields, Austin, Texas set a similar spin-off process in motion. In the 1990s, SUNY Stony Brook encouraged its molecular biology faculty to explore the practical implications of their research through seed research grants and an incubator facility for those who wished to take further steps toward product development. Linkoping University in Sweden adopted a similar strategy, first for information technology and then for medical devices. Building on a base of 80 colleges and universities, six medical schools and 24 teaching hospitals, Philadelphia, an older industrial region, has renewed its economy by creating health care and pharmaceutical clusters. Niche formation in regional innovation, linked in one way or another to academia, has become a third local development strategy, transcending the two traditional methods of either attempting to improve the business climate by cutting taxes or inducing existing firms to relocate by providing direct or indirect subsidies.

The symbiotic connection between academia and knowledge-based industry is exemplified by the growth of California molecular biology companies emanating from the state's universities; similarly there have been changes in the nature and structure of technology firms, some of which are moving toward an academic mode, especially in the biotechnology industry, where firms advertise for post-doctoral fellows. The biotechnology industry also has a strong interest in producing fundamental research which can immediately be used to make a product. Thus, as universities become a bit more like firms, firms are also becoming a bit more like universities—consider the phenomenon of companies sharing research projects as they realize that they cannot always do alone, all the research needed for the development of new products. As anti-trust laws have been changed to accommodate this need an increasing number of firms operate as quasi-universities, sharing knowledge that formerly would have been kept secret, in the interest of their fundamental goal, which is profit.

Despite significant convergences, there is a continuing distinction among the institutional spheres, with universities retaining their fundamental interest in developing new knowledge. Thus, the founders of Karlskronner-Ronnerby, a new regional university in Sweden specialized in computer science, noted that they do not take on any projects that industry wishes, but rather select those that will both solve a practical problem and lead to a fundamental advance in knowledge. Working on these projects under the guidance of their professors is the basis for the training of their students, a contemporary reinterpretation in high technology of the agricultural research model.

The U.S. land grant universities are the model for an "endless transition" that is currently emerging in new areas of technology-based sciences. The two-way flow between industry and academia in computer science, materials science, and molecular biology is not well captured by the "endless frontier" model. New scientific disciplines have recently been created which simultaneously exhibit theoretical and practical implications, rather than the latter emerging after a long time delay. Thus scientific disciplines themselves today—especially the most important emerging ones, molecular biology, computer science, and materials science—do not fit the earlier model.

Potential products are often invented as a normal part of the research process, especially as software becomes commonplace in collecting and analyzing data. As a faculty member commented in the mid-1980s, "In universities we tend to be very good at producing software, [we] produce it incidentally. So there is a natural affiliation there. My guess is a lot of what you are going to see in university-industry interaction is going to be in the software area." In the 1990s this phenomenon spread well beyond the research process, with software produced in academia outside of the laboratory, and start-ups emerging from curriculum development and other academic activities (Kaghan and Barnet 1997).

Incentives for Academic Firm-Formation

The appearance of commercializable results in the course of the academic research process is the necessary cause for the emergence of entrepreneurial science. The "meandering stream of basic research" produces such results even without scientists directing their research programs to this end or universities requiring their disclosure. In addition to the opportunity presented by the appearance of research results with commercial potential, entrepreneurial science has several sufficient causes, both proximate and long-term, that encourage academic scientists to utilize these opportunities themselves rather than leaving them to others. Entrepreneurial incentives in U.S. academic science include stringency in federal research funding, a culture of academic entrepreneurship originating in the seeking of government and foundation funds to support research, examples of colleagues' successful firms, and government policies and programs to translate academic research into industrial innovation.

A necessary condition has been the transition of much of science and engineering from empiricism to codified knowledge resting on good theory. Capable of application through computer modeling and simulation, it has enabled "innovation on command" or "invention by design" in computational chemistry and automated molecular biology. This is not a question of faculty motivation so much as a new accessibility of opportunity that is not blocked by the need for years of "embedded learning by doing" (L. Branscomb, pers. com. 1998). Scientific and technological opportunities open up the path to initiatives that are, at least in part, driven by the demands of the technology itself, such as when the difficulty of maintaining large software programs in an academic environment leads to firm-formation.

Effects of Research Funding Difficulties

Although federal investment in academic R&D increased during the 1990s, academic researchers strongly perceived a shortfall of resources during this period (National Science Board 1996). The explanation of this paradox lies in the expansionary dynamic inherent in an academic research structure, based upon a Ph.D. training system that produces research as a byproduct, and is driven by the ever-increasing number of professors and their universities who wish them to engage in research. Formerly this pressure was largely impelled by the wish to conform to the prevailing academic prestige mode associated with basic research; in recent years, expansionary pressure has intensified awareness of the economic outcomes of basic research, drawing less research-intensive areas of the country into competition to expand the research efforts of local universities as an economic development strategy.

As the ability to conduct advanced research becomes much more widespread, the potential to create new types of academic environments grows in tandem. The limits to growth of the older schools and the increasing flow of research-oriented Ph.D.s to a broader range of universities and colleges are also important factors in building a local research base where none had previously existed. The diffusion of research to a broader range of universities creates the paradox that even as research funding increases, it becomes more difficult to obtain these funds. In response to this perceived shortfall, many state governments have established programs to attract successful faculty researchers by assisting them with research funding and improved infrastructure. For example, through its Eminent Scholars Program, Georgia has funded 17 chairs at six local universities for 1.5 million dollars apiece, with an additional million dollars available to each scholar for research equipment and facilities.

Some faculty have taken the solution of the research funding dilemma into their own hands. For some academic scientists, the intense competition for research funds has become a motivation to found a firm. Some of these professors first started foundations, hoping to find a way to fund a foundation which would in turn fund their academic research projects, but when that stratagem did not work, they started a company, hoping that its success would secure their research funding base.

The increasing scale and costliness of research is also a factor. With the notable exception of a relatively brief wartime and early postwar era, characterized by rapidly expanding public resources for academic research, U.S. universities have always lived with the exigencies of scarce resources. As traditional sources of research funding were unable to meet ever-expanding needs, academics sought alternative sources such as industrial sponsors. A faculty member discussed his involvement with industry: "In some areas we have found it necessary to go after that money. As the experimental needs in computer science [increase], equipment needs build up.

People realize that a small NSF grant just doesn't hack it anymore." Nevertheless, there are cultural and other barriers to overcome before a smooth working relationship can be established. Another professor described the dilemma: "It's harder work with industry funding than federal funding, harder to go through the procurement process, to negotiate the terms of the contract." Industry's expectations for secrecy, for example, are sometimes unreasonable. Dissatisfaction with working with existing companies is another reason professors start their own.

Creating an independent financial base to fund one's own research is a significant motivator of entrepreneurial activity. Stringency in federal research funding has led academic scientists to broaden their search for research support from basic to applied government programs and vice versa. A possible source that has grown in recent years has been research subventions from companies, including firms founded by academics themselves for this purpose, driving industrial support of academic research up from a low of four percent during the early postwar era, to a modestly increased proportion of seven percent. Thus far, these funds are concentrated in a few areas: computer science and the biomedical medical sciences, and in particular research groups. A very few universities, such as MIT, receive as much as 25 percent of their research funding from industry. Concomitant with a shift from military to commercial criteria, new sources of funding for academic research have opened up in some fields that have experienced a rise in practical significance, such as the biological sciences, making the notion of stringency specific to others, such as nuclear physics, that have experienced a decline (Blumenthal 1986a).

Nevertheless, federal research agencies are still the most important external interlocutors for academic researchers. A department chair at Cal Tech noted, "The amount of money from industry is a pittance in the total budget, therefore everybody's wasting their time to try to improve it. It's still a drop in the bucket. We were running about 3% total in our department. We do value our industrial ties, have good friends, interact strongly with them in all kinds of respects and the unrestricted money is invaluable. You wouldn't want to lose a penny of it and would like to increase it a lot, but its impact vis à vis federal funding is almost nonexistent." In contrast to this view based upon the current source of resources, others look toward realizing greater value from the commercial potential of research.

Even without creating a firm themselves, academic scientists can earn funds to support their research by making commercializable results available for sale to existing firms. As a Stanford faculty member described the process, "It's also motivating for us to try to identify things that we do that may be licensable or patentable and to make OTL [Office of Technology and Licensing] aware of that, because according to University policy, 30 percent of the money comes back to the scientist, 30 percent comes back to the Department as well as 30 percent for the University. So, almost all the computing equipment and money for my post docs have been funded by the work that we did. So there's motivation." Earlier in the twentieth century, experiencing difficulties in the French research funding system, the Curies considered exploring the commercial possibilities of their radium discovery for just this purpose (Quinn 1995).

Tapping capital markets through a public offering of stock is an additional source of research funding, especially in biotechnology-related fields, although one not yet recognized in Science Indicators volumes! In response to the increasingly time-consuming task of applying for federal research grants a faculty member said, "Another way to get a whole bunch of money is to start a company." After resisting the idea of starting a firm in favor of establishing an independent non-profit research institute, largely supported by corporate and governmental research funds, two academic scientists returned to an idea they had earlier rejected of seeking venture capital funding. As one of the founders explained their motivation for firm formation, "Post docs who are really good will want to have some place that at least guarantees their salary. And that we were not able to do. It was for that reason we decided to start the company." Other scientists came to the realization that they could combine doing good science with making money by starting a company. As they enhanced their academic salaries through earnings from entrepreneurial ventures, and continued to publish at a high rate, they lost any previous aversion to the capitalization of knowledge (Blumenthal 1986b).

The Industrial Penumbra of the University

The success of the strategy to create a penumbra of companies surrounding the university has given rise to an industrial pull upon faculty members. For example, a faculty member reported that "The relationship with Collaborative is ongoing daily. We are always talking about what project we are going to do next. What the priority is, who is involved, there are probably six projects, a dozen staff members and maybe close to a dozen people scattered around three or four different departments on campus that are doing things with them." Geographical proximity makes a difference in encouraging appropriate interaction. Such intensive interaction sheds new light on the question of industrial influence on faculty research direction and whether this is good, bad, or irrelevant. Thus, the "issue of investigator initiation is much more complicated because I am bringing my investigator-initiated technology to their company-initiated product. It is a partnership in which each partner brings his own special thing. That is the only reason they are talking. Do your thing on our stuff." Previous conflicts based on an assumption of a dividing line between the academic and industrial sides of a relationship are superseded as divisions disappear. A more integrated model of academic-industry relations is emerging along with a diversified network of transfer institutions. Indeed, the very notion of technology transfer, or at least transfer at a distance, is superseded as universities develop their own industrial sector.

Not surprisingly, a receptive academic environment is an incentive to entrepreneurship while a negative one is a disincentive. MIT and Stanford University are the exemplars of firm-formation as an academic mission. In the 1930s, Karl Compton, the president of MIT, persuaded the leadership of the New England region to make the creation of companies from academic research the centerpiece of their regional economic development strategy. With Vannevar Bush's Raytheon, founded in the 1920s, as a model, this strategy was put into practice just after the Second World War through a project to provide the missing links of business advice and capital to potential technical entrepreneurs. In 1946, MIT, the Harvard Business School, and the Boston financial community jointly founded the American Research and Development Corporation, the progenitor of the contemporary venture capital industry. Even earlier, Frederick Terman, Stanford dean of engineering, provided some of the funds to help two of his former students, Hewlett and Packard, to form their firm just prior to the Second World War. There had been a recognition from the turn of the century that a local technical industry was a prerequisite for academic advance and vice versa.

An entrepreneurial dynamic was established in the San Francisco Bay area which had its origins in the realization that to build a great engineering school required strong relationships to local technical industry. This was a missing element in the region earlier in the century, when technology was almost entirely imported from the East coast. A faculty member commented that "Because it has been encouraged here from its inception, it makes it easy to become involved [in firm-formation]. [There are] more opportunities, people come in expecting it." On the other hand, at a university noted for its opposition to entrepreneurial activity an administrator noted that despite the disfavor in which it is held, "There have been some [firms founded]. It's frowned upon. It takes a lot of time and the faculty are limited in the amount of time they have." Under these conditions, procedures that could ameliorate conflicts are not instituted and faculty who feel constricted by the environment, leave.

A surrounding region filled with firms that have grown out of the university is also a significant impetus to future entrepreneurial activity. The existence of a previous generation of university-originated start-ups provides consulting opportunities, even for faculty at other area universities. A Stanford professor noted, "In the area there's a lot of activity and that tends to promote the involvement of people." From their contact with such companies, faculty become more knowledgeable about the firm-formation process and thus more likely to become involved themselves. Faculty who have started their own firms also become advisors to those newly embarking on a venture. An aspiring faculty entrepreneur recalled that a departmental colleague who had formed a firm "gave me a lot of advice. He was the role model." The availability of such role models makes it more likely that other faculty members will form a firm out of their research results, when the opportunity appears.

Once a university has established an entrepreneurial tradition, and a number of successful companies, fellow faculty members can offer material, in addition to moral, support to their colleagues who are trying to establish a company of their own. A previous stratum of university-originated firms and professors who have made money founding their own firms creates a potential cadre of "angels" that prospective academic firm founders can look to in raising funds to start their firms. Early faculty firm-founders at MIT were known on campus for their willingness to supply capital to help younger colleagues.

The Normative Impetus to Firm-Formation

In an era when results are often embodied in software, sharing research results takes on a dimension of complexity well beyond reproducing and mailing a preprint or reprint of an article. Software must be debugged, maintained, enhanced, translated to different platforms to be useful. These activities require organizational and financial resources well beyond the capacity of an academic lab and its traditional research supporters, especially if the demand is great and the software complex. As one of the researchers described the dilemma of success, "We had an NSF Grant that supported [our research] and many people wanted us to convert our programs to run on other machines. We couldn't get support (on our grant) to do that and our programs were very popular. We were sending them out to every place that had machines available that could run them." The demand grew beyond the ability of the academic laboratory to meet it.

Firm-formation is also driven by the norms of academic collegiality, mandating sharing of research results. When the federal research support funding system was unable to expand the capabilities of a laboratory to meet the demand for the software that its research support had helped create, the researchers reluctantly turned to the private sector. They decided that "Since we couldn't get support, we thought perhaps the commercial area was the best way to get the technology that we developed here at Stanford out into the commercial domain." This was a step taken only because the lab failed to receive support from NSF and NIH to distribute the software. "The demonstration at NIH was successful, but they didn't have the funds to develop this resource." The researchers also tried and failed to find an existing company to develop and market the software. As one of the researchers described their efforts, "We initially looked for companies that might license it from us, none were really prompted to maintain or develop the software further." Failure to identify an existing firm to market a product is a traditional impetus to inventors, who strongly believe in their innovation, to form a firm themselves to bring it to market.

Chemists involved with molecular modeling, previously a highly theoretical topic, have had to face the exigencies of software distribution as their research tools increasingly became embodied in software. Since the interest in the software is not only from academic labs but from companies who can afford to pay large sums, the possibility opens up of building a company around a program or group of programs and marketing them to industry at commercial rates while distributing to academia at a nominal cost. Academic firm-founders thus learn to balance academic and commercial values. In one instance, as members of its board, the academics were able to influence a firm to find a way to make a research tool available to the academic community at modest cost. An academic described the initial reaction to the idea, "The rest of the board were venture capitalists, you can imagine how they felt! They required we make a profit." On the other hand, "It was only because we were very academically oriented and we said, "Look, it doesn't matter if this company doesn't grow very strongly at first. We want to grow slow and do it right and provide the facilities to academics." The outcome was a compromise between the two sides, meeting academic and business objectives at the same time, through the support of a government research agency to partially subsidize academic access to the firm's product.

There is some evidence that firms spin out of interdisciplinary research or, at least, that some such collaborations are a significant precursor to firm-formation. As one academic firm-founder described the origins of his firm, "If it had not been for the collaboration between the two departments [biochemistry, computer science], intimate, day to day working [together], it never would have happened. Intelligenetics and Intellicorp grew out of this type of collaboration. We had GSB [the Graduate School of Business], the Medical School, and Computer Science all working together." In this model, the various schools of the university contributed to the ability of the university to spin out firms, providing specialized expertise well beyond the original intellectual property.

Academic Entrepreneurial Culture

An entrepreneurial culture within the university encourages faculty to look at their research results from a dual perspective: (1) a traditional research perspective in which publishable contributions to the literature are entered into the "cycle of credibility" (Latour and Woolgar 1979) and (2) an entrepreneurial perspective in which results are scanned for their commercial as well as their intellectual potential. A public research university that we studied experienced a dramatic change from a single to a dual mode of research salience. A faculty member who lived through the change described the process: "When I first came here the thought of a professor trying to make money was anathema, really bad form. That changed when biotech happened." Several examples of firm-formation encouraged by overtures from venture capitalists led other faculty, at least in disciplines with similar opportunities, to conclude, "Gosh, these biochemists get to do this company thing, that's kind of neat, maybe it's not so bad after all." Although some humanists and science policy experts remain concerned that the research direction of academic science will be distorted, serious opposition dissipated as leading opponents of entrepreneurial ventures from academia, such as Nobel Laureate Joshua Lederberg of Rockefeller University, soon became involved with firms themselves, in his case, Cetus.

A research group within an academic department and a start-up firm outside are quite similar despite apparent differences represented by the ideology of basic research, on the one hand, and a corporate legal form on the other. As an academic firm-founder summed up the comparison, "The way [the company] is running now it's almost like being a professor because it is all proposals and soft money." There is an entrepreneurial dynamic built into the U.S. research funding system, based on the premise that faculty have the primary responsibility for obtaining their own research funds. As a faculty member described the system, "It is amazing how much being a professor is like running a small business. The system forces you to be very entrepreneurial because everything is driven by financing your group." At least until a start-up markets its product or is able to attract funding from conventional financial sources, the focus of funding efforts is typically on a panoply of federal and state programs that are themselves derived from the research funding model and its peer review procedures. As a faculty entrepreneur viewed the situation, "What is the difference between financing a research group on campus and financing a research group off campus? You have a lot more options off campus but if you go the federal government proposal route, it is really very similar." The entrepreneurial nature of the U.S. academic research system helps explain why faculty entrepreneurs typically feel it is not a great leap from an on-campus research group to an off-campus firm.

Until the past two decades a skeptical view of firm formation was taken for granted as the perspective of most faculty members and administrators at research universities. A typical trajectory of firm formation is the transition from an individual consulting practice, conducted within the parameters of the one-fifth rule, to a more extensive involvement, leading to the development of tangible products. A faculty member described his transition from consulting to firm formation, "It got to the point where I was making money consulting and needed some sort of corporate structure and liability insurance; so I started [the company] a couple of years ago. From me [alone, it has grown] to eight people. We're still 70 percent service oriented, but we do produce better growth media for bacteria and kits for detecting bacteria." The firm was built, in part, on the university's reputation but was symbiotic in that its services to clients brought them into closer contact with on-campus research projects.

In another instance, an attempt was made to reconcile the various conflicting interests in firm formation and make them complementary with each other by the university having some equity in the company and holding the initial intellectual property rights. Despite the integrated mode arrived at, some separation worked out on technical grounds was still necessary to avoid conflicts. "There is no line. It's just a complete continuum. It is true that I have a notebook that says [university name] and a notebook that says [firm name] and if I make an invention in the [company] notebook then the assignment and the exclusive license goes to [the firm] and if I make an invention in the university notebook then the government has rights to the invention because they are funding the work. [Interviewer: How do you decide which notebook you are going to write in?] We have ways of dividing it up by compound class. In the proposals that I write to the government I propose certain compound classes. There is no overlap between the compound classes that we work on on-campus and the compound classes that we work on off-campus, so there is a nice objective way of distinguishing that." The technical mode of separation chosen, by compound classes, suggests that while boundaries have eroded as firm and university cooperate closely to mount a joint research effort, a clear division of labor persists.

Once the university accepted firm-formation and assistance to the local economy as an academic objective, the issue of boundary maintenance was seen in a new light. An informant noted, "When the university changed its attitude toward entrepreneurial ventures, one consequence was that the administration renegotiated its contract with the patent management firm that dealt with the school's intellectual property. ...If the university chooses to start an entrepreneurial new venture based upon the invention then the university can keep the assignment and do what ever it wants. [Interviewer: Why did the university make that change?] Because the university decided that it wanted to encourage faculty to spin off these companies." Organizational and ideological boundaries between academia and industry were redrawn, with faculty encouraged to utilize leave procedures to take time to form a firm and entrepreneurial ventures cited in university promotional literature as contributing to research excellence.

A Company of Their Own

There has been a significant change of attitude toward industrial collaboration among many faculty members in the sciences, a shift away from reliance on federal funding as a given. Three styles of participation in technology transfer have emerged, reflecting increasing degrees of industrial involvement. These approaches can be characterized as (1) hands off, leave the matter entirely to the technology transfer office; (2) knowledgeable participant, aware of the potential commercial value of research and willing to play a significant role in arranging its transfer to industry; and (3) seamless web, integration of campus research group and research program of a firm. Of course, many faculty fit in the fourth cell of "no interest" or non-involvement. These researchers are often referred to under the rubric of the federal agency that is their primary source of support as in, "She is an NIH person." The approach of leaving it up to the technology transfer office to find a developer and marketer for a discovery precisely met the needs of many faculty members, then and now, who strictly delimit their role in putting their technology into use. A faculty member delineated this perspective on division of labor in technology transfer: "It would depend on the transfer office expertise and their advice. I am not looking to become a business person. I really am interested in seeing if this could be brought into the market. I think it could have an impact on people's lives. It is an attractive idea." This attitude does not necessarily preclude a start-up firm but it does exclude the possibility that the faculty member will be the entrepreneur.

A stance of moderate involvement is becoming more commonplace, with scientists becoming knowledgeable and comfortable operating in a business milieu while retaining their primary interest and identity as an academic scientist. A faculty member exemplifying this approach expressed the view that "In science you kind of sit down and you share ideas. There tends to be a very open and very detailed exchange. The business thing when you sit down with somebody, the details are usually done later and you have to be very careful about what you say with regard to details because that is what business is about: keeping your arms around your details so that you can sell them to somebody else, otherwise there is no point." Faculty are learning to calibrate their interaction to both scientific and business needs, giving out enough information to interest business persons in their research but not so much so that a business transaction to acquire the knowledge becomes superfluous. Another researcher said, "I am thinking about what turns me on, in terms of scientific interest, and the money is something, if I can figure out how to get it then it is important but it is certainly not the most important thing to me." The primary objective is still scientific; business objectives are strictly secondary.

The relatively new existence of regularized paths of academic entrepreneurship, as a stage in an academic career or as an alternative career, is only of interest to some faculty; others prefer to follow traditional modes. However, for some members of the professoriate, participation in the formation of a firm has become an incipiently recognizable stage in an academic career, located after becoming an eminent academic figure in science. For others, typically at earlier stages of their career, either just before or after being granted permanent tenure, such activity may lead to a career in industry outside the university. As one faculty member put it: "Different people I have known have elected to go different ways: some back to their laboratories and some running technology companies. You can't do both." This difficulty has not prevented other professors from trying to do both, with or without taking temporary leaves of absence.

The Intensification of Academic-Industry Relations

A more direct role in the economy has become an accepted academic function, and this is reflected in the way universities interact with industry. There has been a shift in emphasis, from traditional modes of academic-industry relations oriented to supplying academic "inputs," to existing firms, either in the form of information flows or through licensing patent rights to technology in exchange for royalties. Utilizing academic knowledge to establish a new firm, usually located in the vicinity of the university, has become a more important objective. Indeed, the firm may initially be established on or near the campus in an incubator facility sponsored by the university to contribute to the local economy. Regional economic development is in tension with the traditional technology transfer objective, the sale of intellectual property rights to the highest bidder wherever located. Universities increasingly have to balance between two competing goals, the earning of funds to support the university in an era of constricting external support and a longer-term goal of expanding the regional economic base. The conduct of academic science is also affected by a heightened interest in its economic potential.

From an industrial perspective, relations with universities have traditionally been viewed as a source of human capital, future employees, and, secondarily, as a source of knowledge useful to the firm. In this view, the academic and industrial spheres should each concentrate on their traditional functions and interact across distinct, strongly defended, boundaries. The hydraulic assumptions of knowledge flows include reservoirs, dams, and gateways that facilitate and regulate the transmission of information between institutional spheres with distinctly different functions (academia: basic research; companies: product development). In this view what industry wants and needs from academic researchers is basic research knowledge; therefore, universities should focus on their traditional missions of research and education, their unique function. A focus on technology transfer makes the university into a barrier between university and industry, displacing the free flow of knowledge through publication and informal discussion, with a wasteful effort to capitalize knowledge and reap a return for the university (Faulkner and Senker 1995).

This classic industrial perspective of academia is expressed in Europe by the industrial group (IRDAC) in the Research Directorate of the European Union and in the United States by the Industry-University-Government Roundtable. These organizations primarily represent large multinational firms, whether of U.S. or European origin. Such firms represent the first sector in a typology of firm perspectives on relations with industry. In these companies R&D is typically internalized within the firm and the window on academic research obtained through consultation and participation in liaison programs is sufficient for corporate needs. In a second group of companies with little or no R&D capacity, relations with academia, if any, will also be informal through engaging an academic consultant to test materials or troubleshoot a specific problem. More intensive relationships occur with a third group of firms that have grown out of university research and are still closely connected to their original source, and with a fourth group that, given the rapid pace of innovation in their industrial sector, have externalized some of their R&D and seek to import technologies or engage in joint R&D programs to develop them (Rahm 1996).

In these latter circumstances, traditional forms of academic-industry relations, such as consulting and liaison programs that encourage "knowledge flows" from academia to industry, become less important as an increasing number of companies look to external sources for R&D or are themselves based upon academic knowledge. As industrial sectors and universities move closer together, informal relationships and knowledge flows are increasingly overlaid by more intensive, formal institutional ties that arise from centers and firms. As companies externalize their R&D they want more tangible inputs from external sources such as universities. As one close observer from the academic side of the equation put it, "From the point of view of the company, they tend to want a lot of bang for the buck. They tend not to get involved in Affiliates programs precisely because they can't point to anything." The growth of centers and the formation of firms from academic research have had unintended consequences that have since become explicit goals: the creation of an industrial penumbra surrounding the university as well as an academic ethos among older firms that collaborate more closely with each other through joint academic links (Etzkowitz and Kemelgor 1998).

Academic-industry interaction is reciprocal and recursive. If academia becomes subject to industrial influence, for example through boards of research centers where academics and their industrial counterparts set research direction, industrial enterprises also admit academic influence. Such effects may be broader than a particular firm and can extend to an industrial sector or even the creation of a new sector, such as biotechnology, that will affect existing ones such as pharmaceuticals and agriculture. Firms formed by academics have been viewed in terms of their impact on the university but they are also a "carrier" of academic values and practices into industry and, depending on the arrangements agreed upon, a channel from industry back to the university.

Another reason for the weakening of barriers between the two cultures (academic and industrial) is the industrial trend, set by

Bell Labs, DuPont, IBM, and GE (in prior years) to allow a certain number of highly visible researchers to live "academic" lives. The number of U.S. Nobel prizes won by people in industry is impressive. Although many of these scientists have since moved to academia, a new wave of firms (for example Microsoft and Intel) are establishing laboratories, less "ivory tower" in format than their predecessors but more in line with contemporary academic peers attuned to the value of the intellectual property that they create. When both sides are equally aware and ready to act upon the theoretical and practical implications of findings: What is academic? Who is the industrialist?

The boundaries among institutional spheres can be modeled by the classic "Venn" diagram of separate circles becoming overlapped, with independent and interlocking segments. An increasing number of firms run extensive high-level training programs, for example "Motorola University." A company such as Rockwell Inc. holds conferences on "decision theory" virtually indistinguishable from academic events. The Rand Institute, a not-for-profit think tank, within which is imbedded an FFRDC (federally funded research and development center) called Project Airforce—a captive contractor to the airforce that now also does civilian research—offers the Ph.D. degree in the policy sciences. As universities such as the North Carolina State "Centennial campus" and the University of Bochum, Germany, are built with corporate R&D facilities interspersed among academic buildings as part of regional economic development strategies, a new world of "hybrid organizations" is created.

Conclusion: Bridging the Gap

A transformation is underway in government's role in industrial policy, moving in opposing directions in different countries depending upon previous levels of involvement, toward a middle ground of partnership with academia and industry. In socialist societies, where government was the total controlling force and incorporated production and research activities, there was a sharp reduction to a lesser role after 1989. Although government at first eschewed a direct science and industrial policy, some former socialist societies have recently taken the first steps to transforming government research institutes into science parks and/or shifting research assets to universities. On the other hand, in laissez-faire societies such as the United States where government played almost no direct role until quite recently, the SBIR (Small Business Innovation Research) program and other initiatives represent a substantive if not ideological commitment to a greater role for government in innovation (Brody 1996). In the United Kingdom, in the absence of a written constitution, the pendulum has swung even further, from aversion to industrial links to espousal of "wealth creation" and competitiveness both as an academic objective and as a tenet of public policy (Mandelsohn 1998).

The university is increasingly a partner with government and industry in initiatives targeted at science-based economic development. This is the case even in a government-industry collaborative program such as the U.S. National Institute of Standards and Technology's Advanced Technology Program (ATP) (Hill 1998: 162–163). Even though universities are officially not allowed to take the lead in projects supported by ATP, university research groups are typically brought in as subcontractors since their knowledge is useful for advanced product development. These changes in the role of university, industry, and government into a "triple helix" mode includes transformations within each of the spheres as well as more integrated relationships among them. Each of these spheres has a distinct purpose separate from the others but also "takes the role of the other." These transmutations make it difficult to clearly discern the shape of this new institutional configuration in which previously distinct institutional boundaries are elided.

The institutionalization of industry-university links, invented over the better part of a century, has revised the mission of the university. Like the six blind men in the fable, we circle around this new "elephant" as half-blind people but have difficulty discerning its contours. We have universities which in the late 19th century underwent the first academic revolution, integrating research with teaching but with a continuing tension between the two. There are always complaints from parents that professors are spending too much time on research and not enough in the classroom. Since the Second World War we have had a second academic revolution, one which, like the first, has its origins in adding research to teaching as an academic mission, and in which universities play a leading role in the economic development of their surrounding region.

These practices represent a potentially fundamental modification of the traditional view of universities as institutions supported by governmental, ecclesiastical, and lay patronage. The new arrangements open the possibility that universities will become, at least in part, financially self-supporting institutions, entities obtaining revenues through licensing agreements and other financial arrangements for the industrial use of new knowledge discovered in their laboratories. At present, this possibility is little more than that, but it certainly represents a novel idea in the history of universities—at least on the scale on which it is envisaged. Although an entrepreneurial dynamic has been part of the U.S. research university system virtually from its inception, its intensity has increased and become more widespread in recent years. Academic-industry relations have been transformed from an earlier, less intensive, "mediated" mode relying upon "knowledge flows" across strongly defended boundaries to a new, more intensive, "integrated" mode based upon rechannelling knowledge flows through institutional ties.

A formal system of academic-industry relations that had existed from the late 19th century in Japan was repressed during the U.S. occupation, due to its connections to military industry. Reduced to informal relations, these ties are currently being re-created on a more formal basis (Kneller, chapter 12). During the postwar period, a former student would send a researcher from his company to a university lab to interest a professor in the research problems of the firm. If there was interest, then a larger project would be funded within the university lab and in a parallel unit in the firm (according to Michiyuki Uenohara, chairman of NEC Research Institute, in a 1998 interview with Henry Etzkowwitz and Kenneth Pechter). Consulting relationships and small grants from companies to academic researchers were also a traditional informal mechanism in the United States, especially in chemistry. As science becomes more important to future economic growth, academic institutions are increasingly looked to as a source of discontinuous innovation. Civil Service academic formats in Japan,

Germany, and Sweden are being modified to insure that formal ownership is transferred so that a larger proportion of potential intellectual property is commercialized. There is little that is culturally unique to any of these linkage modes, but their use is contingent on historical circumstances. The direction of academic change toward entrepreneurial science is constant even if the rate is variable.

Defining and maintaining an organization's relationship to the external environment through "boundary work" has different purposes, depending upon whether goals are static or undergoing change. A defensive posture is usually taken, buttressed by arguments supporting institutional integrity, in affirming a traditional role. The reworking of boundaries around institutions undergoing changes in their mission occurs through a "game of legitimation" that can take various forms. One strategy is to conflate new purposes with old one to show that they are in accord. For example, universities legitimize entrepreneurial activities by aligning them with accepted functions such as research and service. In addition, new identifications of compatibility with other institutions are made as they move closer together. Thus, universities and corporations are found to share a common interest in putting knowledge to use (Langfitt et al. 1989). How far can this convergence process go without institutional identities being merged? Will a future step in the capitalization of knowledge be the listing of universities on the stock exchange? Perhaps Stanford and MIT are already on the bourse, by proxy, through companies spun from the brow of academe.

The university is no longer the university of the Middle Ages, an isolated community of scholars (Rashdall, 1896). Nor is it the university of the late 19th century, whether constituted according to the land grant or the basic research model. Translating the "land grant" model for a new era, the university is currently taking up a more fundamental role in society, one that makes it crucial to future innovation, job creation, and economic growth and sustainability. The university is an increasingly significant social institution, playing as important a role as Wright Mills argued that the military played in relation to the federal government and industry during the cold war (Mills, 1958). The university is currently moving into a more central role in society as it becomes the basis for the industry of the future.

Acknowledgment

Data on university-industry relations in the United States are from studies conducted by the author with the support of the U.S. National Science Foundation Ethics and Values Studies Program and the Andrew Mellon Foundation. Unless otherwise noted, quotations are from the more than 150 in-depth interviews conducted since 1982 with faculty and administrators at universities, both public and private, with longstanding and newly emerging industrial ties.

References

Blumenthal, David, et al. 1986a. "Industrial Support of University Research in Biotechnology." *Science* 231: 242–246.

Blumenthal, David, et al. 1986b. "University-Industry Research Relations in Biotechnology." *Science* 232: 1361–1366.

Brody, R. 1996. "Effective Partnering: A Report to Congress on Federal Technology Partnerships." Washington D.C.: U.S. Department of Commerce, Office of Technology Policy.

Etzkowitz, Henry. 1983. "Entrepreneurial Scientists and Entrepreneurial Universities in American Academic Science." *Minerva* 21(2–3): 198–233.

Etzkowitz, Henry. 1999. *The Second Academic Revolution: MIT and the Rise of Entrepreneurial Science.* London: Gordon and Breach.

Etzkowitz, Henry, and Carol Kemelgor. 1998. "The Role of Research Centers In the Collectivisation of Academic Science." *Minerva* 36(3): 271–288.

Faulkner, Wendy, and Jacqueline Senker. 1995. *Knowledge Frontiers: Public Sector Research and Industrial Innovation in Biotechnology, Engineering Ceramics, and Parallel Computing.* Oxford: Oxford University Press.

Feller, Irwin. 1986. "Universities as Engines of Economic Development: They Think They Can." *Policy Studies Journal* 19: 335–348.

Geiger, Roger. 1986. *To Advance Knowledge: The Growth of American Research Universities, 1900–1940.* New York: Oxford University Press.

Gibbons, Michael, et al. 1994. *The New Production of Knowledge.* Beverly Hills: Sage Publications.

Hill, Christopher T. 1998. "The Advanced Technology Program, Opportunities for Enhancement." In Lewis M. Branscomb and James Keller, eds., *Investing in Innovation: Creating a Research and Innovation Policy That Works.* Cambridge, MA: MIT Press, pp. 162–163.

Jencks, Christopher, and David Riesman. 1968. *The Academic Revolution.* New York: Doubleday.

Kaghan, William, and G. Barnet. 1997. "The Desktop Model of Innovation in Digital Media." In Henry Etzkowitz and Loet. Leydesdorff, eds., *Universities and the Global Knowledge Economy: A Triple Helix of Academic-Industry-Government Relations.* London: Cassell.

Krimsky, Sheldon, James Ennis, and R. Weissman. 1991. "Academic-Corporate Ties in Biotechnology: A Quantitative Study." *Science, Technology & Human Values* 16(3): 275–287.

Langfitt, Thomas, et al. 1989. *Partners in the Research Enterprise.* Philadelphia: University of Pennsylvania Press.

Latour, Bruno, and Stephen Woolgar. 1979. *Laboratory Life.* Beverly Hills, CA: Sage Publications.

Mandelsohn, Peter. 1998. *Our Competitive Future: Building the Knowledge-Driven Economy.* London: The Stationery Office Limited.

Matkin, Gerald. 1990. *Technology Transfer and the University.* New York: Macmillan.

Mills, C. Wright. 1958. *The Power Elite.* New York: Oxford University Press.

National Science Board. 1996. "Science and Engineering Indicators." Washington D.C.: National Science Foundation.

Powell, William, K. Koput, and L. Smith-Doerr. 1996. "Interorganizational Collaboration and the Locus of Innovation: Networks of Learning in Biotechnology." *Administrative Science Quarterly* 1(41): 116–145.

Quinn, Sally. 1995. *Marie Curie: A Life.* New York: Simon & Schuster.

Rahm, Diane. 1996. "R&D Partnering and the Environment of U.S. Research Universities." In *Proceedings of the International Conference on Technology Management: University/Industry/Government Collaboration.* Istanbul: Bogazici University.

Rashdall, Hastings. 1896. *The Universities of Europe in the Middle Ages.* Oxford: Oxford University Press.

Rosenberg, Nathan, and Richard Nelson. 1994. "American Universities and Technical Advance in Industry." *Research Policy* 23(3): 323–348.

Shimshoni, Daniel. 1970. "The Mobile Scientist in the American Instrument Industry." *Minerva* 8 (1).

Weber, Max. 1922 [1948]. "Science as a Vocation." In Hans Gerth and C. Wright Mills, eds., *From Max Weber.* New York: Oxford University Press.

9

The Hesitant Relationship Reconsidered: University-Industry Cooperation in Postwar Japan

Takehiko Hashimoto

Introduction

The need for more intensive collaboration between universities and industry has been well-recognized in recent years in Japan, and various attempts are being made to construct a closer relationship between the two sectors. Although such collaboration existed on a much smaller scale in the past, the student protests of the late 1960s inhibited it. However, university-industry collaboration was strong in prewar Japan, when universities were funded by foundations as well as the government. The present paper explores the differences before and after the Second World War, first surveying funding from industry to universities in the prewar period. It will then analyze three important events that may have influenced the relationship: the Occupation, technology importation, and the student protests.

Prewar Background

A symbolic episode showing the close ties between universities and industry was the donation of the Furukawa Family to Kyushu and Tohoku Imperial Universities. Furukawa owned a large mining corporation and had made a fortune during the Sino-Japanese and Russo-Japanese Wars, but his mines caused serious pollution in Ashio, which developed into a political scandal. The Furukawa family decided to donate a large sum of money for the establishment of new departments of the above two universities at the suggestion of Takashi Hara, Minister of Interior, who was serving as an advisor for the Furukawa Corporation (Tachi 1983).

Although less dramatic than this example, many large *Zaibatsu* and industrialists, some of whom made fortunes during the above two wars and the First World War, donated handsome sums to assist the establishment and consolidation of universities and their attached research laboratories. For example, the Sumitomo Corporation helped support a major expansion of the Metallurgical Research Laboratory of the Tohoku Imperial University, and the industrialist Ginjiro Fujiwara helped establish the Engineering Department of Keio University.

Donations from industrialists and corporations were an important financial source for the universities, though they also received government funding. As in the United States, philanthropic foundations were also important financial supporters of university investigators in prewar Japan. The following foundations provided actively for social welfare as well as for the advancement of research and development: Morimura Homei Kai (1913), Toshogu Sanbyakunensai Kinenkai (1914), Keimei Kai (1918), Harada Sekizen Kai (1920), Saito Hoon Kai (1923), Taniguchi Kogyo Shorei Kai (1929), Hattori Hoko Kai (1930), Asahi Kagaku Kogyo Shorei Kai (1934), and Mitsui Hoon Kai (1934). The word "Hoon" in the names of some of these foundations means the requital of kindness, indicating that their mission was nationalistic and patriotic as well as philanthropic; financial support for research and development at universities was considered a patriotic deed in these periods (Hayashi and Yamaoka 1984: 43–130).

Just as many corporations contributed financially to universities, so many university faculty served as technical consultants for corporations in prewar Japan. At one Diet meeting, it was suggested that university professors were too frequently engaged in consulting work for private corporations, to the detriment of the education of their students. It was, however, argued that the raison d'être of the universities was to serve the nation and that professors' consulting work should be considered national service (Bartholomew 1989: 230).

Because of this consulting work, the quality of research work at universities appeared lower that those at Western universities. Hidetsugu Yagi, the father of Japanese electronics research and the director of the Board of Technology—the Japanese counterpart of the OSRD (and so highly regarded that Karl Compton, then director of the OSRD, called him "the Vannevar Bush of Japan")—recalls the following prewar episode when his German mentor Heinrich Barkhausen visited Japan. Barkhausen, a leading electronics engineer, told him that he had expected large corporations such as Mitsui and Mitsubishi to have excellent research laboratories. Surprised to find none, he pointed out the need for such research laboratories if the companies were to be competitive in the world, and further remarked that university professors were instead engaged in the kind of R&D activities that would be more naturally pursued at industrial corporations (Yagi 1953: 130–31). Yagi could only answer that Japanese shareholders expected high dividends and did not respect corporate R&D efforts which, they considered, would not increase their dividends.

The above two episodes tell us that there were generally very few R&D efforts at industrial corporations, and therefore that the engineering faculty at Japanese universities took the role of pursuing industrially oriented research and development.

The Wartime Mobilization of Science and Technology

After the war began, university and industrial research was geared to military purposes. The government took measures for rapid expansion of scientific and technological departments at universities and a variety of new research laboratories were set up to conduct war work. At the University of Tokyo, the Second School of Engineering was established to meet the sharply increased demand for young engineers for military R&D. The number of engineering faculty and students was doubled.

Other universities also established various research laboratories before and during the war. The following is a list of such academic research institutes established in the 1930s and 1940s.

1930s: Architectural Materials (TIT), Telecommunications (Tohoku), Resource Chemistry (TIT), Industrial Science (Osaka), Precision Machinery (TIT).

1940s: Ore Dressing (Tohoku), Cryogenics (Hoku), Fluid Engineering (Kyushu), Ceramics (TIT), Scientific Measurement

(Tohoku), Elastic Engineering (Kyushu), Ultra Short Wave (Hoku), Catalysis (Hoku), High Speed Dynamics (Tohoku), Aeronautical Medicine (Nagoya), Acoustics (Osaka), Wood Materials (Kyoto and Kyushu), Glass (Tohoku).[1]

Some research laboratories were planned but not built during the war. The Research Institute for Aerial Electricity at Nagoya University was designed to investigate the technology of meteorological forecasting in the southern areas by measuring disturbances in atmospheric electricity. Its research began during the war, but it was not established until afterward.

Two points should be mentioned on the prewar and wartime origins of many of these university research laboratories. First, although some of them had long traditions of relevant research and many continued to pursue significant investigations in their own technological fields, the nationwide arrangement of research laboratories at Japanese universities was not designed to adapt to the postwar Japanese economy. Because of the tenacity of the national university system and their wartime origins, some university laboratories found it difficult to adjust to postwar industrial conditions.

Second, some of the university research laboratories were beneficiaries of adjunct foundations, which formed a financial conduit between industry and the university in order to fund their parent research institute. The Foundation for the Advancement of Telecommunications Technology, for instance, was established in 1944 to support its parent, the Telecommunications Research Institute of Tohoku University. The foundation, however, did not take an active role as a financial conduit until around 1953. In 1956, seventeen percent of the total budget of the Telecommunications Research Institute came from this foundation (Tohoku University 1960: 1651–53). These adjunct foundations served as financial supporters of postwar university laboratories.

Another notable feature of the Japanese wartime mobilization was the establishment of research groups called the Kenkyu Tonarigumi (Research Neighborhood Groups). These groups, each of whose missions was intended for investigation in a specific technological field, consisted of academic and industrial engineers and facilitated the exchange of fresh technical information among its members. The discovery of a new dielectric characteristic of a titanium alloy was reported at the 8012nd Kenkyu Tonarigumi, which consisted of engineering professors at Tokyo, Kyoto, Nagoya, and Osaka, and engineers at Toshiba, Hitachi, and others. The information network developed by this wartime arrangement would serve university-industry collaboration in postwar Japan.

The Occupation and the Reorganization of the University System

With the end of the war, the institutional system of research and development in Japan was forced to change radically. The emerging postwar framework of the university-industry relationship was deeply influenced by the educational and industrial policy implemented by relevant sections of the Occupation's General Headquarters (GHQ) under the Supreme Commander of the Allied Powers. Their initial and primary purpose was to demobilize and democratize all of Japan. For this purpose, they ordered the Japanese leaders to abolish old institutions and organize new ones. Accordingly, the military R&D activities in wartime Japan were thoroughly investigated, and most of them stopped and disbanded. The experimental cyclotron of the Physico Chemical Institute (Riken) was destroyed and thrown into the sea, and all aeronautical research was forbidden. While the GHQ tried to build up a new national system of research and development at universities and industries, their policy was not entirely consistent among their sections and throughout the period of the Occupation. These policies and their inconsistencies cast a shadow on the postwar relationship between Japanese universities and industry.

Reparations were the main issue in the first phase of the Occupation, and the original policy of the GHQ was to reduce Japan's economic level to that of other Asian countries that had been exploited by military Japan during the war, for which they planned to confiscate half the equipment and facilities still in Japanese factories. But around 1947, they significantly moderated their former economic policy owing to the rapid advent of communism in China and the rise of a Marxist movement inside Japan. The National Security Council announced a change of policy toward

Japanese economic recovery due to the Communist threat; instead of reducing the economic power of postwar Japan to the prewar level, they began to promote Japan's economic activities.

One of the organizations chiefly responsible for the reorganization of Japanese scientific and technological R&D was the GHQ's Economic and Scientific Section (ESS). It included a Scientific Division, later renamed the Scientific and Technological Division (ESS/ST), which was in charge of scientific and technological matters in relation to economic problems. Many American New Dealers, it is said, joined the ESS, and they were active in disbanding large Japanese concerns. But some officials at the ESS tried to reorganize and reestablish a new Japanese R&D system in order to help the restoration of the postwar Japanese economy. Harry Kelly, at ESS/ST, was one of them.

Kelly, a young physicist who had worked on the development of radar at the U.S. Radiation Laboratory during the war, was invited to assist with technical details in investigating Japanese devices and facilities. Talking to Japanese scientists, Kelly found that GHQ did not need to concentrate on the demobilization of wartime research activities, but rather should pay more attention to restoring the devastated Japanese economy and reforming its research and development system (Yoshikawa 1987).

The ESS/ST accordingly emphasized the importance of industrial applications of academic research, and attempted to strengthen the tie between the universities and industries. In 1947, under Kelly's initiative, it sponsored a "science advisory group" organized by the American National Academy of Sciences, which turned in a report on the "Reorganization of Science and Technology in Japan" and suggested that an Advisory Council on Higher Education and Research be established, following its own model. This eventually led to the establishment of the Science Council of Japan. At the same time, the report mentioned the condition of university research. It criticized its orientation toward basic research, saying that too much emphasis on basic research was detrimental to applied research (Hata 1995). The ESS/ST proposed this report as a basic policy text, and emphasized the importance of education in science and technology for the sake of the economic recovery of Japan.

However, the ESS/ST proposal was strongly opposed by the Civil Information and Education Section (CIE) which was responsible for the reformation of the newly democratized educational system. The CIE worked to reform the Japanese educational system, beginning by implementing the American 6-3-3 system in primary and secondary education. In reforming the universities, they emphasized the importance of a liberal education at the college level.

In opposition to the policy of the ESS/ST, the CIE argued that scientific research was only one of the many functions of the university, and that the university's primary goal was to "train and produce all kinds of leaders needed for a free society." It therefore proposed that the report should be accepted only insofar as it was not against the policy and plan already adopted by the CIE. In this way, the CIE encouraged the separation of university professors and university education from industrial goals. Scientists and engineers were certainly necessary for the economic recovery of postwar Japan, it said, but their number had grown so greatly during the war that there were enough of them for postwar Japanese needs.

In the end, the ESS/ST policy was not implemented in Japanese universities, which were largely shaped by the policies of the CIE.

The Importation of Technology

In the early 1950s, the San Francisco Peace Treaty ended the Occupation, and the Japanese government started to implement its own economic and industrial policy. As foreign capital regulation was greatly relaxed, Japanese corporations began vigorously to introduce the advanced technologies of foreign companies. To introduce a foreign technology, a Japanese company had first to submit a proposal and to receive permission from the government. The proposals were examined by government officials and academic engineers. The above-mentioned Yagi was one of these examiners. Between 1950 and 1960, over a thousand technical innovations had been introduced (most of them from U.S. companies), at a total expenditure of more than 100 billion yen.

There were various ways to purchase foreign technologies. The Japanese companies could simply buy patent rights relevant to a specific industrial product, but in many cases they obtained not only such patent rights and design drawings but also the services of engineers to provide on-site technical know-how. This was especially helpful in introducing technologies relating to new production systems such as oil refining and automobile production (Arai 1995: 166).

The introduction of Western technologies has been the most important concern of Japanese industrial leaders from the Meiji Restoration onward; the direct introduction of cutting-edge technologies from foreign to Japanese companies significantly influenced the relationship between university researchers and industrial corporations. University researchers had no doubt played an indispensable role as technological consultants in earlier technological transfers, being able to understand the foreign technologies, explain them to industrial engineers, and even occasionally to transform them for domestic industrial needs, but now the technical staff of Japanese companies were in touch with those at their counterpart foreign companies, and they learned advanced technologies directly from U.S. or European companies. Because of this cooperation, the previous role of the university researchers as technology transfer agents became minimized.

Between 1955 and 1963, in an attempt to introduce foreign technologies, Japanese firms established a total of 108 "central research laboratories," and invested heavily in new facilities. Especially in the electrical and chemical industries, a substantial research and development effort was required to digest advanced technologies from foreign countries. These "central laboratories" were established during the last phase of the investment boom. Many were the result of the reorganization of small laboratories within the same company, and aimed at introducing and digesting foreign technology. But others aimed at developing their own new technologies.

During this period large corporate research laboratories recruited many university researchers, and some corporate researchers even boasted that the university was unnecessary for industrial research and development in Japan. While certainly an exaggeration, this implies the superiority of research facilities at corporate laboratories to those at universities. By the mid-1960s, however, the boom was over; the recession forced companies to cut the budget of their research laboratories.

The Research Institute as a National Engineering Forum

Some industries soon recognized the need for research and development of their own technologies, and the need for domestic technology, in which academic engineers played critical roles. Episodes from the two leading Japanese industries—steel and shipbuilding—will serve as examples.

In the Japanese steel-making industry, corporations adopted advanced foreign technologies—machines, processes, and designs. However, such technological transfers impeded the further development of the industry: As the Japanese corporations grew larger and more competitive in the international iron and steel market, foreign corporations became less willing to share their advanced technologies with their potential competitors. Secondly, and more important, Japanese metallurgical engineers gradually became aware of the impediments deriving from technical reliance on foreign companies. They recognized that because the users of steel products also relied on other foreign technologies, they lacked information on the physical properties of the materials they bought and were thus unable to develop reliable new products.

For these reasons the National Research Institute for Metals was established in 1956 under the newly established Science and Technology Agency. Its official function was to perform (1) research on the production and processing of metallurgical materials which required equipment outside the scope of one corporation; (2) basic research on a consistent production process from materials to products; (3) research on the production and processing of metals appropriate to the natural resources of Japan; (4) research on metallic materials for nuclear and aeronautical technologies; (5) research on the production of pure metals; and (6) materials testing requiring large-scale facilities, and the development of metallic materials that private corporations were unable to produce.

That same year, the Science Council of Japan requested that the Science and Technology Agency establish a new research laboratory on the physical properties of metallic materials. And the next year, in 1957, an Institute for Solid State Physics was established at the University of Tokyo. Its first director was Seiji Kaya, a notable metallurgical physicist. At the Research Institute for Industrial Technology, a blast furnace for experimental purposes was built in 1955 and metallurgical research and development began. Research laboratories were also established at private corporations. In 1959 and 1960, such research laboratories were set up at Fuji Seitetsu, Yawata Seitetsu, Sumitomo Kinzoku, Nihon Kokan, and Kobe Seikojo. From these corporate research laboratories important industrial innovations emerged, some of which were transferred to foreign corporations.

The Japanese shipbuilding industry also recognized the need for cooperative research, especially for the construction of supertankers of greater than 50 thousand tonnes capacity. For this purpose a Shipbuilding Research Society was established. In the postwar development of the Japanese shipbuilding industry, both prewar and wartime naval engineering played an indispensable role, and in postwar Japan it provided new key technologies as well as a niche for those leading naval engineers who had gone into the private sector due to postwar restrictions on bureaucratic positions. In 1955 a 45,000-tonne supertanker was built by Mitsubishi, and the shipbuilding companies received orders for still larger tankers. But the design and construction of ships larger than 60,000 tonnes required technological breakthroughs on the welding, structure, and performance of such large vessels: As an example, thicker steel plates were harder to weld, which required the development of both special steels and special welding techniques (Kaneko 1964: 474).

Because of this necessity to develop new technologies, a new Japanese Society for the Research of Naval Architecture was established as a forum for representatives from shipbuilding companies, ship-owners, academic laboratories, and the government. Their three-year research on the subject brought new technological information which was fully utilized for the subsequent construction of the new class of ships. Among others, *Nissho-maru*, a 130,000-tonne tanker, attracted worldwide attention.

The above two industries, based on a prewar accumulation of relevant technologies, were perhaps exceptional. But they show how academic engineers were able to contribute to important developments in industrial technologies. The national and academic research institutes served as a forum for the exchange of technological information between engineers from both academia and industry; annual meetings of professional engineering societies also provided an important forum for information exchange, not only between academia and industry but also between industries, through academia. Japanese corporations were a fairly closed world, and these public forums were catalysts for one corporation to provide technical information to another.

Collaboration: Encouraged and Discouraged

As Sputnik in 1957 had a great impact on scientific and technological research and education at American universities, the Ministry of Education in Japan also decided to promote scientific and technological education at the universities. In that year, Chuo Kyoiku Shingikai, the main advisory board of the Ministry of Education, proposed measures to promote scientific and technological education. The report pointed out the poor financial and material conditions at universities and the need to improve both their research facilities and the quality of their graduates. As to university and industry cooperation, the report stated that the universities should "make closer cooperation with industry by taking into account requests from industry and by sending students to industrial factories," and that "attached university laboratories should be able to re-educate engineers in industry" (Chuo Kyoiku Shingikai 1957). The University of Tokyo accordingly changed its rules, and began to encourage industrial engineers to visit its laboratories.

Industrial leaders, too, acknowledged the urgent need to cooperate more closely with the universities in recruiting competent engineers and collaborating on industrial research. "Keizai Doyukai," a group of such industrial leaders, actively promoted such a collaborative relationship between industry and universities, and planned a center for industry and university collaboration (YKDGKK 1973: 201–202).

Thanks to these policies and efforts, in the 1960s the engineering departments of the universities began to receive more financial support from relevant industries. Such donations allowed the

University of Tokyo, for instance, to build many research facilities, and the Engineering School responded by reorganizing and expanding its departments to adapt to contemporary industrial needs (TDHHI 1987: 648–49).

The movement toward closer ties between universities and industry was blocked by the student protests of the late 1960s: While American universities were excoriated for their military connections, Japanese students criticized the close relationship between the universities and private corporations. I will cite two statements responding to this protest movement, one by industrial leaders and the other by leaders at a university.

Industrial leaders responded to the student protests by emphasizing the social need for such collaboration and proposing to reform the university administration. The Keizai Doyukai group proposed a document, "The Institution of Higher Education for an Advanced Welfare Society," as their response to the student movement (YKDGKK 1973: 246–64). Concerning university-industry cooperation, it stated:

> Keizai Doyukai has been proposing the urgent need to promote university-industry cooperation since 1959. We consider that such cooperation is necessary for the universities to play a leading role in the coming advanced welfare society, and repeat our view below.
>
> University-industry collaboration coincides with modern educational thought to encourage the shift from "learning for the sake of learning" in an ivory tower to "academic research opened for society." The history of the relationship between universities and industry in the United States, the USSR, and Japan, which accomplished rapid industrialization, would tell us that both grew and developed stimulating each other and that university-industry collaboration is a historical trend.
>
> Some delimit university-industry cooperation as a simple exchange of money and between a specific researcher at a specific university and an individual corporation, and regard it as a wrong effort to make university researchers industrial subcontractors and to distort research freedom. But the functions of universities and industry interact with each other to drive social development, and their cooperation exists in various fields. Furthermore, the advancement of industrial research and development, the emergence of big science, the increasing social demand for adult education—all will increase the importance of their cooperation.
>
> The reason we dare to promote university-industry cooperation is that the institutional structures of universities do not respond to tendencies in real society. So-called academic freedom would be diminished, if applied

research did not open new fields for basic research or if theoretical research did not relate to real processes. We oppose the attitude that criticizes a part of university-industry collaboration and denies it altogether. We agree with the present state of the collaboration and propose a new organization and rule to correct present errors. The question will arise as to the financial distribution between technological areas relating to production and humanities areas relating to the national and traditional culture and values. This question will be solved by the redistribution of a specific proportion of funding from industry to the latter research areas or by establishing large multi-purpose foundations. (YKDGKK 1973: 255)

After stating its view on university-industry cooperation, Keizai Doyukai proposed a plan to reform the university organization, including the recommendation to recruit on a contractual basis and to reorganize all universities as non-profit corporations.

In contrast to the industrialists' statement, a statement from the university community clearly showed its more restrictive view on industry-university cooperation. The following is an excerpt from the first report of the Investigative Committee for the Preparation of University Reform at the University of Tokyo, concerned with the university-industry relationship. It shows that universities were expected to distance themselves from industry, and probably contributed to the further alienation of university researchers from industrial needs:

In relation to the research at the University of Tokyo, there is the problem of so-called "university-industry cooperation." Although the concept of "university-industry cooperation" is not necessarily clear, it would be defined here in the following relatively limited sense: that a university does research through making official or unofficial contracts with private corporations, governmental agencies, foundations, and the like for certain commissioned researches and receiving financial assistance for research.

University-industry cooperation in this sense has been performed at the University of Tokyo, but its details are not clearly known and are to be duly investigated. At the discussion of the present special committee, the following points have been proposed for industry-university cooperation.

First, the issue of industry-university cooperation lies in the danger that scholarly research would be subservient to the interest of the investor. It is felt, for instance, that the publication of data and research results might be restricted, and that the preconditions of research and, in an extreme case, the conclusion itself might be circumscribed. In other cases, the content of research is regarded as secondary, and the commissioned research is utilized for authorization under the name of the University of Tokyo. To avoid such misconduct, the plan and results of the research and its accounting data should be opened for various commissioned research.

Second, how to place emphasis among various topics during research (and education) at universities is clearly influenced by the prosecution of commissioned research and the like. But we should absolutely avoid the possibility that the selection of research planning be determined by factors in conflict with the prerequisites of academic pursuits. It is certainly desirable that university research receive stimulation from the practical needs of society and play an active role as an intellectual center of society. ... But if such factors as the interest of a specific corporation, which are in conflict with the prerequisites of academic research, influence the policy and planning of research and education, it would result that the autonomy of scholarship is lost. For industry-university cooperation, therefore, careful consideration and institutional regulation are needed to protect academic freedom. Furthermore, if industry-university cooperation is done over a long period and a close connection emerges between commissioned faculty and such investors as corporations and governments, the autonomy of researchers would be lost without their being aware. This is a problem of "the decadence of a researcher," and each faculty member must always be careful.

And third, one of the reasons large-scale industry-university cooperation is under way in certain specific departments is that the expenditures of the government for research and education are very low. It must be said, for instance, that because of this extremely meager financial condition at universities, it is necessary to rely on commissioned research fees and others from corporations in order to guide and educate graduate researchers. In other words, because university faculty are able to receive such commissioned research fees from industry, ... the meager financial situation of the universities is covered up. It is necessary to improve the financial situation of the universities drastically in order for them to perform scholarly research and education in an autonomous way. (DKJC 1969: 99–100)

The statement seems to have accurately represented the thinking of the faculty of national universities. It constituted the formal stance of a leading national university toward collaboration with industry. Despite the strong requests and demands from industry, Japanese universities tended toward restriction of collaboration with industry in the 1970s. These restrictions began to be moderated only in the 1980s; as the editors of an educational journal put it, in a special issue on university-industry collaboration, "the allergic response to university-industry collaboration is recently diminished" (Tachi 1983: 5).

In this history of the hesitant relationship between universities and industry in postwar Japan, the role of university engineers in industrial development was in general implicit and indirect. Most of them did not commit themselves to making industrial innovations or selling their technological ideas. Perhaps one of their important roles was to mediate technological ideas between academic and industrial engineers or among industrial engineers. National and university research laboratories as well as annual meetings of academic societies provided occasions for such engineers to exchange important technical information. On a less formal level, university professors took the initiative in constructing a network of academic and industrial engineers, and made information on new technological ideas flow through that network.[2]

The history of an instrument manufacturer shows the ways in which such networking worked in an almost ideal manner. This company, Murata Seisakusho, developed from being a small ceramics manufacturer into a leading high-tech electronics instrument company. In the process of its postwar development, its president, Akira Murata, received pivotal advice from Tetsuro Tanaka, of the engineering faculty at Kyoto University. Tanaka knew the interesting dielectric characteristics of a certain titanium alloy, and at a meeting of the "Kenkyu Tonarigumi" he recommended that Murata manufacture some products using this material. Murata responded by producing a titanium alloy condenser through close collaboration with Tanaka's laboratory at Kyoto University (Murata Seisakusho 1996: 15–20). When Murata was later at a loss as to what other products might use the piezo-electric characteristics of this titanium alloy, Tanaka again gave him the crucial advice that it could be used as a filter for radio waves; Murata Seisakusho became a leading manufacturer of radio wave filters, which eventually has led to their production for mobile telephones (Murata Seisakusho 1996: 60). The history of Murata Seisakusho tells us that one of its managerial philosophies was "to borrow wisdom from experts," and that such a policy has worked throughout its history (Murata Seisakusho 1996: 47).

The fruitful collaboration between Murata and Tanaka may not be a typical case in the history of university-industry collaboration in postwar Japan. But their give-and-take relationship seems to reflect well the postwar situation of university-industry cooperation.

Conclusion

The year 1945 marked a watershed in the history of the university-industry cooperation in Japan. In prewar Japan, such a cooperative relationship was encouraged by the government and enhanced by the nationalistic environment. Not only those corporations that had made fortunes through war-related transactions but also beneficent industrialists who had succeeded in business were potential funding sources for prewar Japanese universities in expanding their research facilities and organizations. Large corporations donated money to establish foundations whose mission was to promote academic research activities, but did not establish their own research laboratories. As Barkhausen noted, there were no industrial laboratories at firms under Mitsubishi or Sumitomo. Many research efforts necessary for industrial innovation were, consequently, conducted at university laboratories.

The close prewar relationship between universities and industry was broken in 1945, and a new relationship constructed. The GHQ, most of whose influential members were Americans, attempted to disband the old university system and to implement a more democratic one. Although there were some discrepancies in their views of the role of the university in postwar Japanese society, university researchers lost their corporate patrons and had to keep doing their investigations in extremely poor facilities for more than a decade. During that period, Japanese industrial corporations vigorously imported foreign technologies, bypassing Japanese university researchers. Large corporations began to establish their own central research laboratories and even recruited competent engineers from universities.

In the decade after Sputnik, the need for university-industry cooperation was increasingly recognized and budgets for scientific and engineering departments were greatly increased to remedy material and institutional deficiencies at universities. University faculty began to receive financial support from industrial firms through nonprofit organizations. However, the university reform caused by student protests in the late 1960s again restricted the cooperative relationship between university faculty and industry. It was only in the 1980s that the close cooperation between the universities and industries was generally encouraged by government agencies, including the Ministry of Education.

It is now typical that industrial firms offer experimental equipment or research funds to university faculty, who in return offer industry crucial technical information which may eventually lead to marketable products. Such symbiosis based on an informal give-and-take relationship seems to have existed and worked fairly well in at least some quarters of postwar Japanese industry. The informal but nonetheless significant give-and-take relationships between Japanese universities and industrial firms are certainly worth a thorough examination.

Notes

1. TIT: Tokyo Institute of Technology. Tohoku: Tohoku University. Hoku: Hokkaido University. Osaka: Osaka University. Kyushu: Kyushu University. Kyoto: Kyoto University.

2. Some fifty networks in various technological and biomedical fields are introduced in (NSS 1989).

References

Arai, Katsuhiro. 1995. "Gijutsu Donyu (the Importation of Technology)." In *Tsushi Nihon no Kagaku Gijutsu (The Social History of Science and Technology in Contemporary Japan)*, eds. Shigeru Nakayama et al. Tokyo: Gakuyo Shobo. Vol. 2.

Bartholomew, James R. 1989. *The Transformation of Science in Japan: Building a Research Tradition.* New Haven: Yale University Press.

Chuo Kyoiku Shingikai. 1957. The Fourteenth Report, "Measures to Promote Scientific and Technological Education."

Daigaku Kaikaku Junbi Chosakai (DKJC). 1969. (The Investigative Committee for the Preparation of the University Reform). The First Report.

Hata, Takashi. 1995. "Shinsei Daigaku to Riko Kyoiku (New-system Universities and Education in Science and Technology)." In *Tsushi Nihon no Kagaku Gijutsu*

(The Social History of Science and Technology in Contemporary Japan), eds. Shigeru Nakayama et al. Tokyo: Gakuyo Shobo. Vol. 1.

Hayashi, Yujiro and Yoshinori Yamaoka. 1984. *Nihon no Zaidan (Foundations in Japan)*. Tokyo: Chuo Koron Sha.

Kaneko, Eiichi ed. 1964. *Gendai Nihon Sangyo Hattatsu-shi (The History of Modern Japanese Industries)*. Vol. 9. Tokyo: Kojunsha.

Murata Seisakusho Gojunenshi Hensan Iinkai. 1996. *Fushigi na Ishikoro no Hanseiki: Murata Seisakusho Gojunenshi (A Half Century of Wonderful Stones: The Fifty Year History of Murata Seisakusho)*. Kyoto: Murataseisakusho.

Nikkei Sangyo Shinbun (NSS). 1989. *Nihon Gijutsu Jinmyaku* (Human Networks of Technologies in Japan). Tokyo: Nihon Keizai Shimbunsha.

Tachi, Akira. 1983. "Kigyo to Daigaku: Senzen no Sobyo (Corporations and Universities: A Sketch of the Prewar Period)." IDE 244: 3–11.

Tohoku University. 1960. *Tohoku Daigaku Goju nen shi (The Fifty Year History of Tohoku University)*. Sendai: Tohoku University.

Tokyo Daigaku Hyakunenshi Henshu Iinaki (TDHHI). 1987. *Tokyo Daigaku Hyakunenshi (A Hundred Year History of the University of Tokyo)*. Vol. 3. Tokyo: Tokyo Daigaku Shuppankai.

Yagi, Hidetsugu. 1953. *Gijutsu-jin Yawa (Night Stories of an Engineer)*. Tokyo: Kawade Shobo.

Yokohama Kokuritsu Daigaku Gendai Kyoiku Kenkyujo (YKDGKK). 1973. *Zoho Chukyoshin to Kyoiku Kaikaku: Zaikai no Kyoiku Yokyu to Chukyosin Tosin (Zen) (The Central Educational Board and Educational Reform Policy: Pedagogical Requests from Industry and the Reports of the Board)*. Tokyo: Sanichi Shobo.

Yoshikawa, Hideo. 1987. *Kagaku wa Kokkyo o Koete (Science has no National Boundaries)*. Tokyo: Mita Shuppan Kai.

10

University-Industry Collaboration in Japan: Facts and Interpretations

Hiroyuki Odagiri

Introduction

It has become common, especially in the United States, to measure the extent of university-industry (UI) collaborations by the amount of funds spent for UI joint projects or donated by the industry to universities; by the number of patents jointly applied for by university faculty members and company researchers; by the number of citations made by company patents to the patents or papers produced by university members; by the number of venture businesses started by current or former university members, or by a positive correlation between the number of venture businesses and the research activity of universities in the same or neighboring regions.

In any of these accounts, Japan appears to lag far behind the United States. Concerned by this situation and by Japan's apparent lag in biotechnology and other science-based fields, the Japanese Government has been eager to promote basic research and UI research collaborations. Thus, the Science and Technology Basic Law passed the Diet in November 1995, the Industrial Structural Council and Industrial Technology Council of the Ministry of International Trade and Industry (MITI) published a report titled "Promotion of Creation and Utilization of 'Intellectual Assets': The Way toward a Nation Based on Creation of Science and Technology," and the Ministry of Education (Monbusho) published a report on the alliance and collaboration between industry and university (Monbusho 1997).

However, does the weak quantitative performance indicated above necessarily imply that UI collaborations are infrequent or ineffective in Japan? The answer is by no means simple because, first, historically speaking, there were many instances where Japanese professors actively and effectively involved themselves in starting up new businesses or conducting (or guiding) industry R&D efforts. Secondly, both in the past and at present, Japanese professors have been frequently collaborating with industry, albeit less formally, by, for instance, giving advice through committees, conferences, and workshops; allowing company researchers (often former students) to work in their university laboratories; mediating between companies or between a company and national laboratories or other organizations; or diffusing information through former students.

This paper considers the diversity of such collaborations and will show that a simplistic international or intertemporal comparison of UI collaborations, whether by statistics or by a single case, is misleading and that the form of UI collaboration is embedded in the social, economic, and institutional factors of the nation and of the period. Discussion of the early stage of Japan's development, when universities played important roles in technological and industrial development, is followed by an overview of the current situation of UI collaborations and the government policies for this purpose. Two such policies—ERATO and the Fifth-Generation Computer Project—will be discussed in detail, to indicate in what ways university members were involved and how much these ways can vary across cases.

How It Was in the Early Days

Let me start with an old story—the role of engineering schools in the early stage of Japan's industrial development, that is, in the latter half of the 19th century, a period described in detail in Odagiri and Goto (1996), and Odagiri (1998).

Investment in education was probably the single largest contribution of the government in the late 1800s, followed by its investment in a communications and transportation network and other infrastructure. One of the higher-education institutions built by the government was a technical college named Kogakuryo, founded in 1873. The government emphasized practical technical education at a time when European universities tended to regard practical studies as inferior to pure science. This college, which later became the Engineering Department of the University of Tokyo, actively collaborated with industry.

Probably most prominent was the case of I. Fujioka, who studied electrical engineering with a British professor at the college and, after graduation, became an assistant professor there. While teaching at the college, he designed the first domestically-manufactured power generator and then, upon returning from his first overseas trip during which he visited Edison's company in the United States, he left the college and began to develop and manufacture light bulbs. Even with equipment imported from the United Kingdom, the enterprise met many difficulties. It took him a year of experiments and trial and error before he succeeded in the development phase and started a company to commercialize it. The company, therefore, was a venture business, if one uses the current terminology, founded by a former professor. It eventually evolved into today's Toshiba.[1]

Another interesting case is found in the medical field. When Japan's first drug-making company, Dai Nippon Seiyaku, was founded in 1885 with government support, N. Nagai, then studying in Germany, was asked to join the company to lead the technological aspect of the business, and offered at the same time a professorship at the just-established Pharmacy Department of the University of Tokyo. This episode indicates a surprising flexibility in the university administration of the time, in contrast to the later policy environment in which employees of national universities would be prohibited from taking positions in private companies.

It was common for professors to give advice and guidance to industries; a typical case was that of K. Noro, a professor in metallurgy at the University of Tokyo. He solved technical problems at the two major iron mills, Kamaishi and Yawata, both built by the state, by redesigning their imported furnaces and supervising their startup. Subsequently, his students became the chief engineers in these mills and helped improve their operation. Although Noro later had to resign from the University of Tokyo for an unrelated incident, he remained a private consultant and maintained a close contact with both universities and industry, playing a major role in the establishment of the Iron and Steel Institute, an academic association, for which he served as the first president. This institute helped foster the exchange of technological information among the members, which included not only university professors but also engineers from companies and arsenals.

Many entrepreneurs asked for technical advice from academics. For instance, when K. Toyoda started the production of automobiles, he sought advice from a number of university professors, some of whom had been his fellow students at the University of Tokyo: He took advantage of the network of the acquaintances he gained at universities, a practice that would help many other entrepreneurs to acquire technical information and to recruit people with technological knowledge. This was not uncommon, but a continuation of traditional Japanese business practices; in discussing today's collaboration of industry and universities, it should always be kept in mind as a precedent.

How It Is Today

Unfortunately, much of the flexibility of the early period has been lost, partly because of growing criticism that some faculty members spent more time in outside consulting jobs than in teaching and, more importantly, because of the postwar limitations on the outside activities of professors in the national universities (see Hashimoto, in this volume). For instance, they could not establish new businesses or become directors or employees of private firms, a regulation relaxed in 1997 in response to the argument in favor of more UI collaborations and of professors' active involvement in starting up new high-tech businesses. Now professors at national universities can work for companies on a part-time basis if the purpose is to conduct or guide the latter's R&D. This arrangement is still uncommon, although a gradual increase is expected.

The extent of joint research conducted by national universities (including national technical colleges) with private firms or local governments is shown in Figure 1. In 1994, 1,488 joint research

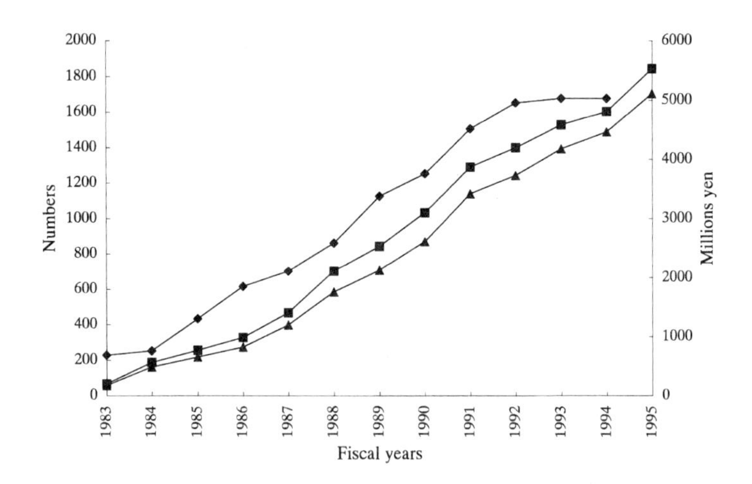

Figure 1 University-Industry Joint Research, 1983–1995. Source: Ministry of Education (1997).

projects were ongoing, in 89 universities with 883 firms (duplications allowed), together spending about ¥5 billion. Although ¥3 million (about $25,000) per project on average is certainly a help to university researchers, this amount is modest indeed; Hitachi and Toyota each spent nearly ¥400 billion for in-house R&D in the same year. The number of these projects is increasing and was 1,704 in 1995.

The figure may give an impression that UI collaborations have been substantially increasing. However, the apparent increase probably is due to the fact that the UI Joint Research Scheme was formalized only in 1983 and more collaborations are now undertaken under this scheme and reported to the Ministry of Education as such. Thus, some of the research funds that used to be commissioned by companies (or government agencies) to universities (*jutaku kenkyu*) or donated by companies to universities (*shogaku kifukin*) are now provided to universities under the joint research scheme. The increase in commissioned or donated research funds is slower, but these funds still reached ¥14 billion and ¥49 billion respectively in 1995, far larger than the industry funds provided under the joint research scheme.

I emphasize that these numbers hardly give a whole picture of UI collaborations in Japan because they cover only national universities, which account for less than a quarter of all universities in terms both of the number of institutions and the number of students. Although most of the prestigious universities are national in Japan, there are several private universities, such as Tokai, Ritsumeikan, and Waseda, that are known to be active in promoting joint research with industry. Furthermore, as will be discussed below, much faculty guidance and collaboration is performed in an informal, unquantifiable manner.

It is difficult to measure the performance of these UI collaborations. Those made under the joint research scheme produced only 20 patent applications in 1996 made jointly by the universities and the firms while, as shown in Figure 1, 1,488 collaborative projects were reported. In other words, nearly 99% of the projects did not result in joint patent applications; this need not imply poor performance of these collaborations because the professors may not have aimed at producing patents or they may have avoided making joint applications. Most probably, they opted to avoid the costs and trouble of applying and administering the patents and instead had the firms apply for patents, with research funds from these firms to be received as compensation. For more details on this issue, see the paper by Yoshihara and Tamai in this volume.[2]

Needless to say, patent acquisitions by universities can be undesirable because it may be better for basic knowledge that the universities are expected to produce to be made public and widely used.[3] This reservation notwithstanding, it is unfortunate if, as I speculate, the sluggish patenting activity of Japanese university researchers has been the consequence of poor support and a lack of incentive. It seems urgent that the Japanese universities be equipped with patent-specialist staff to help the researchers deal with the legal aspects of patenting and licensing.

Owing to the Hashimoto and Obuchi Administrations' enthusiasm for promoting UI collaborations, the menu of the government support for basic research and/or cooperative research has expanded. In addition to the above-mentioned joint research scheme of the Ministry of Education, several ministries offer similar but slightly differentiated programs. For instance, the Ministry of

International Trade and Industry (MITI) has several programs, offering research grants, low-interest loans, or other help. The Ministry of Health and Welfare (MHW) offers a program targeted at basic research in human science while the Ministry of Agriculture, Forestry and Fisheries (MAFF) offers a program targeted at agricultural science. In addition, a number of local governments offer programs for collaborative research. For instance, Kitakyushu City expended ¥46 million in 1997 to support six joint research projects between local firms and universities (together with prefectural or municipal laboratories in two projects). And, finally and perhaps most importantly, the Science and Technology Agency (STA) carries out several policies to encourage basic and/or collaborative research. The oldest and probably the most successful among these is ERATO, which will be discussed below in detail.[4]

The Case of ERATO[5]

The Exploratory Research for Advanced Technology (ERATO) program was initiated in 1981 "for the purpose of fostering the creation of advanced science and technology while stimulating future interdisciplinary scientific activities and searching for better systems by which to carry out basic research."[6] It is administered by the Japan Science and Technology Corporation (JST), an organization fully controlled and supported by STA. A unique characteristic of ERATO is that JST selects its project directors from among a wide variety of researchers based on its own survey, interviewing hundreds of young researchers and asking them to offer names of potential research leaders. Based on this survey, JST makes a list of about 15 candidates for the directorship, asks them to write proposals, and selects four from among them as directors. Each director is given research funds of around ¥2 billion (approximately $15 million) for a period of five years, and a free hand (but with JST's advice) in choosing researchers and locations for the project. The project is non-renewable; this finite research period is a unique feature in Japan, requiring the researchers (who typically number between 15 and 20 per project) to seek new jobs after the five-year period because except for directors all the research positions are full-time. This is quite unusual in Japan, where lifetime employment is supposedly the norm. Among the 995 past and present researchers, 167 are from 30 countries outside of Japan and 421 are from industry (Kusunoki 1998). The researchers from industry are paid by JST and are asked to take a leave of absence from their companies, so that their research is independent of the companies' interests.[7] Most of them return to their companies on completion of the research projects.

Directors can concurrently maintain positions elsewhere and, among the directors of 20 current projects, 12 are university professors (including one Japanese professor at Stanford), three are researchers at national laboratories, and five are from industry. The researchers' backgrounds are also diverse. Many of them are, as shown above, company researchers on leave. But they participate as individuals and not as representatives of their companies; two or more people from the same company rarely join a single project. Others may be post-doctoral students, researchers at various laboratories or universities, or foreigners. Therefore, although the aim of ERATO is to promote basic and interdisciplinary research and not UI collaboration itself, the diversity of researchers in a project ensures that such collaboration actually takes place in various ways.

The results of ERATO projects are impressive. By August 1996, they had produced 1,107 patents (925 in Japan and 182 overseas) and 5,672 papers and presentations (3,335 in Japan and 2,337 overseas) (Kusunoki 1998). Through a questionnaire study, Kusunoki (1993) found that their researchers made more publications and presentations, particularly abroad, than those in a comparable MITI laboratory, and argued that "the dynamic network organization could enable researchers to make full use of external information, as well as to communicate more frequently with outside professionals. In this sense, the findings in this analysis emphasize the effectiveness and the possibility of dynamic network organization like ERATO more than expected" (Kusunoki 1993: 56).

The Case of the Fifth-Generation Computer Project[8]

Even in a more conventional project, like those sponsored by MITI, UI collaboration has taken place in various ways, albeit less formally. Consider the Fifth-Generation Computer Project (FGCP). This project was started in 1982 with funds entirely provided by MITI and continued until 1994. When MITI started planning the project, MITI and its Electrotechnical Laboratory (ETL) organized a committee and three subcommittees to discuss the feasibility of FGCP and the direction it should take. T. Motooka, a professor of the University of Tokyo, chaired the committee; the three subcommittees were headed by a professor, an ETL researcher, and a company director. Therefore, from the beginning, collaboration among universities, ETL, and companies was very much in the scopc of the project.

The main part of the research was conducted at the Institute for New Generation Computer Technology (ICOT), a new institute founded specifically for the project, consisting of researchers mostly seconded from industry and ETL. In the beginning ICOT attempted to persuade professors and other university researchers to join but found this extremely difficult. In Japanese universities, basically every faculty member has tenure and his mobility is more limited than in the United States: Leaving a university to join a temporary organization like ICOT creates uncertainty as to future job availability. Taking a temporary leave from a national university is also difficult because the university hiring policy is constrained by the number of faculty posts set by the Ministry of Education, which does not allow the university to hire temporary substitutes in place of faculty on leave. Consequently, professors cannot buy teaching time with the money provided by outside sources, such as FGCP.[9]

Still, universities played important roles in the project on at least two accounts. First, the director of ICOT and other members sometimes learned of talented young researchers from the professors they knew, for instance those who had just completed their doctorates with these professors in topics related to FGCP and were interested in pursuing further research at ICOT. ICOT recommended these people to be hired in the project's member companies on the mutual understanding that, after, say, six months, the company(ies) would second them to ICOT.

Secondly, ICOT organized a number of working groups, each with a separate research theme. In the beginning, there were only six working groups with 72 outside researchers participating. However, as the interest in FGCP rose and the research at ICOT progressed, more working groups were set up. At the peak year of 1991, there were 16 working groups with 259 outsiders involved, half of them university faculty members. ICOT, being unable to recruit them on a full-time basis, tried to learn from them through these working groups, which typically met once a month. To university members, participation in the working groups provided an opportunity to learn of the progress at ICOT and of research done by other working group members. Furthermore, they could occasionally utilize ICOT's research facility. As Fransman (1990: 208) observed, "through this arrangement university researchers get access to funding and equipment, link into the ICOT network, which as noted serves as a clearing house for information in the area of applied artificial intelligence, and take part in larger mission-oriented cooperative research. On the other hand, ICOT research benefits from university expertise in more fundamental and specialized areas. In addition, many ICOT-related ideas and applications receive further research in university laboratories where professors and postgraduate students are involved in the experimentation and testing processes."

The link between ICOT and universities was further enhanced when 24 of the former ICOT researchers, seconded either from companies or ETL, quit their home companies or ETL to join university faculties. For instance, K. Fuchi, a former member of ETL, who played a central role in the start of the project and served as ICOT's research director, joined the University of Tokyo (later moving to Keio University). These people now teach ICOT-related subjects, such as Prolog, parallel processing, and artificial intelligence, at their universities, arousing interest on these subjects among fellow professors and students, and supporting the students doing research in the field.

Conclusion

In any country and at any time, incentives for UI collaborations abound; however, these collaborations must be conducted within the surrounding social, economic, and institutional conditions. Thus, the way collaborations were made in the early period was much influenced by the fact that knowledgeable persons were few at the time, the industry had not yet accumulated capabilities, and institutional rigidity had not yet developed. All these conditions changed subsequently, particularly after the war, when institutional or regulatory rigidity became common, forcing industry to collaborate with universities only informally.

Recent changes have been also significant. Now that Japan is among those in the forefront of the global technological race, the need for more basic and original research is stressed. Yet many university laboratories suffer from shortage of funds, obsolete equipment, and the lack of support personnel while, by contrast, major companies have grown to invest a huge amount of money for R&D. New fields have appeared, such as biotechnology and artificial intelligence. To accommodate these changes, UI collaborations have to be promoted; several policy initiatives have already been taken. The Ministry of Education has been increasing research funds (Grants-in-aid) that are allocated on a competitive basis and is relaxing the limitations on outside consultation for faculty. MITI has shifted its research focus from applied or development research, such as the VLSI Project, to more basic research, such as the Fifth-Generation Computer Project. STA has been successfully experimenting with the ERATO program and has since added a few more related programs. And other government branches have also started a number of programs.

Similarly, the way U.S. universities and industries collaborate has been influenced by the economic and institutional conditions of that country. The Japan-U.S. difference in capital and labor markets is consistent with the fact that more professors start venture businesses in the United States than in Japan. The dominance of private universities in the United States is consistent with their active patenting. And the huge amount of research funds provided by the U.S. government for defense purposes has exerted a considerable influence on the way research is conducted there.

As we have emphasized elsewhere (Odagiri and Goto 1996), technology cannot be free from the environment in which it is developed and the environment in which it is to be adopted. Furthermore, the process of technological progress and institutional change is bound to be path-dependent, which will ensure that, while global competition tends to force the innovation systems of countries to converge, significant differences nevertheless remain. In addition, innovation will be made not only in technology but also in institutional arrangements and policy devices. ERATO appears to be one such innovation and, as Kusunoki (1998) argues, an effort to achieve organizational flexibility in a Japanese setting has produced an organizational form and research environment in ERATO that will not be realized in the United States, where the allocation of scientists is predominantly made according to market principles.

Indeed, international comparison of innovation systems is not a simple matter at all (also see Nelson 1993), and learning from others' experience and yet adapting it to one's own environment are crucially important to design better institutional settings for university-industry collaboration.

Notes

1. Toshiba was born when Tokyo Denki, the descendant of Ichioka's enterprise, was merged with Shibaura Seisakusho founded by a master of mechanical engineering from the pre-Restoration, Edo era. In this sense, one may say that Toshiba has roots in both pre-modernization and post-modernization technologies.

2. Odagiri and Kato (1998) studied university-industry joint research in the field of biotechnology, but found that hardly any of the projects produced joint patent applications, suggesting again that patent applications are neither beneficial nor easy for university researchers. They found, however, that a firm conducting joint research with a university(ies) has a higher marginal propensity to patent out of its biotechnology R&D expense, suggesting that UI collaboration fosters the firm's patent applications through its spillover to other research projects within the firm, in addition to the direct contribution of the joint research project itself. Unfortunately, the sample is small and the reliability of the estimation result is limited. Zucker and Darby (1998) also studied Japanese biotechnology firms and found that collaborations with university star scientists increase the average firm's U.S. biotech patents.

3. This was a consideration in the Fifth-Generation Computer Project's general policy of not making patent applications unless patenting was needed to prevent others from making the applications and monopolizing the use of invented knowledge. Still, it made 382 patent applications and 163 registrations during its 12-year lifetime.

4. Details on the various programs listed in the text are available on the Internet; for instance, for MITI's programs, see <http://www.miti.go.jp/topic-j/e-menu2j.html#indust> (in Japanese) and for STA's programs, see <http://

www.jst.go.jp/EN/>. Eto (1998) has made a questionnaire study of various research projects sponsored by these programs and found, among other things, that the programs under STA's ERATO are targeted at the basic end of research, as are the programs supported by MHW. Research projects under MITI and MAFF are, by contrast, more biased towards developments in technology or products.

5. The description here is based on ERATO's Homepage (<http://www2.jst.go.jp/erato/index.html>) and Kusunoki (1993, 1998).

6. The quotation is from the statement in ERATO's Homepage: see <http://www2.jst.go.jp/erato/index.html>.

7. This independence was not sought in the joint research projects sponsored by MITI, including the Fifth Generation Computer Project to be discussed later. In fact, these projects were made in close cooperation with the companies.

8. This discussion is based on Nakamura and Shibuya (1995) and Odagiri, Nakamura, and Shibuya (1997).

9. This should explain why university professors joining ERATO projects as directors are allowed to do so on a part-time basis. Since university members wishing to join these projects as non-directors have to resign from their universities, they tend to be the young who expect to have good job opportunities once they publish good papers in the course of their research at the projects.

References

Eto, Manabu. 1998. "New type of cooperative research in Japan." Presented at the conference on the "Triple Helix of University-Industry-Government Relations," State University of New York at Purchase.

Fransman, Martin. 1990. *The Market and Beyond: Cooperation and Competition in Information Technology Development in the Japanese System.* Cambridge: Cambridge University Press.

Kusunoki, Takeru. 1993. "Organizational innovation in the Japanese basic research: Challenges and problems." *Hitotsubashi Journal of Commerce & Management* 28:37–59.

Kusunoki, Takeru. 1998. "Dynamic network of basic research: Organizational innovation of ERATO" (in Japanese). In *Case Book, Nihon Kigyo no Koudou, Vol. 3: Innovation to Gijutsu Chikuseki,* eds. Hiroyuki Itami, Tadao Kagono, Matao Miyamoto, and Seiichiro Yonekura. Tokyo: Yuhikaku, pp. 253–284.

Monbusho. 1997. *Japanese Government Politics in Education, Science and Culture 1997* (in Japanese). Tokyo: Printing Office of the Ministry of Finance.

Nakamura, Yoshiaki and Minoru Shibuya. 1995. "Japan's technology policy: A case study of the research and development of the Fifth Generation Computer Systems." *Studies in International Trade and Industry* 18. Tokyo: Research Institute of International Trade and Industry, Ministry of International Trade and Industry.

Nelson, Richard (ed.). 1993. *National Innovation Systems.* Oxford: Oxford University Press.

Odagiri, Hiroyuki. 1998. "Education as a source of network, signal, or nepotism: Managers and engineers during Japan's industrial development," in *Networks, Markets, and the Pacific Rim,* ed. W. Mark Fruin. Oxford: Oxford University Press.

Odagiri, Hiroyuki and Akira Goto. 1996. *Technology and Industrial Development in Japan: Building Capabilities by Learning, Innovation, and Public Policy.* Oxford University Press.

Odagiri, Hiroyuki and Yuko Kato. 1998. "University-industry research collaboration in the biotechnology industry: An empirical analysis" (in Japanese). *Business Review* 45:62–80.

Odagiri, Hiroyuki, Yoshiaki Nakamura, and Minoru Shibuya. 1997. "Research consortia as a vehicle for basic research: The case of a Fifth Generation Computer Project in Japan." *Research Policy* 26: 191–207.

Zucker, Lynne G. and Michael R Darby. 1998. "Capturing technological opportunity via Japan's star scientists: Evidence from Japanese firms' biotech patents and products." Working Paper No. 6360, National Bureau of Economic Research.

IV

Incentives and Barriers: Technology Transfer Dynamics

University-industry business collaboration is not a natural or an easy relationship in any country, given the radical differences in culture, structure, and goals of the two kinds of institutions. Universities and science-based industry do, however, share a core interest: intellectual property. Incentives to collaboration and barriers to dissemination come together in the copyright and patent systems through which intellectual assets are protected and sold. Part IV explores in some detail the processes through which balances in interest are sought by the parties, presenting a clear picture of the mechanisms by which intellectual property is transferred in each nation. With the background of Part III in mind, the nature of each system becomes clear.

11

The Effects of the Bayh-Dole Act on U.S. University Research and Technology Transfer

David C. Mowery, Richard R. Nelson, Bhaven N. Sampat, and Arvids A. Ziedonis

Introduction

The U.S. research university and the organized pursuit of R&D in industry both originated roughly 125 years ago and have grown in parallel throughout the 20th century (Mowery and Rosenberg 1998). Although this linkage has a long history, recent developments, especially the growth in university patenting and licensing of technologies to private firms, have attracted considerable attention. In particular, the expanded licensing activities of U.S. universities have occasioned both expressions of enthusiasm by some for the enhanced contributions of university research to U.S. economic growth, and expressions of concern by others over the effects of such activities on the culture and norms of academic research.

The recent increases in university patenting and licensing are widely assumed to be the direct consequences of a particular federal policy initiative, known as the Bayh-Dole Act of 1980. Although the Act's importance is widely cited, its effects on U.S. research universities and on the U.S. innovation system have been the focus of little empirical analysis (Henderson, Jaffe, and Trajtenberg 1998 is an important exception). In this paper, we undertake such an analysis, focusing on three academic institutions that have been the leading recipients of licensing and royalty income for much of the 1990s: Columbia University, the University of California, and Stanford University,[1] which are among the most important practitioners of the new approach to university technology transfer. Two of the three, Stanford and the University of California, were active in technology licensing well before the passage of the Bayh-Dole Act, as were such universities as the University of Wisconsin and MIT, while Columbia and many other research universities (for example, Harvard and Yale) were not. A combined analysis of data from Columbia, Stanford, and the UC system thus allows us to consider the effects of these new Federal policies on both universities such as Columbia, which became large-scale patenters and licensors only after 1980, and those such as UC and Stanford, active in patenting and licensing well before 1980.

We use previously unexploited data from each of these universities to compare their patenting and licensing activities and address the following broad questions:

1. How has the level and the mix among fields of university patenting and licensing activity been affected by Bayh-Dole and other factors?

2. Has Bayh-Dole affected the content of academic research at these institutions?

Our evidence suggests that the direct effects of Bayh-Dole on the content of academic research have been modest. The most significant change in the content of research at our three universities, one that is associated with increased patenting and licensing, has been the rise of biomedical research and inventive activity, but Bayh-Dole had nothing to do with this development. Indeed, the rise in biomedical research and the growth of its associated inventions predate the passage of Bayh-Dole in both of the universities (UC and Stanford) for which we have reliable pre-1980 data, and more fragmentary evidence indicates a similar trend at Columbia.

Although our evidence suggests little if any change in the content of academic research, the effects of Bayh-Dole on the marketing efforts of U.S. universities appear to have been considerable. The passage of this law hastened or caused the entry by many universities (such as Columbia) into patenting and licensing activities that they formerly avoided as a matter of policy. Our evidence also suggests that even at universities long active in patenting and licensing of faculty inventions, administrators intensified their efforts to gain access to and/or market these inventions. At the end of the first decade of the Bayh-Dole Act, these three universities display remarkable similarities in their patenting and licensing activities.

The growing importance of biomedical research, much of which relied on federal support that expanded significantly during the 1970s, was at least as important as Bayh-Dole in explaining increased university patenting and licensing after 1980. But other factors also encouraged the growth of university patenting in this and other research fields. Judicial decisions declared that "engineered molecules" were patentable, the U.S. Congress passed a series of laws strengthening intellectual property protection, and the U.S. government expanded its efforts to gain stronger international protection for intellectual property. Nevertheless, Bayh-Dole was an important catalyst, and its provisions are interesting in their own right. Among other things, the Act represents an application of the "linear model" to science and technology policy, assuming that if basic research results can be purchased by would-be developers, thereby establishing a clear "prospect" for the commercial development of these results, commercial innovation will be accelerated.

Immediately below, we discuss the background to the Bayh-Dole Act, by way of underscoring the point that university-industry linkages, university patenting, and university licensing of these patents are not new features of the U.S. innovation system. We then discuss our data for these three universities, present the comparative analysis, and consider the implications of our findings for the academic research enterprise and the U.S. innovation system.

Historical Background

The historic involvement of publicly funded universities in the United States with agricultural research, much of which was applied in character, and the involvement of these universities with the agricultural users of this research, are well-known aspects of U.S. economic history. But throughout this century, the decentralized structure of U.S. higher education and the dependence of public and private universities on local sources of funding also meant that in a broad array of nonagricultural fields, ranging from engineering to physics and chemistry, collaborative research relationships between university faculty and industry were common (Rosenberg and Nelson 1994). Indeed, the Research Corporation, which served for many years as a leading "broker" and licensor of university inventions for many U.S. universities, was founded in 1912 by Frederick Cottrell, a U.C. Berkeley professor, to commercialize his electrostatic antipollution innovations. (The Corporation's licensing activities now are managed by an independent organization, Research Corporation Technologies, founded in 1987).

Thus it is a fallacy to think of U.S. university research as traditionally "basic" and conducted with no attention to practical objectives. Many important advances in applications have emerged from academic research in the United States, and much industry-university collaboration historically has focused on the engineering and applied sciences. University researchers have made important contributions to innovations in scientific instrumentation, medical devices, and computer software, reflecting the fact that university researchers are demanding "users" of these technologies, and their research activities frequently create new advances in applications in these and other areas (Rosenberg 1992 discusses the contributions of academic researchers to innovation in scientific instruments).

Through much of the 1900–1940 period, U.S. universities, especially public universities, pursued extensive research collaboration with industry. Indeed, the academic discipline of chemical engineering was largely developed through such collaboration between U.S. petroleum and chemicals firms and MIT and the University of Illinois (Rosenberg 1998). The Second World War transformed the role of U.S. universities as research performers, as well as the sources of their research funding. Universities' share of total U.S. R&D performance grew from 7.4% in 1960 to nearly 16% in 1995, and universities accounted for more than 61% of the basic research performed within the United States in 1995 (National Science Foundation 1996).[2] The share of industry funding declined within the greatly expanded research budgets of postwar U.S. research universities through the 1950s and 1960s. Much university research nevertheless retained an applied character, reflecting the importance of research support from such federal "mission agencies" as the Defense Department.

Beginning in the 1970s, the share of industry funding within academic research began to grow again. In the early 1970s, federal funds accounted for more than 65% of university-performed research, while industrial support accounted for 2.3%; by 1995, federal funds accounted for 60% of total university research, and industry's contribution had tripled to 7%. Most of the increase in the industry funding share occurred during the 1980s, remaining roughly constant after 1990.

In view of the applied character of a good deal of their research, it is not surprising that a number of U.S. research universities were active in patenting and licensing faculty inventions long before 1980. Beginning in 1926, the University of California required all employees to report patentable inventions to the university administration. Other universities, such as MIT and the University of Wisconsin, developed administrative units to help patent and license inventions resulting from research. During the pre-1940 period, most of this patenting and licensing activity was conducted directly with industrial sponsors of academic research and through organizations such as the Research Corporation.

Expanded federal research funding during the postwar period rekindled the debate over the disposition of the results of academic research. (See Eisenberg [1996] for a review of the history of these policy debates.) During the 1960s, both the Defense Department and the Department of Health, Education and Welfare (now Human and Health Services, the agency housing the National Institutes of Health), which were among the leading sources of federal academic research funding, allowed academic institutions to patent and license the results of their research under the terms of Institutional Patent Agreements (IPAs) negotiated by individual universities with each federal funding agency. IPAs eliminated the need for case-by-case reviews of the disposition of individual academic inventions, and facilitated licensing of such inventions on an exclusive or nonexclusive basis, but tensions between some major IPA participants, such as the University of California, and federal sponsors remained.[3] These debates intensified in the late 1970s, when HEW in particular began to question the use by some U.S. universities of exclusive licenses under IPAs, and proposed limiting the ability of some universities to adopt such policies.

The Bayh-Dole Patent and Trademark Amendments Act of 1980 provided blanket permission for performers of federally funded research to file for patents on the results of such research and to grant licenses for these patents, including exclusive licenses, to other parties. The Act facilitated university patenting and licensing in several ways. First, it replaced the web of IPAs that had been negotiated between individual universities and federal agencies with a uniform policy. Second, the Act's provisions represented a strong Congressional expression of support for the negotiation of exclusive licenses between universities and industrial firms for the results of federally funded research. Finally, it constituted a Congressional endorsement of the argument that failure to establish patent protection over the results of federally funded university research would limit the commercial exploitation of these results. The Bayh-Dole Act responded to a belief by policymakers (based on little evidence) that stronger protection for the results of publicly funded R&D would accelerate their commercialization and the realization of these economic benefits by U.S. taxpayers.[4] These new policies of support for patenting of the results of public R&D programs also appeared to promise increased economic returns for little additional investment of public funds, an attractive feature for policymakers dealing with severe fiscal constraints.

The passage of the Bayh-Dole Act was one part of a broader shift in U.S. policy toward stronger intellectual property rights.[5] Among the most important of these policy initiatives was the establishment of the Court of Appeals for the Federal Circuit (CAFC) in 1982. Established as the court of appeal for patent cases throughout the federal judiciary, the CAFC soon emerged as a strong champion of patentholder rights. Even before the establishment of the CAFC, however, an important U.S. Supreme Court decision in 1980, *Diamond v. Chakrabarty*, upheld the validity of a broad patent in the new industry of biotechnology, opening the door to patenting the organisms, molecules, and research techniques emerging from biotechnology. The origins and effects of Bayh-Dole must be viewed in the context of this larger shift in U.S. policy toward intellectual property rights, and the effects of Bayh-Dole per se are confounded with those of other policy initiatives of the 1980s.

Although this point was ignored in the political debate over Bayh-Dole, the argument that stronger protection of intellectual property accelerates its commercialization conflicted with an important strand of the economic analysis of the social returns to scientific research, which stressed that scientific knowledge was "not rivalrous in use" (Nelson 1959; Arrow 1962). When a good is nonrivalrous in use, use of that good by additional parties and or in additional applications imposes no real economic costs. The denial of access to a scientific discovery by any party that can make good use of it thus imposes costs on that party and on the economy as a whole. Where private investors are the primary source of financial support for scientific research, as is the case with industrial R&D, granting a patent on the results of such work may be necessary in order to induce private R&D investment. But the economic theory of scientific research that was ignored in Bayh-Dole argues that patenting the results of publicly funded research is unnecessary to induce the research investment and that restrictions on use associated with patents reduce the social returns to this public investment.

In contrast, the advocates of Bayh-Dole argued that the findings of publicly financed university research require considerable additional R&D and other investments before they can be commercialized; such investments are more likely if a firm is granted an exclusive license to do that work. But it also can be argued that if the findings of publicly funded university research are placed in the public domain, or are inexpensively licensed to anyone who wants to use them, competition alone may stimulate their widespread application. These issues are complex, and cannot be resolved with the data presented in this paper. But our data provide a useful starting point for such an analysis.

The Effects of Bayh-Dole

The Bayh-Dole Act is contemporaneous with a sharp increase in U.S. university patenting and licensing activity. The data in Table 1 reveal a large increase in university patenting after the passage of the

Table 1 Number of U.S. Patents Issued to 100 U.S. Academic Institutions with the Highest 1993 R&D Funding, 1974–1994.

Year	Number of U.S. patents
1974	177
1979	196
1984	408
1989	1004
1994	1486

Source: National Science Board (1996).

Bayh-Dole Act in 1980: The number of patents issued to the 100 leading U.S. research universities (measured in terms of their 1993 R&D funding) more than doubled between 1979 and 1984, and more than doubled again between 1984 and 1989. Trajtenberg, Henderson, and Jaffe (1994) noted that the share of all U.S. patents accounted for by universities grew from less than 1% in 1975 to almost 2.5% in 1990. Moreover, the ratio of patents to R&D spending within universities almost doubled during 1975–1990 (from 57 patents per $1 billion in constant-dollar R&D spending in 1975 to 96 in 1990), while the same indicator for all U.S. patenting displayed a sharp decline (decreasing from 780 in 1975 to 429 in 1990). In other words, universities increased their patenting per R&D dollar during a period in which overall patenting per R&D dollar was declining.

In tandem with increased patenting, U.S. universities expanded their efforts to license these patents. The Association of University Technology Managers (AUTM) reported that the number of universities with technology licensing and transfer offices increased from 25 in 1980 to 200 in 1990, and licensing revenues of the AUTM universities increased from $183 million to $318 million in the three years from 1991 to 1994 alone (Cohen et al. 1998).

Columbia University

Although Columbia University had no technology licensing office prior to Bayh-Dole, this observation does not imply that Columbia administrators and faculty were unconcerned with patenting and licensing issues during this period. The essence of Columbia University's pre-Bayh-Dole patent policy dates back at least to its 1944 statement on Research and Patent Policy and Procedures, which states that "[w]hile it is the policy of the Faculty of Medicine to discourage the patenting of any medical discovery or invention, and to forbid the patenting or exploitation of such discoveries by members of the staff," staff members in other divisions of the University were "free to patent any device or discovery resulting from their personal researches." In many cases, inventors wishing to patent their inventions did so with the help of the Research Corporation.

During the pre-1980 period, Columbia generally did not assert claims to faculty patents. There were some exceptions to this policy, however, such as cases where a research grant or contract required that the university be party to any patent arrangements. Where patents did result from Columbia research (for example, in wartime research projects funded by the federal government), they were assigned either to the individual inventor or to the sponsoring agency, rarely to the University; and with a few exceptions, Columbia earned no royalty income from patents resulting from faculty research. Although Columbia considered changing its patent policy and organization to "get into the patent business" several times during this period, the university's laissez-faire patent policy remained essentially unchanged from 1944 until 1981. The 1944 policy was revised numerous times over the next several decades, but its spirit remained substantially intact until 1975, when the stipulation against patenting medical inventions and discoveries was dropped.

Columbia's patent policy was significantly altered in response to the passage of Bayh-Dole. The new policy, which took effect on July 1, 1981 (the effective date of Bayh-Dole), reserved patent rights for Columbia and shared royalties with the inventor and department. In 1984, a new policy statement clarified and codified the rules: It mandated that faculty members disclose to the University any potentially patentable inventions developed with university resources. In 1989, Columbia's policy on reserving rights to the University for faculty inventions created with University resources was extended to cover software. Inventions were to be disclosed to

Columbia's technology transfer office, the Office of Science and Technology Development, which was founded in 1982.

Figure 1 shows the rapid "ramping up" of Columbia invention reports during the 1980s. Since most academic research programs change only gradually, the initial surge of reports almost certainly reflects increased identification by university administrators (based on a more intensive canvassing of the faculty) of potentially valuable inventions derived from research projects already under way. Almost 75% of the 877 invention reports disclosed between 1981 and 1995 originated in the medical school, and biotechnology figured prominently in these biomedical inventions. Biotechnology inventions account for 60% of these biomedical inventions, 45% of the biomedical inventions that result in patents, and nearly 70% of the biomedical inventions that are licensed. Although our Columbia data on inventions and patenting do not extend back into the pre-1980 period, we believe that at this university, like Stanford and UC, the growth of post-1980 inventive activity in biomedical technologies was not directly affected by Bayh-Dole. The surge of inventive activity in the biomedical and biotechnology fields had been building up for some time, as a result (among other things) of the long-term expansion of federal biomedical research funding from the National Institutes of Health and other sources. Indeed, Columbia filed for a patent on what became its most profitable single invention, a biotechnology research tool, before the passage of Bayh-Dole.

Outside of the medical school, Columbia's "inventing" is concentrated in a few departments and research institutes that, like the medical school, rely heavily on federal R&D funding. Over 60% of the non-biomedical invention reports, and over 65% of the patenting associated with non-biomedical invention reports during the 1981–1995 period, emanate from two departments, electrical engineering and computer science, and two research centers, the Center for Telecommunications Research and the Lamont-Doherty Earth Observatory. Much of this research is associated with electronics and software. Software inventions also account for a significant share of Columbia faculty disclosures, increasing to more than 10% of disclosures by the 1990s; software inventions also account for a large share of Columbia licensing agreements. As the data in

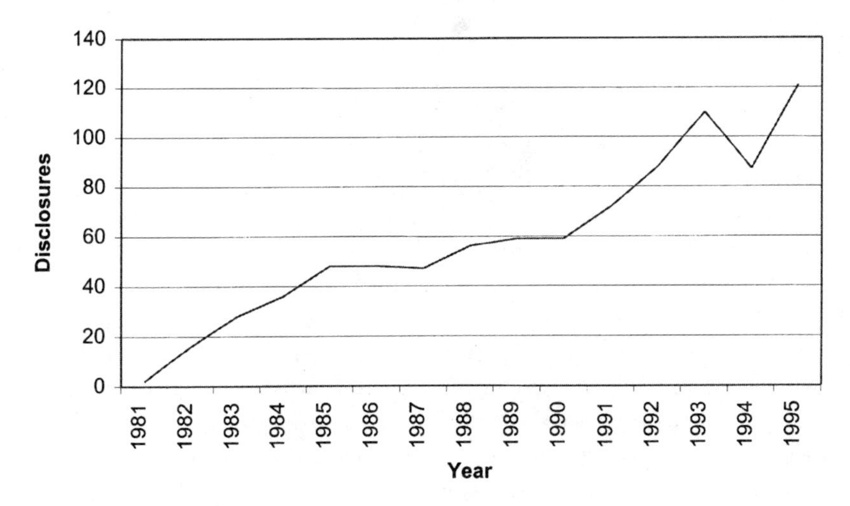

Figure 1 Columbia University Disclosures, 1981–1995.

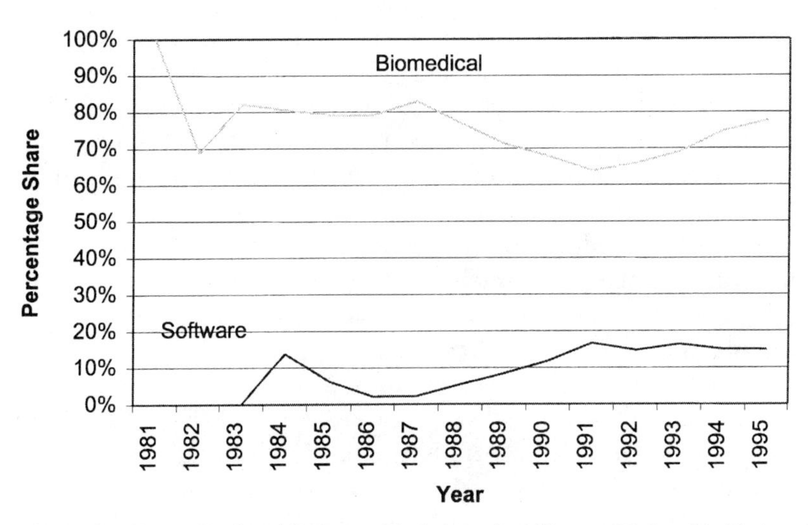

Figure 2 Biomedical and Software Technologies' Share of Columbia University Disclosures, 1981–1995.

Figure 3 suggest, increased invention reports generate increased patenting activity with a slight lag.

As Figure 4 and the comparative data in Table 2 indicate, Columbia University's technology licensing activities have been associated with a surge in gross licensing revenues, which grew almost 60-fold (measured in 1992 dollars) in the decade between 1985 and 1995. This income was highly concentrated among a small number of inventions: The "top 5" accounted for more than 90% of gross revenues throughout this period. Turning to the number of licenses (a measure that heavily weights inventions licensed on a nonexclusive basis), Figure 5 illustrates the growth in importance (in terms of numbers) of licenses for software inventions, which account for well over 50% of Columbia licensing agreements after 1988; the majority of these licenses, however (420 of a total of 648), are associated with one software invention.[6] Nevertheless, biomedical inventions accounted for a large and growing share of the revenues of the "top 5" inventions throughout the period covered by these data (Table 2). In 1985, 1990, and 1995, respectively, 5, 4, and 3 of the top 5 money-earning disclosures were biomedical, which throughout the 1985–1995 period accounted for more than 80% of the income earned by the "top 5" inventions.

In assessing the effects of Bayh-Dole on Columbia University, we lack a compelling counterfactual: What would have happened in the absence of this federal law, given the other trends operating in university finances and research after 1980? We believe that industrial interest in Columbia's research results, especially in the biomedical area, combined with the prospect of large licensing revenues, would have led Columbia to develop some administrative machinery for patenting and licensing in the absence of this federal law. As we noted earlier, Columbia filed for a patent on its most lucrative single biotechnology invention before the effective date of Bayh-Dole. The change in federal policy embodied in Bayh-Dole nevertheless hastened a shift toward more intensive patenting and licensing of Columbia faculty inventions.

There is less evidence of any effects of Bayh-Dole on the underlying research activities of Columbia faculty and staff, although our data on this point are incomplete. It appears that a good deal of

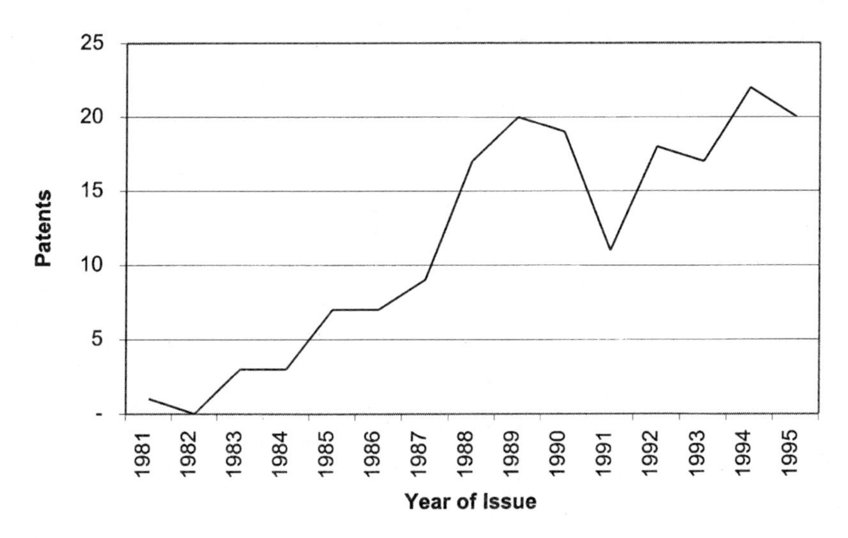

Figure 3 Columbia University Patents, 1981–1995.

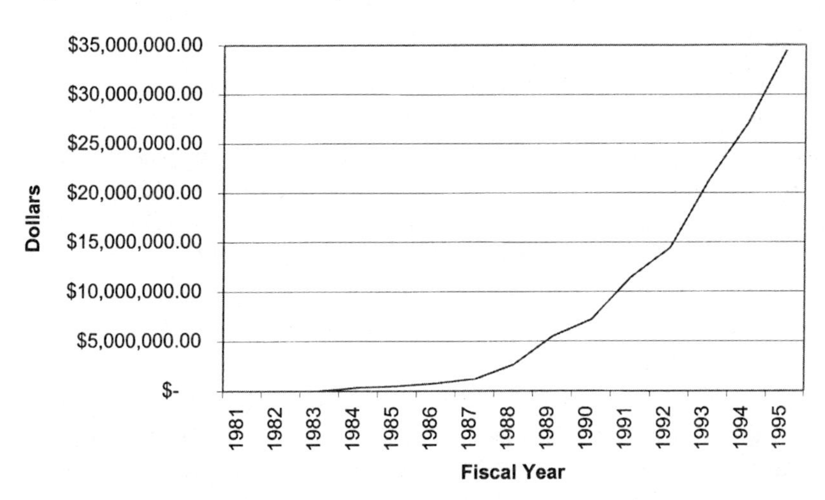

Figure 4 Columbia University Licensing Revenues, 1981–1995.

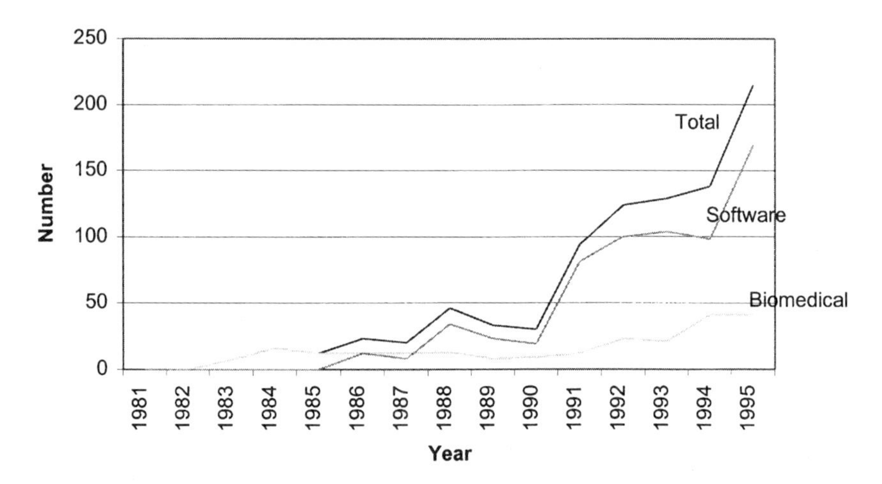

Figure 5 Columbia University License Agreements, 1981–1995.

Columbia research that in the pre-Bayh-Dole era would have been published now results in invention reports, patent applications, patents, and sometimes licenses, in addition to published papers. The consequences of the new regime are reflected in the sharp increase in the ratio of Columbia patents to research spending, from 6 patents on a research budget of $118 million in 1984 to 21 on a research budget of $229 million in 1994.

We have argued that the growing importance of biomedical research at Columbia University probably would have produced some shift in University policy to favor greater patenting and licensing of faculty inventions without the Bayh-Dole Act. Another piece of evidence relevant to an assessment of the effects of Bayh-Dole at Columbia concerns the significant role of software in Columbia's post-1980 licensing activities. Virtually all of the software inventions licensed by Columbia are protected by copyright, a form of intellectual property never affected by Bayh-Dole, rather than by patents, the focus of this federal law. Software licensing is a new form of technology marketing that has resulted from the University's creation of a technology "marketing" operation and (like biomedical research) from the growth of a new academic research area, rather than from the specific policy shifts embodied in Bayh-Dole. Moreover, although the arguments for Bayh-Dole

Table 2 Selected Data on University of California, Stanford University, and Columbia University Licensing Income, FY 1970–1995

	FY1970	FY1975	FY1980	FY1985	FY1990	FY1995
University of California						
Gross income (1992 $: 000s)	1140.4	1470.7	2113.9	3914.3	13240.4	58556.0
Gross income from top 5 earners (1992 $: 000s)	899.9	1070.8	1083.0	1855.0	7229.8	38665.6
share of gross income from top 5 earners (%)	79	73	51	47	55	0.66
share of income of top 5 earners associated with biomedical inventions (%)	34	19	54	40	91	1
share of income of top 5 earners associated with agricultural inventions (%)	57	70	46	60	09	0
Stanford						
Gross income (1992 $: 000s)	180.4	842.6[a]	1084.4	4890.9	14757.5	35833.1
Gross income from top 5 earners (1992 $: 000s)		579.3[a]	937.7	3360.9	11202.7	30285.4
share of gross income from top 5 earners (%)		69[a]	86	69	76	85
share of income of top 5 earners associated with biomedical inventions (%)		87[a]	40	64	84	97
Columbia						
Gross income (1992 $: 000s)				542.0	6903.5	31790.3
Gross income from top 5 earners (1992 $: 000s)				535.6	6366.7	29935.8
share of gross income from top 5 earners (%)				0.99	0.92	0.94
share of income of top 5 earners associated with biomedical inventions (%)				0.81	0.87	0.91

a. FY1976.

stressed the importance of exclusive licensing in effective technology transfer and commercialization, many of Columbia's licensed inventions, including its biggest single source of revenues, have been licensed on a nonexclusive basis.

The University of California

Unlike Columbia, the University of California established policies requiring faculty disclosure of potentially commercially useful research results long before Bayh-Dole. Mechanisms for supporting the commercial exploitation of any resulting patents were put in place in 1943, and assignment by faculty of their inventions to the university was determined on a case-by-case basis. Patenting and any licensing were the responsibility of the UC General Counsel's office, which oversaw the creation and gradual growth of the UC Patent Office. The UC Board of Regents established the "University Patent Fund" in 1952 to invest the earnings from University-owned inventions in the UC system's General Endowment Pool: Earnings from the Fund also supported the expenses of UC patenting activities and faculty research.[7] In 1963, the UC Board of Regents adopted a policy stating that all "Members of the faculties and employees shall make appropriate reports of any inventions and licenses they have conceived or developed to the Board of Patents,"[8] that latter being a committee of UC faculty and administrators charged with oversight of the Patent Office.

In 1976, responsibility for patent policy was transferred from the General Counsel to the Office of the President of the University, and the Patent Office was reorganized into the Patent, Trademark, and Copyright Office (PTCO). Only in 1980, however, was the PTCO staffed with experts in patent law and licensing, as part of a broader expansion in UC patenting and licensing activities. The Board of Patents was abolished in 1985, and new policies allowing for sharing by campuses in patent licensing revenues were adopted by the Office of the President and the campus Chancellors in 1986. Staff employment in the PTCO grew from 4 in 1977–1978 to 43 in 1989–1990. In 1991 the PTCO was renamed the Office of Technology Transfer (OTT); but even before this date, in 1990, UC Berkeley and UCLA had established independent patenting and licensing offices, relying on the system-wide Office of Technology Transfer selectively for expertise in patent and licensing regulations. By 1997, four UC campuses (in addition to Berkeley and UCLA, UC San Diego and UC San Francisco) had established independent licensing offices.[9]

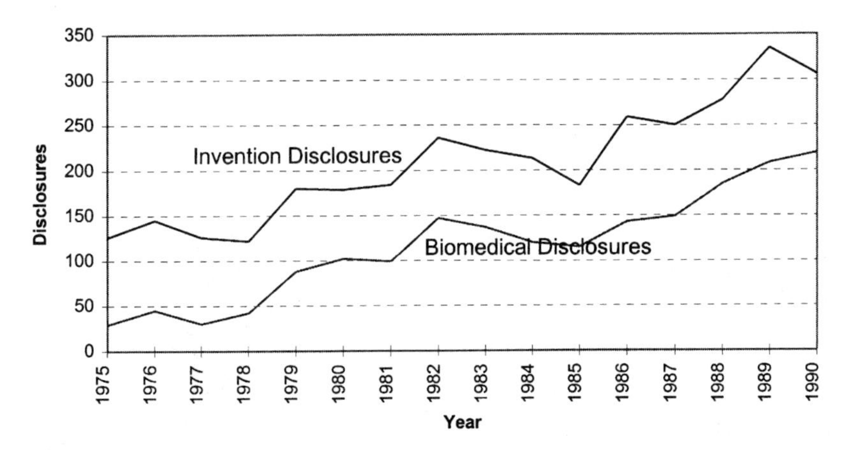

Figure 6 University of California Invention Disclosures, 1975–1990.

Since the University of California was active in patenting and licensing well before the passage of the Bayh-Dole Act, a comparison of the 1975–1979 period (prior to Bayh-Dole) and 1984–1988, following the passage of the bill, provide a "before and after" test of the Act's effects. The average annual number of invention disclosures during 1984–1988, following passage of the Bayh-Dole Act, is almost 237, well above their average level (140 annual disclosures) for the 1975–1979 period. The period following the Bayh–Dole Act thus is associated with a higher average level of annual invention disclosures (confirmed in Figure 6); but the timing of the increase in annual disclosures suggests that more than the Bayh-Dole Act affected this shift.

Figure 7 displays a 3-year moving average for annual invention disclosures by UC research personnel for the 1975–1990 period. The increase in the average annual number of invention disclosures in Figure 7 predates the passage of the Bayh-Dole Act; indeed, the largest single year-to-year percentage increase in disclosures during the entire 1974–1988 period occurred in 1978–1979, before the Act's passage. This increase in disclosures may reflect the important advances in biotechnology that occurred at UC San Francisco during the 1970s, or other changes in the structure and activities of the UC patent licensing office that were unrelated to Bayh-Dole; in view of the longstanding requirement in the UC system that faculty disclose all potentially patentable inventions, we

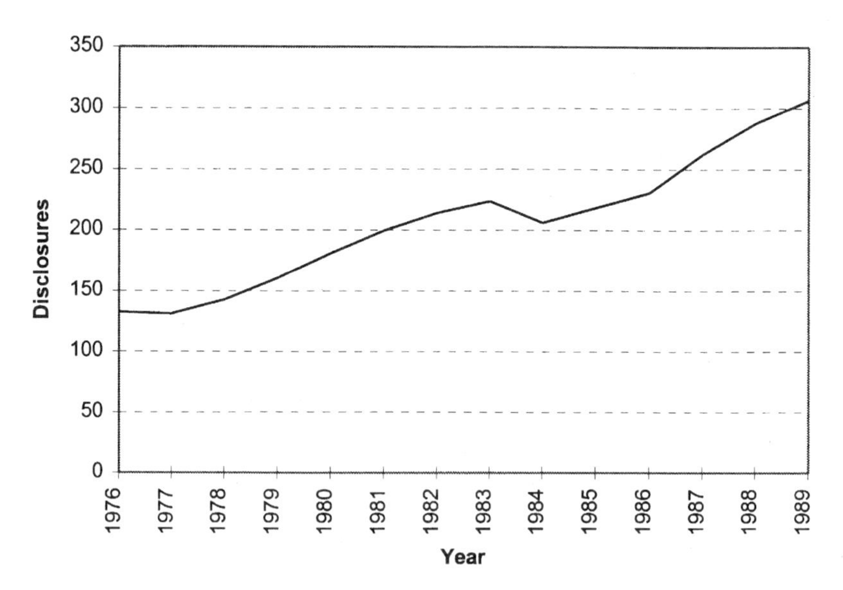

Figure 7 University of California Invention Disclosures, 1975–1990: 3-year moving average.

believe that this increase does not reflect an intensified search by administrators for such faculty inventions. For example, the Cohen-Boyer DNA splicing technique, the basis for the single most profitable invention licensed by the UC system and Stanford University, was disclosed in 1974 and the first of several patent applications for the invention was filed in 1978, well before the passage of Bayh-Dole (this patent issued in 1980).

Since biomedical inventions account for the lion's share of UC patenting and licensing after 1980, our assessment of trends "before and after" Bayh-Dole focuses on biomedical inventions, patents, and licenses. Figure 8 reveals that the share of biomedical inventions within all UC invention disclosures began to grow in the mid-1970s, before the passage of Bayh-Dole. Moreover, these biomedical inventions accounted for a disproportionate share of the patenting and licensing activities of the University of California during this period: Biomedical invention disclosures made up 33% of all UC disclosures during 1975–1979 and 60% of patents issued to the University of California for inventions disclosed during that period.[10] Biomedical patents accounted for 70% of the licensed

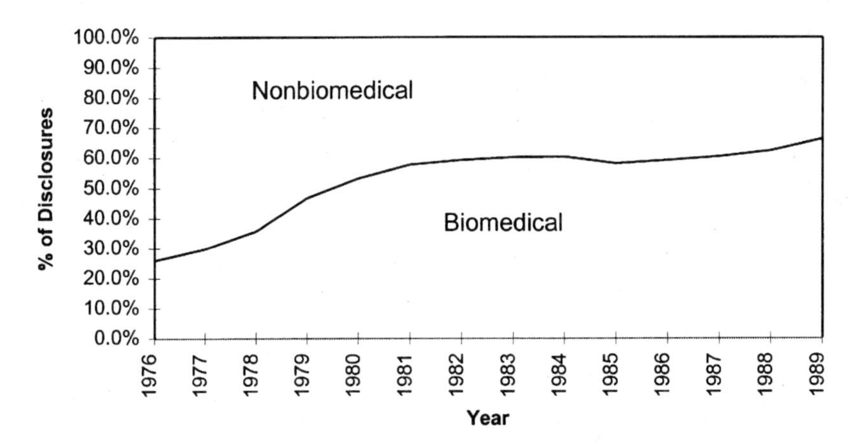

Figure 8 University of California Biomedical Disclosures as a Percentage of Total Disclosures, 1975–1990 (3-year moving average).

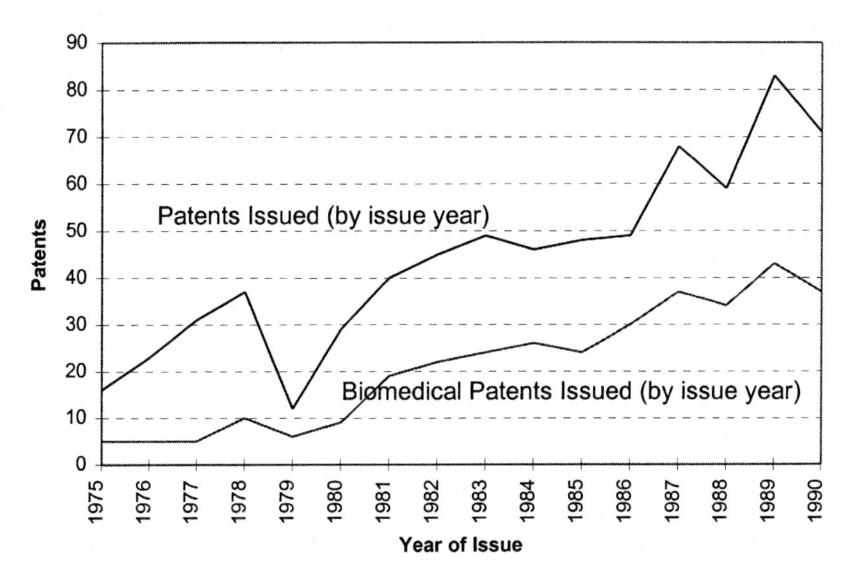

Figure 9 University of California Patents, 1975–1990, by Year of Issue.

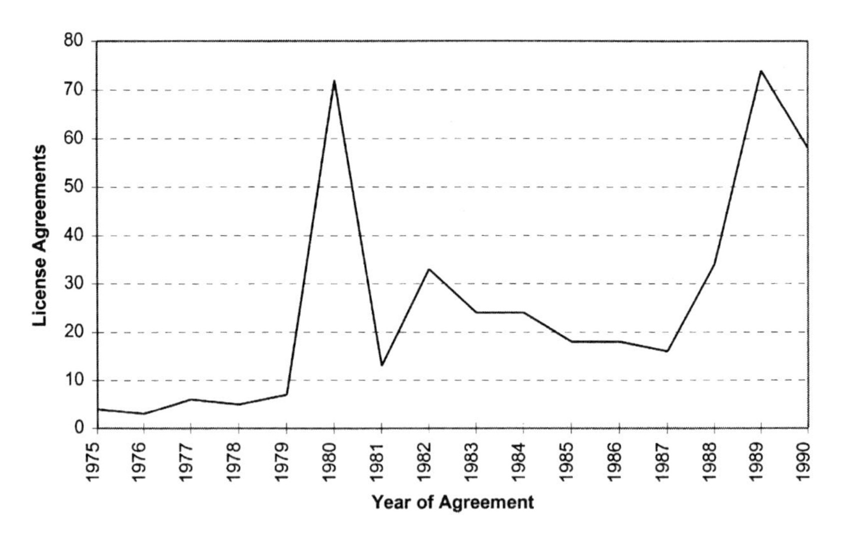

Figure 10 University of California Licenses, 1975–1990, by Year of Agreement.

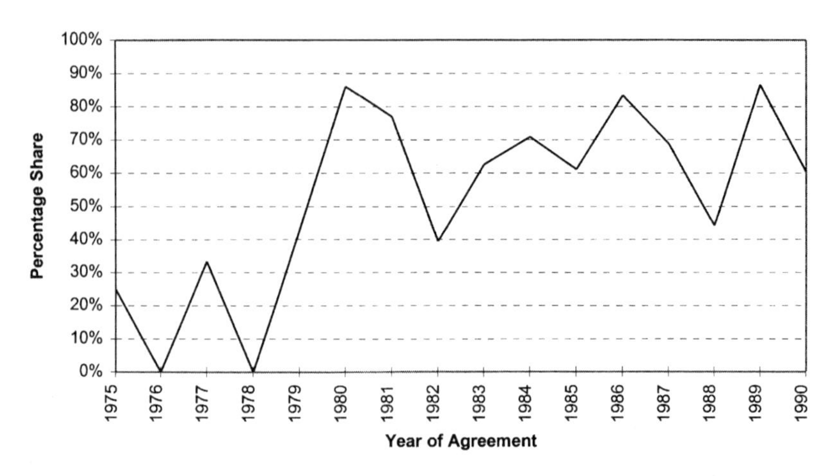

Figure 11 Biomedical Technology Share of University of California License Agreements, 1975–1990.

patents in this cohort of disclosures, and biomedical inventions accounted for 59% of the UC licenses in this cohort that generated positive royalties. Biomedical inventions retained their importance during the 1984–1988 period, as they accounted for 60% of disclosures, 65% of patents, 74% of the licensed patents from this cohort of disclosures, and 73% of the positive-income licenses for this cohort of disclosures.

Growth in the number of biomedical disclosures, which dominated UC patenting and licensing throughout 1975–1990, thus predates the passage of the Bayh-Dole Act. The apparent effects of Bayh-Dole on UC invention disclosures, patenting, and licensing are confounded with those of shifts in the underlying research agenda, as biomedical research funding and scientific advances grew rapidly during the 1970s and 1980s.

Additional evidence on the shifting composition of the University of California technology licensing portfolio is displayed in Table 2. The UC data in Table 2 reveal the high concentration of licensing revenues among a small number of inventions throughout the pre-Bayh-Dole period, as well as indicating remarkable growth (more than 50-fold) in constant-dollar gross revenues during 1970–1995. The share of gross licensing revenues accounted for by the UC system's "top 5" inventions actually decreases throughout the 1970–1995 period, from nearly 80% in fiscal 1970 to 66% in fiscal 1995, having reached a low point of 47% in fiscal 1985.

Equally remarkable is the shift in the UC system's "top 5" inventions from agricultural inventions (including plant varieties and agricultural machinery) to biomedical inventions. Among the three universities, only the University of California maintained a large-scale agricultural research effort. During the 1970s, agricultural inventions accounted for a majority of the income accruing to the "top 5" UC money earners. Beginning in fiscal 1980, however, this share began to decline, and by fiscal 1995, 100% of the UC system's licensing income from its "top 5" inventions, accounting for almost $40 million in revenues (in 1992 dollars), was derived from biomedical inventions, up from 20% in fiscal 1975. Moreover, and consistent with the discussion of the previous paragraph, this share increased sharply before the passage of Bayh-Dole in late 1980: the share of "top 5" licensing revenues associated with biomedical inventions jumped from less than 20% in fiscal 1975 to more than 50% in fiscal 1980.

Stanford University

Stanford University's Office of Technology Licensing was established in 1970, and Stanford was active in patenting and licensing throughout the 1970s. Stanford's patent policy, adopted in April 1970, stated that "Except in cases where other arrangements are required by contracts and grants or sponsored research or where other arrangements have been specifically agreed upon in writing, it shall be the policy of the University to permit employees of the University, both faculty and staff, and students to retain all rights to inventions made by them." (Stanford University Office of Technology Licensing 1982: 1). Disclosure by faculty of inventions and their management by Stanford's OTL thus was optional for most of OTL's first quarter-century.

In 1994 Stanford changed its policy toward faculty inventions in two important aspects. First, assignment of title to the University of inventions "developed using University resources" was made mandatory.[11] Second, the University established a policy under which "Copyright to software developed for University purposes in the course of employment, or as part of either a sponsored project or an unsponsored project specifically supported by University funds, belongs to the University." (Stanford University Office of Technology Licensing 1994a).[12] This policy on software goes beyond anything adopted by the University of California, and appears to be more comprehensive than policies in place at Columbia.

Stanford University's pre-1994 policies toward faculty inventions thus occupy a middle ground between those of Columbia University prior to the mid 1980s on the one hand, and the University of California, on the other. Prior to 1994, faculty disclosure of inventions to university administrators was no more mandatory at Stanford than at Columbia prior to the post-Bayh-Dole reforms there. Nevertheless, especially during the 1970–1980 period, Stanford operated a much more elaborate administrative apparatus for the patenting and licensing of inventions than did Columbia. The expanding scale of Stanford's licensing operations during the 1970s and 1980s also suggests that a substantial fraction of faculty inventions in fact were disclosed to the OTL.

Data from the Stanford OTL provide some insight into the patenting and licensing activities of a major private research

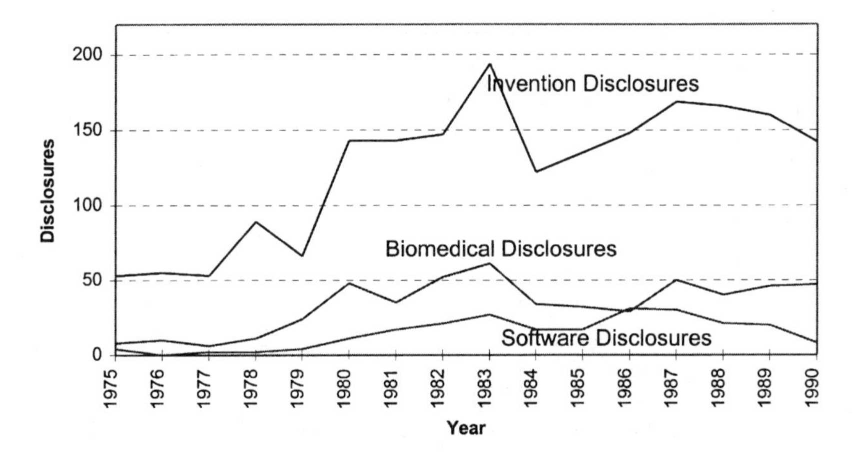

Figure 12 Stanford University Invention Disclosures, 1975–1990.

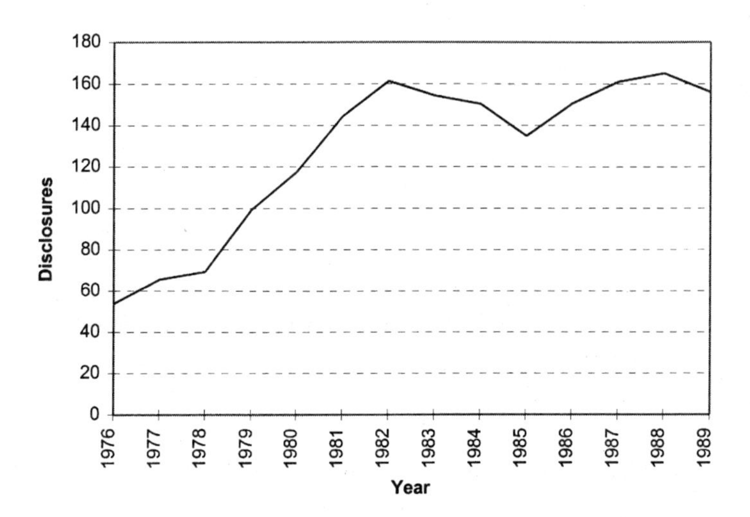

Figure 13 Stanford University Invention Disclosures, 1975–1990 (3-year moving average).

university before and after Bayh-Dole. Similarly to the situation at the University of California, these data suggest that the growth of Stanford's patenting and licensing activities was affected by shifts in the academic research agenda that reflected influences other than Bayh-Dole. Figures 12–13 display trends during 1975–1990 in Stanford invention disclosures. The average annual number of disclosures to Stanford's Office of Technology Licensing increased

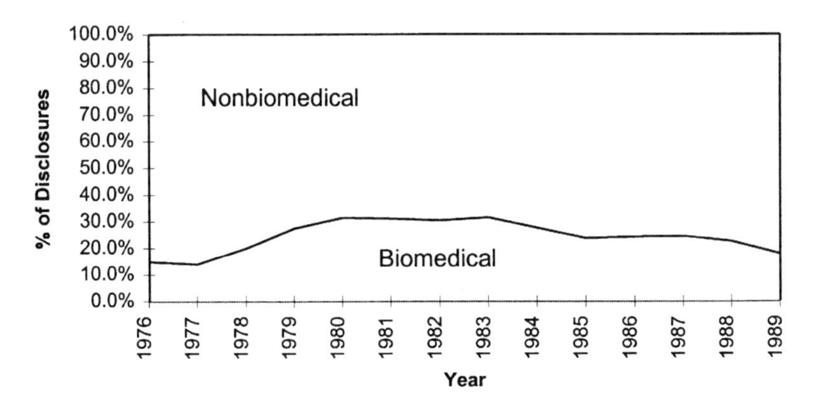

Figure 14 Stanford University Biomedical Disclosures as a Percentage of Total Disclosures, 1975–1990 (3-year moving average).

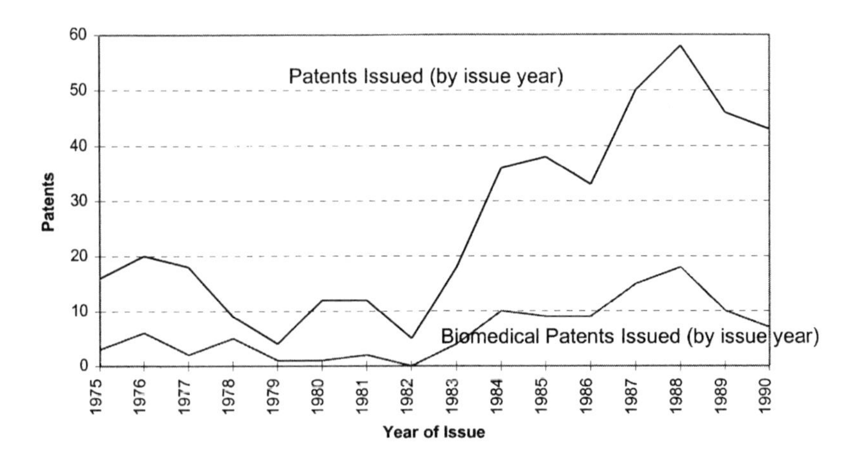

Figure 15 Stanford University Patents by Year of Issue, 1975–1990.

from 74 during 1975–1979, prior to Bayh-Dole, to 149 during 1984–1988. Moreover, the evidence of a "Bayh-Dole effect" on the annual number of disclosures (such as the jump in disclosures between 1979 and 1980 in Figure 12) is stronger in the Stanford data than in the UC data, although the smoothed trends in Figure 13 (computed as a 3-year moving average) suggest that the annual number of invention disclosures was growing prior to Bayh-Dole.

The data in Figures 12 and 14 also suggest that the importance of biomedical inventions within Stanford's invention portfolio advances had begun to expand before the passage of Bayh-Dole.

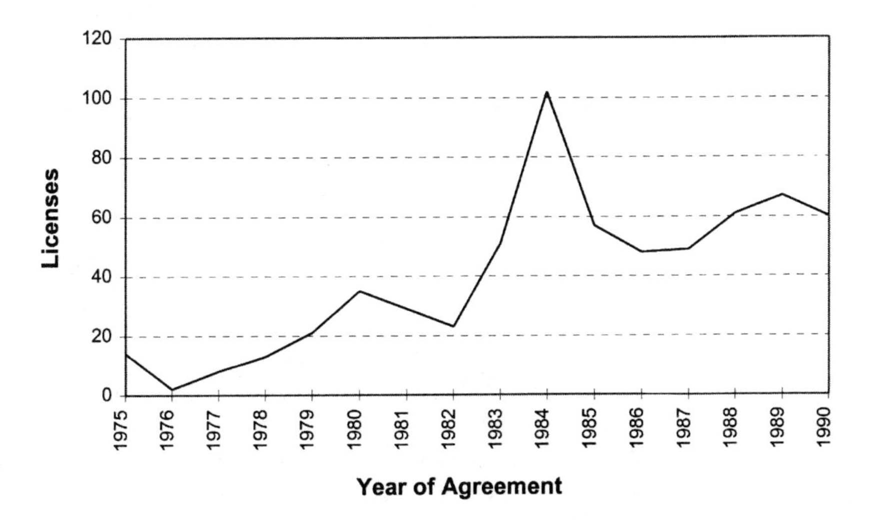

Figure 16 Stanford University Licenses (Excluding Cohen-Boyer and Software Licenses) by Year of Agreement, 1975–1990.

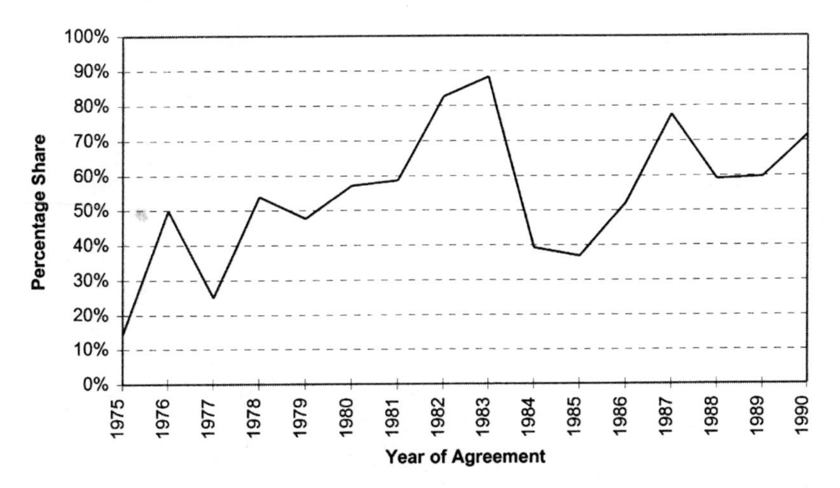

Figure 17 Biomedical Technology Share of Stanford University License Agreements (Excluding Cohen-Boyer and Software Licenses).

Figure 12 indicates that the annual number of biomedical invention disclosures began to increase sharply during the 1978–1980 period, and the share of all disclosures accounted for by biomedical inventions (see Figures 12 and 14) increased steadily from 1977–1980, leveling off after 1980 and declining after 1983. A similar but lagged increase in the share of Stanford patents accounted for by biomedical inventions is apparent in Figure 15. The magnitude of these increases in biomedical inventions prior to Bayh-Dole is more modest than at the University of California, but the trend is similar.

The trends in Figure 17 suggest that similarly to the UC system, biomedical inventions increased somewhat as a share of Stanford's (non-software) licenses during the 1975–1990 period, although the upward trend is less pronounced and fluctuates more widely than in the UC data.[13] Table 2 indicates that as of fiscal 1980, slightly more than 40% of the income from Stanford's "top 5" inventions was attributable to biomedical inventions, suggesting the considerable importance of these inventions prior to Bayh-Dole. This share increases to more than 96% by fiscal 1995. Stanford's licensing revenues grew by almost 200-fold (in constant dollars) during 1970-95, and its "top 5" inventions account for a larger share of gross income for the 1980–1995 period than do the "top 5" UC inventions.

Both Stanford and the UC system thus experienced a shift in the composition of their invention and licensing portfolio toward biomedical inventions prior to Bayh-Dole. Bayh-Dole was an important, but not a determinative, factor in the growth and changing composition of patenting and licensing activity at these institutions.

Like the Columbia University data, Stanford's invention disclosures include a number of software inventions, which account for 10–15% of annual disclosures. As is the case at Columbia during the 1980s, the majority of these inventions were not patented and therefore cannot be traced through annual patent counts. The importance of software disclosures in Stanford's licensing activity has grown over time. Only two of the 41 inventions disclosed during 1974–1979 (less than 5%) that were licensed within eight years of their disclosure were software inventions, but this fraction increased to more than 20% for the 1984–1988 period. Many of these software inventions (for example, the WYLBUR operating system) were licensed on a nonexclusive basis to academic institutions through Stanford's Software Distribution Center during the 1980s. The majority of these licenses involved a small, one-time payment by the licensee institution.[14] Partly because of the large number of such "site licenses," the coverage by our data of Stanford software licensing agreements is spotty and our estimate of the share of all Stanford licensing agreements accounted for by software is less accurate. Nonetheless, like Columbia University, a significant portion of Stanford's licensed inventions (at least 10–20% of annual licensing agreements, and a smaller share of gross revenues) cover nonpatented inventions, Bayh-Dole notwithstanding.

Comparing Invention Disclosures and Licenses at the Three Universities in the Late 1980s

In addition to comparing the periods before and after Bayh-Dole for two of these three universities, we compared the disclosure, patenting, and licensing activities across all three universities for the 1986–1990 period to assess the similarities and differences among them well after the passage of the Bayh-Dole Act.[15] The data in the top panel of Table 3 suggest considerable similarity among these three universities in the characteristics of their invention disclosures: roughly one-fifth of 1986–1990 inventions were patented at all three universities within six years of their disclosure. A larger share of Stanford University's disclosures, however, are licensed and a larger fraction of Stanford's invention disclosures yield positive licensing income than is true of either Columbia or the University of California.[16] The fraction of licenses that yield positive income, however, differs less dramatically among these three universities, keeping in mind that our licensing data for Stanford may omit a number of software licenses.

Restricting the focus to biomedical inventions does little to change the conclusions of this comparison among our three universities (the second panel of Table 3). The share of biomedical inventions resulting in issued patents within six years is remarkably similar across these three institutions, and Stanford once again displays a higher fraction of disclosures that are licensed and a higher fraction of disclosures resulting in licensing agreements that yield revenues. The shares of license contracts that yield positive revenues diverge less among these three universities.

Comparison of software inventions for the two universities within our sample (Stanford and Columbia) that have been active licensors of these technologies suggests a similar contrast between Stanford and Columbia in the share of disclosures that are licensed

Table 3 Comparative Evidence on Invention Disclosures and Licenses at Stanford University, Columbia University, and the University of California, 1986–1990

	Stanford	Columbia	University of California
1986–1990 (6-year "trailing window"): All Technologies			
% of disclosures yielding patents	23.2	18.6	20.4
% of disclosures that are licensed	33.2	16.4	12.3
disclosures with licensing income >0/all disclosures	22.4	12.3	7.4
licensed disclosures with licensing income>0/ number of disclosures that are licensed	67.4	75.0	60.6
% of licensed disclosures that are licensed exclusively	58.8	59.1	90.3
1986–1990 (6-year "trailing window"): Biomedical Technologies			
% of disclosures yielding patents	17.5	15.3	15.7
% of disclosures that are licensed	38.7	17.3	14.8
disclosures with licensing income >0/all disclosures	33.5	13.9	10.0
licensed disclosures with licensing income>0/ number of disclosures that are licensed	86.6	80.0	67.2
% of licensed disclosures that are licensed exclusively	54.9	62.9	90.3
1986–1990 (6-year "trailing window"): Software Technologies			
% of disclosures yielding patents	0	17.6	NA
% of disclosures that are licensed	53.6	35.3	NA
disclosures with licensing income >0/all disclosures	45.5	23.5	NA
licensed disclosures with licensing income>0/ number of disclosures that are licensed	84.7	66.7	NA
% of licensed disclosures that are licensed exclusively	46.3	16.7	NA

and the share of disclosures that result in revenue-producing licenses. As we noted above, software licenses rarely involve patented inventions: 100% of the software inventions disclosed at Stanford during 1986–1990 and 83% of Columbia's software inventions disclosed during this period that were licensed within six years of disclosure were not patented.

A final issue for consideration in our comparative evaluation of licensing in the post-Bayh-Dole era at these three universities concerns the fraction of inventions that are licensed through "exclusive" contracts, which we define here as contracts that are globally exclusive or that contain specified field of use or market restrictions. A large fraction of all inventions that are licensed—as high as 90% for UC licenses and no less than 58.8% for Stanford licenses of "all technologies" during this period—is licensed on a relatively exclusive basis, and these shares are similar for biomedical inventions. Perhaps because of the weaker formal protection for this technology (during much of this period, the lack of patent protection), software inventions are less frequently licensed on an exclusive basis: 46% and 17%, respectively, of software invention disclosures at Stanford and Columbia were licensed exclusively during 1986–1990.

Nevertheless, the most profitable licenses at all of these universities are nonexclusive licenses. The Stanford-UC Cohen-Boyer patents, the single most profitable licensed university invention during the 1980s and 1990s (revenues from these licenses amounted to nearly $250 million during 1981–1997), were licensed widely and nonexclusively.[17] Columbia University's single most profitable invention, the Axel biotechnology patent, also was licensed on a nonexclusive basis.

Although many proponents of patent protection for university inventions argue that would-be commercial developers of these technologies needed exclusive title to this intellectual property in order to obtain a clear "prospect" for their significant investments in this activity, these cases suggest that for inventions of broad promise and potential widespread use, nonexclusive licenses can accommodate both universities' interest in revenues and the needs of commercial users for access to the essential intellectual property, in a way that facilitates competition and limits the risk of monopolization of the commercial development of important technologies.

Conclusions and Concerns

The effects of the Bayh-Dole Act on U.S. research universities have received extensive rhetorical attention but modest empirical analysis. In this paper, we have used an invaluable byproduct of the Act, the systematic records of their faculty inventions, patents, and licenses compiled by three leading U.S. research universities, in an analysis of some of the Act's effects. Our data on the University of California and Stanford University suggest that for universities already active in patenting and licensing, Bayh-Dole resulted in expanded efforts to market academic inventions. The Act also led Columbia, along with many other research universities formerly inactive in this area, to revise its longstanding policies and enter into large-scale patenting and licensing of faculty inventions.

Nevertheless, other factors also affected the upsurge in patenting and licensing at U.S. research universities after 1980, and it is difficult to separate their effects from those of Bayh-Dole. In particular, by the mid-1970s biomedical technology, especially biotechnology, had increased significantly in importance as a productive field of university research, with research findings that were of great interest to industry. The feasibility of technology licensing in biotechnology was advanced by *Diamond v. Chakrabarty*, which in combination with the broader shift in U.S. policy to strengthen intellectual property rights, contributed to the increased post-1980 patenting and licensing activities of U.S. research universities.

An array of developments in research, technology, industry, and policy thus combined to increase U.S. universities' activities in technology licensing, and Bayh-Dole, while important, was not determinative. We believe that even without Bayh-Dole, both Stanford and the UC system would have expanded their patenting and licensing activities, and their licensing revenues would have grown significantly. Similar shifts in its research portfolio led Columbia University, formerly inactive in patenting and licensing academic inventions, to shift toward a more active role in patenting and licensing immediately prior to the passage of Bayh-Dole.

By the end of the first decade of Bayh-Dole, these three universities display remarkable similarities in their patent and licensing portfolios, as illustrated by the data in Tables 2 and 3. For all three, a very small share of all patented inventions account for the majority of gross licensing revenues. Moreover, these leading earners are concentrated in the biomedical area, a technology field characterized by relatively strong patents that are economically significant (Levin et al. 1987). A second important area of licensing at two of our three universities, however, is software, for which formal patent protection is less important.

Our analysis suggests that thus far, Bayh-Dole probably has had minimal effects on the content of academic research at these major research universities. Here as elsewhere, one must specify the counterfactual with care. We have noted earlier that Bayh-Dole had nothing to do with the creation of new areas of academic research, such as biotechnology or computer software, which have provided new financial opportunities for entrepreneurial faculty; in the absence of Bayh-Dole, these new possibilities, and their potential to change faculty behavior, would be no less real. Even without Bayh-Dole, the universities might well have increased their administrative support for patenting and licensing to some extent.

In any case, the post-1980 patenting and licensing activity affects a relatively small portion of the only one of our three universities for which we have reliable data, Columbia University. The bulk of Columbia invention disclosures, patenting and licensing is accounted for by a small number of university departments, primarily the medical school, electrical engineering, and computer science, and most of this University's departments have been unaffected by the post-1980 shift toward more intensive patenting and licensing of faculty inventions. We believe that similar data from Stanford and the UC system would lead to the same conclusion.

This shift in university policies raises two other concerns that are not addressed by the data presented in this paper, but which we wish to discuss here. First, are universities' patenting and licensing efforts increasing or reducing the social returns to the results of the publicly funded research performed within their boundaries? Second, are universities' expanded efforts to patent inputs into the scientific research process impeding progress in this sphere?

The theory behind Bayh-Dole was that companies needed exclusive patent rights to pick up, develop, and commercialize the results of university research, a theory that flies in the face of the position that patents tend to restrict use of scientific and technological information, and that open publication facilitates wider use and application of such inventions and knowledge. Are patents or restrictive licenses necessary to achieve application? Should such licenses be negotiated by universities, institutions not always known for their commercial expertise? Does the presence of a university-assigned patent and the requirement for licensing delay and narrow technology transfer? In future research, we plan to compile case studies of the commercialization of university inventions, examining whether an exclusive license facilitated the transfer of a given technology transfer, or whether technology transfer likely would have proceeded just as fast and widely had the results simply been placed in the open literature. We believe that the answers to these questions will vary considerably among inventions, technologies, and time periods.

As we noted earlier, one widely voiced concern over the effects of Bayh-Dole on academic research asserts that faculty research may be shifted toward applied research topics. But a very different, more recent concern reflects the influence of both Bayh-Dole and the broader shifts in U.S. policy toward intellectual property rights in general and toward intellectual property in biotechnology in particular. Although we have found little evidence of any shift in these universities' research agenda toward applied topics in the wake of Bayh-Dole, more of what universities naturally would have produced and placed in the public domain now is subject to more complex administrative procedures. A number of universities have extended patenting and licensing policies since 1980 to cover the results of scientific research, rather than focusing their patenting on the results of applied research. These policies may raise the costs of use of these research results in both academic and nonacademic settings, as well as limiting the diffusion of these results. These tendencies are most apparent in "research tools" in molecular biology and biotechnology (National Research Council 1997). Universities and private firms alike are now patenting genetic materials far more extensively and requiring licensing and royalty payments for their use, raising the "transactions costs" of conducting research in this area of science.

In our three universities, by and large biotechnology research tools have been licensed widely, and the transactions costs involved in taking out a license appear to have been kept relatively low. This issue merits continuing attention: although universities now can collect revenues from patenting and licensing research tools that in an earlier era likely would have simply been placed in the public domain, this practice does not spur technology transfer. The argument that allowing universities to patent and exclusively license the results of university research will enhance technology transfer does not apply to this class of inventions.

The principal risk posed by the Bayh-Dole Act and related initiatives in U.S. science and technology policy flows from the premise that underpins many of these legislative and policy initiatives. All too often, these initiatives assume that patents and exclusive licensure of the results of federally sponsored research is the best approach to maximize the social returns to the federal R&D investments. We believe this premise understates the effectiveness of publication and other, more open channels for information dissemination and access in enabling society to benefit from publicly funded academic research. Administrators and licensing professionals at all three of the universities discussed in this paper are aware that restrictive licensing terms often are undesirable. But it is important to keep that awareness alive; it may be in danger.

Acknowledgments

Authors' names appear in alphabetical order. An earlier version of this paper was presented at the conference on "The U.S. and Japanese Research Systems," Kennedy School of Government, Harvard University, September 1–12, 1998, and benefited from the comments of conference participants, particularly Roland Schmidt and Lewis Branscomb. We are indebted to the staff of the technology licensing offices of Columbia University, Stanford University, and the University of California for invaluable assistance with the collection and analysis of these data. The research on the Columbia University data underpinning this paper also benefited from the efforts of Michael Crow, Holly Raider, and Annetine Gelijns of Columbia University. Michael Barnes and Lynn Judnich of the University of California assisted in the collection and analysis of the University of California data, and the research on the Stanford data benefited from the assistance of Sandra Bradford. Support for this research was provided by the California Policy Seminar, the U.C. President's

Industry-University Cooperative Research Program, the Office of the Provost at Columbia University, the Alfred P. Sloan Foundation, and the Andrew Mellon Foundation.

Notes

1. According to the annual report of the Association of University Technology Managers (AUTM), fiscal 1993 gross licensing revenues for the UC system, Stanford University, and Columbia University amounted to $45.4 million, $31.2 million, and $21.1 million respectively; the fourth and fifth leading academic institutions in this ranking, the University of Wisconsin-Madison and the University of Washington, earned $15.8 million and $14.8 million respectively (Association of University Technology Managers 1994: 8).

2. Derek Bok (1982) and others have noted that the postwar growth of federal funding of university research, which accelerated in the late 1950s, contributed to some "turning away" by university faculty from collaborative research with industry.

3. According to the "Report on University Patent Fund and University Patent Operations for the Year ended June 30, 1968" of the Board of Regents of the University of California, "The United States Public Health Service (PHS) of the Department of Health, Education, and Welfare is revising its Institutional Agreements under which patent rights can be retained by educational institutions. The PHS intends to make these Institutional Agreements available to many more institutions than at present. At the same time, it is making its patent provisions more restrictive. Most objectionable of the provisions included in the draft under consideration are: (1) a limitation on the amount of royalty the University can share with its inventors, and (2) a requirement that the University and its licensees provide the Government with copies of all licenses, and that the University incorporate into commercial licenses the provisions of the Institutional Agreement" (11/1/68, p. 4).

4. The "evidence" of unsuccessful commercialization of federally owned patents consisted mainly of numerous references to the tiny percentage of the 28,000–30,000 patents owned by the federal government that had been commercialized. Advocates of the Bayh-Dole Act overlooked the fact that title to most of these patents, which resulted from federal defense contracts, had been ceded to the federal government by private contractors who had not invoked their rights under the policies then prevalent in the Defense Department to retain title to the patents. In other words, as Eisenberg (1996) points out, the statistical data cited in support of the Bayh-Dole Act suffered from a serious selection bias.

5. According to Katz and Ordover (1990), at least 14 Congressional bills passed during the 1980s focused on strengthening domestic and international protection for intellectual property rights, and the Court of Appeals for the Federal Circuit created in 1982 has upheld patent rights in roughly 80% of the cases

argued before it, a considerable increase from the pre-1982 rate of 30% for the Federal bench.

6. In addition, and similarly to the situation at Stanford University (see below), more than 300 of the 420 licenses for this software invention are academic licenses.

7. According to a March 10, 1975 letter from UC President Charles J. Hitch to Governor Edmund G. Brown, Jr., "The possibility of developing a formal patent policy and program was first considered in the University in the Thirties. But the idea did not achieve full impetus until the war years when the Federal Government began to sponsor research in the University on a large scale and inventions began to be made under research contracts."

8. As revised in 1973, the "University Policy Regarding Patents" states that "An agreement to assign inventions and patents to The Regents of the University of California, except those resulting from permissible consulting activities without use of University facilities, shall be mandatory for all employees, academic and nonacademic." The policy statement goes on to emphasize that "The Regents is [sic] averse to seeking protective patents and will not seek such patents unless the discoverer or inventor can demonstrate that the securing of the patent is important to the University." This latter sentiment notwithstanding, UC administrators were actively seeking patent protection for faculty inventions by the mid-1970s, as the historical data of the Office of Technology Transfer show.

9. These "independent" licensing offices, which continue to pay a portion of their revenues to the state government, are in charge of invention disclosures (along with any revenues or expenses associated with these disclosures) occurring after their foundation. The centralized database maintained by the UC Office of Technology Transfer accordingly is most reliable for invention disclosures occurring through 1990, and we focus our analysis on the 1975–1990 period.

10. Figures 9 and 11 indicate that biomedical inventions accounted for a growing share of UC patenting and licensing during the entire 1975–1990 period.

11. Almost simultaneously with this shift in University patent policy, an internal study by the OTL Advisory Board in 1993 recommended that "OTL need not be constrained by the principle of 'preference for non-exclusive licensing' ... " Stanford University Office of Technology Licensing, 1993: 2).

12. Reflecting faculty sensitivity over assignment to the University of all ownership of all copyrighted material produced under University sponsorship, Stanford's OTL explicitly exempted ownership of "books, articles, popular nonfiction, novels, poems, musical compositions, or other works of artistic imagination which are not institutional works" from the policy governing software (Stanford University Office of Technology Licensing, 1994b: 1).

13. Figures 16 and 17 exclude licenses for the Cohen-Boyer patents, which were managed by Stanford's OTL on behalf of the UC system and Stanford University. Strictly speaking, since the revenues from these licenses are split between the UC system and Stanford University, the licenses also should be allocated between the

two institutions. Exclusion of this heavily licensed invention understates the growth in the biomedical share of Stanford and UC licensing agreements during the 1980s in Figures 11 and 17.

14. Some indication of the relative magnitudes of licensing revenues from these "site licenses," which for some years were administered by the OTL Software Distribution Center, is given by the following data cited in the 1988–1989 report of Stanford's Office of Technology Licensing, which separated software licensing revenues into those derived from "... direct software distribution through OTL's Software Distribution Center ($453,581 from 515 use licenses) and from royalties paid by commercial distributors ($420,000 from 40 distribution licenses to software firms, computer companies, and publishers)." (Stanford University Office of Technology Licensing, 1990: 4). Unfortunately, we have been unable thus far to consistently separate software licenses between these two channels of distribution.

15. In order to deal with the problems of "truncation bias" while accommodating the fact that our data end in 1997, we have imposed a 6-year "trailing window" on our invention disclosures. In other words, the analysis includes issued patents or licenses only if these events occur within 6 years after the date of disclosure of the invention. This convention is used to avoid unfairly biasing the indicators of "productivity" in favor of older disclosures, which have much longer time periods during which to produce patents or licenses.

16. As we noted earlier, a large fraction of Stanford's software licenses cover low-cost "site licenses" at other academic institutions, which may well raise the shares of Stanford disclosures that yield licensing income without necessarily having a significant effect on overall licensing income. In addition, the Stanford invention disclosures and licensing data contain a large number of agreements covering "clones" of various pieces of genetic material—such agreements are less common in the Columbia or UC data. These licenses are somewhat more formal than Materials Transfer Agreements, and often involve the payment of modest licensing fees. But like the Stanford software licenses, the effect of including these agreements drives up the shares of disclosures that are licensed or that yield licensing income without having much effect on overall licensing income.

17. See the case study of Cohen-Boyer in the recently published summary of a National Research Council workshop on "Intellectual Property Rights and Research Tools in Molecular Biology" (National Research Council 1997).

References

Arrow, K. 1962. "Economic Welfare and the Allocation of Resources for Invention," in R.R. Nelson, ed., *The Rate and Direction of Inventive Activity*. Princeton, NJ: Princeton University Press.

Association of University Technology Managers. 1994. *The AUTM Licensing Survey: Executive Summary and Selected Data, Fiscal Years 1993, 1992, and 1991.* Norwalk, CT: Association of University Technology Managers.

Bok, D. 1982. *Beyond the Ivory Tower.* Cambridge MA: Harvard University Press.

Cohen, W., R. Florida, L. Randazzese, and J. Walsh. 1998. "Industry and the Academy: Uneasy Partners in the Cause of Technological Advance," in R. Noll, ed., *Challenges to the Research University.* Washington, D.C.: Brookings Institution.

Eisenberg, R. 1996. "Public Research and Private Development: Patents and Technology Transfer in Government-Sponsored Research," *Virginia Law Review* 82: 1663–1727.

Henderson, R., A.B. Jaffe, and M. Trajtenberg. 1998. "Universities as a Source of Commercial Technology: A Detailed Analysis of University Patenting, 1965–88," *Review of Economics & Statistics* 80(1): 119–127.

Katz, M.L., and J.A. Ordover. 1990. "R&D Competition and Cooperation," *Brookings Papers on Economic Activity: Microeconomics,* 137–192.

Levin, R.C., A. Klevorick, R.R. Nelson, and S. Winter. 1987. "Appropriating the Returns from Industrial Research and Development," *Brookings Papers on Economic Activity,* 783–820.

Mowery, D.C., and N. Rosenberg. 1998. *Paths of Innovation: Technological Change in 20th-Century America.* New York: Cambridge University Press.

National Research Council. 1997. *Intellectual Property Rights and Research Tools in Molecular Biology.* Washington, D.C.: National Academy Press.

National Science Board. 1996. *Science and Engineering Indicators: 1996.* Washington, D.C.: U.S. Government Printing Office.

National Science Foundation. 1996. *National Patterns of R&D Resources: 1996.* Washington, D.C.: U.S. Government Printing Office.

Nelson, R.R. 1959. "The Simple Economics of Basic Scientific Research," *Journal of Political Economy,* 297–306.

Rosenberg, N. 1992. "Scientific Instrumentation and University Research," *Research Policy* 21: 381–90.

Rosenberg, N. 1998. "Technological Change in Chemicals: The Role of University-Industry Relations," in A. Arora, R. Landau, and N. Rosenberg, eds., *Chemicals and Long-Term Economic Growth.* New York: John Wiley.

Rosenberg, N., and R.R. Nelson. 1994. "American Universities and Technical Advance in Industry," *Research Policy* 23: 323–348.

Stanford University Office of Technology Licensing, 1982. *Thirteenth Annual Report.* Stanford, CA: Stanford University.

Stanford University Office of Technology Licensing, 1990. *Twentieth Annual Report.* Stanford, CA: Stanford University.

Stanford University Office of Technology Licensing, 1993. *1991–92 Annual Report: Office of Technology Licensing.* Stanford, CA: Stanford University.

Stanford University Office of Technology Licensing, 1994a. "Office of Technology Licensing Guidelines for Software Distribution," 11/17/94; processed.

Stanford University Office of Technology Licensing, 1994b. "Copyrightable Works and Licensing at Stanford," Spring 1994; processed.

Trajtenberg, M., R. Henderson, and A. Jaffe. 1994. "University Versus Corporate Patents: A Window on the Basicness of Inventions," *CEPR Working Paper #372*, Stanford University.

12

Intellectual Property Rights and University-Industry Technology Transfer in Japan

Robert Kneller

Introduction

There is a widespread consensus in Japan today that relations between Japanese universities and industry should be improved in order to promote the development of high-technology products and industries. The formation of technology transfer offices (TTOs) and university spin-off companies figures prominently in proposals to facilitate the transfer of university discoveries to industry, including the 1998 Technology Transfer Law, promulgated to promote university-industry technology transfer.

This chapter first takes a step back from the concern with TTOs and start-up companies to clarify how companies have traditionally obtained intellectual property (IP) rights to university discoveries. It analyzes the principal legal documents and procedures that have shaped the system of technology transfer since the 1970s, and how they create incentives for companies, university administrators, and university researchers to transfer technology in ways that are less formal and transparent than the manner in which U.S. university technologies are transferred under the Bayh-Dole Regulations (37 CFR §401.1–401.16). The chapter then analyzes how the new legislation will affect university-industry technology transfer. Finally it examines how technology transfer practices interact with other factors to influence the overall nature of university-industry cooperation and the development of high-technology industries in Japan. This analysis should be of interest to persons concerned with science policy as well as companies and individual investors interested in cooperating with Japanese university researchers.

Overview of Technology Transfer and the U.S. System

The United States and Japan lead the world in government support for non-military research and development (R&D) according to official statistics.[1] The level of this support and that for university R&D are shown in Table 1.

It is important that this research leads to social benefits through improved health and beneficial products and services. However, much government-supported university research tends to be basic, investigator-initiated research. Often much additional development by private companies is needed to translate the results of such research into new products. However, companies are unwilling to commit substantial resources to developing university discoveries if competitors can simply copy and market the finished products. Exclusive IP rights to university discoveries enable a company to prevent this, and are particularly important for small companies, which depend upon exclusive control over technology to attract the capital necessary for product development and commercialization. Exclusive IP rights are also particularly important in certain industries, such as biotechnology and pharmaceuticals, where the costs of developing and proving the safety of new products is high, while the cost of copying the final products is low.

In the United States prior to 1980, there was no uniform policy regarding IP rights to university discoveries made with U.S. government support. Policies differed according to the laws or policies applying to each funding agency.[2] In general, IP rights to such discoveries vested in the U.S. government, unless the funding agency waived its rights, and agencies tended to license discoveries on a non-exclusive basis. In 1980, fewer than 250 patents were issued to universities per year (COGR 1993), only a fraction of which were for inventions made with government support (Latker 1977).

Amendments to U.S. patent law in 1980 and the implementing regulations issued in 1987 (commonly known respectively as the "Bayh-Dole Act" [PL 96-517] and the "Bayh-Dole Regulations" [37 CFR §401.1–401.16]) allow companies and non-profit organiza-

Table 1 Indices of Government Support for R&D in Japan and the United States

	Japan	United States
Gov't Support for Non-military R&D in 1996	$14 billion	$31 billion
Gov't Support for R&D in Universities in 1995	$5.75 billion	$18.8 billion
% of Higher Education R&D Funded by Government	52%	68%
% of Higher Education R&D Funded by Private Companies	2.3%	5.5%

Source: NSB 1998.

tions, including universities, to claim world-wide patent rights on inventions made under U.S. government grants and contracts. The Bayh-Dole Regulations also require universities to establish procedures to ensure that university employees inform them of inventions soon after they are made. Because nearly all U.S. universities require their employees to assign to them rights in employment-related inventions (see Etzkowitz 1994 for a history of these requirements), the Bayh-Dole Act and Regulations allow universities to take full control of IP rights in inventions arising from government support. Therefore, universities have strong incentives to ensure that discoveries by their employees are transferred to companies that will effectively develop and commercialize them. As a result, technology licensing offices (TLOs) have become the focal point for technology transfer activities in U.S. universities.

For a variety of reasons, U.S. universities generally license rather than assign rights to companies (Kneller 1998), retaining ultimate ownership of IP created by their employees. Thus, licenses are the cornerstone of the transfer of IP rights from U.S. universities to industry. This is true even in the case of "sponsored research" or "cooperative research agreements" under which a company agrees to support a particular line of university research, usually in return for scientific information and options to license inventions that may emerge. It is also true concerning the formation of "start-up" or "spin-off" companies, to which universities usually exclusively license the key technologies they expect to develop. In 1995, over 5,000 licenses from U.S. universities and non-profit medical research institutions were generating over $480 million in license royalties. Direct corporate support for R&D in these institutions was over $1.6 billion, largely through sponsored or cooperative research agreements. Over 190 start-up companies were formed that year, bringing to over 1,400 the total number formed since 1980. (AUTM 1998)

This summary of the U.S. technology transfer system provides the basis for our comparison with the Japanese technology transfer system.

The Japanese Technology Transfer System Prior to May 1998

Background: Japanese Universities and Monbusho Rules

Japan's university system comprises 595 institutions, of which 385 offer graduate courses and 274 have doctoral programs (NSF 1997a). Even among these 274 institutions, research excellence is heavily concentrated in a small number of universities (Irvine 1990; Yamamoto 1996; Keii 1984).[3] With a few exceptions,[4] most highly regarded research universities are national universities, which are funded and controlled by the Ministry of Education, Science, Sports and Culture (Monbusho) and whose faculty members are civil servants. Most of the national universities have graduate schools, which account for approximately 75% of total R&D expenditures (Irvine 1990).[5] Therefore, the laws and procedures governing technology transfer from national universities determine how most Japanese university discoveries are transferred to the private sector.

Formal technology transfer procedures for Japanese national universities are set forth in a number of official "Notifications" that have been issued from time to time by Monbusho[6] and internal rules that individual national universities have enacted to implement the Notifications. There is no overarching Diet legislation setting forth technology transfer procedures, nor are such procedures set forth in Monbusho regulations, which trace their authority to Diet legislation and which must go through a process akin to the "public review and comment" necessary for regulations of U.S. government agencies to take effect.

The Basic Principle: IP Rights Vest either with Individual Inventors or with the Nation

Probably the single most important Notification underlying the Japanese technology transfer system is Monbusho Notification No. 117 of 1978. It sets forth the basic principle that IP rights to inventions made in national universities belong to the individual inventors. However, it qualifies this principle by stating that inventions arising either (1) under special funding from the government for a project specified as being for the purpose of developing practical applications, or (2) under a project specified as being for the purpose of developing practical applications, and which utilizes special research facilities and equipment (such as nuclear power research facilities and particle accelerators) that are established for use in government-sponsored research, should be classified as National Inventions and belong to the nation. (See Monbusho 1996 for an authorized translation of these two criteria.) The Japan Society for the Promotion of Science (JSPS), an affiliate of Monbusho, manages the patenting of National Inventions arising under Monbusho support (in effect, the majority of National Inventions from national universities). With certain exceptions, the Japan Science and Technology Corporation (JST)[7] is responsible for licensing such inventions. The offices of both the JSPS and the JST are located in the Tokyo area.

Deciding Ownership: The Role of Invention Committees and the Importance of the Type of Funding That Leads to an Invention

In order to determine whether national university inventions should belong to the nation, Monbusho Notification No. 117 states that national universities should establish internal regulations under which university inventors report their inventions to university presidents. University Invention Committees will then review the inventions and recommend to the university presidents whether any of them should belong to the nation. The type of funding under which an invention is made is an important determinant of whether it will be classified as a National

Invention. The principal categories of R&D funding in national universities are listed below; Table 2 summarizes the amount of funding under each category.[8]

Standard Research and Educational Allowances for Faculty Members. These cover general laboratory operating expenses including supplies, equipment such as computers and laboratory instruments, utilities, upkeep, and general maintenance.[9] Monbusho gives national universities funds for Standard Allowances in amounts determined primarily by the number of laboratories and the number of tenure track researchers in each university,[10] although increased amounts are available for laboratories engaged in "experimental" work (NSF 1997a; Tamura 1989).

Grants-in-Aid for Scientific Research. Grants-in-Aid are proposal-based, competitively evaluated Monbusho awards to individual researchers or groups of researchers. There are several classes of Grants-in-Aid, ranging from special priority area research projects that may involve several laboratories, to small-scale research grants to supplement doctoral or post-doctoral fellowships (NSF 1995).

Donations (including Endowments). These are donations from private organizations, foundations and individuals to support scientific research and educational activities. They can be targeted to a particular laboratory. Unlike other R&D funds, they can be rolled over to the next fiscal year and used flexibly to purchase equipment, support travel, hire temporary secretaries or technicians, and so on. Faculty members must disclose Donations to their university administrators, who then report these data to Monbusho headquarters.

Commissioned Research. Under Commissioned Research Agreements, universities can accept funds from companies, governmental organizations other than Monbusho, and other outside organizations to carry out contract-type research. Special Monbusho "investment funds" for application-oriented research are also awarded under Commissioned Research Agreements.[11] Personnel exchanges do not occur under such agreements.[12]

Joint Research with the Private Sector. Under Joint Research agreements, national universities can accept funds from companies to conduct collaborative research in a defined area. Usually such projects involve an exchange of researchers.[13]

Table 2 Support under Various Funding Categories for R&D in Japanese National Universities in 1996

Type of R&D Funding	Number of Projects or Persons	Amount in billions of ¥ (millions of US $)	Approximate Percent of Funding Total*
Standard Research Allowance	N/A	146 (825)	44
Grants-in-Aid	31,398	101.8 (575)*	30*
Donations	not avail.	52.8 (298)	16
Commissioned Research	3714	28.8 (163)	9
Joint Research with the Private Sector	2001	5.4 (30.5)	2
Bench Fees for Commissioned Researchers	790 researchers	0.5 (2.8)	0.15
Total	N/A	335. (1,893)	100

*The amount of Grants-in-Aid includes grants to local and private universities; thus, the amount and percentage for national university R&D should be less than shown in the table.
Sources: Monbusho 1996; Monbusho Advisory Com. 1998; Committee 1996.

Commissioned Researchers. National universities can accept technicians and researchers employed in companies and other outside organizations to do graduate or post-graduate research. Their employers usually pay a research or "bench fee."

The University of Tokyo's Invention Rules implementing Notification No. 117 state that inventions that occur under Grants-in-Aid, Commissioned Research, and Joint Research with the Private Sector (indicated in bold in Table 2) may qualify as National Inventions under the two criteria set forth in Notification 117, but that inventions arising under standard research allowances and Donations do not. They require that an inventor report to the university president any invention made under Grants-in-Aid, Commissioned Research, or Joint Research, as well as any licensing or assignment documents prior to their execution. They add, however, an important proviso: If the inventor is sure the invention meets neither of the above criteria for belonging to the nation, he does not have to report the invention or related technology transfer agreements. The internal rules of Tohoku University (Rule 58, 1978) and the Tokyo Institute of Technology (Rule 53, 1978) do not contain this exception, and university inventors are obliged to report all work-related inventions to the university presidents. However, officials in these universities recognize that some patentable discoveries are not reported. Officials in JSPS believe that the University of Tokyo's position on invention reporting is followed by a number of other national universities, even though their internal rules may not contain similar explicit exceptions to the reporting requirement.

Patenting and Licensing of National Inventions by JSPS and JST

Until April 1999, JSPS was responsible for filing patent applications on National Inventions on behalf of national universities and other Monbusho institutions. JST assumed this responsibility as of April 1999. Patent application and maintenance fees for national university inventions are waived by the Japan Patent Office (Japan Patent Law, sections 107 and 195), although universities must still pay the patent application costs of JST (as they had for JSPS previously).[14] JST assumes responsibility for licensing national inventions, and can consider requests from companies for "preferential licenses" under which a period of conditional exclusivity is negotiated, although the general principle is that National Inventions belong to the public and should be made available non-exclusively to whoever wishes to use them. JST can transfer a preferential license to another company if the licensee fails to develop the technology. Royalties from National Inventions belong to the nation and are transferred to a common national university "general revenue" account under the control of the Ministry of Finance (MOF). JST can use some of these funds for administrative expenses. Neither individual inventors nor their universities share in the royalty stream, although individual inventors are eligible for license bonuses, which theoretically can be as high as ¥6 million ($35,000) annually, depending upon the royalty stream.

However, as indicated by Table 3, the amount of patenting activity by JSPS and licensing activity by JST has been modest.

Table 3 JSPS and JST Patenting and Licensing of University Inventions

	1996	1982–1996
Japanese patent applications by JSPS	35	711
Japanese patents issued to national universities and other Monbusho institutions	79	482
Licenses issued by JST*	2	122

*Most licenses are non-exclusive and non-preferential according to JST.[15]

Source: JSPS Web site: http://www.jsps.go.jp/j-home.htm.

Most University Inventions Are Transferred Informally by the Inventors

As suggested by the first row of Table 3, the number of university inventions reported to university presidents and reviewed by university invention committees is small, especially in comparison to the nearly 10,000 inventions reported by U.S. universities and medical research institutes in 1996 (AUTM 1998). Table 4 summarizes the decisions of national university invention committees for two recent years.

It is clear that the actual number of university inventions exceeds the number reviewed by university invention committees. A study by the Japan Bioindustry Association estimated that over a ten-year period ending in March 1997, an average of 874 applications were filed annually in the Japanese Patent Office in the field of genetic engineering (Japan Patent Office technology code no. C-12). Approximately 40% of these listed a Japanese university faculty member as an inventor (JBA 1998).[16] Although this study (discussed in more detail in chapter 16) has not been repeated for other technologies, it suggests that many university discoveries are transferred to the private sector informally and are thus unaccounted for in any normal statistics[17] (see also Yoshihara and Tamai, chapter 15).

Informal transfers occur in many ways. Scientists who are well-known or whose research is of interest to companies often talk to companies about their research; graduate students find employ-

Table 4 Disposition of Inventions Reviewed by Invention Committees in all 99 Japanese National Universities in 1995 and 1996

	1995	1996
Individual inventor retains ownership	390 (90%)	382 (85%)
IP rights transferred to nation	34 (8%)	53(12%)
Inventor requested his IP rights be transferred to nation	11 (2%)	13 (3%)
Total	435 (100%)	448 (100%)

Sources: Monbusho Advisory Com. (1998); Monbusho (undated).

ment in companies; corporate researchers working in university laboratories communicate research results back to their companies. Sometimes the transfer of IP rights occurs without any written agreement. Sometimes it is accomplished through a short document known as an *obo-e-gaki*, which usually functions as an assignment agreement.[18] If a company believes a university researcher's discovery is important, it may decide to apply for a patent.[19, 20] An official of a major national university said that many of the university's most productive faculty members do not know how many of the various discoveries that they have reported to companies have been patented.

Conversations with university scientists and corporate officials suggest that university inventors rarely negotiate for personal royalties in exchange for transferring their discoveries to companies. However, companies often give Donations to individual professors' laboratories to maintain an amicable relationship with the laboratories and to help ensure the professors' cooperation in providing access to new technologies and promising graduates. Donations tend to be under ¥1 million ($6000), because division directors in many corporations can authorize Donations up to this amount, whereas executive board approval is usually needed for larger ones. Monbusho discourages and must approve Donations above ¥5 million ($30,000) a year to an individual professor's laboratory. Professors often say that the value of the research they perform for companies far exceeds the amounts they receive in Donations.

Donors are prohibited from making advance contracts to receive IP rights (Monbusho 1996).[21] Nevertheless, professors and companies often perceive Donations to be ideal forms of research support, provided both parties trust and cooperate with each other. As noted above, Donation funding is more flexible than other forms of research support. Until Japanese tax authorities recently began to scrutinize more closely the contractual aspects of Donations, companies could automatically deduct from taxable income the entire amount of their Donations, as if they were purely charitable gifts. In contrast, corporate support for Commissioned or Joint Research has been subject to scrutiny by Japanese tax authorities, who can challenge the deductibility of specific expenditures in support of such research. Perhaps most importantly, inventions arising under Donations do not face the delays and uncertainties associated with reporting to university presidents and review by invention committees, because there is a presumption that such inventions do not belong to the nation. This presumption is implicit in the criteria to determine which inventions are National Inventions (Notification 117) and the internal rules of the three national universities discussed above. The low levels of total funding under Commissioned and Joint Research compared with Donations (see Table 2) are consistent with university researchers and companies favoring Donations as a means for companies to support university R&D in return for IP rights to university discoveries and access to university researchers.[22]

Special Provisions Relating to Commissioned and Joint Research Agreements

Prior to the 1970s, there were no formal mechanisms permitting national universities to accept research funds from companies in return for IP rights to discoveries made using such funds. Monbusho Notification No. 260 of 1970 specifies that when a company sponsors Commissioned Research in a national university, it is possible to transfer to the company a "portion" of the nation's rights in any patents that arise under such research. Monbusho Notification No. 195 of 1983 established the system of Joint Research Agreements allowing personnel exchanges in the context of industry-sponsored research. Monbusho Notification No. 172 of 1984 gave national university presidents the authority to issue to sponsoring companies a seven-year "preferential license" to university inventions arising under such research. In 1997, this period was extended to ten years (Notification 186). The following are the principal aspects of Commissioned and Joint Research Agreements:

1. These agreements can be negotiated directly between national universities and companies. Approval from Monbusho is not needed, and the time it takes to conclude such agreements has diminished substantially over recent years so that now relatively uncomplicated agreements can be concluded in one or two months. However, the beginning of research projects must coincide with the start of the Japanese fiscal year, April 1: If a company and a national university decide in July to begin a Commissioned or Joint Research project, the company cannot begin disbursement of funds until the following April. In addition, corporate funds must be disbursed on an annual basis through the MOF, and funds for one year may not be rolled over to the next year. One company reported that because of delays in disbursement, funding for a project it sponsored was not available for use by the university laboratory until the end of September, nearly half a year after the company sent the funds to the MOF.

2. It is imperative that Joint Research inventions be reported to the university president and reviewed by the university's invention committee (Notification 195), and there is a strong presumption that the same is true with respect to Commissioned Research inventions (Notifications 260 & 172).

3. There is a strong presumption that Commissioned and Joint Research inventions by university researchers belong to the nation (Notifications 172, 195 & 260).[23] However, it is possible for the university to contract with the sponsoring company to transfer a portion of its rights to the company. In addition, if an invention committee decides an invention "is not for application purposes," it can recommend that the inventor retain the university's ownership share.

4. There are two methods of transferring IP rights in Commissioned and Joint Research Inventions to the corporate sponsor:

(a) At the time the agreement is made, the sponsor and the

university agree to contract terms that give the sponsor a partial ownership interest (usually 50%) in inventions emerging from the research. When inventions arise, they are reviewed by the invention committee and (unless there is no interest on the part of the sponsor) determined to be National Inventions. The sponsor and the university then become co-applicants on the Japanese patent application. The sponsor pays patent prosecution costs in proportion to its ownership interests and is thereby able to control patent prosecution.[24] JSPS does not become involved in the process. As a co-owner of any patents along with the university, neither the university nor the sponsor can transfer its rights to a third party without the other's permission.[25, 26] The sponsor must pay royalties to the government (Notifications 195 & 260).

(b) If there is no contract regarding co-ownership of patent rights, the university president can approve a preferential license to the collaborating company or another company specifically designated by the collaborating company. The duration of such a license is 10 years from the date of the patent application, although this period can be extended. If the collaborating company or its designee does not, absent sufficient justification, develop or practice the invention within two years after patent filing, or if remarkable public harm is recognized as resulting from the license, the university president can transfer license rights to another party. The government (i) has a right to royalties, and (ii) must approve any sublicensees (Notification 172) (see Notes 25 and 26). Method (a) gives the sponsor stronger IP rights than method (b) and, not surprisingly, is much more commonly used.[27]

Despite their complexity and lack of clarity, the Notifications governing IP rights to Commissioned and Joint Research inventions permit a more efficient and flexible technology transfer process than if JSPS and JST would have had to manage patenting and licensing. By giving universities authority either to apply for patents jointly with the sponsoring companies or to license inventions preferentially to such companies, they represent a decentralization of authority over the management of IP rights to National Inventions.[28]

Most inventions reviewed by invention committees probably

arise under Commissioned Research or Joint Research Agreements. Data for the ten-year period 1988–1997 show that 26 of the 46 inventions (57%) designated as National Inventions by the University of Tokyo's Invention Committee were either Commissioned or Joint Research Inventions. For the most recent five-year period, 1993–1997, the predominance of Commissioned or Joint Research inventions is even greater: 23 of 32 (72%), due to an increase in the number of such inventions (Invention Com. 1998).[28]

Summary of the Basic Framework

University invention committees are often criticized for impeding technology transfer from Japanese universities to industry. In many national universities, the invention committee meets only once a year.[29] Committee members are appointed from various science and engineering departments, but usually have little expertise in evaluating discoveries outside their field of expertise, much less their commercial potential and need for IP protection. They have full-time academic responsibilities, and thus little time to devote to committee work; furthermore, they may be reluctant to antagonize their peers by ruling that an invention should become national property. University administrators and some committee members may also believe that it is better for the invention committees to be inactive, if being more active might result in the committees deciding that more inventions should be National Inventions, thereby diminishing the prospects of commercialization.

In summary, an "official" technology transfer system, focused on the need to identify inventions that should belong to the nation and that should be patented and licensed (usually on a non-exclusive basis) by central government agencies, co-exists with a system which allows university researchers to transfer rights to their discoveries directly to companies. The boundary between these two systems is unclear, but because the criteria for classifying inventions as National Inventions emphasize "practical applications," the inventions that are vulnerable to being so classified are also likely to be of interest to industry. However, because of the remoteness of JSPS and JST from the scientists and engineers who understand the details and utility of their discoveries, the delays inherent in official procedures, and the barriers to companies receiving the strong exclusive or preferential license rights that many companies need, there is little likelihood that discoveries classified as National Inventions will be developed and commercialized by private companies. It follows that university researchers, administrators, and industries all have considerable incentives, whenever possible, to bypass official technology transfer procedures and let university inventors transfer their discoveries informally to the private sector. Available evidence suggests that this is precisely what has been occurring.

This "informal" or "bypass" system of technology transfer can be very efficient if there is a good match between the inventor and company, that is, if the company is willing and able to develop the inventor's discoveries. Transaction costs are low, at least for universities: No funds need to be paid to support the operations of a TTO or TLO; marketing and licensing have already been done by the inventor and the company. However, the informal system lacks a mechanism to select the most appropriate development partner from among potentially interested companies, and relies instead on preexisting relationships between individual researchers and companies. It lacks a mechanism to ensure that the company will actually develop and market the technology, unless the inventor is sufficiently knowledgeable and highly regarded to insist on a formal contract containing enforceable due diligence clauses (see Note 39). There is no mechanism to educate university researchers about the commercial value of their research and the importance of protecting IP rights, even though recent surveys indicate that awareness of these issues is low (Arai 1998), and that IP rights to a large percentage of patentable discoveries are lost because research results are published without filing patent applications.[30] There is no mechanism to collect data on technology transfers as they occur, much less on the outcome of such transfers. Finally, because official procedures are bypassed, companies may be concerned whether they have obtained legally enforceable rights to university discoveries.[31]

Technology Transfer from Private Japanese Universities

Except when private universities receive R&D support from the Japanese government for specific projects, they are generally free to set their own technology transfer policies. Nihon University, Ritsumeikan University, and Tokai University have recently required their employees to assign to them any employment-related inventions and have established active TTOs; other private universities, however, still adhere to the principle that IP rights remain with the inventor. As in national universities, most transfers still occur informally with no reporting of inventions or licensing/ assignment agreements, and R&D funds that private university researchers receive from companies tend to be low by U.S. standards, although several major research projects in Waseda University's Advanced Research Institute for Science and Engineering (RISE) are supported by companies. Waseda established a TTO in 1998. Keio University plans to formally open its TTO in the year 2000, although a precursor organization has been in existence since the 1960s.

Approximately one-third of Keio's R&D budget comes from the national government, largely in the form of Grants-in-Aid from Monbusho. Equivalent data are not available for Waseda, although compared with Keio, a larger portion of the government contribution is for Commissioned Research (primarily in RISE) and is funded by agencies other than Monbusho. As indicated above, these other agencies often specify in advance the disposition of IP rights to any inventions arising from the Commissioned Research they support. Neither Keio nor Waseda has an invention committee to determine whether inventions arising under Monbusho Grants-in-Aid should belong to the nation. At least in the case of small Grants-in-Aid, this possibility seems to be overlooked by all parties concerned. However, in the case of large application-oriented Monbusho-funded projects such as referred to in Note 11, Monbusho asserts rights to private university inventions that arise from such projects.[32]

Changes Likely to Result from the 1998 Technology Transfer Law

The Technology Transfer Law, legislation drafted by the Ministry of International Trade and Industry (MITI) and Monbusho to improve university-industry technology transfer, was officially enacted on 6 May 1998. It authorizes the establishment of TTOs to patent and license university inventions. Even in the case of national universities, the TTOs can be legally independent private corporations or publicly chartered corporations. National universities can thus avoid complications that would arise if they had internal offices collecting royalties from private companies. Private universities can establish TTOs within their official structures. Professors who have made commercially relevant discoveries are being encouraged to donate some of their inventions to their universities' TTOs. National university professors with technology transfer experience can serve as part-time advisors to TTOs, but not as full-time officials. Most of the former Imperial Universities, the Tokyo Institute of Technology, the University of Tsukuba, and several private universities are establishing TTOs, however in many cases it is expected that some universities will share one TTO, or use the services of another university's TTO. Universities have flexibility in shaping TTO operations.

The law leaves unchanged many of the technology transfer procedures described above. In particular, it does not alter the basic principle that inventors retain ownership rights in their inventions, nor the criteria for deciding which inventions should belong to the nation. It does not change the role of the university invention committees, although a number of proposals call for the committees to be more active and to review inventions more quickly. Some of these proposals envisage the committees as reviewing all new inventions (putting more force behind reporting obligations) and deciding quickly (within a few months) whether an invention should belong to the nation.

Article 4 of the law calls for persons or organizations planning to establish a TTO to draw up plans describing how the TTO will operate and to submit these plans for approval to Monbusho and MITI. TTOs with approved plans are allowed to manage the patenting and assignment/licensing of university inventions that do not belong to the nation and to receive licensing royalties. Thus, the new TTOs will provide university inventors with an alternative method of technology transfer. However, inventors will still have the option of transferring their IP rights directly to companies. Most TTOs will require inventors to assign their IP rights to the TTOs in order to use their services. The TTOs will then assume responsibility for patenting and the transfer of IP rights to companies.

Under Article 12, TTO organizers must submit special plans to Monbusho in order for the TTO to have authority to transfer National Inventions. Under such plans, the organizers and Monbusho will need to agree on procedures for Monbusho to assign rights to such inventions to the TTO. University officials hope that such procedures will provide for automatic rather than case-by-case assignments, but as of September 1998 this issue was unresolved. One point of contention is Monbusho's and the MOF's position that, since these inventions are the property of the nation, the national government should receive payment from the TTOs that reflects the value of each National Invention in order to comply with the National Properties Law (§§2, 3 & 21) and the Financial Law (§9). Even if Monbusho agrees to assign rights in National Inventions to TTOs, the TTOs cannot begin patenting or licensing such inventions before April 1999.[33]

Details of funding for TTO operations were also still under consideration at the end of 1998, although it is recognized that they will not be self-supporting through assignment or licensing revenues for at least several years. University officials estimate that annual initial operating costs will be at least ¥50 million ($300,000). University TTOs can apply to MITI for up to ¥20 million ($120,000) per year for five years to supplement their budgets. TTOs are seeking private funding to make up the shortfall. The following brief descriptions illustrate how the TTOs of three major National Universities, the Tokyo Institute of Technology, Tohoku University, and the University of Tokyo, are approaching this issue.

Two of these TTOs[34] have decided to constitute themselves as independent for-profit companies free from direct supervision by Monbusho; the other plans to constitute itself as a non-profit organization within the overall structure of its associated university. The two for-profit TTOs are encouraging science and engineering faculty members to purchase TTO stock. One will allow individuals to purchase up to ¥1 million ($6000) in stock and hopes to attract 25 faculty stockholders. Faculty contributions will be matched by equity investments by financial institutions, particularly local banks interested in promoting the local economy. The other TTO will permit stock purchases up to ¥3 million ($17,000) and is encouraging science and engineering faculty members to purchase at least

¥50,000 ($300) in stock. In both cases, equity ownership will provide rights to future TTO profits, but not authority to manage the university's technologies. All three TTOs are also establishing a system of privileged access based on payment of membership fees. One plans to charge individuals ¥50,000 ($300) and corporations ¥100,000 ($600) for annual membership. It hopes to have approximately 200 corporate members, mostly local companies. Another plans to fix membership charges at between 1 and 2 million yen ($6,000–12,000) and to have 50 to 100 members. Membership charges for small businesses will be reduced by an as yet undecided amount. This TTO also has close contacts with local government organizations whose goal is to encourage technology diffusion to local industries. The third TTO is hoping that 20 companies will each pay ¥5 million ($30,000) in annual fees. In the case of all three TTOs, membership confers the right to see lists of available technologies, sign confidentiality agreements to obtain detailed information about these technologies, and have first rights of refusal in licensing. After 6 to 9 months, if no member takes a license, nonmembers will be able to sign confidentiality agreements and negotiate licenses. The two TTOs charging lower membership fees say that any company, including a foreign company, can become a member. They are also trying to encourage faculty to assign all their inventions to them. The TTO charging $30,000 membership fees welcomes applications from any company, but certain companies from a cross-section of high-technology industries are being encouraged to become members. It is hoped that at least one of these will serve as a broker of technologies to small companies. In the case of all three TTOs, licensing royalties will be used to cover the TTOs' operating and patent prosecution costs, with any remaining funds to be allocated between the inventor, the inventor's laboratory, the inventor's department, the university as a whole, and "profits" to be distributed among shareholders (in the case of the two for-profit TTOs). As of October 1998, none of these TTOs had reached its membership goals. If a family of promising technologies needs further development to be attractive to a company that will commercialize them, these TTOs hope to attract venture funding to bridge this development gap.

The recently formed TTO of a major private university has entered into a cooperative relationship with an investment subsidiary of a major bank. This bank subsidiary will review new technologies that inventors assign or bring to the attention of the TTO and select those technologies it intends to develop or license to other companies. The bank subsidiary will then receive assignments or exclusive licenses to these technologies from either the TTO or the individual inventors. In return, the TTO will receive a percentage of the royalties that the bank subsidiary receives from the transfer of these technologies to third parties. The TTO itself will take responsibility for transferring inventions not selected by the subsidiary. Any remaining funds remaining after paying TTO expenses will be shared between the inventors and the university. Although this arrangement reduces TTO staff requirements and patent prosecution costs, it substantially diminishes the TTO's control over the university's technologies.

Besides the need for financial support, TTOs also appear to be seeking partnerships with private companies in order to obtain access to expertise on evaluating the commercial potential of their technologies. Ironically, many Japanese investment and venture capital firms say that they themselves lack sufficient valuation expertise. In light of this concern about appropriate valuation of technologies, it is curious that many TTOs appear to be eschewing a strategy of active marketing to as many interested companies as possible so that "value" can be better approximated by competitive bidding and royalties based on a percentage of sales.[35]

Summary and Discussion

There are essentially four paths for transferring IP rights to Japanese university inventions to industry so that they can be developed:

National Inventions that are neither Commissioned nor Joint Research Inventions. The average number of such inventions (on which patent applications are usually automatically filed by JSPS or JST for inventions) is probably less than twenty.[36] The number that are licensed by JST is significantly less. Despite these small numbers, the possibility that inventions will be classified by university inventions committees as National Inventions hangs like a Damoclean sword over university-industry technology transfers. This is because many inventions can theoretically be classified as National Inventions, especially if they arose under Grant-in-Aid support, the main source of project-specific R&D funds in national universities. Classification as National Inventions means that such technologies face substantial legal and bureaucratic obstacles to development and commercialization by private companies. Although many Grant-in-Aid inventions are passed to companies informally, often in exchange for Donations, the questionable legitimacy of using Donations in this contractual manner casts a degree of uncertainty over such informal transfers. The possibility that many inventions could be classified as National probably has significantly discouraged university inventors from taking a more proactive role in developing the commercial potential of their inventions and has contributed to a culture of informal, undocumented transfers to companies. This has left university researchers in a weak bargaining position with respect to companies and deprived universities of data on what technologies they create and the fate of these discoveries. A Monbusho official responsible for university-industry cooperation stated publicly in early 1998 that the process of deciding which inventions are National Inventions and the system of patenting and licensing such inventions are without public benefit.

Commissioned and Joint Research Inventions. In 1996, the number of university inventions arising under Commissioned and Joint Research Agreements that were reported to university invention committees and slated to be the subject of patent applications was probably about 40–53. The number of Commissioned and Joint Research Agreements is growing and the number of inventions will likely increase. Although these Commissioned and Joint Research Inventions are almost automatically classified as National Inventions, their transfer to industry does not face the overwhelming barriers faced by other National Inventions. If sponsoring companies contract with universities for joint ownership rights to such inventions before research begins, corporate sponsors can substantially control the patent application process. However, the following factors continue to mitigate against the more widespread and efficient use of this path: (1) Incentives for universities to insist on development commitments in co-ownership or license agreements are low, because universities and inventors do not have access to the royalty stream or any other benefits associated with successful commercialization. (2) Funding under these agreements can only begin on April 1 each year, and funds for one fiscal year cannot be rolled over to the next. (3) Disbursement of funds sometimes takes a long time. (4) Sponsoring companies cannot license or sublicense their rights without permission from the university presidents and perhaps other government agencies, potentially a major concern for small business and investment companies.

Assigning inventions to TTOs. TTOs will not diminish transaction costs associated with having to obtain government permission to patent or license inventions, nor will they give universities control over the inventions made by their employees. Therefore, the two principal factors underlying the success of the U.S. Bayh-Dole system are absent from the new TTO system.[37] As in the case of U.S. TLOs, there will be incentives to concentrate scarce resources on licensing a few promising inventions while ignoring inventions with less immediate commercial potential. In Japan, however, these financial constraints are exacerbated because university inventors are under no obligation to assign inventions to their TTOs, but can instead transfer them directly to companies. This raises the possibility that inventors will transfer their most commercially attractive discoveries to companies with which the have longstanding relationships, leaving their less attractive inventions to the TTOs.

The membership fee priority system that most TTOs are adopting threatens to exacerbate weak TTO finances and undercut their fundamental technology transfer mission. If membership fees are high, companies will have strong incentives to deal directly with individual scientists in order to avoid paying the fees and the royalties that the TTO will charge. Inventors who have long-standing relationships with companies that are not TTO members may also want to bypass the TTOs. The facts that as of April 1999 the TTO that charges ¥5 million membership fees was far short of its membership goals and only one license had been concluded by all the officially approved TTOs, suggest that these concerns are warranted.

More fundamentally, the basic goal of technology transfer is to find the best companies to develop university inventions and to ensure that such companies do their best to follow through on development commitments. High membership fees restrict the pool of potential developers and may compromise this fundamental goal. High membership fees may be particularly burdensome for small companies, and thus may perpetuate the tendency for university technologies to flow to large companies, opposite to recent trends in U.S. universities. If a subsidiary of a large financial institution that is a core member of a "keiretsu"[38] group of companies is given a monopoly on attractive university inventions, that subsidiary may be more likely to license to companies within the same keiretsu group, even though these companies might not be the best candidates to develop certain technologies. Finally, if a limited number of companies have privileged access to a university's technologies, they may be able to obtain these at lower royalty rates than if the TTOs marketed them to an unrestricted number of companies.

Nevertheless, the new TTOs should increase awareness of the importance of technology transfer among university researchers and encourage some universities to develop expertise in managing their technologies. Whether the system will be sustainable depends upon the willingness of inventors to transfer their inventions to their TTOs, the interest of companies in TTO inventions, and the effective use by the TTOs of provisions in their license agreements that require companies to develop the technologies they receive.[39]

Inventors Retain Rights

Although it is risky to speculate on the number of patent applications filed by companies on inventions transferred directly from university inventors, the anecdotal surveys mentioned in our discussion of informal transfers suggest that this number is probably over 600 annually and may even exceed 1,000. In 1996, U.S. and Canadian universities and major non-profit research institutions filed slightly over 3,261 patent applications and negotiated slightly over 2,740 licenses, of which 53% were exclusive licenses (AUTM 1998). Considering that (1) the population of Japan is roughly half that of the United States, and (2) government support for biomedical university R&D in Japan is much less than in the United States (see Kneller, chapter 16) while biomedical inventions account for the majority of patents and approximately two-thirds of active licenses by most U.S. universities (see Mowery, chapter 11; AUTM 1998), the number of patentable Japanese university discoveries transferred to industry is probably not remarkably less than in the United States.

If the Inventors Retain Rights development path is deficient, the reason is probably not lack of contact between university researchers and industry, nor an inability of industry to obtain IP rights to university discoveries. Rather the nature of university-industry contacts and the lack of incentives for Japanese companies to use effectively the technologies they receive from universities are probably at issue. About two-thirds of patented technologies are neither being developed nor used by the patent holder, nor licensed to other companies. The most common reasons companies give for not exploiting patented technologies are (1) certain of these technologies are less effective than alternatives; (2) low profits are expected from their commercialization; and (3) patents were filed for defensive purposes (JPO 1996). The discussion in chapter 16 suggests that the proportion of unexploited informally transferred university inventions may be even higher, at least in the field of biomedical research.

It is doubtful that the new TTOs or the Commissioned/Joint Research System will successfully address this problem in the near future. Considering in addition that (1) the Inventors Retain Rights path is the most prevalently used development path today, (2) incentives for inventors to switch to use TTOs are not clear, and (3) widespread adaptation of a Bayh-Dole system is unlikely in the near future (especially in national universities); the Inventors Retain Rights path will probably remain the most frequently used system of technology transfer for the near future. Therefore, Japan's best strategy may be to try to improve this system. This might be done through outreach to faculty members and by providing them with information and support services to enable them to transfer their discoveries more effectively. Specific steps might include:

• seminars to raise faculty members' awareness of the potential commercial value of their research and the potential benefits of cooperating with industry;

• guidance on how to make contact with companies and negotiate with them;

• information on the basic components of any license agreement, including: specification of the field of use, development commitments, duration, exclusivity, provisions concerning royalty payments or non-monetary research support, and restrictions (if any) on publication and other aspects of academic freedom; and

• information on various resources to help inventors find appropriate corporate development partners, negotiate with prospective partners, find sources of outside capital to develop their discoveries, and acquire additional R&D manpower. Such resources could include the university's TTO, independent technology brokers, consultants, and various business or technology databases.

The basic objective would be to empower individual university inventors to be more effective technology transfer agents; to change the Inventor Retains Rights path from an "informal," "bypass" system to a system that enables inventors to act as effective technology transfer agents. This would require more contractually based transfers. Although this would entail adjustments to traditional relations between university faculty and industry, it would increase the transparency of technology transfers and make collection of data on such transfers considerably easier. However, it would be important to clearly limit the criteria for classifying inventions as National Inventions and to ensure that, except in rare cases, such classifications be made *prospectively* at the beginning of research projects rather than *retrospectively* once inventions have occurred and their commercial potential is apparent.[40]

Japanese Technology Transfer in Its Larger Social and Business Context

Changes in other areas besides IP rights may also be necessary to improve university-industry technology cooperation in Japan. The immediately preceding discussion assumes that faculty members are ready to bargain for increased support and firm development commitments from companies. However, conversations with Japanese university researchers and administrators suggest that university inventors rarely bargain for royalties and development commitments, even in private universities where restrictions on cooperation with industry are less than in National Universities.

One reason professors do not negotiate with companies for royalties and development commitments may be that the transfer of IP rights from themselves to companies is just one aspect of their multifaceted long-term relationships with the companies. Professors depend upon companies to employ their graduate students; corporate contacts provide professors with interesting information, ideas for research with practical applications, and annual Donations usually not tied to any particular invention or line of research. Under these circumstances, it may seem ungrateful or counterproductive for professors to bargain hard over licensing terms when promising inventions do arise.

Another reason is that increased royalties or contributions from industry may be of limited usefulness in view of difficulties in recruiting young researchers to work on temporary research projects. Lifetime employment, with promotion opportunities and retirement benefits depending greatly upon length of service within a particular organization, is still the norm in Japan (Aoki 1990). It is generally not considered to be a career-enhancing step for a promising new M.S. or Ph.D. recipient to spend a couple years working on a temporary university research project (Coleman 1999). Moreover, the number of doctoral, postdoctoral, and assistant professorship positions in Japanese universities are limited (see Note 3), and companies still prefer to hire graduates with M.S. rather than doctoral degrees, whom they will train in-house to suit their needs (Coleman 1999). Research assistantships or stipends to defray tuition and living expenses are rarely given to Ph.D. students,[41] and most national university faculty and administrators interviewed for this study believe it is not permissible to use corporate research support for such purposes. In fact, however, it appears that it is permissible to use Donations and corporate support for Commissioned and Joint Research for stipends and research assistantships for Ph.D. students and post-doctoral researchers, although there are limits on the hourly pay such persons can receive and on the amount of time Ph.D. students can spend per week on industry-supported projects that are not their own thesis research.

National university researchers who dare to form their own companies must resign from their university positions if they hold a management position in the company.[42] In Japan, the infrastructure of individual donor networks and venture capital companies to support such start-up companies is not nearly as developed as in the United States. Japanese venture capital companies, most of which are subsidiaries of large financial institutions or major companies, tend to fund research that is at a more advanced stage than do their U.S. counterparts and to direct much of their investment overseas. It is common to hear officials of Japanese venture capital companies say that the quality of Japanese university research is inferior to that in the United States, and to decry a lack of entrepreneurial spirit among Japanese university researchers. However, the Japanese financial services industry itself lacks the U.S. expertise to evaluate new technologies and to advise on the management of start-up companies, expertise that depends in large part on former scientist-entrepreneurs who have been through the process of forming and managing their own companies. If a Japanese scientist-entrepreneur persists against these financial obstacles and the lack of job security and then his company fails, the personal cost is much higher than in the United States. His credit record is ruined, it is hard for him to find employment, his family members face ostracism, and he often must move because of failure to make mortgage payments.

Therefore—whether because of limited numbers of researchers, greater concern about job security, real or perceived labor immobility, limited access to capital and management advice, or the greater cost of failure—very few university researchers form companies. For example, as of 1995 in all Japan, approximately 89 new companies had been formed that had significant R&D activities related to biotechnology-based pharmaceuticals. Only four of these were started independently rather than as offshoots of large companies or as spin-offs from government-organized research consortia. Of these four, only one had been formed in the 1990s, and this company was capitalized by personal funds of the founding scientist's family (Kano, chapter 14). In the United States, small companies formed to develop promising technologies play an important role in developing university inventions to the point where large companies are willing to pursue final development and commercialization (AUTM 1998; Nelsen 1998; Atkinson 1994). Small companies account for over half the licensees of university technologies (AUTM 1998; GAO 1998). Without such companies, technology transfer from Japanese universities to industry may be more difficult than in the United States, no matter what rules govern the control and transfer of IP rights.

In order to encourage the sort of risky entrepreneurial dedication to the development of potentially valuable discoveries that occurs in the United States, some large Japanese companies are allowing some of their scientists and engineers to form their own "in-house" start-up companies capitalized by funds from the parent companies. Thus, the possibility exists that Japanese universities will never become more integrated into a national or international technology development system, but instead will be bypassed as the private sector develops unique ways to exploit the capabilities of talented scientists and engineers interested in fundamental research topics of importance to industry.[43] On the other hand, it appears that there may soon be a substantial pool of young scientists and engineers who can find neither lifetime employment in major companies nor tenure track academic positions and who are willing to work as post-doctoral researchers or join small companies. As mentioned in Note 3, the government is funding the training of more Ph.D.-level researchers. At the same time, at least in the pharmaceuticals industry, hiring of new university science and engineering graduates has fallen dramatically, with no company hiring more than 10 graduates in 1998 and many companies hiring fewer than five or even none.

In this situation, there might be opportunities for American or European individual investors or venture capital companies to work with Japanese university researchers or recent university graduates to develop promising early-stage technologies. Under the Inventor Retains Rights path (and under the Commissioned/Joint Research path, provided close relations are maintained with university leaders), the transfer of IP rights to foreign investors should be manageable. Whether the new TTOs will assist this process depends upon their openness to working with small companies and independent Japanese and foreign investors.

Acknowledgments

Dr. Kneller's research was assisted by a grant from the Abe Fellowship Program of the Social Science Research Council of the American Council of Learned Societies, with funds provided by the Japan Foundation Center for Global Partnership, and also by the cooperation of many Japanese scholars and government and business officials who kindly provided information for this study.

Notes

Unless otherwise indicated, throughout this paper U.S. dollar equivalents of Japanese yen amounts are estimated using the Organization for Economic Cooperative Development's (OECD's) "purchasing power parity indices." For 1996, this index was 177 yen/dollar. This index is much steadier than market exchange rates which have fluctuated between 85 and 145 yen/dollar from 1994 to 1998.

1. It has been suggested that Japanese government statistics overstate the amount of government support for university R&D, compared to the way expenditures are generally accounted for in other industrialized countries (Ohtawa chapter 6; Irvine 1990). See also chapter 16.

2. The 1995 expenditures for university R&D for the principal U.S. government science agencies were: NIH, $6,271 million; NSF, $1,734 million; DOD, $1,592 million; NASA, $708 million; DOE, 594 million; USDA, 435 million; and others, 599 million (NSB 1998).

3. Most Japanese science and engineering graduates take jobs in industry after receiving either a B.S. or M.S. degree. In 1995, only 5,453 persons received Ph.D.s in science and engineering from Japanese universities, compared with 26,207 in the United States (NSB 1998). Approximately one third of these degrees are "thesis" doctorates awarded by Japanese universities to persons who do not enroll as Ph.D. candidates but instead submit for evaluation the results of research performed at their place of work (NSB 1996). The Japanese government is increasing the number of doctoral and postdoctoral researchers, awarding 2,300 doctoral and 3,800 postdoctoral fellowships in 1996, while planning to increase the number of awards to 3,600 and 6,400, respectively, in the year 2000 (Normile 1995; NSF 1997a).

4. Keio, Waseda, and Sophia Universities among private universities, Osaka Prefectural and Osaka City Universities among local government universities, and a small number of private or local government medical schools have been listed as examples of non-national universities with significant research activities (Yamamoto 1996; Keii 1984).

5. Prominent among the 99 national universities are the seven former "Imperial Universities"—the Universities of Tokyo, Kyoto, Osaka, Kyushu, Nagoya, Tohoku,

and Hokkaido—the Tokyo Institute of Technology, and the University of Tsukuba (formerly the Tokyo University of Education) (Yamamoto 1996; Keii 1984).

6. Most recent notifications setting forth technology transfer procedures have been issued by Monbusho's Science and International Affairs Bureau. Some of the principal technology transfer notifications (for example, Notification 117 discussed below) require approval of Monbusho's Science Advisory Council. This Advisory Council, composed primarily of prominent university professors, serves as an independent advisory body for Monbusho.

7. The JST is a branch of the Science and Technology Agency (STA), one of the principal Japanese government science agencies. Patenting and licensing inventions on behalf of other government ministries and agencies, as well as publicly chartered corporations and local government laboratories, the JST handles the patenting and licensing of all publicly supported research discoveries, unless individual government agencies establish alternative technology transfer procedures.

8. Separate Monbusho accounts, distinct from any of the listed categories of R&D support, cover salaries, bonuses, and health and retirement benefits for full-time government employees (which includes tenured faculty and full-time support personnel); scholarships and stipends for graduate students and post-doctoral researchers; and building and construction costs.

Separate Monbusho accounts also cover special projects to enhance university-industry cooperation, such as the establishment in 1996 and 1997 of 24 fully equipped "Venture Business Laboratories" (VBLs) in universities to provide space for cooperative research between university and industry researchers. The initial 1995 budget to establish these VBLs was ¥25 billion (about $150 million) (ATIP 1998). Inventions made in VBLs are not National Inventions simply by virtue of being made in these laboratories.

9. Although it is theoretically possible to cover the costs of temporary support personnel with these funds, they are rarely sufficient for this purpose, since their total annual amount has remained nearly constant since the early 1980s (Ohtawa chapter 6; Irvine 1990; Tamura 1989).

10. A typical laboratory in a Japanese National University consists of a single "kouza," or tenured group of civil servant academics, typically consisting of one professor, one associate professor, and one or two research associates; plus support personnel.

11. One of the principal Monbusho programs using these investment funds is called "Research for the Future." It provides large-scale competitive grants to groups of university researchers for application-oriented research. Projects are funded for five years, with annual funding levels ranging from ¥50 to ¥300 million ($0.3 to $2.0 million). In Japanese fiscal year 1997 ¥20.6 billion ($120 million) was budgeted for 204 projects under this program (NSF 1997b; NSF 1998). University inventions arising under these grants are National Inventions.

12. The University of Tokyo's Research Center for Advanced Science and Technology had 31 Commissioned Research projects in 1996, six with private

companies, eight with foundations or non-profit corporations, five with local governments, and 12 with national government agencies or corporations. The median annual funding level (including overhead, but excluding travel and administrative expenses and honoraria) was ¥2.3 million ($13,000).

13. Examples of such research include the following three recent or ongoing projects involving scientists at the University of Tokyo's Research Center for Advanced Science and Technology: (1) Basic research to develop drugs to treat atherosclerosis. The collaborator is a major pharmaceutical company. Funding from 16 December 1996 to 31 March 1997 was ¥5.41 million ($45,000 over 3.5 months). (2) Development of high-efficiency photosynthetic bacteria and algae. The collaborator is a foundation concerned with the development of environmental technology. Funding from 28 August 1996 to 31 March 1997 was ¥12.47 million ($100,000 over 7 months). (3) Development of high-speed Josephson memory devices. The collaborator is a major telecommunications company.

14. Decisions by national university invention committees and the respective university presidents are, in effect, decisions on whether patent applications should be filed on behalf of the universities. In practice, JSPS does not question these decisions. Decisions on foreign patent applications (and payment of maintenance fees) are made primarily by JSPS in consultation with JST. University presidents can, if they wish, have considerable influence on such decisions. Monbusho has a budget ceiling for reimbursing universities for patent application expenses, but according to officials in JSPS and JST, this ceiling is rarely approached.

15. Although an exact breakdown between non-exclusive and preferential licenses is not available, approximately one-third involve the transfer of a technology to two or more licensees (implying non-exclusive licenses), while two-thirds involve only a single licensee for a particular technology (suggesting that these could preferential licenses). However, officials at JST say it does not often issue preferential licenses.

16. It has been suggested that the true proportion should be even higher than 50%, because a company may substitute the name of one of its own scientists for the name of the actual university inventor. This may not be a frequent practice, however, because many professors take pride in the fact that they are listed as inventors on patents. In addition, if a company anticipates later applying for a U.S. patent, it must name the inventors accurately or risk invalidating any future U.S. patent (Kayton and Kayton 1995).

17. A 1997 article in the *Nikkei Industrial Times* stated that over a recent five-year period 317 of 359 inventions (88%) made in the graduate engineering research department of Tohoku University were assigned informally to companies, and that 185 of 197 inventions (94%) developed in the University of Tokyo's Institute for Industrial Technology were similarly passed informally to companies.

18. One factor that might prompt a company to request a written assignment from a university inventor is that, under U.S. patent law, a company cannot apply for a U.S. patent unless there is a written assignment from the inventor(s) (37 CFR §§1.41 & 3.71).

19. A few university inventors actually apply for patents on their own and then formally either assign or license their rights to companies. However, because of the expense of patent prosecution, this option is available only to researchers who already have substantial sources of outside support.

20. Although in principle the right to apply for a Japanese patent derives from inventorship, an inventor's assignee may file an application on the invention, even without a written assignment agreement (Japanese Patent Law, §§ 33 & 34). There is a rebuttable presumption that assignments are valid. An inventor can challenge this presumption during examination proceedings at the JPO. However, publication of the application 18 months after the applicant's filing is a novelty bar against the inventor or any alternative assignee filing its own application. Therefore, if an inventor opposes a particular company applying for a patent on his invention, unless he successfully challenges the application prior to its publication, he has little hope of depriving the purported assignee of ownership rights.

21. Nevertheless, companies sometimes attach letters of understanding to their Donation documents stating an expectation that they will have IP rights to inventions arising from research supported by the Donations. Such letters of understanding are probably not legally enforceable.

22. Nevertheless, funds received under Commissioned Research agreements increased almost threefold from 1994 (¥6.6 billion) to 1996 (¥23.3 billion) (Monbusho Advisory Com. 1998), due in part to pressure from the Japanese government to encourage the use of Commissioned or Joint Research Agreements for large research projects, instead of Donations. In certain fields, such as cooperative drug trials, the government has severely restricted the use of Donations.

Also, some companies are concerned that the contractual use of Donations may be improper, at least in certain circumstances. An official of a Japanese subsidiary of an international biomedical corporation reported the dilemma his company faces in obtaining experimental reagents from a professor at a national university. The company would like to pay the professor's laboratory the equivalent of a non-exclusive, commercial-use license fee for the reagent, and a Donation of approximately a million yen is a mutually acceptable way to do so. However, the company is concerned that if Monbusho were to scrutinize this transaction, there might be negative consequences both for the professor and the company. The company is not sure who owns the reagent. If it is National Property, then, at least in theory, licensing would have to be handled by JST and payments made to MOF (National Property Law (Law No. 73 of 1948) §§2, 3 & 21; Financial Law (Law No. 34 of 1947) § 9). This would entail a long bureaucratic process. On the other hand, if the reagent is treated as analogous to an invention of the professor, then the professor should be able to distribute it as he sees fit, unless it meets one of the criteria for being a National Invention. Even if the reagent is regarded as the professor's own property, the company still is concerned that such a contractual use of Donations might be considered improper by Monbusho. These concerns take on added urgency because corporate research deductions are occasionally audited by Japanese tax authorities, who are employees of the MOF. At least so far, when tax authorities determine that Deductions are made in consideration for university technology, they apparently do not report such determinations to Monbusho or to authorities within the MOF responsible for managing National Properties. This particular company seems to be quite conscientious in trying to comply with the law, but it is not sure what the law is. If it turns out that such reagents are National Properties, either industry access to research materials from national universities would effectively cease, or donors and Donation recipients would be in violation of the National Property and National Finance laws.

23. If a Commissioned Research invention arose under funds from another Japanese government agency or under special Monbusho investment funds (Note 11), that agency or Monbusho may require that the invention be automatically designated as a National Invention and that the funding agency make patenting and licensing decisions. For example, the Science and Technology Agency (STA) usually stipulates in advance that any inventions arising under Commissioned Research that it sponsors will belong to the nation and that STA, working through JST, will determine to whom and how they are assigned or licensed. The Ministry of International Trade and Industry (MITI) or one of its affiliated organizations usually draws up a special contract with the university that specifies in advance the disposition of rights to any inventions arising under Commissioned Research that it sponsors.

24. Controlling patent prosecution is often important for companies receiving IP rights, because it allows them to shape the patent specifications and claims to meet their interests, to decide when to ask the Japan Patent Office to review the application, to decide how to respond to actions from the Patent Office, etc.

25. Japanese Patent Law states that the consent of all joint holders of a patent right is necessary to license the patent right exclusively to anyone or non-exclusively to a third party (Japanese Patent Law §73). In contrast, a co-holder of a U.S. patent does not need permission from the other co-holders to license or assign his or her rights.

26. None of the Notifications address sublicensing. In practice, the university president (and by extension, the inventor himself) has substantial influence over any sublicensing decisions and his approval must be obtained (personal communication with patent attorneys familiar with technology transfer issues). In the case of Commissioned Research funded by an agency such as MITI or STA, the funding agency also has considerable authority over sublicensing decisions. According to officials at JST and JSPS, although JST and JSPS theoretically can block sublicenses, in practice they are unlikely to block proposals backed by a university president.

27. The President of the University of Tokyo has issued no licenses for the 26 Commissioned and Joint Research inventions patented on behalf of the Univer-

sity of Tokyo between 1988 and 1997. All but one of the 18 Joint Research inventions were filed as joint applications by the University of Tokyo and the corporate sponsor. Three of the eight Commissioned Research inventions were also jointly patented. The remaining five Commissioned Research inventions were reviewed by the Invention Committee in 1997 and joint patent applications may yet be filed (Invention Com. 1998).

However, in the case of Commissioned Research, while Notification 172 of 1984 clearly sets forth the procedures for (b), Notification No. 260 of 1970 provides only vague authority for the procedures described in (a). In the case of Joint Research, Notification No. 195 of 1983 provides somewhat clearer authority for the procedures described in (a).

28. In the interest of completeness it should be mentioned that in the case of an invention made by a Commissioned Researcher, the university or the Commissioned Researcher's home institution will apply for patents jointly or independently, according to the level of contributions. The university president and the invention committee have primary authority to decide upon the allocation of inventorship rights (Monbusho 1996).

29. This infrequency is primarily due to the small number of reported inventions. If necessary, most committees can be convened once a month. Even a month's delay, however, could be detrimental for researchers in competitive fields who may need to obtain quick patent protection for their discoveries.

30. By identifying university employees whose names appeared as inventors on patent applications within selected technical areas, and also the articles authored by these university inventors in selected applied science journals, an expected ratio of patent applications to journal articles was calculated. In a personal communication, a patent attorney familiar with technology transfer told me that when compared to the total number of articles in these applied science journals, this ratio predicted that the number of patent applications based on university inventions should be about five times the actual number of such applications. Of course, part of this discrepancy may be explained by the fact that professors whose names appear on patents are a self-selected group, whose ratio of patent applications to publications would be higher than that of other professors publishing in the same journals.

However, interviews with officials in pharmaceutical companies who are responsible for seeking out potentially valuable university technologies also suggest that important technologies are frequently lost in Japan due to publication of research results without timely filing of patent applications.

31. At least one Japanese company that serves as an intermediary between foreign venture capital and Japanese university researchers routinely asks that all technologies that are candidates for licensing by its clients be reviewed by the appropriate university invention committees. Although in other circumstances the inventors might not bring the inventions to the attention of the committees, it insists on this step to diminish the possibility that an invention that is licensed to one of its clients might later be ruled to be a National Invention.

32. In one recent case, a private university inventor, who was working under a Research for the Future grant but unaware that he had to transfer rights in his invention to the nation, "transferred" his invention to a company with which he had been collaborating. JSPS invalidated the transfer. The company was forced to write a letter of apology. But then the company negotiated a co-ownership agreement with JSPS under which it agreed to pay royalties to the government. Neither the inventor nor the university obtained rights to the royalties.

33. If blanket assignment agreements can be worked out, at least one major national university is hoping that most of its faculty members will agree to assign all of their inventions to the nation. After assignment from Monbusho to the university's TTO, the TTO will manage all university inventions, and technology transfer will occur under a de facto Bayh-Dole system.

34. Because final plans are still being formulated and because these descriptions are for illustrative purposes only, attributions to the specific universities are omitted from this description.

35. Some senior officials and scientists advising TTOs say that they do not want their TTOs to be in a situation where two or more companies would be bidding for the same technology, preferring more managed and less competitive access to TTO technologies.

A competitive marketing system would require some way to prevent companies that receive confidential information about university technologies from duplicating the technologies in their own laboratories without licensing from the TTOs. One TTO abandoned consideration of competitive marketing, under which potential licensees would have had to sign confidentiality agreements containing promises not to use the technologies disclosed to them if they did not take a license, because large companies had objected to such "no use" clauses. The IP departments of these large companies maintained that they could not know what research was being conducted in all their companies' laboratories and therefore could not agree to such clauses.

However, another TTO will address this problem by filing Japanese patent applications on a large number of inventions before marketing them to prospective licensors. In the case of inventions with questionable licensing prospects, it will not use outside patent attorneys and will not request examination of the application, but will pay only the basic application fee of ¥21,000 ($120). This will establish, at minimal cost, a priority date that will prevent potential licensors from copying the inventions that are disclosed to them.

In the United States, with its "first to invent" system of patent priority, this problem is less acute because inventors and their universities can claim priority back to the date of invention. Under the "first to file" priority system that exists in Japan and most other countries, this is not possible.

36. Data are not available on how many of the national university inventions recently patented by JSPS (see Table 3) or designated to be National Inventions (see Table 4) were Commissioned or Joint Research inventions. The fact that the number of Commissioned and Joint Research inventions from the University of

Tokyo is growing and accounts for approximately three-fourths of all University of Tokyo National Inventions suggests that a large proportion of those from the other national universities are also Commissioned or Joint Research inventions.

37. Although there is an ongoing debate in the United States about the effectiveness of the Bayh-Dole system (see Mowery chapter 11; Nelsen 1998), the principal benefit most often attributed to this system is to eliminate these transaction costs and thereby give universities strong incentives to ensure that a large number of promising technologies are commercially developed. The Bayh-Dole Regulations have thus positioned U.S. university TLOs as effective technology transfer and innovation agents (Kano chapter 14). For a detailed account of the policy debates and legislative history related to the Bayh-Dole Act and subsequent U.S. technology transfer laws, see Eisenberg (1996).

38. "Keiretsu" is a term to describe associations between large numbers of companies based primarily upon a common major bank which provides financial support to the various companies in its keiretsu in return for equity in these companies. Cross-holding of stock is also common among companies within a particular keiretsu (Miyashita and Russell 1996).

39. A common practice of U.S. TLOs is to require prospective exclusive licensees to submit business development plans for the inventions they want to license and then to include "benchmark" or "due diligence" clauses in the license agreements enabling the TLO to revoke the license if the licensee is not meeting the development milestones set forth in these clauses. Japanese public institutions appear to have very little experience using such clauses. Some Japanese TTOs say they intend to use them in most exclusive licenses. Others say they are afraid that licensees will object and are still considering whether to use them. The author knows of no case of individual inventors including such clauses in *obo-e-gaki* or other agreements they negotiate directly with companies.

40. Germany is the only other major industrialized country that, as a general principle, allows university inventors to retain rights to their inventions. Since there is no need to determine retrospectively whether rights belong to the government, there are fewer incentives to hide transfers to the private sector, and technology transfer is more transparent. However, if the inventor is not a permanent or full-time faculty member, or if an invention is made using special government or European Union grants, patenting or licensing often becomes the responsibility of either the university or the funding agency. In these cases, it has often been difficult for the university or funding agency to issue exclusive licenses (Max-Planck-Gessellschaft 1994). The present German system may provide clues as to what the Japanese system would be like if the determination of National Invention status were made prospectively (except in exceptional, e.g., national security, cases) based upon clearly limited criteria concerning which projects will give rise to inventions that the government should control.

41. Annual tuition for doctoral students in national universities is ¥470,000 (about $3000). It is considerably higher in most private universities.

42. Whether they can hold a substantial equity interest has not been clarified.

43. Technology transfer from non-university government research institutes will be covered in a forthcoming companion study. In summary, although overall intramural R&D expenditures are roughly equivalent in these two sectors (Statistics Bureau 1996), technology transfer from government research institutes generally faces higher hurdles. Depending on the policies of the particular ministry or agency, inventions made in government research institutes either automatically become National Inventions or the government retains a half-ownership interest in them. This means that patent and licensing decisions are often made by central government agencies, and it is difficult for a company to obtain exclusive license rights, as the funding ministry or agency must approve licensing decisions. Some ministries and agencies are trying to streamline these procedures.

References

Aoki, Masahiko. 1990. "Toward an Economic Model of the Japanese Firm." *Journal of Economic Literature* 28:1–27.

Arai, Hidehiko. 1998. *Actor Analyses on Cases of Successful Transfer of Technology from Universities and National Institutions.* Tokyo: Second Theory-Oriented Research Group, National Institute of Science and Technology Policy, Japan Science and Technology Agency (in Japanese).

Asian Technology Information Program (ATIP). 1998. *Report: ATIP98.084: Tohoku University Venture Business Laboratory (VBL)* (distributed by Internet by ATIP, 12 Oct. 1998).

Association of University Technology Managers, Inc. (AUTM). 1998. *AUTM Licensing Survey FY 1996* (AUTM).

Atkinson, Stephen. 1994. "University-Affiliated Venture Capital Funds." *Health Affairs,* Summer 1994:159–175.

Coleman, Samuel. 1999. *Japanese Science: Hierarchy, Mobility and Control. London:* Routledge (In press).

Committee. 1996. Committee for the Study of Scientific Research Funding (ed.) *List of Approved Projects for Monbusho's Grants-in-Aid for Scientific Research, Part 1 or 2.* Tokyo: Gyousei (in Japanese).

Council on Governmental Relations (COGR). 1993. *University Technology Transfer: Questions and Answers.* Washington D.C.: COGR.

Eisenberg, Rebecca. 1996. "Public Research and Private Development: Patents and Technology Transfer in Government-Sponsored Research." *Virginia Law Review* 83:1663–1727.

Etzkowitz, Henry. 1994. "Knowledge as Property: The Massachusetts Institute of Technology and the Debate over Academic Patent Policy," *Minerva* 32:383–421.

Government Accounting Office (GAO). 1998. *Technology Transfer: Administration of the Bayh-Dole Act by Research Universities* (GAO/RCED 98–126). Washington DC: GAO.

Invention Com. 1998. University of Tokyo's Invention Committee. *Japanese Patents Belonging to the Nation: Applications and Acquisitions* (in Japanese).

Irvine, John et al. 1990. *Investing in the Future: an International Comparison of Government Funding of Academic and Related Research—Report of a study sponsored by the UK Advisory Board for Research Councils and the US National Science Foundation.* Hants, UK: Elgar.

Japan Bioindustry Association (JBA). 1998. *The Usefulness of University Research for the Promotion of Our Country's Bioindustry* (in Japanese).

Japan Patent Office (JPO). 1996. *Survey on the Appropriate Promotion of Patents: Fact Finding Survey of Unused Patents: Information Report: Summary* (Industrial Technology Laboratory. Tokyo: Japan Technomart Foundation (in Japanese); and Japanese Patent Office. 1996. *Outline of Results of Status Survey on Unused Patents* (English summary of above).

Japan Science and Technology Corporation (JST). 1997. *JST 1996–1997* (English brochure).

Japanese Patent Law (various cited articles). At http://www.jpo-miti.go.jp/index.html (in Japanese and English).

Kayton and Kayton (eds). 1995. *Patent Practice 4th Ed.* Charlottesville NC: Patent Resources Institute, Inc. (Chapter 16: Inventorship).

Keii, Tominaga. 1984. *A Study Evaluating Universities.* Tokyo: Tokyo University Press. (in Japanese).

Kneller, Robert. 1998. "University-Industry Technology Transfer in the United States", *Tokugikon* (a publication of the Japan Patent Office) 198:15–46 (in English and Japanese).

Latker, Norman. 1977. Statement before the Subcommittee on Science, Research and Technology of the U.S. House of Representatives, 26 May 1977. Also: public remarks by Niels Reimers, founder of Stanford University's Technology Transfer Office (Recruit Conference Center, Tokyo, 23 March 1998).

Max-Planck-Gessellschaft. 1994. *E2/94. European Research Structures—Changes and Challenges: The Role and Function of Intellectual Property Rights* (compilation of papers presented at Ringberg Castle, Tegernsee, January 1994). Also, personal communications with Dr. Joseph Straus, Professor of Law, Max Planck Institute for Foreign and International Patent, Copyright and Competition Law, Munich.

Miyashita, Kenichi & Russell, David. 1996. *Keiretsu: Inside the Hidden Japanese Conglomerates.* New York: McGraw-Hill.

Monbusho. 1996. (Science and International Affairs Bureau of the Ministry of Education, Science, Sports and Culture). *Research Cooperation Between Universities and Industry in Japan.* Tokyo: Monbusho (in English and Japanese).

Monbusho undated. *Pertaining to the Handling of Research Results (Patents) Arising in Universities* (undated Monbusho data sheet). Tokyo: Monbusho. (in Japanese).

Monbusho Advisory Com. 1998. *Report of the Monbusho Advisory Committee for the Promotion of University-Industry Cooperation: Aiming at the Construction of a New Technology Transfer System Related to Patents* (in Japanese).

National Science Board (NSB). 1998. *Science and Engineering Indicators: 1998.* Washington D.C.: National Science Foundation (Appendix Tables 2–30, 4–2, 4–41, 4–45, 4–46, 5–8).

National Science Board (NSB). 1996. *Science and Engineering Indicators: 1996.* Washington D.C.: National Science Foundation (Appendix Tables 2–30, 2–31, 5–5.

National Science Foundation (NSF). 1998. *Report Memo 98–01: The 1997 Monbusho White Paper.* Tokyo: National Science Foundation Regional Office.

National Science Foundation (NSF). 1997a. *Report Memo 97–11: Japanese Government Organization for Science and Technology.* Tokyo: National Science Foundation Regional Office.

National Science Foundation (NSF). 1997b. *Report Memo 97–06: Japanese Government Science and Technology Budget—Fiscal Year 1997.* Tokyo: National Science Foundation Regional Office.

National Science Foundation (NSF). 1995. *Report Memo 95–18: Japanese Ministry of Education's Grants-in-Aid for Scientific Research.* Tokyo: National Science Foundation Regional Office.

Nelsen, Lita. 1998. *"The Rise of Intellectual Property Protection in the American University."* *Science* 279:1460–1461.

Nikkei Industrial Times. 1997. "University Industry Cooperation: New Industry Creation, Part 1 of 3." (24 December 1997, page 4) (in Japanese).

Normile, Dennis. 1995. "Japan Expands Graduate, Postdoc Slots." *Science* 269:1335–1336.

Notification 117. 1997. Monbusho, Science and International Affairs Bureau and the Director of the Accounting Branch, "Notification 117: Pertaining to the Handling of Patents Related to Inventions by Faculty, of National Universities" (issued 25 March 1978, amended by Notifications No. 138 on 20 May 1987 and No. 163 on 27 March 1997), in The Study Committee for Handling Outside Funding for National Universities, ed. *Administrative Handbook for Research Cooperation between Universities and Industry, 2nd Ed.* Tokyo: Gyou-sei (pages 319–322) (in Japanese).

Notification 172. 1984. Monbusho, Science and International Affairs Bureau and the Director of the Accounting Branch, "Notification 172: Pertaining to the Licensing of Patents Related to Commissioned Research and Joint Research with the Private Sector" (issued 8 May 1984, amended by Notification No. 186 on 31 March 1997). In The Study Committee for Handling Outside Funding for National Universities, ed. *Administrative Handbook for Research Cooperation Between Universities and Industry, 2nd Ed.* Tokyo. Gyou-sei (pages 322–323) (in Japanese).

Notification 186. 1998. Monbusho, Director of the Science and International Affairs Bureau and the Director of the Accounting Branch of the Minister's Secretariat, "Notification 186: Revisions to 'Licensing of Patents Related to Commissioned Research and Joint Research with the Private Sector'" (in Japanese).

Notification 195. 1983. Monbusho, Science and International Affairs Bureau and the Director of the Accounting Branch. "Notification 195: Pertaining to Handling Joint Research with the Private Sector" (Issued 11 May 1983, amended by Notification No. 172 on 8 May 1984). In The Study Committee for Handling Outside Funding for National Universities, ed., *Administrative Handbook for Research Cooperation Between Universities and Industry.* Tokyo: Gyou-sei (pages 179–182); (reissued under the same title with revisions on 31 March 1997). In The Study Committee for Handling Outside Funding for National Universities. ed., 1997. *Administrative Handbook for Research Cooperation Between Universities and Industry, 2d Ed.* Tokyo: Gyou-sei, Tokyo (pages 205209) (in Japanese). See also Notification No. 186 of 1997.

Notification 260. 1970. Monbusho, Director of the Accounting Branch and Director of the Bureau, "Notification 260: Pertaining to the Handling of Commissioned Research" (issued 30 April 1970, amended by Notifications No. 115 on 1 April 1988) in The Study Committee for Handling Outside Funding for National Universities, ed., 1997. *Administrative Handbook for Research Cooperation Between Universities and Industry.* Tokyo: Gyou-sei (pages 252–253) (in Japanese).

Public Law (PL) 96-517. 1980. Amending Title 35 of the U.S. Code (USC) by adding Chapter 18, Sections 200–212.

Rule 53. 1978. "Tokyo Institute of Technology Invention Rules," in *Vol. 7: Cooperative Research* (approved 8 September 1978) (in Japanese).

Rule 58. 1978. *Tohoku University Rules for Handling Faculty Inventions* (approved 21 November 1978) (in Japanese).

Statistics Bureau. 1996. Statistics Bureau, Management and Coordination Agency, Government of Japan. *Report on the Survey of Research and Development.* Tokyo: Japan Statistical Society Foundation (in Japanese and English).

Tamura, Kenzi. 1989. "Funding Research in Japan." *Science and Public Affairs* 4: 31–41. London: Royal Society.

Technology Transfer Law. 1998. Full title: Law on the Promotion of the Transfer to Private Enterprises of Research Results Pertaining to Technologies Arising in Universities. No. 52. Tokyo: Ministry of International Trade and Industry. (in Japanese).

University of Tokyo Invention Rules. In *Vol. 4: Commissioned Research and the Like* (approved 17 April 1979 by the Board of Trustees; amended 21/4/81, 17/7/84 (completely revised), 22/4/86, 19/7/88, 2/3/91, and 18/5/93); and Regulations Implementing the University of Tokyo Invention Rules (enacted 17 July 1984, amended 2 March 1989) Tokyo: University of Tokyo (in Japanese).

Yamamoto, Shinichi. 1996. "The System of Mass Higher Education and Research Universities." *Research Reports*, No. 91. Chapter 8, pages 271–284. Tokyo: Broadcast Education Development Center (in Japanese).

13

Lack of Incentive and Persisting Constraints: Factors Hindering Technology Transfer at Japanese Universities

Mariko Yoshihara and Katsuya Tamai

One of the major characteristics of Japan's innovation system is the apparent absence of academic contribution. Unlike in the United States where universities are viewed as an "active and important participant" in research and development (Warshofsky 1989: 383), the Japanese universities are not perceived as a key source of innovation. The advocates of the conventional view readily cite official statistics that demonstrate the scarcity of university-based patents. Nevertheless, there are indications that the public data do not accurately represent the practicality of university-based research and the magnitude of university-industry collaboration in Japan.

Technology transfer at Japanese universities takes place through channels that are significantly different from those in the United States. This article aims to point out some characteristics of the current practice of technology transfer in Japan and to examine the mechanism behind its peculiarities. It starts by reviewing the popular perception that universities play only a minimal role in Japan's technological progress, then demonstrates that the official statistics downplay the magnitude of the inventive activities of the Japanese academic community, and finally analyzes the factors that discourage patent applications by academic researchers.

Perceptions of Technology Transfer in Japan

There is a popular perception, shared both by Westerners and Japanese, that the main conveyor of Japan's technological development has been the private companies rather than the universities. Unlike the academic community in the United States (Kennedy 1986, Warshofsky 1989), the Japanese universities are not perceived as the key source of innovation.[1] Anderson's comment (1984: 72) may well represent the Western view on Japan's research environment: "The general standard of research is still lower than in the West; each of the best universities contains only a few research groups of international standard. ... It is not in universities but in industry ... that really good research facilities are found." Similarly, Rohlen (1992: 339) contends that fundamental research in Japan has "occurred most typically in private corporations, rather than universities." Bloch (1992: 224) maintains that Japan has demonstrated "an abysmal neglect of its university research establishment." Swinbanks (1994) urges that the university researchers in Japan should "increase their interaction with the outside world." Finan and Williams (1992: 137) assert that the universities have acted as a "bottleneck" in the development of Japan's software industry. A bold censure of Japanese higher education has been expressed by Cutts (1997: 84), who proclaims that the Japanese universities are "disconnected" from their research responsibilities and have become "an extreme, almost grotesque form of an ivory tower." Although the degree of criticism may vary, there is one thing on which nearly all seem to agree: The universities have played only a peripheral role in Japan's technological development.

Accordingly, the practice of technology transfer from universities to private sector has been seen as minimal in Japan.[2] In their study of the international competition in high technologies, Brandin and Harrison (1987: 98) argue that there is a striking difference between the roles played by faculty members in the United States and in Japan,[3] maintaining that the Japanese professors have only "a very indirect influence over industry." Uenohara et al. share this view: In their study of the American and Japanese semiconductor industries (1984: 20), they state that the cooperation between universities and industry in Japan is "generally poor." Mukaibo (1991: 6) maintains that the linkage between universities and the private sector in research and development in Japan is "very tenuous, to the point of being negligible."

Patent statistics are considered to reflect inventive activity and innovation (Holman 1978; Pavitt 1985);[4] official data on patent applications seem to support the view that the academic community does not play an active role in Japan's technological development. According to a study conducted by the Ministry of International Trade and Industry (MITI), in 1994 only 129 patent applications were submitted by Japanese universities, amounting to only 0.04% of the nation's total patent applications.[5] In contrast, the total number of patents registered by the American universities in 1994 was 1,862, 1.83% of all patents filed in the United States during that year.[6] The Japan Patent Office reports that there were 137 patents filed by universities in 1995, 90 in 1996, and 107 in 1997, and it is probable that the corresponding figures in the United States were significantly higher: For example, the Japan Society for the Promotion of Sciences, a special legal entity under the jurisdiction of the Ministry of Education (Monbusho), reports that there were only eight technologies licensed by universities to private companies in Japan during 1994, compared to 2,049 in the United States in the same year. These numbers have been widely cited in the Japanese media,[7] reenforcing the image of the Japanese universities as a minimal player in technology transfer.

The Problem of Inadequate Data

The popular perception of the private sector as the main conveyor of Japan's technological development may be a legitimate characterization of the innovation system in postwar Japan. However, the practicality of academic research and the magnitude of university-industry collaboration in Japan seem to have been overlooked in the above-mentioned patent statistics. The following issues suggest that the official data do not properly reflect the nature of university-based research in Japan.

First, the increase of collaboration between universities and private companies is not adequately represented simply by the number of university-based patents. During the past decade, Japan has witnessed an increasing interaction between the academic and business communities.[8] Fujita (1997) portrays emerging university-industry collaboration in Japan.[9] He reports that ten out of the thirteen interviewees representing the Japanese major electronics companies[10] assert that the Japanese universities are playing an increasing role in research and development. The number of joint research projects between the national universities and private companies grew seven-fold between 1985 and 1994,[11] with a comparable increase in the number of industry researchers engaged in commissioned research at national universities.

However, university-industry collaboration has produced few visible results; according to the Ministry of Education, the total number of patent applications generated out of the joint research was only 326 through fiscal 1994. During 1994, only 20 patent applications were filed jointly by academic and industry researchers, although there were 1,488 on-going joint research projects (Odagiri 1997: 6). In biotechnology, Odagiri and Kato (1997) found no patent applications submitted jointly by academic and industry researchers as a result of their collaborative research projects.[12]

Second, the generally cited statistics on university-based patents are an inadequate and misleading measure of the magnitude of academic research. As mentioned earlier, although the Patent Office data (Japan Patent Office 1997) show that there were 124, 137, 90, and 107 patent applications filed by universities in Japan in 1994, 1995, 1996, 1997 respectively, a monthly report of Diamond Management Development, a Japanese private institution which specializes in the analysis of the domestic patent market, presents a very different picture. According to that report (Diamond Management 1997), there were 1,149, 911, and 941 patent applications filed by individual faculty at Japanese universities in 1994, 1995, and 1996 respectively, a significant difference from the Japan Patent Office figures.

We will discuss the source of these discrepancies below, but first let us examine the data referring to only one university: the University of Tokyo, using the same three years cited above. In 1994 its official number of patent application was two; Diamond lists 73. For 1995 and 1996, the official count was 13 and 3 respectively; Diamond counts 99 in 1995, and 98 in 1996.

Moreover, a 1996 survey by the Department of Engineering of the University of Tokyo[13] showed that of the patents filed that year approximately 150 listed department faculty as the inventors, while Diamond's assessment counted only 69 for the entire university. If three patent applications were filed by the University of Tokyo, and 69 by its engineering faculty, there remain at least 78 innovations unaccounted for, filed for neither by the University nor by the inventors themselves. What became of them? Typically, the applications would have been filed by private companies. This single case is a demonstration that statistics do not accurately represent the magnitude of academic involvement in innovation: We must not take these data literally.

The Current System of Technology Transfer in Japan

To explain the mechanism behind the current practice of technology transfer in Japan, we will argue that a significant proportion of the inventions made within the Japanese academic community are filed for patents by the private sector. The argument proceeds in two stages. First, it outlines the legal basis of the technology transfer practice in Japan in which the majority of the titles to inventions go back to the individual inventors. Second, it demonstrates that the current system creates no incentive for the faculty to file intellectual property rights themselves, making acquisition of the titles to inventions by private companies a common practice.

The Legal Basis of the Technology Transfer Practice in Japan

The notice issued in 1978 by the Ministry of Education (Monbusho) sets the basic rules concerning inventions developed at national universities.[14] It states that the faculty members of national universities are required to report to the university's president if they develop an invention. According to the Monbusho Notice, the title to the invention, or the right to file for a patent, belongs to the state if the invention was made under a special government research fund or using special research facilities and equipment established for the use of government-sponsored research. All other inventions become privately owned by the inventors. Table 1 highlights the relationship between the funding sources and the ownership of the title to inventions at all Japanese universities.

Table 1 Ownership of the Title to Invention According to Source of Funds[a]

Types of Funds	Funding Sources	Owner of Title to Invention
Regular Research Allowance to Each Chair[b]	Government (Monbusho)	Individual Inventor
Grants-in-Aid	Government (Monbusho)	State
Grants and Endowments	Private Sector	Individual Inventor
Commissioned Research	Government (Monbusho)	State or Individual Inventor
Joint Research Funds	Government (Monbusho)	Co-owned by State and Private Sector
Funds to Sponsor Commissioned Researchers	Government (Monbusho)	Co-owned by State and Individual Inventor

a. Funding for academic research in Japan is classified into: (1) the regular research allowance for faculty, (2) Grants-in-Aid, and (3) external funding, including grants and endowments from the private sector, funding for commissioned research, and funding for joint research with the private sector.
b. At the Japanese national universities, research is organized in a "chair" system. Each chair consists of a professor, an associate professor, assistants, and technical assistants. Funds are allocated according to a standard formula based on the number of researchers in a chair, and whether or not the chair holds graduate courses.

The Science Council of Japan, an independent advisory body to the government,[15] released an "administrative guidance"[16] in 1977, stating that each university should establish its own policies for handling inventions. The 1978 Monbusho Notice specifically encourages the national universities to set up an internal council of faculty members called an "Invention Committee," which assesses appropriate ownership of the title to an invention developed at the university and makes a recommendation to the university's president, who makes the final decision.[17] If the title to the invention is assigned to the state, the Japan Society for the Promotion of Science files the patent on behalf of the university. If the university's president determines that the title to invention does not belong to the state, it will be assigned to the inventor. It is up to the inventor whether or not to file for a privately-owned title to the invention.

Currently, a large proportion of the titles to university-based inventions are assigned back to the individual inventors. For example, of the 435 inventions assessed by the Invention Committees at national universities in 1995, 390 (89.7%) became privately owned by their inventors; 45 (10.3%) became state-owned, of which 11 were voluntarily transferred (Monbusho 1996).

The Mechanism behind the Current Practice: Informal Exchanges

Faculty and Firms. In many cases, technology transfer in Japan takes place through a channel that differs significantly from that in the United States. In the United States, intellectual property rights to inventions made by professors most commonly belong to the universities;[18] in Japan, they generally go back to individual inventors, who most commonly transfer the title to their inventions to private companies for little or no fees without filing for patents. This practice forms the basis of technology transfer at Japanese universities which is often described as "usually ... arranged rather indirectly" (Anderson: 107–108). University faculty, especially engineering professors, have close ties with the firms where their former students are employed. When these companies decide to file patent applications for the acquired intellectual property, the faculty inventors are listed either as joint applicants or only as inventors. In the cases in which the companies file unilaterally, without listing the faculty member's name,[19] the patents are officially listed as having been filed by the private sector, and the inventive activity within the academic community does not appear in the public data.

In return for the rights to the intellectual property, the companies provide the faculty with grants and endowments as Donations.[20] (According to Monbusho, grants and endowments from the private sector to national universities amounted to as much as ¥87.7 billion in 1996, accounting for 26% of total university income.[21] An executive of Oki Electronics maintains that as much as forty percent of Oki's academic investment was in the form of gifts to universities and to academic associations (Fujita 1997). Such unrestricted income is administratively much less complex to handle and is thus preferred by faculty.)

Disincentives for Faculty Ownership of Patents. Why don't the Japanese professors file for patents on their own discoveries? First, because of economic disincentives: The Japanese system imposes high transaction costs upon the inventors. As outlined earlier, currently as much as 90% of the titles to inventions go back to the individual inventors who are responsible for the initial application and for subsequent maintenance of the patents once they own the inventions. The administrative fee is at least ¥105,300 ($752) to file a single patent in Japan. In addition, it usually costs from ¥200,000 ($1,430) to ¥300,000 ($2,143) per case to hire a patent agent.[22] Furthermore, owners of the intellectual property must pay an annual fee to maintain their rights. The patent maintenance fee increases exponentially, further increasing the burden on the patent holders, as shown in Table 2.[23]

Moreover, in order to generate income from a patent, the owner of the intellectual property needs to spend a significant amount of time negotiating for licensing, warning violators, and filing lawsuits against them. Such a financial burden is particularly onerous for the underpaid Japanese professors.[24] In addition, a scarcity of patent attorneys in Japan further makes the transaction cost higher if an inventor seeks professional help. All these costs create a major disincentive for faculty inventors to file patents themselves. In their study of biotechnology-related research, Odagiri and Kato (1997: 17) suggest that the university researchers prefer not to file their intellectual property right due to costs associated with the application process.

A second inhibiting factor is the legacy of the university disturbances of the late 1960s, when radical students led protests against university collaborations with the private sector. Only recently has this collaboration between academics and the business community come to be viewed in a positive light, but academic allergy to university-industry collaboration still prevents many academics from participating in technology transfer at their universities.[25]

Nor is there academic incentive to begin with, for the Japanese professors to file patent applications. Unlike in the United States, where patent ownership is considered along with publications when recruiting faculty, patents are not important to the academic community in Japan.

Incentives for Private Firms. In contrast, the current system encourages private companies to file as many patents as possible. As Helfand (1993) points out, the number of patents serves as a

Table 2 Transactions Costs for Japanese Patents

Years	Annual Fee per Claim
1–3	¥14,000 ($100) plus ¥1,400 ($10)
4–6	¥20,300 ($145) plus ¥2,100 ($15)
7–9	¥40,600 ($290) plus ¥4,200 ($30)
10–12	¥81,200 ($580) plus ¥8,400 ($60)
13–15	¥162,400 ($1,160) plus ¥16,800 ($120)
16–18	¥324,800 ($2,320) plus ¥33,600 ($240)
19–21	¥649,600 ($4,640) plus ¥67,200 ($480)
22–25	¥1,299,200 ($9,280) plus ¥134,400 ($960)

Source: Japan Patent Law, Article 107.
Note: Article 3 of the Japan Patent Law was modified in 1998 by Legislation No. 51 to exempt all state-owned titles to inventions from fees associated with patent application, effective April 1, 1999.

barometer of business activities in Japan. The Japanese companies encourage patent filings so that they can be used as a part of advertising. There is thus a strong incentive for individual employees to produce as many patents as possible, not only to maximize their financial rewards[26] but also to be promoted.

In sum, the essence of Japanese technology transfer lies in the informal exchange between faculty inventors and private companies. Companies have incentives to maintain the current mechanism, which allows them to acquire ownership of university-based inventions at low cost; professors have incentives to maintain it, since it ensures stable funding for their research.[27]

Structural Problems of the Current System

The current practice of technology transfer in Japan is inadequate from a purely technological point of view, as it ignores supply and demand and encourages informal exchanges between faculty and firms. This means that new technologies are not always transferred to the company that most needs the inventions. The recipients of the new inventions are typically large companies, which may leave new inventions undeveloped in order to eliminate potential competition with their existing products. At the same time, a small start-up company may be a better recipient of new niche technologies that are unsuited to the economies of scale practiced by large firms. However, they are often left out of technology transfer from Japanese universities, which is practiced typically between faculty and larger firms.

The current system may be economically damaging as well, because patents could be used to provide comparative advantages to smaller-scale companies (Ziman 1984). According to Swanson,

> [U]nder the umbrella of patent, a new company can compete against larger, older, and more entrenched corporations. This, in turn, stimulates the older companies to increase their own R&D efforts, and to lower prices on older products now challenged in the market place by new products. I believe that both of those results are desirable from a public policy point of view, in that they provide incentives to continually enrich and extend the nation's research and technology base. (Swanson 1983: 182)

Patent protection ultimately leads to a healthier economy by strengthening competition. However as discussed earlier, the university-based inventions are in most cases acquired by large firms in Japan.

And finally, under the current Japanese mechanism there is no apparent correlation between quality of faculty research and financial compensation by the recipient firms: Firms appear to contribute, whatever the quality or the content of the research. Clearly, there is no reason under this system for faculty researchers to develop more useful technologies.[28]

Conclusion

This article has pointed out some characteristics of technology transfer in Japan and explained the mechanism behind its current practice. A major aspect of Japan's technology transfer practice is that faculty members have little incentive to secure their intellectual property. The majority of titles to inventions return to the individual inventors under Japan's patent policies. However, the current system discourages university faculty from filing for patents. Professors most typically transfer the title to their inventions to private firms, and in return, the recipient firms provide financial contributions to the faculty as unrestricted gifts.

Since the mid-1990s, scholars and policymakers in Japan have discussed the system's shortcomings. As a result, the Japanese government has introduced a series of reforms to promote technology transfer at Japanese universities.[29] Most recently, the Ministry of Education and the Ministry of International Trade and Industry introduced the Technology Transfer Promotion Law (TTPL)[30] to help create an infrastructure that will facilitate transfer of university-based technology to the private sector, particularly to small businesses. While the policy has brought up some intriguing changes at the Japanese universities,[31] its value has not yet been proved.

Notes

1. The issue of Japan's innovation dynamics has long concerned scholars and policymakers. It is now well established that the sustained increase in research and development expenditure has been one of the major contributing factors to technological progress in postwar Japan (Kodama 1991). Concurrently, the nation witnessed progressive growth in the number of patent applications since the 1970s. According to the Science and Technology Agency (1998: 23), patent applications increased more sharply in Japan than in any other industrialized country between 1975 and 1995. Pointing out that the number of patents received by the Japanese had almost quadrupled between 1967 and 1984, Okimoto and Saxonhouse (1987) contend that the rise in Japan's patent productivity can be partially attributed to the sustained increases in R&D spending.

2. In this article, technology transfer is defined as the process through which inventions made in universities are transferred into the private sector for further development.

3. Various scholars argue that the role of higher education has been defined differently in Japan than in the United States. Rohlen (1992: 339) characterizes the essential role of the university system in Japan as "a combination of sorting talented recruits for companies and ministries and serving as an important conduit and legitimizer for foreign ideas." Anderson (1984: 72) contends that the Japanese universities are "better thought of as suppliers of a large number of generally educated people … to industry." Feigenbaum and McCorduck (1983: 144) argue that "industry uses the university as a filtering device, acting on the assumption that the strict entrance exams will identify the brightest and most tenacious." In the Joint Symposium of the American Association for the Advancement of Science's Annual Meeting held in February of 1993, the Japanese participants expressed similar perceptions of the role of universities in Japan. Imai (1993: 2) maintains that the "general view of industrial firms is that

universities need to provide good students who are well prepared for the future of industrial development ... industries seek 'raw material' from the universities—not 'finished parts.'" Similarly, Nishi and Kobayashi (1993: 34) contend that "[i]n Japan, the universities are viewed as a resource supplier to industry; industry provides funds to universities to support education, but not research in most cases."

4. Patents are often cited as a barometer of technological development. See, inter alia, Kamien and Schwartz (1982: 50), Anderson, (1984: 23), Stoneman, (1987: 12), and Basberg (1988). Basberg (1988: 457–58) categorizes the literature on patents into the following three groups: the first deals with the legislation and functioning of the patent system, the second deals with the rationale of the patent system, and the third uses patents as an indicator of technological information. The present article falls within the second category.

5. The data discussed in this article mainly concern national universities. Accordingly, the analysis centers around the Japanese national universities unless mentioned otherwise.

6. Personal communication with Monbusho officials.

7. See, for example, *Nihon Keizai Shinbun* (January 7, June 30, and July 5 of 1997, and January 9, February 14, July 5, and July 27 of 1998) and *Tokyo Yomiuri Shinbun* (July 18, 1997).

8. Monbusho has since the early 1980s created a series of policies to promote joint research between universities and private companies, establishment of endowed chairs, privately-funded departments specializing in research, and collaborative research centers.

9. The report delineates an increasing enthusiasm of the Japanese companies for academic research on the one hand, and an emerging awareness among Japanese engineering professors of the importance of collaboration with the private sector, on the other.

10. These thirteen companies were Hitachi, Toshiba, Mitsubishi, Matsushita, NEC, Fujitsu, Sony, Sharp, Oki, NTT, Sanyo, IBM Japan, and Sharp.

11. Odagiri (1997) points out that the official statistics concerning joint research may overlook the collaborative research that had taken place prior to the 1980s. Odagiri and Goto (1996) challenge the popular perception of Japanese universities as a minimal player in Japan's technological development, arguing that there existed various forms of university-industry collaborations in prewar Japan.

12. Odagiri and Kato (1997: 17) admit that the nature of biotechnology-related research tends to be too fundamental to be qualified as patentable. However, they further suggest that the university researchers prefer not to file their intellectual property rights due to costs associated with the application process. This point is discussed later in this article.

13. The members of the survey group reviewed all the patents filed in 1996, counted the ones that had the names of University of Tokyo engineering faculty listed as the inventor(s), and contacted the faculty for confirmation. The survey

may not be the most accurate representation of the inventive activity at the Department, in that some of the faculty did not respond.

14. The rules and procedures of the Monbusho Notice are applicable to the faculty and staff engaged in science-related research at national universities.

15. The Science Council of Japan was established in 1948 with the objective of contributing to the peaceful reconstruction of the country, promoting the welfare of mankind, and advancing science, through the joint will of the scientists throughout Japan (Anderson 1984).

16. Johnson (1989: 265) defines administrative guidance as "the authority of the government, contained in the laws establishing the various ministries, to issue directives (*shiji*), requests (*youbou*), warnings (*keikoku*), suggestions (*kankoku*), and encouragement (*kanshou*) to the enterprises or clients within a particular ministry's jurisdiction." Although an administrative guidance is not legally enforceable, it can impose strong restrictions upon universities.

17. University presidents typically follow the Invention Committee's recommendations in deciding ownership of the intellectual property of university-based inventions.

18. The U.S. government added a new chapter to the U.S. Patent Code in 1980. The amendment, which is commonly known as the Bayh-Dole Act, provided that nonprofit organizations and small businesses, particularly universities, may retain worldwide titles to inventions made under U.S. federal funding. The Act is argued to have influenced the way the U.S. universities manage their technologies.

19. In some cases, the faculty inventors themselves prefer to remain anonymous.

20. As discussed earlier, the Japanese companies contribute money in order to ensure a personal tie with faculty, expecting to acquire their best and brightest students. This article focuses on a less discussed component of the exchange, intellectual property.

21. An engineering professor at the University of Tokyo, in a personal communication, maintains that more than one-third of his research is funded by grants and endowments from the private sector. During the fiscal year of 1997, this professor's laboratory acquired approximately ¥40 million from Monbusho, including ¥15 million in Monbusho Grants-in-aid for Scientific Research, ¥70 million from other government funds, and ¥40 million from the private sector. At its peak, the professor's laboratory received as much as ¥60 million in a year from the private sector.

22. The dollar value is calculated using the foreign exchange rate indicated in the Management and Coordination Agency Statistics Bureau's (1997) *Japan Statistical Year Book 1998*.

23. To promote patent applications, the Japanese government passed legislation in February 1998 which prevents the annual fees associated with patent applications from increasing after the 10th year.

24. This is particularly true for the professors at national universities. According to the scheme for 1997 defined in the Public Servant Salary Law, the monthly salary for a full-time tenured professor started around ¥373,200 ($2,666). Even the highest-paid faculty received only ¥608,000 ($4,343) per month.

25. In personal communications, several Monbusho bureaucrats maintained that they still encountered persistent opposition from faculty members when the ministry tried to introduce new policies to promote collaboration between universities and industry. Kuwata (in Fujiwara Press 1995) contends that some students display discontent toward faculty's active involvement with industry even today.

26. Some companies set up a reward structure to encourage their researchers to file patents.

27. This position is also held by several authors in the present volume.

28. This may pose a problem regarding academic freedom. A Japanese professor in applied physics who is currently pursuing research in the United States points out that the Japanese system may be more geared toward producing genuinely innovative studies because the scholars are less constrained to cater to industrial needs than they are in the United States (personal communication).

29. Particular emphasis on reform was placed in the following four areas: (1) development of human resources, (2) promotion of collaborative research between universities and industry, (3) promotion of the transfer and commercialization of university-based technologies, and (4) increase in public spending on research and development.

30. The TTPL has three main pillars. First, it grants approvals and provides subsidies and guaranteed loans to universities that intend to establish an office specialized in handling university-based inventions. Second, it provides financial assistance to small business in commercializing university-based inventions. Third, approved university offices are exempt from fees in the maintenance of state-owned intellectual property.

31. For example, the Research Center for Advanced Science and Technology (RCAST) of the University of Tokyo launched an incorporated company in August 1998 to handle its university-based inventions. The Company acquires titles to inventions made by faculty members, files for patents on behalf of the inventors, and then licenses these intellectual property rights to private companies.

References

Anderson, Alun M. 1984. *Science and Technology in Japan.* Essex: Longman.

Basberg, Bjorn L. 1988. "Patents and the Measurement of Technological Change." In *Innovation: A Cross-Disciplinary Perspective,* eds. Kjell Gronhaug and Geir Kaufmann. Oslo: Norwegian University Press.

Bloch, Erich. 1992. "Comments on Policy Implications." In *Japan's Growing Technological Capability: Implications for the U.S. Economy*, eds. Thomas S. Arrison et al. Washington, D.C.: National Academy Press.

Brandin, David H. and Michael A. Harrison. 1987. *The Technology War: A Case for Competitiveness*. New York: Wiley-Interscience.

Cutts, Robert L. 1997. *An Empire of Schools: Japan's Universities and the Molding of a National Power Elite*. New York: M.E. Sharpe.

Diamond Management Development Information (Daiamondo Keiei Kaihatsu Jyoho). 1997. Analysis on Intellectual Property. (November).

Feigenbaum, Edward A. and Pamela McCorduck. 1983. *The Fifth Generation: Artificial Intelligence and Japan's Computer Challenge to the World*. Reading, Massachusetts: Addison-Wesley.

Finan, William F. and Carl Williams. 1992. "Implications of Japan's 'Soft Crisis': Forcing New Directions for Japanese Electronics Companies." In *Japan's Growing Technological Capability: Implications for the U.S. Economy*, eds. Thomas S. Arrison et al. Washington, D.C.: National Academy Press.

Fujita, Kaori. 1997. "Capturing Universities in R&D." In *Nikkei Electronics* 679 (January 6).

Fujiwara Press Editorial Board. 1995. *The Cutting Edge of University Reform*. Tokyo: Fujiwara Press.

Helfand, Michael Todd. 1993. "How Valid Are U.S. Criticisms of the Japanese Patent System?" In *Hastings Law Journal* 15: 93.

Holman, M. A. 1978. "An Analysis of Patent Statistics as a Measure of Innovative Activity." In *The Meaning of Patent Statistics*, L. J. Harris et al., eds. Washington, D.C.: National Science Foundation.

Imai, Kaneichiro. 1993. "Technology Management in Japan." In *Technology Management in Japan: R&D Policy, Industrial Strategies and Current Practice*, ed. Robert S. Cutler. Boston: Joint Symposium, American Association for the Advancement of Science (AAAS) Annual Meeting.

Japan Patent Office. 1997. *Annual Report of Patent Administration 1998*. Tokyo: Hatsumei Kyokai.

Johnson, Chalmers. 1989. *MITI and the Japanese Miracle: The Growth of Industrial Policy, 1925–75*. Stanford: Stanford University Press.

Kamien, Morton I. and Nancy L. Schwartz. 1982. *Market Structure and Innovation*. Cambridge: Cambridge University Press.

Kennedy, Martin. 1986. *Biotechnology: The University-Industrial Complex*. New Haven: Yale University Press.

Kodama, Fumio. 1991. *Analyzing Japanese High Technologies: The Techno-Paradigm Shift*. London: Pinter Publishers.

Management and Coordination Agency, Statistics Bureau. 1997. *Japan Statistical Yearbook 1998*. Tokyo: Management and Coordination Agency.

Monbusho. 1996. *Research Cooperation Between Universities and Industry in Japan.* Tokyo: Ministry of Education, Science and International Affairs Bureau.

Monbusho. 1998. *White Paper of Japan's Education Policies 1997.* Tokyo: Ministry of Education.

Mukaibo, Takashi. 1991. "Engineering in Japan." In *Engineering in Japan: Education, Practice and Future Outlook,* ed. Robert S. Cutler. Washington, D.C.: Joint Symposium, American Association for the Advancement of Science (AAAS) Annual Meeting.

Nihon Keizai Shinbun (Nikkei). January 7, 1997; June 30, 1997; July 5, 1997; January 9, 1998; February 14, 1998; and July 27, 1998.

Niki, Etsuo. 1997. "A New Paradigm of University-Industry Linkages: The Method of the Research Center of Advanced Science and Technology (RCAST) at the University of Tokyo." (Unpublished draft proposal).

Nishi, Yoshio and Hisashi Kobayashi. 1993. "A Comparison Between Japanese and American Technology Management Practices." In *Technology Management in Japan: R&D Policy, Industrial Strategies and Current Practice,* ed. Robert S. Cutler. Boston: Joint Symposium, American Association for the Advancement of Science (AAAS) Annual Meeting.

Odagiri, Hiroyuki. 1997. "The Japanese Innovation System and the Role of Universities." Tokyo: Conference Paper for the Research Institute of International Trade and Industry.

Odagiri, Hiroyuki and Akira Goto. 1996. *Technology and Industrial Development in Japan: Building Capabilities by Learning, Innovation, and Public Policy.* New York: Oxford University Press.

Odagiri, Hiroyuki and Yuko Kato. 1997. "University-Industry Research Collaboration in the Biotechnology Industry." Tokyo: Discussion Paper Series of the Research Institute of International Trade and Industry.

Okimoto, Daniel I. and Gary R. Saxonhouse. 1987. "Technology and the Future of the Economy." In *The Political Economy of Japan, Volume 1: The Domestic Transformation,* eds. Kozo Yamamura & Yasukichi Yasuba. Stanford: Stanford University Press.

Pavitt, K. 1985. "Patent Statistics as Indicators of Innovative Activities: Possibilities and Problems." In *Scientometrics* 7 (1–2).

Rohlen, Thomas P. 1992. "Learning: The Mobilization of Knowledge in the Japanese Political Economy." In *The Political Economy of Japan,* Vol. 3: *Cultural and Social Dynamics,* eds. Shumpei Kumon and Henry Rosovsky. Stanford: Stanford University Press.

Science and Technology Agency. 1997. *Indicators of Science and Technology [Kagaku Gijyutsu Yoran] 1998.* Tokyo: Science and Technology Agency.

Stoneman, Paul. 1987. *The Economic Analysis of Technology Policy.* Oxford: Clarendon Press.

Swanson, Robert A. 1983. "Policies to Stimulate Growth: The View from a New Industry." In *The Race for the New Frontier: International Competition in Advanced Technology: Decisions for America.* Panel on Advanced Technology Competition, Office of International Affairs, National Research Council. New York: National Academy Press.

Swinbanks, David. 1994. "Japan's universities 'need to strengthen links to industry'." In *Nature* 371, October 27.

Uenohara, Michiyuki, Takuo Sugano, John G. Linvill, and Franklin B. Weinstein. 1994. "Background." *In Competitive Edge: The Semiconductor Industry in the U.S. and Japan,* eds. Daniel I. Okimoto, et al. Stanford: Stanford University Press.

Warshofsky, Fred. 1989. *The Chip War: The Battle for the World of Tomorrow.* New York: Charles Scribner's Sons.

Ziman, John. 1984. *An Introduction to Science Studies: The Philosophical and Social Aspects of Science and Technology.* Cambridge: Cambridge University Press.

14

The Innovation Agent and Its Role in University-Industry Relations

Shingo Kano

Technology transfer and collaboration between university and industry has been a growing policy issue and of increasing economic interest since the end of 1990s in Japan, in the creation of new industry, in growing existing businesses, and in new job creation due to deregulation and changes in the university system. Biotechnology and information science, especially, are expected to become sources through which universities can contribute directly to industry.

This paper reports on a conceptual study which sought to propose a general definition of resource-coordinating functions between universities and industrial companies. The concepts of the "three-phase, three-mode hypothesis" and the "innovation agent" (IA) and their functions will be introduced to examine these issues. Based on this general concept of IA, three forms of resource coordination in university-industry relations (UIR) provide a model whose characteristics will then be analyzed.

Next, the dominant presence of a particular IA form in the biopharmaceuticals field in Japan is confirmed by empirical data. Discussion follows to explore reasons for this dominance and to elaborate on a UIR resource coordination agenda for Japan's science-based industry.

Characteristics of Science-Based Industries

"Science-based" industries, such as biotechnology and information science, are characterized by the following features:

Scientific Research as a Direct Source of Innovation. Science-based industries seek sources of innovation through research conducted by research universities and other public scientific research organizations. Consequently the companies in the science-based industries must secure relationships with relevant academic institutions, which is an obvious distinction from engineering-driven industry, which does not achieve innovation directly through scientific research.

Limited Understanding of the Applicability of Research. Since the trends in scientific research are constantly changing, only a handful of people can decide whether the embryonic basic research in question would be industrially applicable; in fact, no one may be available in the established companies who can understand the issue. And yet science-based industry must involve commercial entities to push forward with the basic research into application levels.

Fuzzy Differentiation between Basics and Application. Companies engaged in science-based industries find it hard to set boundaries between basics and application and therefore may not be able to determine a suitable extent of in-house fundamental research. This fuzziness also lowers the probability of successful outsourcing. How to differentiate the kind of research that will provide commercially applicable results from research for scientific infrastructure in the public domain, tops the management agenda for established science-based companies if they are to gain advantages in the next-generation competition.

These characteristics of science-based industry call for enhanced UIR and, at the same time, invite difficulties in their rational design. The incomplete understanding of embryonic technology, the vague boundaries between basics and application, and the cost of research, all act as barriers to beneficial association between businesses and academic researchers, and sometimes lead to excessive investment in basic research. They may also have contributed to the births of a large number of venture businesses dedicated to R&D in the gap between academic institutions and established private companies, taking over research projects from universities and other entities and acting as bridge organizations toward product development and marketing. This is often seen in biotechnology fields such as genetics for gene therapy and human genome studies. Science-based industry as represented by these examples relies for its foundation on scientific research results whose industrial applicability may not be easily recognized.

The Importance of Morphological Analysis of UIR

University-industry relations for science-based industry entails technology transfer from universities, which provide innovation sources in their immature states, to industrial companies. For established businesses, this technology transfer may upgrade emerging innovations either through their own efforts or in collaboration with universities. Universities are a fundamental resource for science-based industry, which must coordinate resource allocation among universities and companies involved in the innovation process. At issue is what kind of coordination mechanism would best enhance UIR.

It is not as easy to measure UIR activities in Japan as in the United States, because patent policies in the national universities are still ambiguous and statistics such as an AUTM survey are lacking. Moreover, official statistics of UIR reported by the Ministry of Education do not cover all UIR activities. The choice of UIR indications is more difficult than for R&D activity itself, because they must measure the interactions among many types of R&D organizations.

Conventional UIR studies have measured relationships based on university research results as input and patents and/or products as output, leaving the real mechanism of UIR hidden. This approach, based on the way economics has measured business activities by using production function, puts resource allocation mechanisms in a "black box." It can quantify papers, patents, or products as indices in place of prices for input/output analysis, but it cannot refer to the exact resource allocation mechanism that science-based industry applies between universities and companies. Even though it is important to know precisely which paper, patent, or product has formed an effective linkage, the black-box approach ignores linkage and hence cannot decipher the factual coordination that actually took place. In order to offset this shortcoming, Zucker and Darby (1996; 1998) applied the same type of approach, focusing on those researchers who have enjoyed a certain reputation as "star scientists." Levels of performance, however, are hard to define properly, and since any such definition will change in time it may not be applicable over the long term. Moreover, the immaturity of innovation sources creates uncertainty about their practical values, making resource allocation impossible on quantified bases such as price mechanisms. Consequently the black-box approach will not yield the realities of resource allocation and coordination.

Since UIR in science-based industry is actually conducted based on quality rather than quantity, its analysis may be made more efficient by focusing on the forms of coordination instead of depending on the approaches of economics or production-function-like approaches. Coase (1937) focused on the uniqueness of resource allocation mechanisms within businesses, as opposed to conventional approaches that analyzed resource coordination by price mechanism, classifying the natures of transactions to explain why certain ones were conducted in the market and driven by price mechanisms, while others were conducted inside business organizations; similarly, inter-organization coordination between universities and companies requires the clarification of the contents of the UIR black-box, based on the assumption that there is an inherent resource allocation mechanism other than the price mechanism. In analyzing coordination mechanisms, the premise that the number of personnel in the recipient company who could recognize potential innovation sources is limited to very few invokes the concept of "bounded rationality" discussed by the researchers of Comparative Institutional Analysis: UIR planners hope to select sensible alternatives to meet their interests, but are in reality only allowed partial understanding of the various interests surrounding the issue. As a result they tend to be only boundedly rational and are unable to make optimal decisions.

The planning of UIR is tantamount to designing the relationship between providers and recipients of technology. As the outcome of any UIR is unknown at its inception, its primary use would be to determine how many resources were available. An analysis of the resource coordination mechanism should ask first what kind of medium or bridging method would serve the purposes of the university providing technology and/or those of the company, to be able to secure more R&D resources under the "boundedly rational" situation. This leads the present article to discuss the morphology of industrial-academic relations. Harmon et al. (1997) mapped the five technology transfer processes of 23 different technologies developed at one university, suggesting our morphological approach to UIR analysis.

Coordination Mechanisms of University-Industry Relations: A Morphological Approach

To prepare to discuss resource coordination issues in science-based industry, we first introduced a basic problem of in UIR by hypothesizing a three-phase, three-mode model and defining mismatches; next we considered the general concept of "an innovation agent," representing the function that avoids mismatches, mediates both sides, and executes R&D activities according to the maturation of innovation phase through appropriate management; finally, we examined the classification of innovation agent forms and features to analyze UIR morphology.

The Three-Phase, Three-Mode Model of the Innovation Process

In innovation research, the innovation process is generally divided into phases. There are many concepts of the classification method, including Furukawa (1993), who has classified the process into eight phases, Rothwell (1994), who defined the innovation process as "generations" based on the premise that the innovation process is non-linear, and others, who have classified the process into many different phases. However, as Amable (1998) pointed out, it is clear that no classification concept corresponds to all situations, since there are many different fields and each depends on the researcher's viewpoint. As university researchers have multiple viewpoints in setting their research agendas, from basic to applied, it is difficult to categorize each collaboration project or licensing deal in the same manner: To do so would require dividing and subdividing the innovation process into cumbersome detail.

However, by limiting the subject to who will take responsibility for the innovation process, the innovation process in university-industry collaboration can be classified in the same manner from several points of view. In this paper the innovation process will be classified into three phases to discuss the correlation between the innovation process and university-industry relations: the individual creation phase (basic research phase), the evaluation phase (intermediate phase), and the advanced development phase (product development). The approximate images of activity in the three phases are as follows:

Individual Creation Phase. This phase includes generating a concept, making a scientific discovery, and creating a theory/hypothesis, which is then evaluated by conducting small-scale experiments, followed by obtaining intellectual property rights (such as basic patents). This phase is usually carried out by individual researchers and small groups.

Evaluation Phase. Large-scale experiments are conducted in order further to evaluate the hypothesis and concepts. Compared to the individual creation phase, the evaluation content will be advanced, since there will be combinations with many other technological factors. Here, obtaining intellectual property rights (such as peripheral patents) may take place. This is the intermediate phase between individual creation and advanced development, which aims to prove both concept and hypotheses.

Advanced Development Phase. Here R&D activities are conducted and the innovation readied for practical application. This phase includes production technology development, final evaluation of the product concept, marketing survey, and test marketing.

This classification is expedient for two reasons. Since research and development are difficult to divide at the mid-phase (evaluation phase) and cross each other in non-linear fashion, the mid-phase activity bridges the other two and thus should be distinguished from both. The other reason is that the roles of the players in the technology transfer process are distinctly different. During the full life-span of R&D activities, especially in biotechnology and information technology, a three-party relationship often exists between a university, a venture company, and a large company; it is thus reasonable to divide the process into three phases, according to who is responsible for the innovation process, and who are players in the university-industry relationship as a total technology transfer process.

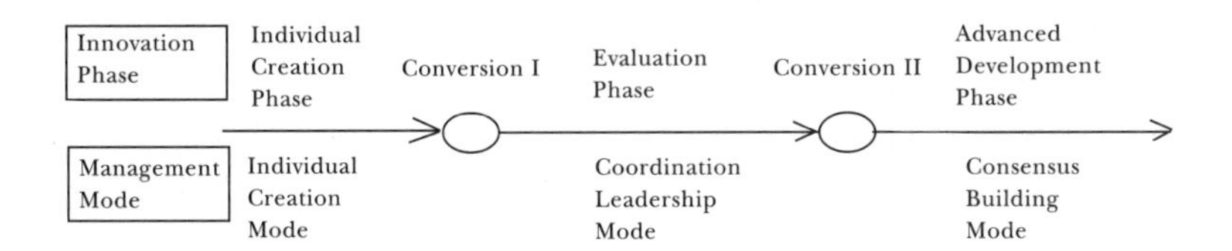

Figure 1 Innovation Phases and Management Modes (three-phase, three-mode hypothesis).

Three Management Modes

Management characteristics differ in each phase of R&D, just as R&D management in a university laboratory, in a venture company, and in a large corporation must be different from each other. This is an inevitable result of the maturation of innovation demanded by each type of management. Their characteristics are sketched in Figure 1.

Individual Creation Mode. The individual creator sets a goal at his/her own discretion and it is managed at the level of individual researchers with a small investment of R&D resources. In the final stage of the individual creation phase, the individual creator should demonstrate the possibilities of the technology as well as its uniqueness and explain the maturity level of innovation so that the corporate-side evaluator can decide whether the technology should be funded.

Coordination Leadership Mode. Significant resources are often required to study the possibilities of the technology in the evaluation phase. Following this expansion of scale, multiple technological factors will be more necessary than in the individual creation phase. Therefore, powerful resource coordination skills and strong leadership, as well as a good understanding of technology, are required. This may call for a high-risk investment at a stage where many uncertainties exist; in the evaluation for setting the target, the project leader must have an agreement with the provider of R&D resources and must keep in close communication with the scientists and engineers conducting the experimental and development work.

Consensus Building Mode. The possibilities of the technology having been evaluated, the number of persons involved to put the technology into practical use should be determine by consensus, rather than through one individual's taking a powerful leadership position. This is important in order to maintain the overall teamwork needed in the advanced development phase.

Mismatches in University-Industry Relations

As an innovation matures, the strict definition of "phase" and "mode" shown above must be kept in mind. If a mismatch occurs during the process, the results could be disastrous, for example if a consensus-building management mode is used during the evaluation phase, there will be problems defining R&D aims and securing and allocating R&D resources; in a high-risk R&D project, only a small number of people understand the subject and can evaluate the possibilities of specific advanced technologies. Premature consensus-seeking will rouse opposition from people who do not understand, and the R&D theme may be dropped or may result in situations in which the necessary R&D resources will be denied. Similarly, if a promising R&D theme that has already reached the evaluation phase is still managed in the individual creation mode, the opportunity may be lost.[1]

For these reasons, it is crucial to change management modes as the innovation matures. Problems in university-industry relations may also arise when the management mode changes from individual creation to coordination leadership. If the company has been involved in more basic research with the university and the mode conversion occurs inside the company, the issue arising in UIR will be isometric, as it involves a change of resource coordination from small-scale investment to mid-scale investment.

The Innovation Agent as a Problem Solver

As we have seen, mismatches between innovation phase and management mode often occur during the intermediate, evaluation phase. This is where successful management becomes crucial for the success of an innovation. By focusing on this point, we can define the activity that works as a medium to link the individual creation phase and the advanced development phase as the "innovation agent." I propose the following definition:[2]

The "innovation agent" is a coordinator in the evaluation phase, mediating between the individual creation phase and the advanced development phase, having the interface function to convert the management mode from individual-creation to coordination-leadership (Conversion I) and from coordination-leadership to consensus-building (Conversion II) along with the maturation of the innovation (the mediating aspect of the agent's role).

The "innovation agent" takes various management forms, whether as part of a large corporation or of a venture company, and has the function to propose and evaluate R&D strategies, secure R&D resources (people, material, money, intellectual property, and so on) and to conduct R&D activities (the cultivating aspect of the agent's role).

This general definition of innovation agent leads to a comparison of various forms of UIR morphologies. In this context, resource coordination activities required in the mid-phase of innovation are essentially common among the various types of UIRs, which makes general definition possible. The role of innovation agent, defined above as having both a mediating aspect and a cultivating aspect, has the following six functional elements:[3]

Structuring the management scheme. This function organizes management schemes for R&D activities or establishes a mechanism for resource investing, and selects a leader with the initiative and authority to make decisions. The forms of these schemes vary, ranging from establishing an R&D project within an existing company to founding a venture company with a new form of management. The person having the decision-making authority in the coordination leadership mode is the advocate for the management scheme and can be the individual creator him/herself, the one responsible for R&D in the company, an entrepreneur, or a venture capitalist. Powerful leadership is indispensable for a massive investment of resources. It is crucial that the management mechanism be established in order successfully to convert the management mode from the individual creation mode (usually in the university) to the coordination leadership mode (usually centered in the industry side).

Planning of R&D strategies. This function sets a course for the link between the individual creator's idea and the R&D activity of the corporation, converting basic research to applied research. If the university researchers devise a development strategy, the corporate side is represented during strategy development by a knowledgeable appraiser; if the university researchers cannot form a strategy, the corporate side is able to comprehend the idea and to form an original R&D strategy.

Evaluation. This involves evaluation of the research content of the university and its R&D strategies. Evaluators must understand the technology itself and its comparative analysis with other technological options, and thus must have analytic and networking skills. The evaluation ability should be within the innovation agent, otherwise it cannot conduct an effective evaluation that does not link it to a decision-making process.[4]

Networking. This involves developing many different channels to find an individual creative researcher and incorporate him/her into the management scheme of an enhanced research activity and thus speed the maturation of innovation. It is also necessary to form research networks in the evaluation phase.

Securing R&D resources. This involves securing human resources, R&D funds, research materials, research space, intellectual property, and so on. Inside an established company it means assigning research priorities and securing in-house R&D resources; in a newly established company, it means securing new resources from outside sources.

Rights and contracts. This involves managing intellectual property, allocating different roles under R&D, negotiating to secure R&D resources, and producing contracts. It provides the legal basis for innovation development, especially management mode conversions. If there is going to be a mode conversion within the company, its function is to produce necessary documents for the decision-making process.

Innovation Agent Forms

Generally innovation agents in collaboration between university and industry take three basic forms. In this chapter, we will call these forms the Existing-IA Model or Corporate-IA Model (IA model inside an existing company); the Venture-IA Model (IA model in a "new" start-up company); and the Consortium-IA Model (Consortium-based IA model). In the case where there is no joint project between university and industry, we will call it an "In-house R&D Model."

In the Existing-IA Model, the IA function is performed inside the organization and "Conversion II" is conducted within the company. This is different from the Venture-IA Model, in which the innovation agent itself is newly formed and Conversions I and II take place with outside organizations. University spin-off companies are basically derivatives of the Venture-IA model. It can be said that, although management control by researchers from universities is higher than in normal venture companies, it is inside this model's boundaries. In the Consortium IA Model, the researchers consist of delegated corporate researchers and some university researchers. It is different from other models in a sense that Conversion I and mid-phase activities become major activities, since part of Conversion II is conducted by each participating party.

Mode conversion within the same organization is classified as "informal conversion," and mode conversion in different organizations is classified as "formal conversion." This perspective focuses especially on the contract-drafting function of the innovation agent. For example, an in-house conversion is a decision-making process using a common language and tacit corporate-specific methods in a common corporate culture. However, a more formal process is necessary in inter-organization conversion and has different transactions costs. Generally, Japanese companies tend to choose the tacit approach, decreasing their transactions costs.[5]

Table 1 describes the features of each IA by using six IA functional elements. Since the six functions exhibited are completely different among these three IAs, they are also based on different resource coordination mechanisms. Three of the six functions, Networking, Evaluation, and Planning R&D Strategy, are related to the functions that evaluate the opportunity for UIR; positive results of these three functions assures realization of UIR. The other three, Structuring Management Scheme, Securing R&D Resources, and Rights and Contracts, are related to the probability that UIR will occur or the probability that resources for UIR will be secured. Whether UIR is possible or not will be based on the "Opportunity Evaluation" and its positive result as a necessary condition, and

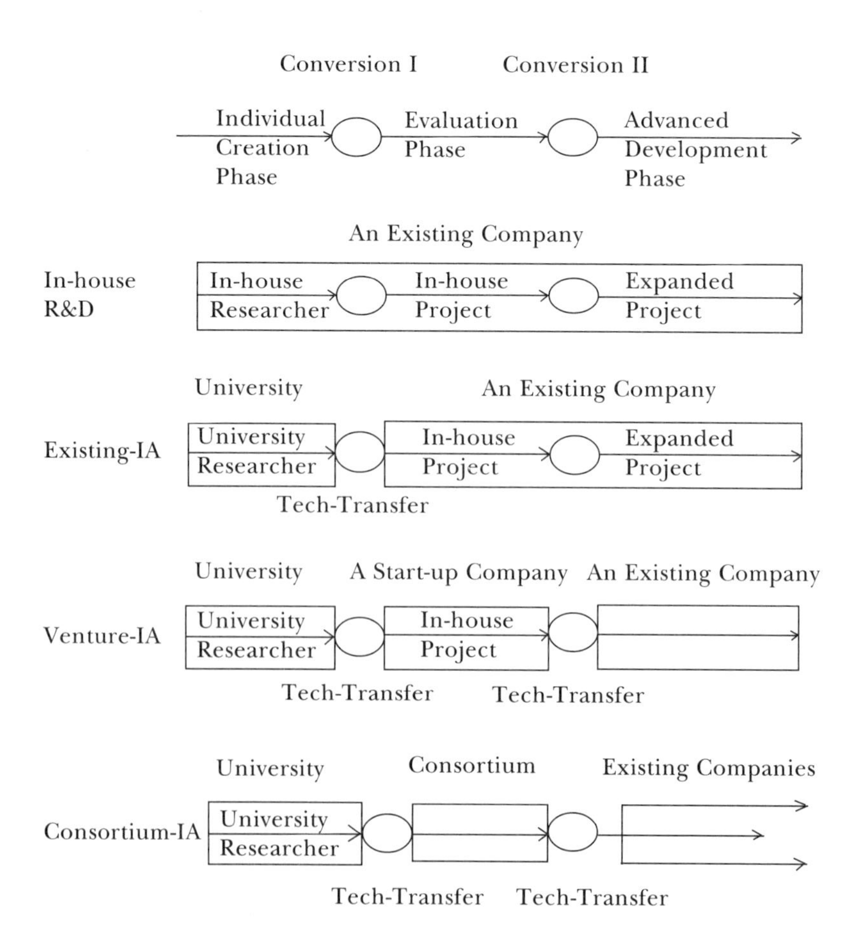

Figure 2 Innovation Agent Forms.

"Resource Securability" as a sufficient condition. The reason "Resource Securability" is regarded only a sufficient condition is that without "Opportunity Evaluation" it is assumed there will be no investment.

This integrated "Opportunity Evaluation" of the University's R&D theme by an Existing-IA depends on the company's ability to check the technology's relevance to existing businesses, to compare it with in-house resources and to evaluate its absolute value. On the contrary, an "Opportunity Evaluation" by a Venture-IA is an evaluation of its pure potential without respect to the existing

Table 1 Resource Coordination Features of IAs by Six IA Functions

	IA Function	Existing-IA	Venture-IA	Consortium-IA
Opportunity Evaluation				
I	Networking	Corporate Researchers	University Researchers, TLO, SAB, VC	Government
II	Evaluation	Corporate Researchers	University Researchers, Entrepreneurs	Case-by-Case
III	Planning R&D Strategy	Corporate R&D Planning Division	University Researchers and/or Entrepreneurs	Consensus-based
Resource Securability				
IV	Structuring Management Scheme	Corporate Project	Newly Established	Newly Established
	Securing R&D Resources	—	—	—
	- Human Resources	University Researchers, Corporate Researchers	Outside Company	University Researchers, Participant's Researchers
V	- Funds	Corporate R&D Budget	Angel, VC	Participating Companies, Government
	- Space	University/ Company	Outside University	Outside University
	- Intellectual Property	Corporate IPD, TLO (+/–)	TLO (+/–)	Corporate IPD, TLO (+/–)
VI	Rights & Contracts	Corporate IPD	TLO (+/–)	A Managing Company

TLO: Technology Licensing Office; SAB: Scientific Advisory Board of the newly established company. VC: Venture Capitalist; IP: Intellectual Property; IPD: Intellectual Property Division.

company's resources and capabilities. In other words, an "Opportunity Evaluation" by an Existing-IA is a relative evaluation of the company's core competence; by a Venture-IA it is an absolute evaluation of the R&D theme. It is more complex in a Consortium-IA than in the other two models because evaluation capability partially depends on the ability of government policy-makers, who use a case-by-case approach.

The integrated "Resource Securability" depends on the possibility of investing in existing in-house R&D resources in an Existing-IA, but in a Venture-IA it means the possibility of securing R&D resources from outside. The other difference is where resources are invested. Existing-IAs invest in both in-house and university projects, but Venture-IAs mainly invest in themselves as an investing vehicle. In Consortium-IAs, since human resources usually come from the participant companies, "Resource Securability" relies on the ability of existing companies.

IA Forms in the Japanese Biopharmaceuticals Field

Biotechnology-based pharmaceuticals provide good opportunities to analyze the features of UIR resource coordination by using the IA-form method in a science-based industry: The percentage of IA forms in a particular industry might reflect the UIR features of its home country.

Table 2 shows the business structures of 145 Japanese companies that entered the biotechnology-based pharmaceuticals field (excluding diagnostics) in 1995. These 145 companies represent nearly all the samples in Japan, but do not include foreign companies or their subsidiaries. In the "science-based" biopharmaceutical field, UIR activity is necessary to build up R&D activities, so all entities could be automatically regarded as conducting some sort of UIR such as collaborations, research funding, donations, simple licensing, and recruiting researchers from universities. For this reason the identification of IA forms can be focused on the business structure of each entity. An existing company from another industry which has a R&D-specific business structure in the pharmaceuticals industry can be classed as a Venture IA-like Existing IA form, even though it is an existing company. However, a company from another industry which not only has R&D activity but also market-

Table 2 Entry Data in the Japanese Biopharmaceuticals Industry

	A	%	B	%	B/A	C	C/A	D
Existing-IA Total	125	86.2	69	77.5	55.2	27	21.6	75
Pharmaceuticals	49	33.8	0	0.0	0.0	27	38.8	75
Other Industries Total	76	52.4	69	77.5	90.8	8	10.5	26
Chemical	37	25.5	33	37.1	89.2	4	10.8	11
Food	25	17.2	22	24.7	88.0	4	16.0	14
Fiber	6	4.1	6	6.7	100.0	0	0.0	0
Petro & Energy	2	1.4	2	2.2	100.0	0	0.0	0
Electronics & Machines	1	0.7	1	1.1	100.0	0	0.0	0
Steel	3	2.1	3	3.4	100.0	0	0.0	1
Paper & Pulp	1	0.7	1	1.1	100.0	0	0.0	0
Others	1	0.7	1	1.1	100.0	0	0.0	0
Venture-IA	4	2.8	4	4.5	100.0	2	50.0	0
Consortium-IA	16	11.0	16	18.0	100.0	—	—	—
Total	145	100.0	89	100.0	61.4	29		75

Sources: Nikkei Bio96, Japan Bioindustry Association Company Book 1996
A: Number of companies that entered into biopharmaceutical field by industries before 1995, without respect to the actual year entered. Entities from other industries are regarded as derivatives of the Existing-IA model.
B: Number of companies with R&D-specific activities among the Entered Companies (A). Pharmaceuticals companies have a marketing capability and are thus not regarded as R&D-specific.
C: Number of companies that launched biotech-based pharmaceutical products by acquiring rights from foreign companies or by discovering original sources. The C/A Ratio indicates the extent of success in utilizing biotechnology by industry.
D: Companies that launched traditional chemical drugs. The D/A Ratio indicates industry's success in entering the pharmaceuticals business.

ing capability could be regarded as a pure Existing IA. There are 86.2% Existing IAs in a broad sense, 52.4% other industry entities, 11.0% Consortium IAs, and 2.8% Venture IAs. Venture-IAs have an extremely small presence: Only one company was established in the 1990s.

Historical View of Business Diversification

Entry patterns from other industries to the pharmaceuticals industry are divided into three periods. The first period was in the 1950s and 1960s, and was chiefly in the antibiotics market, mainly among food companies that had microorganism-related technologies derived from amino acid fermentation. The second period was in the 1970s, when mainly chemical and synthetic fiber companies tried to diversify from declining industry fields with higher profit potentials. Their technological bases are mainly chemical synthetic technologies. The third period was in the late 1980s, when genetic engineering experienced a boom. While entities other than chemicals and foods also embraced the diversification trend of all industries in Japan, genetic technologies were regarded as a dream technology; their introduction was also attempted by the first two groups. There were only four entities in the chemical industry, and three in the food industry, that could establish marketing capability in pharmaceuticals. Among these seven companies, only one food company entered in the late 1980s, others having entered in the first and second periods. Entities from other industries also appeared on the stage as new players who provided R&D resources independently from established pharmaceuticals companies, such as U.S. biotech venture companies.

Since the companies based on fermentation and chemical synthetic technology sought synergistic effects with their existing technologies and pharmaceuticals R&D efforts, and most of the researchers in charge were company employees who sought additional new recruits from the pharmaceutical departments of universities, their approaches were closed to Existing-IA (or Pseudo-EIA). On the other hand, the entities from fiber, steel, pulp, machines, and energy had newly set up their R&D activities, so that their approaches were closed to Venture-IA (or Pseudo-VIA). However, both approaches secured monetary resources from the internal R&D budgets of their companies, which restricted their opportunity evaluation capabilities. Their investment activity was not independent of the technological opportunity, like that of a Venture-IA.

Percentages of IA Forms and Performances

Table 2 lists 49 pharmaceutical companies plus 7 other industry entities that established marketing capabilities, 56 companies that can thus be regarded as Existing-IAs. As mentioned above, among the 69 companies with R&D-specific activities, 55 chemical and

Table 3 Percentages of IA Forms and Performances

	Number of Entities	Percentage	Success Ratio in Biopharmaceuticals (%)
Existing-IA	56	38.6	38.8
Pseudo-EIA	55	37.9	12.9
Venture-IA	4	2.8	50.0
Pseudo-VIA	14	9.7	0.0
Consortium-IA	16	11.0	NA

food companies can be regarded as Pseudo-EIA players and 14 others as Pseudo-VIA players. Table 3 shows the distribution of numbers and percentages of IA forms.

Although chemical and food companies had the advantages of complementary technologies in entering the pharmaceuticals business, their existing technological assets also imposed restrictions on them as Pseudo-EIA players. It was difficult for them to evaluate new paradigms having no connection with their technological assets. In fact, there were only 34 IAs newly established for the purpose of UIR, including 14 Pseudo-VIAs, 16 Consortium IAs, and 4 Venture-IAs. Especially among Venture-IAs, only one University spin-off company was established in early 1990s. This means that Japanese university researchers were unable to secure R&D resources for their research unless the research was on topics of interest to Existing-IAs.

In terms of business performance, food and chemical companies were relatively successful. Chemical drugs were launched by 30% of the chemical companies and 56% of the food companies. By 1995 entities other than chemical and food had generally failed, since they had no complementary technologies and sought technological bases only for the genetic engineering technology that was similar to that of the pharmaceuticals companies. Biotechnology-based pharmaceutical products were launched by 38.8% of pharmaceuticals companies (half of them licensed-in from foreign companies) and 12.9% of chemical and food companies, showing that in Japan pharmaceuticals companies were better able than food and chemical companies to introduce biotechnology applications.

UIR Features in the Japanese Biopharmaceuticals Industry

The dominant IA forms in Japan that had sufficient resource coordination capabilities for UIR were Existing-IA and Pseudo-EIA. This means that there were very few ways for university researchers to acquire R&D resources to accelerate their research topics which could not be absorbed by existing companies. Due to the low mobility of researchers and poor social infrastructures such as the financing system, capital market, technology transfer system from universities to industry, and other business incubation systems, Venture-IA forms that could adapt to the situation were not well developed. Consortium-IA was the only form that could make up for the low mobility of researchers, by sending researchers from existing companies and effectively coordinating resources. The consequence of the lack of a specific IA form that could bridge the gap between research topics and industrial applications was that Japanese biotech-related industries became less competitive. While in the United States over a thousand biotech start-up companies emerged, even though not all succeeded, in Japan there was a distinct lack of innovation agents that could capitalize on the explosion of research diversity in the biotechnology field.

Conclusion

In this study, the general concept of "innovation agent" (IA) was introduced for the analysis of resource coordination mechanisms between universities and companies, and IAs were classified into three forms: Existing IA, Venture IA, and Consortium IA, whose differences were analyzed according to six IA functions. Using this morphological typing as a UIR analysis method, the IAs in the Japanese biopharmaceutical field were analyzed by using empirical data. It is obvious that UIR in Japan is mostly conducted with already existing companies, including new entries from other industry branches. As opposed to the situation in the United States, IA forms for resource coordination in the gap between university research fields and existing companies in Japan are few and fragile.

Although this morphological approach in UIR analysis provides only indirect evidence, it is thought to be useful for international and inter-industry comparisons.

Notes

1. There are many explanations for resistance to a novel, innovative idea inside a large company. For example, the Nomura Research Institute (1990) proposed an "Idea Killer" in its "IGKP model"; Pinchot (1985) proposed a "Corporate Immune System"; and Shane (1994) focused on an "innovation champion" which overcomes resistance inside the company, and explained its five functions. These sources also discuss the difficulty of managing a large corporation in the coordination leadership mode. The coordination leadership mode can also be explained as decision-making by a small group whose members understand the potential of technology, but do not need to reach a consensus. The resistance phenomenon becomes serious when the source of innovation comes from outside the company. Unless an enabler (equivalent to the "Idea Promoter" or "Coordination Leader") exists during the shift from the individual creation phase to the evaluation phase, the recipient of the university's technology in the company will be unable to manage the resources through the coordination leadership mode.

2. Technically, an agent is a helper who acts for another and is not a player itself. However, here "Innovation Agent" means not only one who mediates, but also one who cultivates innovation activities in case the ultimate receiver of the technology transfer cannot absorb the technology in its immature state. In this context, Innovation Agent does not simply mean an information agent.

3. Roberts and Malone (1996) proposed four functions (or players)—technologist, entrepreneur, licensing office, and venture investor—for analyzing university spin-off companies. Evan and Olk (1990) proposed six functions: recruiting personnel, obtaining resources, recruiting new parties, decision-making, legal issues, and evaluation, for analyzing consortiums. In this paper, the six IA functions are designed to analyze the three types of institution in general while containing the functions proposed by the two papers cited.

4. The evaluators in Conversions I and II are always receivers of the technology transfer. An Innovation Agent must possess evaluation ability in Conversion I and accountability in Conversion II. Bandt (1995) pointed out that evaluation of innovation is always based on multiple, partial, and insufficient criteria. Thus, the inevitable insufficiency of evaluation should be complemented by the risk-taking ability of the decision-maker. This linkage of evaluation and decision-making abilities is indispensable in the course of conversion to the Coordination Leadership mode.

5. This is an issue related to the "inevitable insufficiency of evaluation." The company should translate the innovation process to an approximate linear process and set some sort of step-wise milestones, even in the case where innovation process is non-linear and difficult to evaluate in a multiple-criteria situation. Especially, a venture company should set the milestones linked to the maturity of the innovation and explain the present value of the project related to a given milestone for the purpose of financing or alliance with another

company. This approximate translation from the non-linear to a linear model is a conversion from tacit knowledge to explicit knowledge, as Nonaka and Takeuchi (1996) wrote. Culturally, Japanese business people seem to be good at this intra (or informal)-translation process but not good at inter (or formal)-corporate translation.

References

Amable, T.M. 1998. "A model of creativity and innovation in organization." In B.M. Staw and L.L. Cummings, eds. *Research in Organization Behavior* 10: 123–67. Greenwich, CT: JAI Press.

Coase, R. 1937. "The Nature of the Firm." *Economica* 4: 386–405.

de Bandt, Jacques. 1995. "Research and innovation: Evaluation problems and procedures at different levels." *International Journal of Technology Management* 10: 4–6, 365–377.

Evan, William M. and Paul Olk. 1990. "R&D Consortia: A New U.S. Organization Form." *Sloan Management Review* (Spring): 37–46.

Furukawa, Kiminari. 1993. "Diversification and internationalization of corporate management and its technological management." *Yokohama Keiei Kenkyu.* XIV(1): 1.

Harmon, Brian et al. 1997. "Mapping the University Technology Transfer Process." *Journal of Business Venturing* 12: 423–434.

Nomura Research Institute. 1990. *Creating Creativity.* Tokyo: NRI Press.

Nonaka, Ikujiro and Hirotaka Takeuchi. 1996. *A Knowledge Creating Company.* Tokyo: Toyo Keizai Shinpo-sa.

Pinchot, G. 1985. *Intrapreneuring.* New York: Harper & Row.

Rothwell, Roy. 1994. "Towards the fifth-generation innovation process." *International Marketing Review* 1(1): 7–31.

Roberts, Edward B. and Denis E. Malone. 1996. "Policies and structures for spinning off new companies from research and development organizations." *R&D Management.* 26(1): 17–48.

Shane, Scott A. 1994. "Are champions different from non-champions?" *Journal of Business Venturing* 9(5) (September): 397–421.

Zucker, Lynne G. and Michael R. Darby. 1996. "Star Scientists, Institutions, and the Entry of Japanese Biotechnology Enterprises." *NBER Working Paper* No. 5795, National Bureau of Economic Research.

Zucker, Lynne G. and Michael R. Darby. 1998. "Capturing Technological Opportunity via Japan's Star Scientists: Evidence from Japanese Firm's Biotech Patent and Products." *Working Paper* No. 6360, National Bureau of Economic Research.

15

Venture Capital and the Commercialization of Academic Technology: Symbiosis and Paradox

Josh Lerner

Introduction

Venture capitalists have long been involved in the commercialization of academic research. The first modern venture capital firm, American Research and Development (ARD), was designed to focus on technology-based spin-outs from MIT. As envisioned by its founders, who included MIT President Karl Compton, Harvard Business School Professor Georges F. Doriot, and Boston-area business leaders, this novel structure would be best suited to commercialize the wealth of military technologies developed during World War II. As Doriot noted (Liles 1977):

> Scientific intelligence does not always bring business modesty. Success would be attained more often if good idea men would entrust their ideas to good operating men. The discovery of a beauty of an expense account does not always suggest good controllership.

From its inception, the venture capital industry has sought to bundle capital with the effective oversight of university-based enterprises, a symbiosis that continues today.

While the tradition of interactions between venture funds and universities has been a long one, particularly at MIT and Stanford (see, for instance, Roberts 1991), the relationship has undergone a fundamental change in the past decade. Universities have become increasingly interested in venture capital-backed spin-outs as a mechanism to commercialize early-stage technologies and to produce the greatest returns for the institution. This trend has been manifested in numerous ways, such as increases in the staff within academic technology transfer offices devoted to working with professors to establish new firms. The most dramatic manifestation, however, has been the proliferation of funds dedicated to investing in new firms spawned from these institutions. Increasingly, institutions are seeing internal venture funds as an avenue to generate more wealth for the university, as a recent set of remarks by the director of Yale's Office of Cooperative Research illustrates (Gardiner 1997):

> It is even more instructive to look at Yale intellectual assets that could have matured into new ventures ... like Human Genomic Sciences or Incyte Pharmaceuticals. Each has a market capitalization in excess of 500 million dollars. Though Yale had the ideas, technology and personnel to form such a company a year or two in advance of HGS or Incyte, it did not happen because our development strategies were limited to licensing.

Table 1 summarizes those university-based efforts that have actually been established. It does not denote the many more organizations that are considering establishing such funds.

Past efforts by academic institutions to sponsor venture capital funds, however, present several grounds for concern. Consider, for instance, the experience of Boston University. The school's venture capital subsidiary invested in a privately held biotechnology company founded in 1979 by a number of scientists affiliated with the institution. As part of its initial investment in 1987, the school bought out the stakes of a number of independent venture capital investors, who had apparently concluded after a number of financing rounds that the firm's prospects were unattractive. Between 1987 and 1992, the school, investing alongside university officials and trustees, provided at least $90 million dollars to the private firm. (By way of comparison, the school's entire endowment at the end of the fiscal year in which it initiated this investment was $142 million.) While the company succeeded in completing an initial public offering, it encountered a series of disappointments with its products. At the end of 1997, the University's equity stake was worth only $4 million.[1]

This essay represents a preliminary exploration of this complex and multi-faceted terrain. Because of the very limited data avail-

Table 1 Academic-Affiliated Venture Capital Funds

Name	Location	Year Begun
Enterprise Development Fund	MIT	1972
Community Technology Fund	Boston Univ.	1974
British Technology Group Venture Capital Fund	Various British universities	1981
Center for Biotechnology Research	Stanford Univ./Univ. of Calif.	1982
BCM Technologies	Baylor College of Medicine	1983
Tennessee Innovation Center	Oak Ridge Nat'l Laboratory	1984
Dallas Biomedical Corporation	Univ. of Texas, Southwestern Medical Center	1985
A/W Company	Washington University, St. Louis	1987
Triad Investors	Johns Hopkins Univ.	1988
Medical Science Partners	Harvard Univ.	1989
ARCH Venture Partners	Univ. of Chicago/Argonne Nat'l Laboratory	1989
Technology Ventures Corp.	Sandia Nat'l Laboratory	1993
Northwestern Univ. Investment Partners [Evanston Business Investment Corp.]	Northwestern Univ.	1993
Thermo Technology Ventures	Three U.S. nat'l laboratories	1994
JAFCO	Two Japanese universities	1997
Southwest One	Virginia Polytechnic Institute and State Univ.	1998

Source: Compiled from press accounts. This table lists venture capital funds sponsored by or targeted toward investing in particular academic institutions. In some cases, the efforts were abandoned before any investments were made; in others, the fund focus ultimately shifted to include other institutions or types of investments.

able, many of the observations about the relationship between academic commercialization and venture capital must be tentative in nature. Nonetheless, the empirical and field research that has examined the venture capital industry more generally over the past decade provides the starting point for some explorations into these issues. The essay begins with a consideration of the reasons why venture capital is a natural mechanism for financing new university-based ventures. It highlights the problems that these early-stage technology-based spin-offs pose for traditional financiers of small businesses, such as banks, and explores the mechanisms that venture investors employ to address them. In the third section, some of the apparent limitations of venture capital financing are considered, in particular the size and substantial geographic and industry-specific concentration of venture investments. I review two contrasting interpretations of these events: On the one hand, they may be an appropriate shift in response to a changing set of opportunities; on the other, they may reflect the inability of the venture industry to restructure speedily in the face of rapid growth.

Finally, I consider one avenue through which universities are addressing the perceived limitations of the independent venture capital sector, beginning with a case study of a pioneering university-based venture capital fund, ARCH Venture Partners. The case highlights both the rationales for encouraging the involvement of venture investors in university-based technology transfer and the challenges associated with targeted funds. While the limited experience of universities with their own funds makes it difficult to demonstrate empirically that the University of Chicago's experiences are representative, similar issues that have emerged during the longer historical experience with corporate-affiliated venture funds are also reviewed.

The essay concludes by pointing to a fundamental paradox. Venture capital has from its inception been designed to solve the difficult problems of the commercialization of academic technologies. But efforts by universities to duplicate the venture capitalist's role by establishing captive funds seem unlikely to be successful.

The Symbiosis between Venture Capital and Academic Technology Transfer[2]

University technology transfer offices typically focus on nascent firms in high-technology industries with tremendous promise. Unfortunately, these firms are also characterized by uncertainty and informational asymmetries, which make it difficult for the investors to evaluate business plans or to oversee the entrepreneurs once the investments are made. The consequences are often unfortunate. In some cases, the idea is commercialized, but the return to the academic institution is small; often these information problems discourage outside investors entirely and the discovery languishes.

To briefly review the types of conflicts that can emerge in these settings, Jensen and Meckling (1976) demonstrate that agency conflicts between managers and investors can affect the willingness of both debt and equity holders to provide capital. If the firm raises equity from outside investors, the manager has an incentive to engage in wasteful expenditures from which he may benefit disproportionately—such as lavish offices; similarly, if the firm raises debt, the manager may increase risk to undesirable levels. Because providers of capital recognize these problems, outside investors demand a higher rate of return than would be the case if the funds were internally generated. Even if the manager is motivated to maximize shareholder value, informational asymmetries may make raising external capital more expensive or even preclude it entirely. Myers and Majluf (1984) and Greenwald, Stiglitz, and Weiss (1984) demonstrate that equity offerings of firms may be associated with a "lemons" problem (first identified by Akerlof [1970]). If the manager is better informed about the investment opportunities of the firm and acts in the interest of current shareholders, then new shares are issued only when the company's stock is overvalued. Indeed, numerous studies have documented that stock prices decline upon the announcement of equity issues, largely because of the negative signal that it sends to the market. These information problems have also been shown to exist in debt markets. Stiglitz and Weiss (1981) show that if banks find it difficult to discriminate among companies, raising interest rates can have perverse selection effects. In particular, the high interest rates discourage all but the highest-risk borrowers, so the quality of the loan pool declines markedly. To address this problem, banks may restrict the amount of lending rather than increasing interest rates.

Described in this manner, these problems may appear to be abstract. But they have very real implications for university technology managers, who may find corporations unwilling to invest the time and resources to examine early-stage technologies, or offering only modest payments in exchange for exclusive licenses to innovations that the technology transfer officers believe to be valuable. University-based start-ups may find it impossible to access bank loans or equity investments.

These problems in the debt and equity markets are a consequence of the information gaps between the entrepreneurs and investors. If the information asymmetries could be eliminated, financing constraints would disappear. Financial economists argue that specialized financial intermediaries, such as venture capital organizations, can ameliorate these problems by intensively scrutinizing firms before providing capital and then monitoring them afterwards.

To address information problems, venture capitalists employ a variety of mechanisms. First, business plans are intensively scrutinized: Of those firms that submit business plans to venture capital organizations, historically only 1% have been funded (Fenn, Liang, and Prowse 1995). The decision to invest is frequently made conditional on the identification of a syndication partner who agrees that this is an attractive investment. Once the decision to invest is made, the venture capitalists frequently disburse funds in stages, forcing the managers of the venture-backed firms to return repeatedly to their financiers for additional capital, in order to ensure that the money is not squandered on unprofitable projects. In addition, venture capitalists intensively monitor managers, demanding preferred stock with numerous restrictive covenants and representation on the board of directors.[3] Thus, it is not surprising that the capital provided by venture capital organizations is the dominant form of equity financing for privately held technology-intensive businesses.[4]

The Limitations of Venture Capital

There are reasons to suspect that despite the presence of venture capital funds, there still might be attractive university spin-offs that cannot raise capital. I will consider two sets of arguments: those based on the finance literature about capital constraints, and those based on observations about the changing dynamics of the venture capital industry.[5]

A growing body of empirical research suggests that new firms, especially technology-intensive ones, may receive insufficient capital. The literature on capital constraints (reviewed in Hubbard 1998) documents that an inability to obtain external financing limits many forms of business investment. Hall (1992), Hao and Jaffe (1993), and Himmelberg and Petersen (1994) show that capital constraints appear to limit research-and-development expenditures, especially in smaller firms.

However compelling the evidence that capital constraints limit investments by small, technology-intensive firms,[6] these studies' relevance for university technology transfer officers and administrators today is unclear. Many of these works examine firms during the 1970s and early 1980s, when the venture capital pool was relatively modest in size. As Table 2 illustrates, the pool of venture capital funds has grown dramatically in recent years. Thus, even if small high-technology firms had numerous value-creating projects that they could not finance in the past, it is not clear that this problem remains today.

The second set of arguments is based on the perceived limitation of the venture capital industry, namely that it funds a modest number of firms each year, with these investments highly concentrated, and invests generally only in firms that need a substantial amount of capital.

Venture capitalists back only a tiny fraction of the technology-oriented businesses begun each year. In 1997, a record year for venture disbursements, 707 companies received venture financing for the first time (VentureOne 1998). (By way of comparison, the Small Business Administration estimates that in recent years close to one million businesses have been started annually.) Furthermore, these funds—as in previous years—have been very concentrated. In 1997, 53% of venture funding went to companies based in either California or Massachusetts; 85% went to firms specializing in information technology and the life sciences (VentureOne 1998).

It is not clear, however, what lessons to draw from these funding patterns. Concentrating investments in such a manner may well be an appropriate response to the nature of opportunities. Consider, for instance, the geographic concentration of awards. Recent models of economic growth—building on earlier works by economic geographers—have emphasized powerful reasons why successful high-technology firms may be very concentrated. The literature highlights several factors that lead similar firms to cluster in particular regions, including knowledge spillovers, specialized labor markets, and the presence of critical intermediate good producers. The theoretical rationales for such effects are summarized in Krugman (1991); case studies of the development of high-technology regions (see, for example, Saxenian 1994) have

Table 2 Summary statistics for venture capital fund-raising by independent venture partnerships, venture capital disbursements, and venture capital-backed initial public offerings. All dollar figures are in millions of 1992 dollars.

	1978	1979	1980	1981	1982	1983	1984
First closing of funds							
Number of funds	23	27	57	81	98	147	150
Size (millions of 1992 $)	407	460	1,186	1,629	1,988	5,189	4,605
Sources of funds							
Private pension funds	15%	31%	30%	23%	33%	26%	25%
Public pension funds	a	a	a	a	a	5%	9%
Corporations	10%	17%	19%	17%	12%	12%	14%
Individuals	32%	23%	16%	23%	21%	21%	15%
Endowments	9%	10%	14%	12%	7%	8%	6%
Insurance companies/banks	16%	4%	13%	15%	14%	12%	13%
Foreign investors/other	18%	15%	8%	10%	13%	16%	18%
Disbursements							
Number of rounds	173	238	360	676	802	1,249	1,372
Size (millions of 1992 $)	353	552	869	1,713	1,959	3,720	3,710
Initial public offerings							
Number of rounds	6	4	24	50	21	101	44
Size (millions of 1992 $)	134	62	670	783	738	3,451	731

	1985	1986	1987	1988	1989	1990	1991
First closing of funds							
Number of funds	99	86	112	78	88	50	34
Size (millions of 1992 $)	3,988	4,212	5,117	3,538	3,290	2,385	1,455
Sources of funds							
Private pension funds	23%	39%	27%	27%	22%	31%	25%
Public pension funds	10%	12%	12%	20%	14%	22%	17%
Corporations	12%	11%	10%	12%	20%	7%	4%
Individuals	13%	12%	12%	8%	6%	11%	12%
Endowments	8%	6%	10%	11%	12%	13%	24%
Insurance companies/banks	11%	10%	15%	9%	13%	9%	6%
Foreign investors/other	23%	11%	14%	13%	13%	7%	12%
Disbursements							
Number of rounds	1,275	1,332	1,508	1,431	1,434	1,321	1,044
Size (millions of 1992 $)	3,243	3,653	3,724	3,603	3,322	2,866	1,963
Initial public offerings							
Number of rounds	35	79	69	36	39	43	119
Size (millions of 1992 $)	819	2,003	1,602	915	1,110	1,269	3,835

	1992	1993	1994	1995	1996	1997
First closing of funds						
Number of funds	31	54	105	72	97	136
Size (millions of 1992 $)	1,913	2,422	4,536	4,388	7,542	10,408
Sources of funds						
Private pension funds	22%	59%	47%	38%	43%	40%
Public pension funds	20%	a	a	a	a	a
Corporations	3%	8%	9%	2%	13%	30%
Individuals	11%	7%	12%	17%	9%	13%
Endowments	18%	11%	21%	22%	21%	9%
Insurance companies/banks	14%	11%	9%	18%	5%	1%
Foreign investors/other	11%	4%	2%	3%	8%	7%
Disbursements						
Number of rounds	1,150	969	961	1,175	1,637	1,852
Size (millions of 1992 $)	2,682	3,541	2,461	3,490	5,027	10,070
Initial public offerings						
Number of rounds	157	193	159	205	284	134
Size (millions of 1992 $)	4,317	4,905	3,408	6,251	10,976	4,270

a. Public pension funds are included with private pension funds in these years.

Note: "First closing of funds" reports the number and ultimate size of funds that had their initial closing in each year. "Sources of funds" reports the sources of capital for all venture funds that had their first closing in a given year. "Disbursements" reports the number and amount of all early- and later-stage venture investments made by private equity organizations in portfolio companies in each year. Buyouts and other investments by venture groups are excluded. 1997 disbursements are from a different data source, and thus not exactly comparable to those in earlier years. "Initial public offerings" report the number of venture capital-backed initial public offerings in each year, and the amount raised in these offerings.

Source: Compiled from the Venture Economics funds and investments databases, the VentureOne investments database, the *Venture Capital Journal*, and an examination of initial public offering prospectuses.

emphasized the importance of intermediaries such as venture capitalists, lawyers, and accountants in facilitating such clustering.

A related argument is that the structure of venture investments may be inappropriate for many young firms. Venture funds tend to make substantial investments, even in young firms. The mean venture investment in a start-up or early-stage business between 1961 and 1992 (expressed in 1996 dollars) was $2.0 million (Gompers 1995). The substantial size of these investments may be partially a consequence of the nature of institutional investors. The typical venture organization raises a fund (structured as a limited partnership) every few years. Because investments in partnerships are often time-consuming to negotiate and monitor, institutions prefer making relatively large investments in venture funds (typically $10 million or more). Furthermore, governance and regulatory considerations lead institutions to limit the share of the fund that any one limited partner holds.[7] These pressures lead venture organizations to raise substantial funds. Because each firm in his portfolio must be closely scrutinized, the typical venture capitalist is responsible for no more than a dozen investments. Venture organizations are consequently unwilling to invest in very young firms that only require small capital infusions.[8] This problem may be increasing in severity with the growth of the venture industry. The U.S. Department of Labor until 1979 essentially prohibited private pension funds from making venture investments. Even after this policy changed, private pension funds did not invest in venture funds in significant numbers until the mid-1980s, and a variety of regulatory and political factors restrained public pension funds from making substantial venture investments until the 1990s. As the number of dollars per venture fund and dollars per venture partner have grown, so too has the size of venture investments; for instance, the mean financing round for a start-up firm has climbed (in 1997 dollars) from $1.6 million in 1991 to $3.2 million in 1997 (VentureOne 1998).

Again, it is not clear what lessons to draw from these financing patterns. Venture capitalists may have eschewed small investments because they were simply unprofitable due to the high cost associated with these transactions or the poor prospects of the thinly capitalized firms.[9]

University-Based Venture Funds

Venture capitalists have developed a variety of mechanisms to address the information problems that young firms pose. But because the scope of venture capital has been limited—as we have seen, highly concentrated in certain industries and regions, and in firms that can accommodate substantial investments—universities are increasingly seeking to raise venture funds of their own.

As the discussion above highlights, there is still a considerable degree of uncertainty about whether the distribution of venture funding reflects a market failure or the nature of investment opportunities. Putting these contentious issues aside, I will next consider the challenges associated with implementing university-based venture capital funds, illustrated by the experiences of one pioneering academic-based venture fund, ARCH Venture Partners, and its affiliated academic institution. I then explore the experience of corporate venture funds, which highlights how difficult such institutionally affiliated funds can be to manage.

The Case of ARCH Venture Partners[10]

The Argonne National Laboratory/University of Chicago (ARCH) Development Corporation was established in 1986. Argonne National Laboratory was a national laboratory operated for the U.S. Department of Energy (and its predecessor organization, the U.S. Atomic Energy Commission) by the University of Chicago since 1946. In 1986, it employed about 4500 people and had a budget of over $400 million. Among its areas of expertise were advanced energy systems, hazardous waste detection and clean-up, advanced materials (especially superconducting materials), and high-performance instrumentation. The University of Chicago was a private institution, whose researchers had been awarded over 50 Nobel Prizes. In 1986, the University's research expenditures were over $88 million; its faculty, about 1,200. The University had pioneered a variety of commercially important discoveries, including the first sustained nuclear reaction, the isolation of proinsulin, and the invention of the scanning-transmission electron microscope. The University's record in profiting from these inventions, however, had been much poorer. Among the most conspicuous failures was erythropoietin, which was first isolated at the University of Chicago but not patented. The drug ultimately generated annual revenues of $1.2 billion for Amgen, which marketed it as Epogen. (Johnson & Johnson also sold erythropoietin.) The University received no royalties from these sales; nor did it receive equity in Amgen or Johnson & Johnson.

A key role in the conception of ARCH was played by two men: Dr. Walter Massey, who at the time was the University's Vice President for Research and responsible for the operations of Argonne National Laboratory (and who subsequently became Director of the National Science Foundation), and Alan Schriesheim, the Director of Argonne and former General Manager of the Exxon Engineering Technology Department. These men fleshed out the concept for ARCH in conjunction with a number of University trustees. Hanna Gray, then President of the University of Chicago, was also a strong supporter of the ARCH concept.

They had several motivations in creating this organization. First, of course, the University stood to generate revenue from licensing revenues and equity in spin-off firms. Second, the effort would contribute to the regional economy by creating new enterprises and jobs. This was especially important because at the time, the bulk of the early-stage venture capital investments—even by venture funds based in Chicago—were devoted to firms based on the East and West Coasts. Finally, Massey believed that ARCH could have a broader impact on the way that scientists looked at business. As Massey noted (Palmer 1993), "We really had in mind changing the culture of both institutions. The biggest problem was, how do you capture research without having the scientist feel he or she is being directed?" It was hoped that ARCH could address these multiple goals. Consequently, in late 1986, ARCH Development Corporation was created as a separate, private, not-for-profit corporation affiliated with Argonne and the University of Chicago.

A critical task for the committee was the recruitment of an appropriate head for ARCH. It was crucial to find an individual who combined familiarity with private-sector practices with an ability to operate in an environment governed by Federal regulations and the formal and informal constraints that characterize academic institutions. The man selected as ARCH's head in late 1986, Steven Lazarus, had a background that spanned both the public and private sectors. A graduate of Dartmouth College and Harvard Business School, Lazarus had served for 21 years in the U.S. Navy. Between 1972 and 1974, he had been Deputy Assistant Secretary of Commerce for East-West Trade. In 1974, Lazarus joined Baxter International. He spent 13 years in this organization, rising to Senior Vice President for Technology and Group Vice President of the Health Care Services Group. The leaders also recruited a blue-ribbon board of directors, drawn from the University's trustees and other Chicago business leaders.

Lazarus had begun with a small office in Walker Museum, the home of the University's Graduate School of Business, and a secretary. He had a broad mandate to develop both the licensing of the University's technology and the promulgation of spinoff companies, but few resources. Lazarus consequently sought an incremental implementation of his charge to develop ARCH. Among his first steps was to develop an understanding of ARCH's mission on the part of researchers and administrators. Such technology transfer organizations often encounter some initial resistance from researchers, as scientists fear that the outsiders will seek to influence the direction of their research. In addition, they may be concerned that they will not be given a chance to be involved in—and to profit from—the commercialization of their discoveries. At the opposite extreme, scientists may push to have favorite projects funded without an understanding of the demands of the commercial marketplace.

The support of the top University and Laboratory administrators was critical in overcoming initial concerns from the organizations' legal staffs. The potential for profits from these enterprises created, in some eyes, the "appearance of a conflict of interest." Their concerns centered on two areas. The first related to the licensing process. Particularly with respect to the Laboratory, the lawyers argued for extensively publicizing technological discoveries, and then soliciting formal proposals for non-exclusive licenses. Their concerns were based in part on an extensive scrutiny of earlier efforts to transfer technology from the federal laboratories. For instance, Martin Marietta, which operated Oak Ridge National Laboratory, had been criticized by the U.S. General Accounting Office and Representative John Dingell of Michigan. Marietta had established a venture capital subsidiary, the Tennessee Innovation Center, which sought to establish new businesses around Oak Ridge and had invested in a business that subsequently received an exclusive license to develop an Oak Ridge technology. After congressional criticism, the contractor had been forced to restructure the relationship with its affiliate in a financially unattractive way. While acknowledging these concerns, Lazarus argued that without rapid diffusion and exclusive licenses, many technologies developed at the University and Laboratory would be unlikely to be commercialized.

A second area of legal concern was the nature of the relationships between Laboratory researchers and spin-off companies. While researchers were allowed to serve as directors of and consultants to spin-off companies, they were not allowed to hold equity in these enterprises. Arguing forcefully that the researchers needed to benefit directly from these investments, Lazarus worked out an arrangement in which ARCH would invest in spin-off firms, distributing 25% of any capital gains to the researchers involved. In addition, he pushed for researchers to receive as large a share of royalties as possible, undiminished by any deductions for overhead (as had been the case at other institutions).

The next step was for ARCH to create a relationship with the University of Chicago's Graduate School of Business (GSB). Under the leadership of Lazarus and GSB Dean John Gould, a program was set up to allow first- and second-year MBA candidates to work with new enterprises as "ARCH Associates." Out of this effort, Lazarus recruited two partners to assist him in this project, Keith Crandell and Robert T. Nelsen, both working on their MBA degrees at the GSB.

A third step was the raising of a venture fund. Lazarus realized that to maximize the return to the University, he should be developing new businesses in addition to simply licensing technologies to existing firms. When technology transfer officials at schools such as MIT identify an innovation that appears to be the foundation for a new enterprise, they contact local venture capitalists with whom they have a long-standing relationship. (MIT has invested in venture capital funds since the 1940s, and many alumni work for venture organizations or venture-backed firms.) ARCH did not have that luxury. The venture capital community in Chicago was a relatively modest one, and did virtually no early-stage investing in local companies during the 1980s. Consequently, Lazarus realized that ARCH would need to raise a fund of its own. The ARCH partners succeeded in raising a total of $9 million. The limited partners (investors) included the University of Chicago, State Farm Insurance, and two venture firms.

In its first five years, the companies begun by ARCH encompassed a broad range of technologies. Many were traditional outgrowths of the University's research programs. An example was NiOptics Corporation. Researchers in the University's Physics Department had developed a method of managing light with several times the efficiency and brightness of traditional lens-based optics, whose immediate application was the development of backlights for the screens of color laptop computers. Other sources of business ideas were less traditional. For instance, Everyday Learning Corporation, a publisher of elementary school mathematics textbooks, grew out of the University's School Mathematics Project, which attempted to design new methods to teach math to grade-school students. Of the dozen firms funded by ARCH, the fundamental technology for eight had originated at the University of Chicago and for four at Argonne National Laboratories. The smaller number of companies to have been spun out of Argonne reflected the difficulty of the technology transfer process there. ARCH officials estimated that the spin-off of a new business, which typically took six to twelve months at a private university, consumed two years at a national laboratory.

The fund's greatest success in its first five years was an Argonne spin-off, Illinois Superconductor Corporation. This Evanston, Illinois firm went public in late October 1993 in an offering underwritten by Gruntal and Co. At the time of the offering, the company had licensed thirteen superconducting patents (or patent applications) from Argonne and Northwestern University. In addition, it had been awarded one patent, and had four applications pending. This firm had used these discoveries to develop several promising technologies, including a process that allowed high-temperature superconductors to be applied to surfaces through a process similar to painting. Previously, superconducting film had been applied through chemical vapor deposition, a far more costly process. A second innovation was a superconducting sensor that provides continuous readings of the temperature of ultra-cold refrigerators that store human tissue; technicians had previously been required to open the units periodically to check refrigerant levels. Ultimately, the firm hoped to develop a variety of signal processing and filtering components for the cellular telephone and wireless communication industry that would employ superconducting materials.

The financing history of this firm was typical of ARCH's firms. In October 1989, in conjunction with its formation, the company issued 136 thousand shares of common stock to ARCH. In its Series A financing (undertaken during 1990 and 1991), it raised $500,000 each from ARCH Venture Partners, Batterson, Johnson, and Wang (a Chicago-based venture capital organization), and the Illinois Department of Commerce and Community Affairs. The firm raised several million more dollars from these same sources and others in two additional financing rounds in 1992 and 1993.

Since one of the crucial goals of the ARCH partners was to add value through the provision of oversight, almost all the firms in its first fund were based in the Chicago area. Lazarus also sought to involve other venture capitalists as investors, although as an outsider to the close circle of venture organizations, he had initially found it difficult to interest venture capitalists. This reluctance was exacerbated by the unwillingness of many in the industry to consider seed investing. Many felt that the time and effort needed to monitor a seed investment was as great as an investment in a much larger firm, and that early-stage investing represented a luxury that could be not afforded in a venture fund with several hundred million dollars of committed capital. Other venture capitalists, insisting that they be located near their early-stage investments, were reluctant to invest in firms based in the Midwest. Finally, ARCH, limited by its charter to investing in deals where the technology had originated at the University of Chicago or Argonne National Laboratory, was unable to invest in almost all transactions initiated by other venture organizations. Since this deal-sharing was an important component of the venture investment process, the prohibition had limited ARCH's ability to build strong ties with the venture community.

Lazarus began aggressive efforts to overcome those barriers. He sought to exploit old connections from Baxter (a disproportionate number of executives of venture-backed biotechnology firms were from the ranks of Baxter executives), as well as more recent contacts from the University of Chicago. Among the syndication partners of ARCH Venture Partners, L.P. in its first four years were Batterson, Johnson, and Wang, Columbine Ventures, Hillman Medical Ventures, Institutional Venture Partners, and Sierra Ventures.

In its first five years, ARCH created 720 jobs directly and promised to create many more. The number of patents filed annually by the University and the Laboratory had increased from about a dozen to between 120 and 150; over 60 licensing agreements encompassing more than 125 products had been signed, generating nearly $2 million in royalties for the University and Argonne. While these revenues were well below the $10 to $20 million in revenues that Stanford and MIT received annually, ARCH had been in existence for a much shorter period.

Soon after its initial successes, however, the relationship between ARCH and the University of Chicago was restructured. By late 1993, ARCH had invested or reserved nearly its entire initial venture fund. Many of the firms were still in the development stage and would require additional infusions of capital. Furthermore, ARCH officials were regularly being contacted by other academic institutions seeking their involvement in technology transfer at their own organizations, activities outside the scope of the partnership agreement establishing ARCH Venture Partners, L.P. Furthermore, there were a variety of problems in the design of the first fund. One of the barriers to ARCH's ability to attract venture investors into its transactions, the partners believed, was their own inability to invest in other fund's deals. In addition, none of the three ARCH principals received a share of the capital gains, or carried interest, in the first fund (traditionally, venture capitalists serve as general partners of their venture fund and receive 20% of the capital gains from the fund). Instead, the University of Chicago served as both a limited and the general partner, while the three received a salary comparable to that of other University officials.

Prompted by these developments, Lazarus and his partners renegotiated their relationship with the University. They were given permission to raise a second, more substantial, venture fund. As part of the agreement, the nature of the relationship between the University and Lazarus and his partners changed. The venture capitalists retained responsibility for ARCH Venture Partners, but relinquished their direct role in the licensing of University technology and ceased to be employees of ARCH Development Corporation. ARCH Venture Partners could, however, continue to finance new enterprises that sought to commercialize University technologies. The University endowment invested in the second fund as well, and was a limited partner alongside the various institutional and individual investors. The role of ARCH Development Corporation in this fund, however, was substantially different. In the earlier fund, ARCH Development had been the sole general partner; now the general partner's carried interest was divided between the three ARCH venture capitalists. In addition to the financial relationship, ARCH Venture Partners' special relationship with the University was preserved through a clause in the agreement with the University formalizing a "right of first look" at the University's technology.

ARCH rapidly expanded after raising the second fund, adding a vice president, three post-graduate fellows, and a number of consultants, and beginning operations at other national laboratories and universities based in Massachusetts, New Mexico, New York, and Washington State. These new locations generated a stream of new investment opportunities and the share of new transactions originating from the University of Chicago and Argonne fell dramatically. Only a little more than a year after the closing of ARCH Venture Partners II, they began seeking a third fund with a broad mandate to pursue early-stage technology investments. They were successful in raising $107 million, primarily from institutional investors and corporations, in 1997.

General Evidence about Affiliated Venture Funds

The experience of ARCH Venture Partners illustrates a variety of challenges faced by university-affiliated funds. With ARCH many things went right. Substantial barriers to the recruitment of personnel were overcome, regulatory barriers designed to forestall conflicts of interest and informal organizational concerns were addressed, and the investments proved to be reasonably successful. Nonetheless, the structure proved unstable.

A natural question concerns the generality of this example. Is it representative of other university-based efforts? It is premature, however, systematically to study the success of funds affiliated with academic institutions. Few of these efforts have reached maturity, and data on their activities is limited.[11] We can gain some useful insights by examining another form of institutionally affiliated venture capital funds: those sponsored by corporations, which face many similar challenges as university funds, but have a much longer and better-documented historical record.

The first corporate venture funds began in the mid-1960s, about two decades after the first formal venture capital funds. The corporate efforts were spurred by the successes of the first organized venture capital funds, which backed such firms as Digital Equipment, Memorex, Raychem, and Scientific Data Systems. Excited by this success, large companies began establishing divisions that attempted to emulate venture capitalists; during the late 1960s and early 1970s, more than 25% of the Fortune 500 firms attempted corporate venture programs. In 1973, the market for new public offerings—the primary avenue through which venture capitalists exit successful investments—abruptly declined. Independent venture partnerships began experiencing much less attractive returns and encountered severe difficulties in raising new funds. At the same time, corporations began scaling back their own initiatives. The typical corporate venture program begun in the late 1960s was dissolved after only four years.

Funds flowing into the venture capital industry and the number of active venture organizations increased dramatically during the late 1970s and early 1980s, once again attracting corporations to the promise of venture investing. These efforts peaked in 1986, when corporate funds managed two billion dollars, or nearly 12% of the total pool of venture capital. After the stock market crash of 1987, however, the market for new public offerings again went into a sharp decline. Returns of and fundraising by independent partnerships declined sharply; corporations likewise drastically scaled back their commitment to venture investing. By 1992, the number of corporate venture programs had fallen by one-third and their capital under management represented only 5% of the venture pool.

Interest in corporate venture capital climbed once again in the mid-1990s, both in the United States and abroad. Much of this interest was stimulated by the rapid growth of independent venture funds and their attractive returns. Corporate funds have been invested directly in a variety of internal and external ventures, as well as in funds organized by independent venture capitalists.[12]

Numerous studies suggest that the instability of earlier corporate venture programs was due to three structural failings. First, these programs suffered from a lack of well-defined missions (Fast 1978; Siegel, Siegel, and MacMillian 1988). Typically, they sought to accomplish a wide array of not necessarily compatible objectives: from providing a window on emerging technologies to generating attractive financial returns, a confusion over program objectives which often led to dissatisfaction. For instance, when outside venture capitalists were hired to run a corporate fund under a contract that linked compensation to financial performance, management frequently became frustrated about their failure to invest in the technologies that most interested the firm. A second cause of failure was insufficient corporate commitment to the venturing initiative (Hardymon, DiNino, and Salter 1983; Rind 1982; Sykes 1990). Even if top management embraced the concept, middle management often resisted. R&D personnel preferred that the funds be devoted to internal programs; corporate lawyers disliked the novelty and complexity of these hybrid organizations. New senior management teams in many cases terminated programs, seeing them as expendable "pet projects" of their predecessors. Even if they did not object to the idea of a program, managers often were concerned about its impact on the firm's accounting earnings; during periods of financial pressure, money-losing subsidiaries were frequently terminated in an effort to increase reported operating earnings. A final cause of failure was inadequate compensation schemes (Block and Ornati 1987; Lawler and Drexel 1980). Corporations have frequently been reluctant to compensate their venture managers through profit-sharing ("carried interest") provisions, fearing that they might need to make huge payments if their investments were successful. Typically, successful risk-taking was inadequately rewarded and failure excessively punished. As a result, corporations were frequently unable to attract managers who combined industry experience with connections to other venture capitalists to run their venture funds. Furthermore, all too many corporate venture managers adopted a conservative approach to investing. Nowhere was this behavior more clearly manifested than in the treatment of lagging ventures. Independent venture capitalists ruthlessly terminate funding to failing firms because they want to devote their limited energy to firms with the greatest promise; corporate venture capitalists have frequently been unwilling to write off unsuccessful ventures, lest they incur the reputational repercussions that a failure would entail.

A recent study (Gompers and Lerner 1998) used the VentureOne database of private equity financings to examine over thirty thousand investments into entrepreneurial firms by corporate and traditional venture capital programs. The empirical evidence suggests that corporate programs were considerably less stable than independent funds, frequently ceasing operations after only a few investments. Instability was notable in corporate funds whose investments lacked a strong strategic focus. When corporations invested in areas that they knew very well, they could effectively select and add value to portfolio firms. These benefits were sufficient to overcome the structural problems faced by corporate venture funds. When there was no fit between the corporate parent and the portfolio firms, the programs almost invariably failed.

Many of these concerns are germane to university-sponsored funds. As the ARCH case illustrates, many of the same structural challenges surface in university and corporate funds. Furthermore, it is unclear in many cases what a university-based fund can "bring to the table" aside from money, as it appears difficult for a university to offer a portfolio company the strategic benefits that a corporation investing in an area related to its core line of business can.

Conclusions

This essay thus ends with an ambivalent conclusion. On the one hand, it has emphasized the importance of venture capital for the commercialization of academic technologies. University spin-outs pose severe information problems for investors, which often can preclude the school from realizing much, if any, value from discoveries made in its laboratories. Venture capital organizations employ a variety of mechanisms to address these problems.

At the same time, venture capital is limited in the geographic and industrial scope of the firms in which it invests. Numerous schools have been tempted to consider the establishment of funds that would duplicate the activities of independent venture funds, but case studies and empirical evidence raise serious questions about whether such efforts are likely to be sustainable. Rather than entering into these treacherous waters, university technology transfer officials and administrators may be better served by investing in developing strong relationships with the venture capital community, both with the significant local organizations and with the leading national funds.

Acknowledgments

I thank participants in the November 1997 Universities and Science-Based Industrial Development pre-conference, the September 1998 conference, and the December 1997 Association of University Technology Managers Workshop on University-Based Venture Capital Funds for helpful comments. Research support was provided by Harvard Business School's Division of Research.

Notes

1. This account is based on Seragen's filings with the U.S. Securities and Exchange Commission. In a 1992 agreement with the State of Massachusetts' Attorney General's Office, the University agreed to make no further equity investments. The school, however, made a $12 million loan guarantee in 1995 (subsequently converted into equity) and a $5 million payment as part of an asset purchase in 1997. The firm was merged in 1998 into another biotechnology concern. Even if all contingent payments associated with the transaction are made, the university will have received far less than the amount invested.

2. This and the following section are based in part on Lerner (1998).

3. Various aspects of the oversight role played by venture capitalists are documented in Gompers and Lerner (1999b) and Lerner (1999); the theoretical literature is reviewed in Barry (1994).

4. While evidence regarding the financing of these firms is imprecise, Freear and Wetzel's 1990 survey suggests that venture capital accounts for about two-thirds of the external equity financing raised by privately held technology-intensive businesses from private-sector sources.

5. For a detailed presentation of these arguments, see U.S. Small Business Administration, Office of Advocacy (1996a, 1996b).

6. A related body of literature documents that investments in R&D yield high private and social rates of return (e.g., Griliches 1986, Mansfield et al. 1977). These findings similarly suggest that a higher level of R&D spending would be desirable.

7. The structure of venture partnerships is discussed at length in Gompers and Lerner (1999b).

8. There are two primary reasons why venture funds do not simply hire more partners if they raise additional capital. First, the supply of venture capitalists is quite inelastic. The effective oversight of young companies requires highly specialized skills that can only be developed with years of experience. A second important factor is the economics of venture partnerships. The typical venture fund receives a substantial share of its compensation from the annual fee, which

is typically between 2% and 3% of the capital under management, and is strong motivation to increase the capital that each partner manages.

9. For a theoretical discussion of why poorly capitalized firms are less likely to be successful, see Bolton and Scharfstein (1990).

10. This section is based on Lerner (1995).

11. The available evidence, however, suggests that several university-based efforts have been disappointing. For case studies of a number of other programs, see Atkinson (1994).

12. Venture Economics estimates that corporate investors accounted for 30% of the commitments to new funds in 1997, up from an average of 5% in the 1990–1992 period.

References

Akerlof, G. A. 1970. "The market for 'lemons': Qualitative uncertainty and the market mechanism." *Quarterly Journal of Economics* 84: 488–500.

Atkinson, S. H. 1994. "University-affiliated venture capital funds." *Health Affairs* 13: 159–175.

Barry, C. B. 1994. "New directions in research on venture capital finance." *Financial Management* 23 (Autumn): 3–15.

Block, Z., and O. A. Ornati. 1987. "Compensating corporate venture managers." *Journal of Business Venturing* 2: 41–52.

Bolton, P., and D. Scharfstein. 1990. "A theory of predation based on agency problems in financial contracting." *American Economic Review* 80: 93–106.

Fast, N. D. 1978. *The Rise and Fall of Corporate New Venture Divisions.* Ann Arbor: UMI Research Press.

Fenn, G. W., N. Liang, and S. Prowse. 1995. *The Economics of the Private Equity Market.* Washington D.C.: Board of Governors of the Federal Reserve System.

Freear, J., and W. E. Wetzel, Jr. 1990. "Who bankrolls high-tech entrepreneurs?" *Journal of Business Venturing* 5: 77–89.

Gardiner, G. E. 1997. "Strategies for Technology Development: A Presentation to the Board of the Yale Corporation." <http://www.yale.edu/ocr/yalecorp.html>.

Gompers, P. A. 1995. "Optimal investment, monitoring, and the staging of venture capital." *Journal of Finance* 50: 1461–1489.

Gompers, P. A., and J. Lerner. 1999a. "The determinants of corporate venture capital success: Organizational structure, incentives, and complementarities." In R. Morck, ed., *Concentrated Corporate Ownership.* Chicago: University of Chicago Press for National Bureau of Economic Research.

Gompers, P. A., and J. Lerner 1999b. *The Venture Capital Cycle.* Cambridge, MA: MIT Press.

Greenwald, B. C., J. E. Stiglitz, and A. Weiss. 1984. "Information imperfections in the capital market and macroeconomic fluctuations." *American Economic Review Papers and Proceedings* 74: 194–199.

Griliches, Z. 1986. "Productivity, R&D, and basic research at the firm level in the 1970's." *American Economic Review* 76: 141–154.

Hall, B. H. 1992. "Investment and research and development: Does the source of financing matter?" Working Paper No. 92–194, Department of Economics, University of California at Berkeley.

Hao, K. Y., and A. B. Jaffe. 1993. "Effect of liquidity on firms' R&D spending." *Economics of Innovation and New Technology* 2: 275–282.

Hardymon, G. F., M. J. DeNino, and M. S. Salter. 1983. "When corporate venture capital doesn't work." *Harvard Business Review* 61 (May–June): 114–120.

Himmelberg, C. P., and B. C. Petersen. 1994. "R&D and internal finance: A panel study of small firms in high-tech industries." *Review of Economics and Statistics* 76: 38–51.

Hubbard, R.G. 1998. "Capital-market imperfections and investment." *Journal of Economic Literature* 36: 193–225.

Jensen, M. C., and W. H. Meckling. 1976. "Theory of the firm: Managerial behavior, agency costs and ownership structure." *Journal of Financial Economics* 3: 305–360.

Krugman, P.R. 1991. *Geography and Trade.* Cambridge, MA: MIT Press.

Lawler, E., and J. Drexel. 1980. *The Corporate Entrepreneur.* Los Angeles: Center for Effective Organizations, Graduate School of Business Administration, University of Southern California.

Lerner, J. 1995. ARCH Venture Partners: November 1995. Case no. 9-295-105 and teaching note no. 5-298-138, Harvard Business School.

Lerner J. 1998. "'Angel' financing and public policy: An overview." *Journal of Banking and Finance* 22: 773–783.

Lerner J. 1999. *Venture Capital and Private Equity: A Casebook.* New York: John Wiley.

Liles, P. 1977. *Sustaining the Venture Capital Firm.* Cambridge, MA: Management Analysis Center.

Mansfield, E., J. Rapoport, A. Romeo, S. Wagner, and G. Beardsley. 1977. "Social and private rates of return from industrial innovations." *Quarterly Journal of Economics* 91: 221–240.

Myers, S. C., and N. Majluf. 1984. "Corporate financing and investment decisions when firms have information that investors do not have." *Journal of Financial Economics* 13: 187–221.

Palmer, A. T. 1993. "Collaboration turns research into jobs." *Chicago Tribune* (June 6), C1, C3.

Rind, K.W. 1982. "The role of venture capital in corporate development." *Strategic Management Journal* 2: 169–180.

Roberts, E.B. 1991. *Entrepreneurs in High Technology: Lessons from MIT and Beyond.* New York: Oxford University Press.

Saxenian, A. 1994. *Regional Advantage: Culture and Competition in Silicon Valley and Route 128.* Cambridge, MA: Harvard University Press.

Siegel, R., E. Siegel, and I. C. MacMillan. 1988. "Corporate venture capitalists: Autonomy, obstacles, and performance." *Journal of Business Venturing* 3: 233–247.

Stiglitz, J. E., and A. Weiss. 1981. "Credit rationing in markets with incomplete information." *American Economic Review* 71: 393–409.

Sykes, H. B. 1990. "Corporate venture capital: Strategies for success." *Journal of Business Venturing* 5: 37–47.

U.S. Small Business Administration, Office of Advocacy. 1996a. *Creating New Capital Markets for Emerging Ventures.* Washington D.C.: U.S. Small Business Administration.

U.S. Small Business Administration, Office of Advocacy. 1996b. *The Process and Analysis behind ACE-Net.* Washington, D.C.: U.S. Small Business Administration.

VentureOne. 1998. *Venture Capital Investment Report: 1997.* San Francisco: VentureOne.

16

University-Industry Cooperation in Biomedical R&D in Japan and the United States: Implications for Biomedical Industries

Robert Kneller

Introduction

This chapter compares university-industry cooperation in biomedical research and development (R&D) in Japan and the United States. As the only chapter in this volume that looks in depth at a particular industry, it seeks to integrate insights from the other chapters and to draw conclusions concerning how university-industry cooperation affects the strength of one particular industry. This analysis focuses on the system of university-industry cooperation, including its financial and legal aspects, and also on evidence for such cooperation. This systemic approach makes analysis of a complex subject more manageable, although it misses many of the nuances and important exceptions that might be revealed by a more descriptive analysis focusing on specific technologies or projects. Nevertheless, an emphasis on the overall cooperative "system" identifies significant differences in the nature of university-industry biomedical R&D cooperation in Japan and the United States which may influence the strength of biomedical industries in both countries. Some of these differences are unique to biomedical R&D, but others also apply to cooperation in other fields. Nevertheless, because of certain unique characteristics of the biomedical industry, notably (1) the influence of the U.S. Food and Drug Administration (FDA) and the Japanese Ministry of Health and Welfare (MHW) in deciding which products can be marketed and (2) the influence of MHW in setting prices for pharmaceutical products and biomedical procedures in Japan, the conclusions of this chapter cannot necessarily be generalized to other industries. Finally, although the scope of this chapter includes all biomedical-related industries, including electronic imaging and biomaterials, its principal focus is the pharmaceutical industry, and its analysis of comparative industrial strength applies mainly to that industry.

Supply and Demand Considerations

The United States and Japan are the world's two largest markets for biomedical products and services. In 1997, Japan's pharmaceutical market constituted 21% of the global pharmaceutical market of approximately $300 billion, second only to the U.S. market at 31% of the global market (SBC 1998). Per capita spending on prescription drugs is higher in Japan than any other industrialized country, followed by France, then the United States (PhRMA 1998). With approximately half the population of the United States, Japan's per capita sales of drugs are considerably higher. Projections that in 2025 the percentage of persons aged 60 and older will increase to 33% in Japan (an increase of 12% over 1996) compared to 24.6% in the United States (an increase of 8% over 1996) (U.S. Bureau of the Census 1996) suggest that the demand for innovative biomedical therapies may rise relatively faster in Japan than in the United States. This increased demand may be particularly high for cancer-related therapies and diagnostics, as cancer is the leading cause of death in Japan, as opposed to circulatory heart disease in the United States. (Utsunomiya 1993).

At the same time, the variety of useful drugs and other biomedical technologies is rapidly increasing, and expanding genetic knowledge is continually suggesting new drugs and diagnostic tests. New classes of drugs that did not exist ten years ago are now on the market or in development. Rapid advances are occurring in diagnostic tests, biomaterials, medical imaging technologies, technologies for non-invasive surgery, and so on. Biomedicine is an area where the technical feasibility of rapid, substantial technical advances, coupled with great demand for these new technologies, predict great R&D and commercial potential.

The Special Importance of University-Industry Linkages in Biomedical R&D

A variety of data sources indicate that cooperation between U.S. universities and industry in biomedical R&D is pervasive and of great importance to industry: Mowery et al. (chapter 11) document how biomedical discoveries account for the majority of reported inventions and licensed technologies at Columbia University and the University of California, and also a large proportion of invention reports and licenses at Stanford University. Moreover, the majority of licensing revenue in all three of these universities is from biomedical inventions, suggesting that industry pays more for university biomedical technologies than for all other university technologies combined. This suggestion is confirmed by the latest survey by the Association of University Technology Managers (AUTM), which found that 86% of gross licensing income received in 1996 by responding institutions was for life science inventions (AUTM 1998).[1,2]

Cockburn and Henderson (1996) examined case histories of 21 drugs introduced between 1965 and 1992 that had great impact upon therapeutic practice. Only five of these (24%) were developed without any input from the public sector. (In most cases, "public sector" meant publicly supported university researchers.) Of these five, all but one were discovered through random screening of large numbers of compounds. Among the 14 drugs whose discovery depended on detailed knowledge of the causes of the target diseases and/or the structure of the compounds being tested, all but one involved significant input from publicly supported researchers. In addition, when Cockburn and Henderson examined patterns of coauthorship involving researchers in major pharmaceutical companies, they found a strong correlation between the proportion of coauthored papers that involved university researchers and a company's research productivity.[3] In other words, considering the set of all publications with multiple authors where at least one author has a corporate address and at least two institutional addresses are listed in the publication, the likelihood that one of the institutional addresses will be a university as opposed to another branch of the same company is strongly correlated with a high ratio of important patents per dollar of research expenditure.

Mansfield's survey (1991) of 76 U.S. firms suggested that the pharmaceutical industry relied even more than other industries on university R&D. Although his survey included only six pharmaceutical companies, these reported that 27% of their products and 29% of their processes would not have been developed (or would have been greatly delayed) without university research, as opposed to 11 and 9%, respectively, for survey respondents as a whole. A survey of 210 life science companies by Blumenthal et al. (1996) showed that 90% of the companies that conducted research in the United States had life science relationships with U.S. academic institutions. Of the respondents 59% actually supported research in such institutions, significantly more than reported by a comparable survey ten years earlier. In 1994, this research support amounted to $1.5 billion, or approximately 12% of all life science R&D expenditures in such institutions (about $12.7 billion), a higher proportion than the 7% average for industry support for U.S. academic research across all fields of science and engineering.[4] Most of the life science companies with academic ties derived patents, products, or sales from such ties (Blumenthal 1996). Nevertheless, these companies reported that the greatest value of academic relationships was to keep their own researchers current with important research, to provide ideas for new products and to assist in recruiting able researchers, rather than to invent new products that these companies would license. Unpublished data suggests that investment in academic research by life science companies produces more patents per dollar of investment than investment in nonacademic (including in-house) research (Blumenthal et al. 1995).

Narin (1997a) has shown that each patent granted by the U.S. Patent and Trademark Office (PTO) for inventions in the field of drugs and medicines cites on average more scientific articles than do inventions in other areas: nine articles compared with three for chemical, 1.3 for instrument and 1.0 for electronic patents in 1995. These citations to open scientific literature, most of which consists of academic journal publications, indicate the degree to which technologies that are commercially valuable to industry are based upon academic science.[5] Over time, the number of scientific citations per patent has increased, a phenomenon which is generally true for all fields of research and also for inventions that originate in Japan and other countries in addition to the United States.[6]

Data showing the extent and nature of university-industry cooperation in biomedical R&D in Japan are scarce compared to the United States. As described in chapter 12, because invention reporting requirements are lax and the majority of technology transfers occur informally with no reporting to university officials, data on university inventions are incomplete and there are no systematically collected data on university technology transfers. Nevertheless, contacts between university faculty and industry are extensive and it is relatively easy for Japanese companies to obtain intellectual property (IP) rights to university inventions.

A survey in the late 1980s of six major Japanese biotechnology firms found that their most important external source of knowledge was joint research with domestic universities, while joint research with foreign universities was probably the second most important (Fransman 1992).[7] In chapter 5, Niwa shows that since 1986 Japanese companies have provided more support for R&D related to drugs and medicines than any other area, both to Japanese universities and to foreign universities. In fact, since 1992 Japanese corporate support for drug and medicine R&D in foreign universities has exceeded their support in this area to Japanese universities. A recent survey by the Japan Bioindustry Association (JBA) estimated that over a ten year period ending in 1997, 38% of Japanese patent applications in the field of genetic engineering (on average 874 of 2,327 applications annually) listed at least one Japanese university faculty member as an inventor. Out of a sample of 252 genetic engineering patent applications, at least 72% were filed by private companies. None were filed by the universities themselves (JBA 1998).[8] These findings support one of the main points of chapter 12, that statistics on numbers of patents filed by Japanese universities vastly understate and have little correlation with the amount of commercially valuable technology that is passed from universities to industry.

Finally, coauthorship data for both the United States and Japan show that industry authors of biomedical papers are more likely to coauthor papers with academic researchers than are industry authors as a whole, while in the United States, academic authors of biomedical papers are less likely to coauthor papers with industry researchers than academic authors as a whole (Pechter and Kakinuma chapter 4; NSB 1998). This suggests that the biomedical industry may be more dependent on academic research than other industries in general, although U.S. academic biomedical researchers may be less likely to cooperate with industry than their colleagues in other academic fields. (Perhaps this is because generous public funding for U.S. academic biomedical research supports large numbers of researchers, many of whom do not work in areas of interest to industry or do not need industry support.)

Less Effective Linkages between Universities and Industry in Japan than in the United States

Despite the findings above, university-industry linkages in biomedical R&D may be weaker or less effective in Japan than in the United States. Chapter 12 describes the traditional system of university-industry technology transfer under which inventions are passed informally and cheaply to companies with whom professors have connections, without any guarantees that the inventions will actually be developed. That chapter hypothesizes that this system has (1) led universities to adopt a passive approach to relations with industry and (2) decreased incentives for companies to work concertedly with universities to exploit the potential commercial benefits of university R&D. Although evidence for this hypothesis is based partly upon the structure of the Japanese technology transfer system and upon personal observations of Japanese scientists and university and business officials, the JBA survey cited above provides statistical support for this hypothesis in the field of biotechnology. Of the sample of 252 patent applications listing a university faculty member as an inventor, only 40 (16%) had actually issued as patents. In the case of 156 applications (62%), the applicant either had not requested that the JPO review the application,[9] or (in one case) had asked the JPO not to issue a patent after the JPO had indicated that it would approve the application.[10] The survey notes that the 16% ratio of granted patents to non-denied applications "does not exceed" the overall ratio for genetic engineering patent applications, suggesting that the likelihood that genetic engineering technologies acquired by companies from universities will not be developed is probably higher than the likelihood that similar technologies obtained from other sources will not be developed.[11] On the basis of the survey by the Japan Technomart Foundation cited in chapter 12, it might be assumed that these technologies are not being developed because the recipient companies believe the discoveries are not sufficiently valuable to them, yet probably also want to block competitors from developing these technologies (see Note 9).

The JBA survey also includes responses from 39 JBA member companies to a JBA questionnaire administered in 1997 on university-industry cooperation and IP management over the preceding five-year period.[12] Of the 39 companies, 92% said they had cooperative relationships with Japanese universities over that time.[13] The average number of cooperative relationships per respondent was 156; the average annual expenditure per respondent to support such relationships was ¥151 million (roughly $1 million). Over this period 116 patent applications were filed on discoveries to emerge as a result of cooperation with universities (slightly less than one per respondent per year). However, only 24 of these applications were for discoveries that were of "practical utility" to the companies. The respondents reported that discoveries originating in universities accounted for 14% of their total development projects that reached the stage of practical utility for their companies. Interestingly, expenditures on cooperative relationships with universities accounted for only about 1.4% of their biology-related R&D expenses. This suggests that, even though the biomedical industry does not develop the vast majority of new technologies it receives from universities, the industry's return on investment in university R&D is very high.

When the 39 JBA survey respondents were asked about various shortcomings of research with Japanese universities, issues of confidentiality, IP protection and exclusive access to research results were the major concerns, followed by difficulties in influencing the direction of university research, slowness of university research, and poor accountability concerning how universities use corporate research support.[14] When asked about reasons for cooperating with U.S. universities, the most frequently checked response (checked by 26 respondents) was the abundance of venture-capital-supported university spin-off companies in the United States. This suggests that Japanese companies view venture capital funding as complementary to, rather than competitive with, their own support for U.S. university researchers. Other responses also emphasized the abundance of research resources, as well as and the ease of university-industry interactions in the United States.[15] When asked about differences between foreign and Japanese universities, "clarity of contracts" was checked more frequently than any other option;[16] and when asked how Japanese university research should be reformed to promote commercialization of biomedical research, the most frequently selected option by far was "universities should establish offices to handle patenting and licensing."[17]

Therefore, the JBA survey suggests that Japanese technology transfer procedures are a significant impediment to university-industry cooperation, at least in biomedical research. It also suggests two other important factors: (1) lack of venture capital and university start-up companies, and (2) low levels of government support for biomedical research.

Although the question of whether small- and medium-size enterprises (SMEs) per se contribute disproportionately to technological innovation is still unresolved, Kortum and Lerner (1998) found that small firms backed by venture capital do have a disproportionate impact on U.S. technological innovation. As Kano notes in chapter 14, before 1995 only four independent Japanese start-up companies had been formed to commercialize pharmaceutical-related research. In contrast, over 1600 new companies have been formed since 1980 based on a license from an academic U.S. institution (AUTM 1998). Blumenthal et al. (1996) estimate that over a thousand science-based SMEs are pursuing development of life science technologies. A substantial portion of these are probably university licensees. U.S. university faculty have participated in founding 24 Fortune 500 companies and over 600 smaller life science companies (Blumenthal et al. 1995). SMEs derive a much higher number of patents from university research than do large companies, and the proportion of their total patents derived from university research is also higher (Blumenthal et al. 1996).[18] This indicates that an important component of the U.S. national innovation system for biomedical technologies may be absent from Japan. (Blumenthal's 1996 estimate that 58% of the patents of large U.S. life science companies are derived from university research is more than four times greater than the JBA survey's estimate that 14% of the useful projects of JBA members were based on university discoveries.)

One reason for the dearth of Japanese biomedical venture companies may be the nature and limited amount of venture capital funding in Japan, an issue that is discussed briefly in chapter 12 and in more depth by Kano in chapter 14. Other factors may be limited research manpower and various social factors that dissuade capable young Japanese researchers from starting their own companies. Underlying low levels of biomedical research manpower are low levels of public funding for biomedical research in Japanese universities, at least in comparison to the United States, and the likely existence of slow-to-change enrollment and funding quotas for university biomedical departments of the type described by Ogura and Kotake in chapter 21. Biomedical or life science research accounts for a substantially smaller proportion of total university research in Japan than it does in the United States. Of approximately 1,145 ongoing Joint Research projects between Japanese National Universities and private companies in 1991, approximately 100 (9%) were classified as biotechnology related (Diamond 1993). In contrast, Note 4 above suggests that biomedical research accounts for the majority of university-industry cooperative research projects in the United States. In Japanese fiscal year 1995, the value of standard life-science-related Monbusho Grants-in-Aid for university research was approximately ¥15.2 billion, or 31% of the ¥49.6 billion total of standard Monbusho Grants-in-Aid for all fields of science and engineering (Committee 1995).[19]

In contrast, 1995 U.S. government support for life-science R&D in academic institutions was $6.7 billion, or 54.6% of the total $11.5

billion federal support for academic R&D (NSB 1996). Moreover, it appears that the absolute level of government support for university biomedical research is substantially less in Japan than in the United States. Total 1995 intramural expenditures for life-science R&D in Japanese national universities from all sources were ¥277 billion, or 24% of the ¥1.19 trillion total for all fields of science and engineering (Statistics Bureau 1996). While this figure excludes Japanese government support for R&D in private and local government universities, it includes private and local government support of R&D in national universities. Depending upon whether market exchange rates or the more stable Organization for Economic Cooperation and Development (OECD) purchasing power parity index is used, this in itself suggests that U.S. government support for life-science university R&D is two to four times the level of Japanese government support. However, as described by Irvine et al. (1999) and by Ohtawa in chapter 6, Japanese methods for calculating government support for university R&D overestimate such support in comparison with methods used in the United States and Europe. Assuming their analyses apply specifically to life-science university R&D, government support for university life science R&D is at least four and perhaps 12 or more times higher in the United States than in Japan.[20] The data of Negishi and Sun (chapter 7) showing that Japanese publication rates on a per population basis in medical fields are far below rates in the United States are consistent with public funding for biomedical research being particularly low in Japan compared to other industrialized countries.[21]

University-Industry Cooperation in Clinical Trials

Much of the development work in biomedical R&D consists of tests on human subjects, usually hospitalized patients. Known as "clinical trials," these tests are to determine the safety and effectiveness of new drugs, diagnostics, and medical devices. Each new drug approved by the U.S. Food and Drug Administration (FDA) requires on average 6.7 years of clinical trials and over 4,200 patients participating in over 68 clinical trials.[22] Clinical trials constitute the most expensive stage of the biomedical development process, accounting for 36% (approximately $6.2 billion) of the $17.2 billion total R&D expenditure of the U.S. pharmaceutical industry in 1998 (PhRMA 1998). Industry-financed clinical trials are shaped largely by the requirements of MHW in Japan and the FDA in the United States. Although only a portion of them occur in university hospitals, such hospitals can be ideal sites for such trials, because they have (a) large numbers of patients, (b) doctors and nurses who are attuned to clinical research needs, and (c) on occasion, special units or wards and trained staff to conduct clinical trials. In addition, clinical researchers (most of whom are physicians in university or government medical centers) often initiate their own clinical trials to test the effects of drugs or combinations of drugs against particular diseases or in particular types of patients. These investigator-initiated trials are sometimes supported by public funds and sometime by funds from the biomedical industry. Indeed, clinical trials account for approximately 13% of the National Institutes of Health's (NIH's) budget to support university research, while all areas of clinical research (most of which are university hospital-based) account for approximately 38% of NIH's extramural research budget. In other words, clinical trials and other forms of hospital-based research account for a substantial proportion of publicly supported biomedical research in the United States.[23] The U.S. National Cancer Institute (the largest institute within NIH) estimates that the cost of study management, data collection, and data monitoring for cancer-related clinical trials is in the range of $2500 to $3000 per patient. This cost does not include the additional physician and nursing care and the additional tests required to monitor and provide extra care for clinical trial participants. MHW supports several major research hospitals (for example, the National Cancer Center Hospital in Tokyo) that are major centers for clinical trials. Monbusho supports medical schools and affiliated hospitals in national universities, and thus the infrastructure for many of the university-based clinical trials in Japan. One of these national universities, Hamamatsu University School of Medicine, has a clinical pharmacology department that used to be supported by the Japan Pharmaceutical Manufacturers Association (JPMA), but was recently taken over by Monbusho.

U.S. university hospitals have experienced a significant rise in industry-sponsored clinical trials over the past ten years, and such trials now are a significant source of revenue for many medical schools and their affiliated hospitals. Usually, a pharmaceutical company interested in sponsoring a clinical trial at a medical school will negotiate a budget with school or hospital officials that includes line items for equipment, laboratory tests, other quality control monitoring, overhead, and salary support for physicians and nurses. Many major university-affiliated hospitals have established clinical trial units staffed with research nurses, data managers, statisticians, and administrative personnel whose jobs depend upon their units' maintaining a certain minimum level of clinical trial business. A few of these centers have been established primarily with NIH funds, although these can conduct industry-sponsored, as well as NIH-sponsored, clinical trials.[24] Many physicians who are medical school faculty members devote considerable time to patients in such centers. Up to 80% of the total salary of some of these physicians is derived from clinical trials, although the average is probably between 30 and 40%.

Pharmaceutical companies have found that hiring well-known academic physicians as consultants, or sponsoring trials for drugs under the direction of such physicians, provides valuable scientific and clinical insights into the development of new drugs. Moreover, such physicians may be able to influence advisory committees convened by the FDA to decide whether new drugs should be approved. Although such physicians cannot sit on the FDA advisory committees reviewing the drugs in which they have an interest, they often know and are respected by committee members. While this raises concerns about pharmaceutical companies being able to exert undue influence on regulatory decisions, this sort of academic input often greatly assists the speedy development of safer and more effective drugs and diagnostic methods. In all their publications and representations before the FDA, academic physicians are expected to disclose the sources of support for their research. Most universities have conflict of interest guidelines, and all are required by federal law to constitute independent institutional review boards (IRBs) to review all clinical trial protocols prior to initiation of such trials. However, only some universities (perhaps a minority) require review by a special ethics committee of a faculty member's stock holdings, outside consulting arrangements, and so on, before permitting him or her to accept industry sponsorship for a clinical trial.

In Japan, the vast majority of clinical trials are financed by industry, although data on the level of this support are not available. A common practice of pharmaceutical companies wishing to sponsor clinical trials is to seek the assistance of an authority in the field of the product to be tested, who then recruits from his network of associates and former students a sufficient number of physicians who could in turn recruit the necessary number of patients for the trial. This results in patients for most trials being dispersed among a large number of hospitals, including some university hospitals but also many non-university hospitals, including some with only 200 to 300 beds. Another factor encouraging the dispersal of trial subjects is the tendency of pharmaceutical companies to involve as many hospitals as possible in order to lock in new markets for their drugs once they are approved. Hospital clinical departments may also have an incentive to become involved in as many trials as possible, because this can provide an important source of funds for their own research (Fukuhara et al. 1997).

At least until recent years, the number of clinical trials and new drug approvals in Japan has been quite high. From 1982 to 1990, Japanese companies led the world in the number of new drugs introduced to market: 113 compared with 97 for U.S. companies, while German companies were a distant third (JEI 1992). Around 1990, up to 160 new Japanese drugs were entering clinical trials each year and 40 were being approved for marketing (SBC 1998), compared with an average of 25 new drug approvals per year by the FDA for 1989 to 1991 (PhRMA 1998). As will be discussed later, most of these drugs were "me-too" drugs: modifications or derivatives of previously approved drugs. Nevertheless, coupled with the practices described in the previous paragraph, the large number of trials exacerbated the tendency for patients in a particular trial to be dispersed among many hospitals as hospitals vied to become involved in trials but found they had the resources to handle only a small number of patients per trial. Some hospitals conduct up to 30 trials at a time, although none have special wards or clinics for such trials (except for a few special units to handle Phase 1 trials, which usually involve fewer than 50 patients and are intended to determine safe dosages of a new drug). One study reported that in a typical Phase 3 trial (which tests the effectiveness of a new drug and often involves over 1000 patients) only two or three patients are enrolled in each hospital (Fukuhara 1997).[25]

Predictably, the quality of trial procedures and trial data has suffered. Japan has had Good Clinical Practice (GCP) rules in effect since 1990. Among other things, these require that all clinical trial protocols be reviewed by IRBs, that clinical trial agreements be contractually-based agreements between the pharmaceutical sponsor and the testing institution (not individual investigators) and that, if a patient gives only oral consent to participate in a trial, a record of the oral consent must be preserved. However, MHW inspections of 41 pharmaceutical companies and 77 national and private hospitals in 1994 revealed inadequate reporting of serious adverse effects, failure to measure clinical test values, involvement of unauthorized institutions in trials, failure to detect protocol violations, and a tendency for IRBs to rubber-stamp proposed protocols. Other MHW studies suggested that about a third of the physicians involved obtain "informed consent" without providing their patients any written materials or receiving their written consent (Fukuhara 1997). Japanese clinical trial data were often regarded as unacceptable by the FDA and European regulatory agencies, hindering the marketing of some Japanese drugs in America and Europe.

These concerns coincided with the Green Cross scandal involving distribution of HIV-contaminated blood products,[26] increased pressure from foreign pharmaceutical companies for greater transparency and fairness in the drug approval process, and the ongoing effort by Japan, Europe, and the United States to harmonize drug approval procedures and standards so that results of appropriately conducted tests in one region will be accepted by regulatory authorities in the other regions. As a result, stricter Japanese GCP guidelines were issued in April 1997 and came into full force a year later (see MHW Ordinance 1997).

The new GCP guidelines require that written informed consent be obtained and impose rigid requirements concerning the consent form. In 1994, Nara Medical University voluntarily mandated written informed consent in all clinical trials, and subsequently the rate of refusal to participate in such trials rose from 5% to about 70% (Fukuhara 1997). Although there are various reasons patients may refuse to participate in trials,[27] the fundamental concern is that under the stricter informed-consent provisions it will be hard to recruit patients for clinical trials. A statistician advising MHW on clinical trial data estimates that fewer than 40% of patients now consent to participate in clinical trials, while the Asahi newspaper (1998) reported that the number of patients participating in clinical trials has fallen to 10% of levels in past years.

Another significant challenge to clinical trials posed by the new GCP guidelines concerns how to compensate physicians and nurses for the extra work associated with such trials, including the new responsibility of obtaining written informed consent from patients. It is illegal for physicians, nurses, pharmacists, and other hospital or medical school employees to receive salary support from clinical trial sponsors. Prior to 1997, a variety of reimbursement mechanisms were used, including generous publication stipends and travel allowances for physicians. The new GCP guidelines reinforce the principle that all industry-sponsored clinical trials be based upon contracts between the sponsor and the hospital, and specify in detail the points such contracts should address. In particular, the guidelines require the sponsor to submit to the hospital directors a document explaining the charges and expenses associated with the trial. This information is to be submitted to the institution's IRB, which has more authority and independence under the new GCP guidelines (MHW Ordinance 1997; Fukuhara 1997). In addition, other recent MHW notifications prohibit a variety of informal or indirect reimbursement schemes. Officials of several pharmaceutical companies and contract research organizations report that it has become difficult to recruit physicians to participate in clinical trials. An official of MHW acknowledged that personnel problems are severe. However, he explained that MHW does not deem participation in clinical trials to be an important duty of hospital personnel, and that employees of public university hospitals (who are employees of Monbusho or local governments) and of public non-university hospitals (who are employees of MHW or local governments) are regarded as public servants who may not receive outside income. Private university hospitals follow suit according to guidelines set forth by the Japan Council of Private Medical Universities. Furthermore, nationwide personnel regulations prevent hospitals from employing temporary medical care staff. However, the new GCP guidelines do permit contract research organizations (CROs), acting as contractors for companies sponsoring clinical trials, to hire study coordinators and data managers to handle some of the additional work associated with the trials. Both the Japan Nurses Association and MHW are starting training programs for pharmacists and nurses to become clinical research coordinators, although as yet there is no MHW-approved job description (that is, no formal position) for clinical research coordinators.

The number of clinical trials has fallen significantly in Japan since 1990, with the decrease especially dramatic since 1997. From the peak levels around 1990, entries of Japanese drugs into the clinical trials process fell to 39 in 1996, while the number of Japanese drugs to successfully complete trials and be approved by MHW fell to eight (SBC 1998). In the first half of 1998, clinical trials for more than 20 drugs were called off (SBC 1998), and the number of patients enrolled in trials may have fallen to 10% of levels in the early 1990s (Asahi 1998). The remaining trials are taking much longer to complete (Asahi 1998). It appears that the decline in the number of clinical trials is due in part to the the new GCP guidelines and in part to fundamental changes in the Japanese pharmaceutical industry and how it is regulated by MHW.[28]

Whatever the reasons, it appears that university hospitals are poised to play a more important role in clinical trials and may have the opportunity to rationalize and improve the clinical trial process. New MHW guidelines requiring at least six patients per trial per hospital should limit the dispersion of clinical trial patients and encourage the conduct of trials in large hospitals. Although statistics on the percentage of trials conducted in university hospitals are currently not available, an official of one of the two largest CROs in Japan stated that most clinical trials are now being conducted in university hospitals, with the shares in public and private university hospitals being approximately equal. Even though salary support for medical staff cannot come from "soft" industry money, university hospitals may have more flexibility than other hospitals to legitimately acquire or shift resources to conduct such trials. Senior clinical researchers of the University of Tokyo along with colleagues in other Japanese universities have proposed the creation of special clinical trial units in major medical institutions which would be staffed exclusively with clinical research nurses and other specially trained personnel. This would further consolidate clinical trial sites and allow for careful review of trials by IRBs, better care for trial participants and more reliable data (Fukuhara 1997). Although some increased government investment would be required, it would appear that industry donations might support some of the some of the physical infrastructure required by such facilities. Perhaps, if there were a reasonable expectation of steady support from industry, MHW and Monbusho might permit some industry funds to be pooled to support full-time permanent employees of such units.[29]

The end result of the current changes, wrenching as they are for the pharmaceutical industry, may be closer and more rational cooperation between industry and universities in the field of clinical drug and medical device development. However, a recent highly publicized scandal involving a well-known professor of pharmacology at Nagoya University Medical School and three Japanese pharmaceutical companies indicates that this process will not be smooth. The professor was arrested and convicted of receiving bribes passed through a dummy company headed by his wife, in return for letting the companies send researchers to national university facilities and exploit those facilities for commercially-oriented research. Corporate officials, including the president of one of the companies, were also arrested. The amounts of money involved were in the range of $50,000 to $200,000 per company per year over a series of several years—sums that are not exceptionally high by U.S. standards, although it appears that much of this money was used for personal rather than research purposes. The companies maintain that the payments were based upon formal contracts and were made in return for technological guidance important to their drug research and development. In the case of one company, Otsuka Pharmaceuticals, the payments were to support collaborative research to develop a new drug to replace Otsuka's successful anti-platelet therapy, which the professor had helped Otsuka to develop and whose patent was to expire in 1999. There are apparently no allegations that the professor gave the companies access to confidential information of other companies or that the payments were an attempt to influence a regulatory decision by MHW. Although the professor should have asked Nagoya University to enter into Joint or Commissioned Research agreements with the companies, in practice it is often difficult for companies to support research under such agreements (chapter 12). Furthermore, it is widely believed that the alleged practices are fairly common. Perhaps the arrests were intended to warn other researchers to cease sponsored research relationships with companies outside of officially sanctioned channels. Nevertheless, there now appears to be widespread uncertainty and caution in the Japanese scientific community about the permissible bounds of cooperation with industry, just a few months after enactment of a new law in May 1998 to encourage such cooperation (see chapter 12 and Saegusa 1998). To make matters worse, public skepticism concerning university-industry cooperation appears widespread in Japan, and this incident may have reinforced the belief of many persons that any form of industry-sponsored university research smacks of corruption.

Conclusion

The principal goal of this paper has been to analyze and compare the nature of university-industry cooperation related to biomedical R&D in Japan and the United States. However, a corollary objective is to provide insights into how the nature of this interaction influences the strength of biomedical industries in each country. Although a detailed comparison of the biomedical industries in each country is beyond the scope of this paper, it appears that, at least with respect to pharmaceuticals, Japanese industry is substantially weaker than U.S. industry in terms of sales and new innovative products (see Note 28). A variety of reasons have been suggested for this situation: failure of Japanese pharmaceutical companies to internationalize their operations (SBC 1998); lack of consolidation in the Japanese pharmaceutical industry (SBC 1998); government overprotection and over-regulation of biomedical industries (SBC

1998; JEI 1992); and price setting under Japan's universal health insurance system that has encouraged the development of derivative drugs for the domestic market and squeezed profits so that companies cannot support the large-scale R&D necessary to develop internationally-competitive innovative medicines (SBC 1998; JEI 1992). However, this chapter, together with others in this volume, suggests that another fundamental reason might be weak or ineffective linkages between the biomedical industry and Japanese universities. The following factors appear to be primarily responsible for this situation: low levels of public support for university biomedical research; limited numbers and mobility of university researchers; a technology transfer system that permits widespread under-exploitation of university discoveries by industry; the absence of university-start up companies backed by independent capital; and difficulties in rationalizing clinical trials in university hospitals. At the risk of passing judgement too quickly on a variety of reforms now under way or under consideration in Japan, only in the case of clinical trials does it appear that effective reforms with a reasonable chance of success in the next few years are being implemented. Thus, effective university-industry cooperation in biomedical R&D may remain problematic in the near future.

With the exception of clinical trials, many of the problem areas just mentioned may also apply to other fields of university-industry R&D cooperation. If this is indeed the case, what is the implication for Japanese industry as a whole? It may be more than just a coincidence that Japanese industry appears weakest in one of the areas where university-industry cooperation is most important for technology creation and development. Effective university-industry cooperation may be sub-optimal in a number of fields, but the negative effects are most evident in areas of technology that require close cooperation between industry and universities (or SMEs based upon university technologies). However, as noted in the introduction, other factors besides ineffective university-industry cooperation might explain the weakness of the Japanese pharmaceutical industry. Biomedical industries are unique in that development of new technologies is significantly influenced by safety regulations and also by health insurance systems. Therefore, analyses of other industries might provide valuable tests of this hypothesis.

Acknowledgments

This research was assisted in part by a grant from the Abe Fellowship Program of the Social Science Research Council of the American Council of Learned Societies with funds provided by the Japan Foundation Center for Global Partnership.

Notes

1. The 173 responding institutions consist of 127 U.S. universities (including almost all major research universities), 27 U.S. hospitals and research institutes, 16 Canadian institutes and 3 patent management firms.

2. Formal transfers of intellectual property (IP) rights are not the only measure of university-industry cooperation. IP protection may be more important for biomedical industries than for other industries, because of the high cost and uncertainty of drug development and the ease with which pharmaceuticals of proven safety and effectiveness can be copied. Also, a single university biomedical invention may often provide more protection than a single invention in other fields, because IP protection for a new drug or diagnostic can often be based on a single principal patent, whereas other high-technology products (e.g., new integrated circuits chips or electronic devices) often incorporate a large number of patented technologies. Therefore, biomedical companies may place greater emphasis than other companies on obtaining IP protection for commercially valuable information obtained from universities.

3. Research productivity was measured as "number of important patents per research dollar," where an "important patent" was defined as a patent granted in at least two of the following three major world markets: Europe, Japan, and the United States.

4. U.S. universities and major U.S. hospitals and research institutions responding to AUTM's annual licensing survey reported that in fiscal year 1994 research expenditures from industrial sources in all fields of research was $1.6 billion (AUTM 1996). It may be that Blumenthal was able to capture a greater proportion of industry support for university research in his survey than AUTM captured in its survey, because the AUTM survey probably does not cover a large number of university-affiliated hospitals. In any case, Blumenthal's data suggest that industry support for university life-science R&D accounts for a substantial proportion of industry support for university R&D in all fields. To put Blumenthal's estimate of industry support in perspective, $1.5 billion of industry support for life science research is about one quarter of the $6.2 billion that U.S. universities received in 1994 from the National Institutes of Health (NIH), the main supporter of academic biomedical research.

5. U.S. patents, unlike those issued by patent offices in Japan and some other countries, are required to cite all prior art that is relevant to the invention.

Usually, most of the citations are to previously issued patents, but, as Narin has shown, the number of citations to open scientific literature has been increasing.

6. However, Narin also found that the average number of citations to scientific literature is higher for inventions that originate in the United States than in Japan. This difference is particularly great for drug and medicine patents (11.6 citations per U.S. invention vs. 3.3 citations per Japanese invention) and is becoming more pronounced over time. Although this may reflect in part the fact that U.S. patent law requires U.S. patent applications to list all relevant prior art, while Japanese patent law does not have this requirement (and therefore U.S. patent applicants are more used to citing more prior art references), Narin suggests that it may also mean that commercial development of biomedical products is more closely linked to academic research (and therefore perhaps is more innovative) in the United States than in Japan, an issue that will be explored in more detail below (Narin 1997b).

7. This interview survey asked Ajinomoto, Kyowa Hakko, Mitsubishi Petrochemical, Sumitomo Chemical, and Takeda Pharmaceuticals, to rank in order of importance the following external sources of knowledge: joint research with competing companies; joint research with noncompeting companies; joint research with Japanese universities; joint research with non-Japanese universities; licensing and other purchases from Japanese companies; and licensing and other purchases from foreign companies. Four of these companies ranked "joint research with Japanese universities" first in order of importance. Of these four companies, two ranked "joint research with foreign universities" second, while another ranked "joint research with foreign universities" third. One of the companies ranked "joint research with foreign universities" first and "joint research with Japanese universities" second.

8. The JBA compiled a list of 2,897 names consisting of its own individual members who are university faculty and also successful 1995 and 1996 university applicants for Monbusho Grants-in-Aid in life-science fields of research. Taking a 10% random sample of 290 names from this list, the JBA found that 252 Japanese patent applications filed between 1 April 1987 and 31 March 1997 in the field of genetic engineering (JPO classification code C-12) listed at least one of these names as an inventor. Adjusting for the ten percent sample and also for the fact that the success rate for Grants-in-Aid applications was 25.6% in 1995 and 1996 (and thus the total pool of potential inventors should be adjusted upwards from 2,897), the JBA study estimated that university faculty members were listed as inventors on 8,743 out of a total of 23,274 patent applications (38%) in the field of genetic engineering filed in the ten-year period from 1987 to 1997. Among the 252 surveyed patent applications, private companies accounted for at least 72% of the named applicants, while individual inventors accounted for 9% of applicants, national government agencies other than universities or Monbusho for 8%, nonprofit organizations for 5%, and local governments for 3% of applicants.

9. Under U.S. patent law, the U.S. PTO automatically reviews all applications. However, under Japanese Patent Law (Section 48), the JPO does not review

applications until the applicant makes a specific request for review. If the applicant does not make such a request within seven years from the date of the application, the application is deemed abandoned. Whether or not a request for review is made, however, Japanese patent applications are published 18 months after they are made. Publication in and of itself makes it impossible for another party to apply for a Japanese patent on the same invention.

10. In the case of 56 applications (22%), the applicant had requested review by the JPO, but a final decision was still pending at the time of the survey.

11. Apparently, the patent applications in this data base, known as PATOLIS, include only pending applications and applications which have been approved by the Japan Patent Office (JPO). Applications denied by the JPO are not included.

12. Fifty-one companies were contacted. These were selected from a list of JBA members recommended by the JBA Study Committee for IP rights and a list of members of the JBA Board on Biology/Biotechnology. Thirty-nine companies responded, for a response rate of 76.5%. Although respondents were not identified, the vast majority of JBA members are large companies (over 1000 employees) or affiliates of large companies. JBA members include pharmaceutical and biotechnology companies, and companies from other industries such as chemicals, environmental technologies, and food products.

13. Of the 39 respondents, 22 reported having a total of 292 cooperative relationships with foreign universities over the five-year period: 79% of these were with U.S. universities, 16% with European universities, and 5% with universities from other locations. The survey did not disclose other information concerning these foreign collaborations.

14. The 39 respondents were presented with a list of 12 potential problems and asked which problems they considered to be important for companies. With respect to IP-related concerns: 18 checked "ineffective control over publications," 10 checked "there are so many collaborating companies that it is hard to obtain exclusive access to research results," three checked "universities do not protect secrets," and two wrote in responses indicating concern over difficulties in obtaining patent protection for university inventions. With respect to non-IP related concerns: 12 respondents checked "difficulties in regulating content of research" and seven checked "unequal cooperative research relationships." Eleven checked "insufficient research speed" and four checked "research deadlines not met." Seven checked "insufficient research records and management of research funds" and five checked "uses of research expenditures not clear," while one respondent wrote in an answer indicating concern about accountability of research funds.

15. Twenty-five respondents cited abundant research funding and 15 cited abundant research personnel. Two respondents specifically praised meritocratic U.S. research support, presumably referring to the U.S. system of research support through peer-reviewed grants. Twenty-five respondents cited a climate

of successful university-industry interactions, and three specifically mentioned the ease with which research results can be commercialized.

16. Tied for second place were "abundance of equipment and personnel," "price of useful research is expensive (for companies)," and "research gets done faster."

17. This suggests a preference by the Japanese biomedical industry for a more contractually based system of university-industry cooperation and technology transfer, despite an awareness that industry will probably have to pay more for university technologies under such a system. (See Note 16.)

18. Among SME and large-company respondents to Blumenthal's survey of patent activity by life-science companies from 1989 through 1993, the median number of patents derived from university research was, respectively, 6.7 vs. 1.7, while the proportion of total patents derived from university research was, respectively, 66% vs. 58%.

19. Monbusho "Grants-in-Aid" is the largest category of project-specific support for Japanese university R&D, totaling ¥92.4 billion in 1995 (see chapter 12). This amount consists of standard awards totaling ¥49.7 billion in 1995, and also several categories of special awards. The ¥15.2 billion total value of standard life-science awards was estimated by summing awards made under the following categories: under "General Sciences": biology; under "Agriculture": agricultural science, agricultural chemistry, interdisciplinary agricultural studies, animal husbandry, and veterinary science; under "Medicine": physiology, pathology, internal medicine, dentistry, pharmacology, and general medicine; and under "interdisciplinary fields": biochemistry, fundamental biology, experimental zoology, medical ecology, and biomaterials. Since special Grants-in-Aid are not classified according to these categories, it is somewhat difficult to estimate the value of special Grants-in-Aid for life-science research. Therefore, the ratio of life-science support to total support is reported for standard Grants-in-Aid awards only.

20. Using the average 1995 market exchange rate of ¥94/$1, ¥277 billion was equivalent to $2.9 billion. Using the more stable OECD purchasing power parity index of ¥176/$1, ¥277 billion was equivalent to $1.6 billion (NSB 1996). However, Japanese data include all labor costs of all full-time academic personnel without adjustments for time spent in teaching and other non-research activities. Also, they include core funding not usually included in statistics from other countries. They include research-related expenditures in the humanities and social sciences, and expenditures in Inter-university Research Institutes such as the National Laboratory of High Energy Physics, whose functional equivalents in other countries are often classified as academically related research institutions rather than universities. Irvine et al. (1990) estimated that these differences require discounting Japanese university expenditure data by about 40% to be equivalent to U.S. and European data, while Ohtawa (chapter 6) estimated that this discount factor should be at least 67%.

21. Indeed, Japanese population-adjusted publication rates in medical fields are below those in all other industrialized countries except Russia, although in most other fields they are closer to or even above U.S. rates.

22. Between 1988 and 1997, the FDA approved on average 29 new drugs per year, suggesting that approximately 120,000 patients are involved in trials for the drugs that are approved for marketing in an average year. However, since only one out of five drugs that enters clinical trials is ultimately approved by the FDA, the actual number of patients involved in trials is considerably larger.

23. In 1995, NIH support for clinical trials was approximately $1.2 billion, about 20% of industry's $6.2 billion level of support for clinical trials. ($1.2 billion may include clinical research in NIH's own Clinical Center as well as support to extramural institutions). An NIH survey of new extramural awards made in 1996 showed that 13% of such awards were for clinical trials. An additional 25% of all NIH extramural awards was for clinical research (i.e., patient-oriented research in which investigators directly interact with human beings, epidemiologic and behavioral studies, and health services and outcomes research) other than clinical trials (NIH 1997). In 1995, NIH supported $6.2 billion of extramural research (NSB 1998).

24. In fact, medical schools usually charge industry lower overhead for industry sponsored trials than they charge NIH for NIH-sponsored clinical trails.

25. On the other hand, a former official of the JPMA estimated that patients in such a study would be dispersed among 50–60 hospitals.

26. One of the allegations in this scandal was that MHW's Pharmaceutical Affairs Bureau was aware of the contamination but failed to take appropriate action—especially to protect hemophilia patients.

27. The Japanese universal health insurance system reimburses physicians at a fixed rate per procedure and therefore creates incentives for physicians to see many patients but to spend little time with each one. Explanations concerning the nature of a patient's illness and treatment are minimal. Many patients are dissatisfied with having to wait hours for a visit that lasts just a few minutes. Also, since written documents are used less often in Japanese society to express assent or personal wishes as, for example, in the case of wills, patients may be put off by a request to sign a formal consent form. In addition, in the case of certain serious diseases, particularly cancer, physicians are reluctant to discuss diagnosis and treatment openly with patients, preferring instead to communicate with family members. In many respects, Japanese physicians feel responsibility as much to a patient's family as to the patient herself. Therefore, obtaining written informed consent very often means the physician must obtain approval from family members. Often in the case of cancer, an open discussion about the disease is out of the question, usually more out of consideration for wishes of family members than of the patient herself. In the case of other diseases, a formal request for written consent can trigger concern on the part of patients and family members that patients are being used as guinea pigs. Even for physicians who are willing and able to explain the nature of clinical trials to their patients, it is hard to convince patients that they stand to benefit from participating in clinical trials when many of the trials are for derivative drugs that offer little benefit over standard therapies. Monetary incentives for patients to participate in clinical

trials are not great. Since 1997, companies sponsoring clinical trials must pay 100% of the costs associated with such trials. Since 1997, under the Japanese universal health insurance system, patients have faced a 20% co-payment charge and an additional graduated surcharge when two or more medicines are prescribed at one time. However, it seems doubtful that patients can be enticed to participate in clinical trials simply by the prospect of not have to pay the co-payments and surcharges associated with a new drug (Fukuhara 1997).

28. In 1961 Japan established a system of universal health insurance that reimbursed patients fully for the cost of medicines. (See Okimoto and Yoshikawa 1993 for a detailed description of this system up until the early 1990s.) MHW assigned a reimbursement price to classes of similar drugs and allowed hospitals and physicians to dispense drugs at the reimbursement prices, pocketing the difference between the reimbursement prices and the wholesale prices at which they bought the drugs. This encouraged overprescription of drugs but was a boon to the nascent and highly protected Japanese pharmaceutical industry (Kimura 1993; JEI 1992). In the 1980s, MHW began to set reimbursement prices for individual medicines rather than for groups of medicines. These prices sometimes incorporated premiums for drugs developed in Japan, new drugs, or therapies for rare diseases (JEI 1992). Similarly, MHW set reimbursement prices for medical procedures, such as CT scans and later MRI scans and lithotripsy treatments (the use of sound shock waves to break up kidney stones). In the case of CT and MRI scans and lithotripsy treatments, reimbursement prices were set high so as to encourage even small hospitals and clinics to purchase the expensive machines, knowing they could recoup the cost and gain a profit if they prescribed these procedures for many patients. Araki (1993) and Yoshikawa et al. (1993) contend that the reimbursement price system provided Japanese manufacturers of high-technology medical diagnostic and therapeutic equipment with strong domestic market demand that helped them boost their international competitive position vis à vis American and European manufacturers who had pioneered the development of such products. Unlike Japanese electronic equipment manufacturers, however, Japanese pharmaceutical companies did not have the opportunity to become internationally competitive and to develop strong innovative R&D capabilities before cost containment pressures began to collide with industrial policy in the 1980s. Faced with rapidly rising pharmaceutical costs, (which accounted for one-third of Japanese medical expenditures in 1980 compared with 8% in the United States), MHW began systematically to cut reimbursement prices at least once every two years so as to maintain a "reasonable price differential" between the reimbursement price and average competitive wholesale prices. Originally set at 15%, MHW reduced this reasonable price differential to 10%, and the aim of the latest round of cuts in 1998 was to reduce it even more to 5%. In general, price reductions for innovative drugs were the same as those for derivatives. As reimbursement prices have been cut, hospitals and doctors have been able to force wholesalers to reduce their prices, and these have in turn forced pharmaceutical manufacturers to reduce their prices. Before 1997, Japanese pharmaceutical companies managed to maintain profits and increase

sales by economizing on distribution costs and reaping profits from new me-too drugs which were awarded an initially high reimbursement price by MHW. However, their budgets for the expensive R&D necessary to develop innovative world class drugs were limited. (See PhRMA 1998, SBC 1998 and JEI 1992 for statistics on R&D expenditures by Japanese and U.S. companies.) In other words, regulatory and economic incentives pushed Japanese pharmaceutical companies to devote most of their R&D to making derivative drugs for the domestic market (SBC 1998; JEI 1992). As U.S. and European pharmaceutical companies merged, invested billions of dollars in R&D, and developed new innovative products, Japanese companies stagnated in terms of innovative new products. Of 152 major international drugs developed between 1975 and 1994, 45% originated in the United States, 14% in the United Kingdom, 9% in Switzerland, and only 7% each in Japan and Germany (PhRMA 1998). As American and European companies have merged, creating companies that can benefit from economies of scale in pharmaceutical R&D (see Henderson and Cockburn 1996), the Japanese industry remained fragmented. The largest Japanese pharmaceutical company, Takeda, ranks 15th in worldwide pharmaceutical sales ($5.2 billion in 1996), while Japan's second-largest company, Sankyo, ranks 18th in worldwide pharmaceutical sales ($4.1 billion in 1996) (SBC 1998). Finally, sales of Japanese pharmaceutical companies are still geared predominantly toward the domestic market, while 70–90% of European companies' sales are outside their home market and 50% of U.S. companies' sales are in foreign markets (SBC 1998).

In the face of continued reductions of reimbursement prices by MHW and the imposition in 1997 of higher patient co-payments and graduated patient surcharges for prescriptions of two or more medicines, the Japanese pharmaceutical industry experienced its first decline in sales and profits, and declines were also forecast for 1998 (SBC 1998). However, MHW has taken additional steps that suggest it is trying not only to reduce health care costs and overprescription of drugs but also to curb the waste associated with developing derivative drugs that offer minimal advantages over existing drugs. In 1998, MHW declared two or three drugs previously approved to treat dementia to have minimal efficacy. As mentioned in the main text, clinical trials for more than 20 drugs were canceled in the first half of 1998. It appears that MHW may no longer be trying to follow a policy that keeps a large number of small non-innovative pharmaceutical companies in business. Instead it may be letting the industry follow what many analysts believe is its natural and (from a long-term perspective) desirable trajectory, to split into a group of companies that can grow and develop innovative world-class drugs and a group that cannot and will inevitably be absorbed by stronger companies.

A variety of evidence suggests that the companies that emerge the strongest at the end of this process will likely be companies that have learned to cooperate closely with foreign, but not necessarily Japanese, universities. For example, Niwa in chapter 5 shows that Japanese companies outsource more R&D related to drugs and medicines to foreign universities than to Japanese universities. The data from the JBA survey on the perspectives of JBA companies regarding cooperation with universities list a variety of advantages in cooperating with American as opposed to Japanese universities. Similarly, senior officials of major Japanese pharmaceutical companies have said their companies prefer to cooperate with foreign, rather than Japanese universities because, for example, it is easier to assemble strong research teams abroad in contrast to Japan, where job mobility is low (Normile 1994) and it takes much longer to perform clinical trials in Japan than in foreign countries (Asahi 1998). Moreover, in the case of clinical trials for new cancer therapies for which there should be great demand in Japan, it is very difficult to perform such trials in Japan since the new GCP guidelines went to effect (see Note 27) but relatively easy to perform such trials in the United States.

29. Constraints on industry funding arrangements for R&D in national universities may make it easier for industry to support clinical trials in private university hospitals. According to a former official of the JPMA, absent a high-level agreement between national university officials and a pharmaceutical company to permit a large Donation by the company, corporate support for clinical trials usually must occur under a Commissioned Research agreement. As described in chapter 12, such agreements can be cumbersome. Recently, a pharmaceutical company and a major national university entered into a Commissioned Research agreement under which the company provided funds for a trial for which one of the university's hospitals would recruit ten patients. The funds had to be disbursed through the Ministry of Finance. When it turned out that the hospital was only able to enroll five patients in the study, the company had to petition MOF to return the unused funds, and as of September 1998 the funds still had not been returned.

References

Araki, Kasuhiro. 1993. "Understanding Japanese health care expenditures; the medical fee schedule." In Daniel Okimoto and Aki Yoshikawa, eds., *Japan's Health System: Efficiency and Effectiveness in Universal Care.* Washington DC: Faulkner & Gray's Healthcare Information Center.

Asahi Newspaper (Shimbun). 1998. "Domestic development takes time." (13 June 1998, evening edition).

Association of University Technology Managers, Inc. (AUTM). 1998. *AUTM Licensing Survey FY 1996* (AUTM).

Association of University Technology Managers, Inc. (AUTM). 1996. *AUTM Licensing Survey FY 1991–FY 1995* (AUTM).

Blumenthal, David, et al. 1996. "Relationships between academic institutions and industry in the life sciences." *New England Journal of Medicine* 334: 368–373.

Blumenthal, David, et al. 1995. "Capitalizing on public sector research investments: the case of academic-industry relationships in the biomedical sciences."

(A paper presented at the NIH Economics Roundtable on Biomedical Research, National Institutes of Health, Bethesda, Maryland, October 19.)

Cockburn, Iain and Rebecca Henderson. 1996. "Public-private interaction and the productivity of pharmaceutical research." *Proceedings of the National Academy of Science. USA* 93: 12725–12730 (colloquium paper).

Committee. 1995. Committee for the Study of Scientific Research Funding, ed. *List of Approved Projects for Monbusho's Grants-in-Aid for Scientific Research, 1995.* Tokyo: Gyousei, Tokyo (in Japanese).

Diamond. 1993. "Patent Applications of University Researchers." *Diamond Management Development Information MDI Impact* (2): 2–11 (in Japanese).

Fransman, Martin and Shoko Tanaka. 1992. *The Strengths and Weaknesses of the Japanese Innovation System in Biotechnology* (JETS Paper No. 3). Edinburgh: Institute for Japanese-European Technology Studies, University of Edinburgh.

Fukuhara, Shunichi, et al. 1997. "Good clinical practice in Japan before and after IHC: problems and potential impacts on clinical trials and medical practice." *International Journal of Pharmaceutical Medicine* 11: 147–153.

Henderson, Rebecca and Iain Cockburn. 1996. "Scale, scope and spillovers: the determinants of research productivity in drug discovery." *Rand Journal of Economics* 27: 32–59.

Irvine, John, et al. 1990. *Investing in the Future: An International Comparison of Government Funding of Academic and Related Research. Report of a study sponsored by the UK Advisory Board for Research Councils and the US National Science Foundation.* Hants, England: Elgar.

Japan Biotechnology Industry Association (JBA). 1998. *The Usefulness of University Research for the Promotion of Our Country's Bioindustry* (Tokyo: unpublished document) (in Japanese).

Japan Economic Institute (JEI) 1992. *U.S.-Japan Competition in Pharmaceuticals: No Contest?* (JEI Report No. 13A, April 3).

Kimura, Bunji, et al. 1993. "The current state and problems of Japan's pharmaceutical market." In Daniel Okimoto and Aki Yoshikawa, eds., *Japan's Health System: Efficiency and Effectiveness in Universal Care.* Washington DC: Faulkner & Gray's Healthcare Information Center.

Kortum, Samuel and Josh Lerner. 1998. *Does Venture Capital Spur Innovation?* Working Paper No. 6846, National Bureau of Economic Research. Washington, DC: National Bureau of Economic Research.

Mansfield, Edwin. 1991. "Academic research and industrial innovation." *Research Policy* 20: 1–12.

MHW Ordinance on Good Clinical Practices. 1997. In Japanese Ministry of Health and Welfare, Pharmaceutical and Medical Safety Bureau, Evaluation and Licensing Division, Review Assessing Group (eds.). *Japan's New GCP and Other Rules on Clinical Trials.* Tokyo: Yakuji Nippo, Ltd. (Ordinance, chapter 2, Article 10(6)).

Narin, Francis, et al. 1997a. "The increasing linkage between U.S. technology and public science." *Research Policy* 26: 317–330.

Narin, Francis. 1997b. "Science dependence of Asian-invented U.S. patents." Paper presented at the conference, Intellectual Property: Japan and the New Asia, in Washington D.C., October 1997, organized by the Japan Information Access Project.

National Institutes of Health (NIH). 1997. *The NIH Director's Panel on Clinical Research Report to the Advisory Committee to the NIH Director December, 1997.* At <http: //www.nih.gov/news/crp/97report.htm>.

National Science Board (NSB). 1998. *Science and Engineering Indicators: 1998.* Washington DC: National Science Foundation.

National Science Board (NSB). 1996. *Science and Engineering Indicators: 1996.* Washington DC: National Science Foundation.

Normile, Dennis. 1994. "Universities and companies learn benefits of teamwork." *Science* 266: 1174–1175.

Okimoto, Daniel and Aki Yoshikawa. 1993. *Japan's Health System: Efficiency and Effectiveness in Universal Care.* Washington DC: Faulkner & Gray's Healthcare Information Center.

Pharmaceutical Research and Manufacturers Association (PhRMA). 1998. *Industry Profile 1998.* At http://www.phrma.org/publications.

Saegusa, Asako. 1998. "Private deal puts Japanese researcher in hot water." *Nature* 19 (November). At http://www.naturejpn.com/newnature/news/news191198b. html.

SBC Warburg (a division of Swiss Bank Corporation). 1998. *Pharmaceutical Sector—Japan.*

Statistics Bureau, Management and Coordination Agency, Government of Japan. 1996. *Report on the Survey of Research and Development.* Tokyo: Japan Statistical Society Foundation (in Japanese and English).

U.S. Bureau of the Census. 1996. "Global Aging into the 21st Century" (pamphlet). International Programs Center, International Data Base. Washington DC: U.S. Department of Commerce, Bureau of the Census.

Utsunomiya, Osamu. 1993. "Health status and patients in Japan." In Daniel Okimoto and Aki Yoshikawa, *Japan's Health System: Efficiency and Effectiveness in Universal Care.* Washington DC: Faulkner & Gray's Healthcare Information Center.

Yoshikawa, Aki, et al. 1993. "High Tech Fad? Medical equipment use in Japan." In Daniel Okimoto and Aki Yoshikawa, eds., *Japan's Health System: Efficiency and Effectiveness in Universal Care.* Washington DC: Faulkner & Gray's Healthcare Information Center.

17

The Impact of the Internet on University-Based Research and Innovation in the United States and Japan

Y. T. Chien

Introduction

Innovation is a springboard for industrial development in a modern society. Many scholars have argued that to create a favorable environment for innovation, a society must possess a dynamic market, sound government policies in finance and regulation, and sustained investment, both public and private, in science and technology research (Branscomb and Keller 1998). Since the late 1980s, both the United States and Japan have seen changes in these areas that will profoundly affect each nation's future and the well-being of its citizens. It is not the purpose of this article to explore these changes, but to examine in some detail one of the most significant developments of the past two decades, the Internet and its related technologies.[1] For many, the Internet is a tool that enables them to pursue new work in various scientific disciplines, or do old work in new ways; others see it as an evolving creation of science and technology that has formed a new discipline in itself. In any event, for better or worse, the Internet and its derivatives such as the World Wide Web are forever changing the ways we do business, work, and live in the society. More fundamentally, it is altering the ways that research is done and innovation achieved. Our discussion and analysis will focus on the activities in and around information technology, briefly examining Internet's past and present in both countries and then characterizing the ingredients for change caused by the Internet. We will look at its effects on university-based research in relationship to industry and government at different levels and then narrow our view to some key R&D programs, both in the United States and Japan, exploring their interconnectedness through the Internet. Finally, we will present an analysis of how these interconnections lead to innovation and draw a few general conclusions from our study of these particulars.

The Internet and the Information Society

The notion of an "Information Society" had its roots in the introduction and penetration of such mass electronic media as telephones, radios, television, and computers which, over a period of several decades, dramatically changed the ways our societies produce, distribute, and use information. Their interconnection through high-speed networks was only the latest phase in a process that began in the late 1970s, when Yasumasa Tanaka presented a lively discussion on the subject of "Information Space" (Tanaka 1978) as fashionable as today's discussion of "Cyberspace." What Tanaka had discovered is still revealing. Using data from a White Paper by the Japanese Ministry of Posts and Telecommunications, Tanaka showed how Japan had moved steadily from being an "Information-poor" to an "Information-rich" society, in a relatively short period of time compared to the United States and four other industrial nations. To measure the "information orientation" of a nation/society, he devised a numeric index defined by a combination of four factors:

Amount of information: The volume of media activity, such as mail sent per capita, telephone calls made per capita, and the number of newspapers circulated per 100 inhabitants;

Information equipment: The number of telephones, the number of television sets per 100 inhabitants, and the number of digital computers per 10,000 inhabitants;

Communication subjects: Variables such as the number of tertiary-industry workers among the total working population, and the number of college students in the general population;

Coefficient of information: The ratio of miscellaneous expenditures to total personal information consumption.

Table 1 Indices of Information Orientation: A Five-Nation Comparison

Country	Year	Index of Information Orientation	Five-Nation Ranking
Japan	1965	[100]	5
	1973	221	2
United States	1965	242	1
	1973	531	1
England	1965	117	2
	1973	209	5
France	1965	110	3
	1973	210	4
West Germany	1965	104	4
	1973	211	3

Adapted from Tanaka (1978). Original Source: White Paper on Communications for 1975, Ministry of Posts and Telecommunications, pp. 30–31. Japan's [100] was used as the base score.

This index of information orientation allowed the Japanese government to track the "growth of information" in the society, by showing how Japan had improved its standing among the world ranks as an "information-rich" society. For example, during the eight years from 1965 to 1973, the index increased at an annual rate of 10%, almost equivalent to the increase of the net GNP of Japan for the same period. Furthermore, this index, if applied to other industrial countries, could be used to make cross-national comparisons. The MPT report cited by Tanaka included the indices for 1965 and 1973 of five industrial nations—the United States, England, France, West Germany, and Japan (see Table 1).

The ranking among the five nations shown in this table is of little significance. However, in examining the trends for Japan and the United States, it can be seen that the latter stood alone as an information "super state," far ahead of the rest in 1965, while Japan ranked at the bottom of the five nations based on the information index. In 1973, the United States again was on top of the measurement scale, followed this time by Japan as the second richest "Information Society." The position of the United States as a leader of the information-rich "appears to be insurmountable," to use Tanaka's words.

Table 2 Summary of 1998 Internet Hosts: An International Comparison

Country	Hosts	Population (Millions)	Hosts per 1000 Population	5-Nation Ranking	10-Nation Ranking
Japan	1,168,956	125	9.4	4	9
United States	17,252,053	270	63.9	1	2
England	987,733	58	17.0	2	7
France	333,306	58	5.7	5	10
Germany	994,926	82	12.1	3	8
Australia	665,403	18	37.0		3
Canada	839,341	30	28.0		5
Finland	450,044	5	88.3		1
Netherlands	381,172	16	24.4		6
Sweden	319,065	9	36.5		4

Source: Network Wizards Survey (See URL: www.nw.com for definitions and other details).

In 1998, twenty years after that survey and analysis, had the Internet changed anything? Here again it might be informative to draw on another statistic that would reflect the changing information orientation. Specifically, let us look at what the Internet connections worldwide tell us about the extent to which various countries are "going on line" to reap the benefits of the information explosion. There are many indices we could use, from at least a dozen U.S. organizations that specialize in collecting Internet data and produce surveys,[2] but since not all of them use the same methodologies or methods of analysis, resulting in uncertainty and sometimes inconsistency, we will use the result of a January, 1998 survey by Network Wizards.[3]

Table 2 shows two sets of data. The top five rows give the statistics of Internet penetration of the five nations (Japan, the United States, England, France, and Germany, as in Table 1) and their relative standing based on the number of hosts per 1000 people. The second set of data, shown as the entire table, gives the statistics of the top 10 nations (inclusive of the first five) on their Internet penetration and their relative standings using the same measure. Although based only on one Internet statistic, the table suggests two interesting conclusions. First, compared to the data in Table 1, Japan's "information orientation" status in the Internet era has slipped from second in 1973 to fourth in the 5-nation comparison. The second and more significant conclusion is that in the 10-nation comparison shown in Table 2, although Japan has the second-highest number of hosts worldwide, it has the second-lowest number of hosts per capita (among the top ten). Finally, looking at a more global picture, we see that the other four nations in the first list of "Information-rich" all have a lower ranking among the top 10 countries in Internet penetration: The United States is second after Finland, England is seventh, Germany eighth, and France last.

The re-shuffling of these positions among the "information-rich" appears to tell much of the Internet story, giving us a sense of where and how the Internet, and the networking technology in general, is changing the "information orientation" of many of the world's countries—small and large, rich and poor—in the information era. With no way of predicting the unpredictable, Tanaka twenty years ago made a prescient comment on Japan' improved status as an "information society":

> As an "information-rich" and "information-affluent" society, Japan has received every blessing of the progress in information-engineering and capital investment in the information industry. But we know very little about the impact of information technology upon the society. Certainly, we have accumulated fairly satisfactory quantitative knowledge, but "subjective" or "human" aspects are still to be explored. We do not know, for example, how the abundance of information is related to the quality of life. We do not know whether or how freedom and equality can be maintained. We do not know whether or not an information-rich society is merely a linear extension of materialism upon which most of capitalistic societies have been built.

These remarks still speak to our desire to characterize and understand the impacts of the Internet on science and technology, which we believe to be critically important in the development of sound public policies for future R&D investments in a wide range of areas. In what follows we will explore the Internet's effects, real and perceived, on one important aspect of the societal and human dimension, how it is transforming the ways in which academe, industry, and government form new relationships for research and innovation.

The Internet Factor: How Networking Technologies Are Changing Research and Innovation

Ingredients for Change

The Internet, while still young and evolving, has changed the landscape of university-based research and industrial innovation. There are at least three technical ingredients associated with the Internet and its related technologies that we believe have contributed to this change.

Collaboration and Coordination across Time and Space. The term "Collaboratories" has been used (NRC 1993) to describe the kinds of new research laboratories and similar environments that have access to people, instruments, and a variety of other information and physical resources which are functionally co-located while remaining geographically separated. These collaboratories can function independently of the time zones around the world. What makes a collaboratory more attractive as a new modality for research and innovation is, among other features, its dual function as both a real-time operation and a delayed access to or playback of laboratory experiments. Such new capabilities, possibly only over the Internet, usher in fresh opportunities for teamwork across disciplines, institutional constraints, and national boundaries. By reducing the burden and overhead of resource sharing, they help provide flexibility to collaborative laboratory experiments and add new meaning to interdisciplinary research. Many forms of collaboratories for scientific research, especially those of an interdisciplinary nature, already exist in various stages of development and operation. One outstanding example is the Upper Atmospheric Research Collaboratory (UARC) at the University of Michigan, with research partners all over the world via the Internet.[4]

Convergence of Computing and Communications Technologies. For many years, the computing and communications industries have been quietly converging. This began with the breakup of AT&T, but since then technical advances in each industry have evolved naturally into a convergent discipline (Messerschmitt 1996). For telecommunications, the question used to be how to improve long-distance telephone connectivity; in computing, it was how to speed up computation and data transfer within a platform. The Internet and its recent commercialization have required that these two industries become increasingly integrated as a result of the common technologies in hardware and software they share, a convergence that has led to a sea change in the ways university research and industrial innovation are done.

Integration, Interoperation, and Interactivity. Here we refer to the many possible combinations of data formats, communication modalities, human interfaces, access platforms, programming and software systems, and the like.[5] As the Internet community attempts to deal with these complex combinations, new ideas, services, and products quite different from traditional forms come online and eventually into the marketplace, either as laboratory prototypes or as commercial offerings. Not infrequently, new and small companies then spin off from university laboratories or large corporations in search of unique outlets for creativity and innovation—the essence of the Internet research cycle.

Characterizing the Impacts: Networked Paths to Innovation

This section will examine the initial signs of change in, and the evolution of, university-based research as it relates to industrial innovation in the context of the Internet revolution. For purposes of analysis, the impact of the Internet can be looked at from several interrelated standpoints, which form the basis of a discussion of specific research modalities that have emerged from the Internet.

Accelerating the Convergence of Content with Computing and Communications. While the convergence of computing and communications provided the seeds of change in the information industry, the impacts of the Internet on research and innovation were first and foremost derived from its ability to allow the media forms of human interaction to be all-inclusive. Digital content, especially in large, distributed, heterogeneous collections of electronic objects—text, voice, images, graphics, video, and others—is fueling the reciprocal growth of computing and communications. While commercial interest in this three-way convergence has been made clear through the activities of recent mergers in these industries, the impact of digital content on academic research will almost certainly grow.

Already, new kinds of literary and scientific data are appearing in digital forms available on the Internet: The digitization effort of the Library of Congress,[6] the Visible Human Project at the National Library of Medicine,[7] and the vast databases from the Human Genome Project, are a few of the many examples. The convergence of content with various human interface modalities, computing platforms, knowledge domains, and their representations and semantic interpretations, will continue to be a driving force for future research and innovation for universities and industries.

Enhancing the Creation and Use of "Social Capital." Like physical capital and human capital—tools and training that enhance individual productivity—"social capital" is also a key enabler of innovation. Social capital, according to Fountain (1998: 85), refers to features of social organizations, such as networks, norms, and trust, that facilitate coordination and cooperation for mutual benefits. The Internet, with its associated information and human resources, has helped in the creation and effective use of social capital in a variety of ways. With improved facilities and various modes of interactivity available on the Internet, university-based research has increased the variety, as well as the number, of partnerships through consortia, for example, with industry and government in jointly advancing research. Sharable resources, whether they be instruments, computer software, or databases, are often proprietary but they are extremely critical in any large-scale, experimental projects.

Achieving New Economies of Scale and Scope. The advent and popularization of the Internet in many ways redefines economies of scale. While traditional concepts of scaling up the information systems focus on such parameters as computer speed and network bandwidth, the Internet environment allows us to look at a number of other critical parameters, particularly those having to do with mass users and their interactivity with data of different forms. In the world of commerce, this new economy of scale makes it possible, for example, for FedEx to allow its customers not only to track their packages but also to learn the entire history of their transactions with the company (Rayport and Sviokla 1995). This, in effect, creates a process of business improvement, thus achieving a new economy of scale across the number of users demanding and using the service. In the area of investment in research and innovation, the Internet provides a similar opportunity for improvement in university-based research and innovation. Further, just as in the new economies of scope for business, which draws on a set of resources to provide benefits for different and disparate markets, the same principle is applicable to scientific research in networked environments. The Internet has in its evolving infrastructure a set of distributed and heterogeneous resources accessible by many disciplines and applicable to different knowledge domains. Redefining economies of scale and scope on the Internet affords a fresh way of looking at our investment in university-based research.

Leveraging Global, National, and Regional Resources. Information infrastructures defy regional and national borders, since they are effective and useful only if they are globally connected and managed. On the other hand, the contents of the information infrastructures, such as the Internet, more likely and even preferably, will be created and deployed by local sources—regional or national entities. Coordinating this creation and deployment as well as the global use of these contents, however, remains one of the challenges of Internet governance now and for the foreseeable future. In the scientific arena, a variety of new partnerships, consortia, and other forms of joint effort among university, industry, and government at all levels have been the subjects of experimentation in both the United States and Japan. The early results of these experiments in the Internet context show that differences in culture, history, and political systems all contribute in varying degrees to the processes of research and innovation in each country.

Connecting Science and Technology to Social and Economic Development. Not since the years of Vannevar Bush, when the compact between the scientific community and society for support of basic research was first firmly established, has the question of societal connection with science been so severely tested. Many of the recent legislative battles in the United States, state as well as federal, have centered around how much of the budget pie should go to the support of science research and technology development. And yet, even putting aside health and medical research, which normally receives popular support, a recent NRC study (Brooks and Sutherland 1995) reveals unequivocal evidence for the government-funded basic research initiative that ultimately became the backbone of industrial innovation and economic growth. Today, the success of the Internet itself serves as a vivid reminder of the earlier relationship between government and industry, between university research and industrial innovation, and between science and technology and economic growth. In the last decade, during which the Internet matured from public funding to public asset, the United States has clearly reaped the benefits in economic terms. By many accounts, Japan too sees the Internet as a key to the way out of the economic doldrums. The Internet as a technology with social impact cannot be measured more fairly than in this connection.

Cases of Change: National and Regional Partnerships for Research

To find the traces of the Internet's effects on university-based research, we will first take a brief look at some of the R&D initiatives in the United States and Japan. We have chosen those initiatives for their relevance to the Internet and to the information technology focus of our discussion. We will then discuss them in a comparative summary of highlights across the five Internet threads.

U.S. Initiatives—Some Examples

The Computing, Information, and Communications Program: The New HPCC. A centerpiece of the U.S. National Information Infrastructure program announced at the beginning of the first Clinton administration is the highly successful, congressionally-chartered 6-year High Performance Computing and Communications (HPCC) initiative. Conceived in the mid-1980s and implemented by the cross-agency National Coordination Office mainly prior to the commercialization of the Internet, it nevertheless provided the impetus and Federal support for the birth and unprecedented growth of the Internet. While HPCC focused its research agenda largely on computing and communications technologies, a fifth component on Information Infrastructure Technology and Applications (IITA), added in 1995, helped define and propel the follow-on initiative in the ensuing years (Chien and Katz 1994).

An outgrowth of HPCC, the Computing, Information, and Communications (CIC) R&D program,[8] has extended and re-focused the research activities into five new Program Component Areas: High End Computing and Computation (HECC); Large Scale Networking (LSN); High Confidence Systems (HCS); Human Centered Systems (HuCS); and Education, Training, and Human Resources (ETHR). All the CIC components except HECC are based on "information," networking, and human-centered systems in the Internet era to provide the benefits of research to society at large, a conceptual shift of investment strategies from a computationally intensive focus to an information-intensive, human-oriented focus. This shift, while not necessarily reflected immediately in the budgets of the participating agencies, is nevertheless an indication of how the United States is redirecting its research investment. It will almost certainly induce near- and long-term changes in both what CIC research is done and how it is done across Federal programs in science.

University-based research, in relation to industrial innovation, has been CIC's first and foremost target for change. Participating agencies now include not just those with major R&D responsibilities but also those with clear social missions and benefits. A Presidential Advisory Committee, consisting of members from academe, industry, and the general public, has been convened to provide the program with the kind of input needed to help the agencies implement an effective and balanced program. Several of the cases discussed below will further illustrate this University-Industry-Government relationship.

The Digital Library Research Initiative (DLI). Launched in 1994, this joint-agency (NSF, DARPA, and NASA) initiative for advanced research in digital libraries is the first of its kind designed for university-based research to harness the Internet technology for future knowledge networks and human-centered information systems. The vision for digital libraries of the future was (and still is) that they would be libraries without walls, open twenty-four hours a day, and accessible where the network is by everyone, anytime. They would be a system of knowledge networks, connecting people, organizations, and varied information resources through a range of digital communication technologies (such as the Internet and the World Wide Web).

The Initiative led to the funding of six university-led research projects,[9] centered at Carnegie Mellon University, the University of California at Berkeley, the University of Michigan, the University of Illinois, the University of California at Santa Barbara, and Stanford University. Each project brings researchers and library users from academic faculty into partnerships with libraries, museums, publishing houses, schools, as well as the computer and telecommunications industries, an arrangement that is one of its distinguishing features. Each project consortium is joined by a diverse group of parties taking part in its research. Together, the collection of these "partners" numbers several hundreds and represents the real strength of this program. On examining the details of this collection, we find that the industrial partners cluster around the "digital triangle" of computer, communications, and content industries, illustrated in Figure 1. Note that beyond their traditional roles, many of these companies would fit equally well into one or the other groups in the triangle relationship. As the three industries converge and business focuses shift towards one another, their research interests in the Digital Library Initiative seem to mirror their present business strategies in the digital economy.

The Digital Library Initiative's second phase began in the fall of 1998,[10] with new government sponsors from the Library of Congress, the National Library of Medicine, and the National Endowment for the Humanities, a clear indication that the Internet is no longer a tool for an elite group, but will benefit the wider society.

Partnerships for Advanced Computational Infrastructure (PACI). This is the new NSF infrastructure program initiated in 1997, to replace the previous program that had funded the four Supercomputer Centers in the United States for a decade. In many ways, as in the case of CIC, this new effort also signifies a major shift of investment strategy from one of supporting multiple, concentrated computing facilities into one that supports fewer (in this phase of implementation, two) but more versatile distributed partnerships or alliances. The earlier program made high-performance computers more accessible to scientists and researchers, by making standalone supercomputers usable from the desktop via high-speed trunk lines. With the emergence of the Internet and the World Wide Web, the goal now is to transform the vast and diverse digital

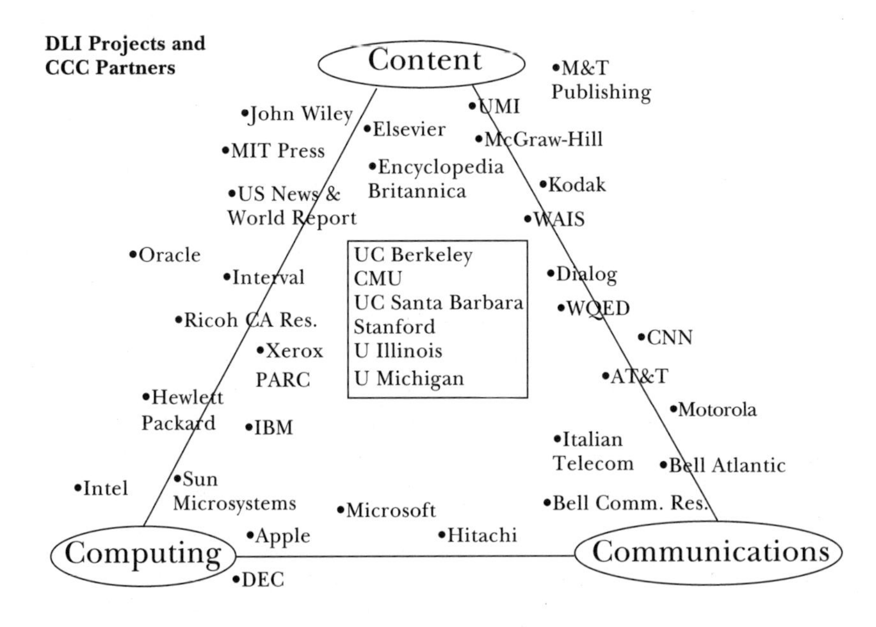

Figure 1 Triangle Relationship for the DLI Industrial Partners

resources into computer-mediated knowledge and make them widely available to scientists and a wider segment of the society.

There are two such knowledge centers under the new PACI program,[11] each led by a university with a unique blend of partners and alliances selected from other universities, industries, and user and professional groups all over the country. The National Center for Supercomputing Applications (NCSA) at the University of Illinois, the birthplace of such icons as MOSAIC and Netscape in the early 1990s, is now the host of the National Computational Science Alliance.[12] The Alliance is a unique partnership among computational scientists, computer specialists, and professionals in education, outreach, and training at more than 50 universities and research institutions, along with industrial partners. A major goal of the Alliance is to build a National Technology Grid. The Grid will integrate instruments, computers, communications, and massive databases via high-speed networks, such as the Internet and its next generation, to form a powerful computational environment for problem-solving and collaborative scientific experiments.

More than providing computing and information resources, however, the Grid will link people and ideas. Researchers and other participants on the National Technology Grid will have at their disposal advanced searching and indexing capabilities, digital libraries, audio and video streaming and virtual conferencing.

The other new partnership is the National Partnership for Advanced Computational Infrastructure (NPACI), led by the University of California at San Diego,[13] linking 37 of the Nation's leading academic and research institutions with industrial laboratories from 18 states. Like NCSA's Alliance, NPACI is developing a knowledge-based information infrastructure to link the highest performance computers, data servers and archival storage systems, providing access to distributed computing power and information resources. Unlike pure, straight research, it focuses more on specific "thrust areas" that join applications with scientists and software developers to ensure rapid deployment, testing, and robustness of the resulting infrastructure. Another unique feature of the San Diego partnership is that it also links up with international groups for collaborative research. Together, NCSA's Alliance and the San Diego NPACI are an outstanding example of how the Internet affects U.S. research and innovation.

vBNS, Internet2, and the Next Generation Internet Initiative. In any discussion of how research and the Internet are intertwined, it is important to look at the developments that followed the successful transition of the Internet into the commercial world. In 1995, the National Science Foundation established the new NSFNET, known as the very high performance Backbone Network System (vBNS).[14] A cooperative project with MCI, the vBNS links NSF's supercomputer centers at the time (NCSA, UCSD, Pittsburgh, Cornell, and NCAR), with potential connections to approximately 100 research institutions, chosen from the normal merit review process. The vBNS currently runs at 622 megabits per second and is expected to operate at 2.4 gigabits per second by the year 2000. This is a Federal investment program intended to keep the nation's information infrastructure for research institutions always a step ahead of commercially available networking. Through a global connections program, it also provides an international link, STAR TAP (for Science, Technology, And Research Transit Access Point) to interconnect international high-performance networks. While vBNS has academic research as its primary goal, its potential impacts on future industrial innovation and societal benefits are enormous, if the history of Internet development is any indication.

A nearly parallel but separate development in the Internet arena was the creation of Internet2,[15] a private effort on the part of the university community in collaboration with industries with a strong commitment to research. Its primary objective is to create broadband network technologies, engineering tools, and applications for research and education. Announced initially by 34 universities in 1996, Internet2 currently has a membership of over 120 leading U.S. universities and more than 30 major corporate sponsors and partners.

Internet2 has three broad, interrelated goals: to provide a leading-edge network capability for the national research community located at universities in every state in the United States; to coordinate the efforts of its members in developing and enabling a new generation of high-performance network applications, including digital libraries, tele-immersion, and virtual laboratories; and to transfer new, advanced network technologies and applications to all levels of educational use and to the broader Internet community. To accomplish these goals, Internet2 has since taken a bold, entrepreneurial step: It has formed a corporate structure, the University Corporation for Advanced Internet Development (UCAID), to manage its assets, projects, and human resources as an educational, nonprofit enterprise.[16] With a board, a CEO, and full-time staff, UCAID is actively moving to implement Internet2's ambitious agenda.

While Internet2 is primarily a private effort between the university community and the information industry, it also complements Federal programs such as the vBNS, which provides a base and initial high-performance backbone network service for Internet2 members to ramp up their networking capabilities. Internet2 also complements the Next Generation Internet (NGI) initiative,[17] a multi-agency federal research and development program designed to create a more powerful and versatile foundation for the networks of the next century. NGI has three goals: to advance research, development, and experimentation in the next generation of networking technologies; to develop a Next Generation Internet testbed, connecting at least 100 NGI sites at speeds 100 times faster than today's Internet and 10 sites at speeds 1,000 times faster, and emphasizing end-to-end performance, to support networking research and demonstrate networking technologies; and to develop and demonstrate novel, high-performance applications.

The NGI goals are clearly related to those of Internet2. More importantly, its implementation as currently carried out across several agencies (and several agency networks—NSF's vBNS; NASA's NREN; the Defense Research and Education Network DREN; the Energy Sciences Network ESnet, and DARPA's Terabit Research Network) relies heavily on the Internet2 constituency for advanced campus-based, local area, and select regional network infrastructure. On the other hand, researchers at Internet2 universities and their industrial partners are developing a wide range of technologies and applications that receive funding from Federal programs, including the NGI. This relationship of overlapping goals and complementary missions contrasts sharply with the recent Abilene project, an implementation of the Internet2 program.

The Abilene Project. As the Internet continues to grow in research, development, and commerce, its impact on the university-industry-government relationship has also been felt in a way not entirely predictable, as demonstrated by the announcement of the Abilene project by the University Corporation for Advanced Internet Development in 1998.

Abilene[18] is an ultra high-performance network being developed by the University Corporation for Advanced Internet Development (UCAID), in partnership initially with three private companies: Qwest Communications, Cisco Systems, and Nortel (Northern Telecom). An important goal of the Abilene project is to provide a backbone network for the Internet2 community. Through access to Qwest's nationwide fiber optic network and technologies provided by Cisco and Nortel, the Abilene network is expected to become the most advanced IP backbone network available to the research and education community. Initially, Abilene will deploy a national backbone capable of operating at OC48 (2.4 gigabits per second) among the network's regional access points, with OC12 (622 megabits per second) and OC3 (155 megabits per second) connections from the Abilene backbone to university aggregation points or to individual institutions; later, this speed is expected to reach OC192 (9.6 gigabits per second). Even in terms only of capacity, this is a tremendous effort and major commitment by UCAID on behalf of the university-industry partnership. It is no coincidence that during the White House ceremony at which the Abilene project was announced, Vice President Gore especially applauded the extraordinary commitments of the three corporate partners. He further predicted, "In the coming years, this investment may enable the medical specialists to give advice to patients in rural hospitals, scientists to use remote supercomputers to predict tornadoes, and adults to get skills through distance learning." This, in plain language, is the goal of the Next Generation Internet Initiative.

Japanese Initiatives—Some Examples

Japanese initiatives, at national or regional levels, have also been put in place in the wake of the Internet and related network technologies. For many in the United States, the Japanese model of inter-relationship among university, industry, and government (central and regional) in support of research and technology innovation has long been a subject of scholarly pursuit as well as practical debate. We will examine this interrelationship in the context of the Internet by describing a few examples, focusing on some of the large-scale initiatives similar to those in the United States.

Real World Computing in the Internet Era. In 1992, the Ministry of International Trade and Industry initiated the Real World Computing (RWC) program[19] as a 10-year project to develop innovative information processing technologies for human-centered, real world applications. In 1997, RWC completed its first phase, basic research and exploratory research in the areas of novel functions, massively parallel systems, and optical information processing, and embarked on its second phase. Renamed the Fundamental Information Technology of the Next Generation, the new RWC project concentrates on work leading to new technologies in the networked environment where people, computers and other systems are interconnected through high-speed networks. Based on the framework developed in phase 1, the new RWC is pursuing research in two complementary fields: Real World Intelligence (RWI) technology and Parallel and Distributed Computing (PDC) technology. The first places emphasis on developing intelligent functions and interfaces for enhancing human-computer-network communications; the second is aimed at research leading to heterogeneous, multi-processor systems across high-speed networks, with the shared goal that the next generation information systems will achieve new economies of scale across large number of users in a connected society and across a variety of computing and communication platforms. As in the first five years, Phase 2 of the program will continue to be a consortium of consumer electronics, computers, and telecommunications companies. Many of Japan's research universities, national and private, are academic partners in this consortium, along with government laboratories.

Like the CIC program in the United States, the new RWC program is a large government R&D investment in information technology, with the Internet as a key motivating, if not driving, force. There are several differences between the CIC and the RWC. First, CIC is a cross-agency program, participated in by many U.S. federal agencies sharing resources, expertise, and outcomes of the research; RWC is a MITI project, although presumably the research outcome will be shared by all of the partners and the society at large. Second, CIC is primarily a university-dominated research program, with industry serving in an advisory capacity, while RWC provides a more direct role for industry. Third, CIC is strictly a national program, but RWC has a sizable international component. Despite these differences, both CIC and RWC are typical cases of government-sponsored R&D programs designed to provide the infrastructure and stimulate work in the post-Internet environment.

Digital Contents and the Next Generation Digital Library Project. In Japan, recognition of an accelerating movement toward the convergence of content with computing and communications seems to have occurred sooner than in the United States. In the last decade, the Japanese government initiated several major programs in multimedia technology, real world computing, the digital library project, and others in an effort to develop new information-based products and services for economic development. Like most of the other national programs initiated by MITI, MPT, and the Ministry of Education, they were designed to encourage a consortium-like relationship between academia and industry. Likewise, recently private sectors also began to promote this collaborative paradigm for research and development.[20] It is the rapid growth of the Internet in recent years, however, that has made the three-way interrelationship more tenable and often a critical ingredient for intellectual or commercial success.

The first Japanese efforts in digital libraries focused more on digitization, particularly the digitization of local content so that it would become more accessible to the Japanese public. Representative projects in this category include the National Diet Library's program to establish a children's electronic library, the all-electronic library at the Nara Institute of Science and Technology, and the digital preservation program at the National Ethnographic Museum. Many of the traditional companies in the computer, communications, and publishing industries also rushed to develop digital contents. IBM Japan's Global Digital Museum, Nikkei's Digital News Service, and Fujitsu's Multimedia Library are just a few of the examples.[21]

As the concept of digital libraries became broader and better understood, new initiatives emerged with more emphasis on research towards future multimedia, interactive, and networked content technologies. The Next Generation Digital Library program currently underway in Japan, a second national project of this kind in recent years, is comparable to the U.S. Digital Library Initiative discussed earlier. The project is funded by MITI, with the Information-technology Promotion Agency (IPA) and the Japan Information Processing Development Center (JIPDEC) as lead agencies.[22] It is a five-year R&D program, started in 1996 with a schedule for completion in the year 2000. In many ways, this program is similar to the U.S. joint-agency Digital Library Initiative (especially DLI phase 1) for its overall vision and objectives. They also have comparable, though different, university-industry collaborative arrangements. For example, the Next Generation DL project, while administratively under the supervision of IPA and

JIPDEC, is actually set up as a private and public consortium, with many universities and industrial laboratories as research partners. Unlike the U.S. approach, which encouraged diversity and alternative architectural designs through a collection of diverse, multimedia, interactive technologies, content-based testbeds, and laboratory prototypes, the Japanese project's technical goal is very much a centrally managed design for a futuristic digital library architecture that can lead to the creation of interactive, multi-media, and distributed information systems accessible over high-speed networks. At a mid-term technical review conference held by JIPDEC and attended by most of the research participants in March 1998, a consensus view of this architecture and its associated functional specifications seemed to have been reached in many of the technical specifications of the next generation Digital Libraries for Japan. That consensus view could very well guide the rest of the research in the remaining years that could lead to innovation.

National Center for Science Information Systems (NACSIS). Perhaps the most comprehensive government-sponsored information infrastructure facility for academic research and education in Japan, NACSIS is an administrative arm of the Ministry of Education, Science, Sports, and Culture (Monbusho), which supports its university constituency and allied institutions. It is charged with gathering, organizing, and providing scholarly informational material and relevant information systems for researchers all over Japan. Since its inception in 1986, and especially in recent years, NACSIS has built up a large number of databases, information services, and experimental systems for managing and disseminating digital contents. Among these resources are an extensive collection of electronic journals representing the humanities, law, science, engineering, and medicine, and search and retrieval services, available in a typical digital library, which are provided to the university community and professional societies. The projected future for NACSIS is both ambitious and convincing: It intends to bring online databases and services potentially consisting of 200 million books and 2.7 million journals held at more than 500 universities and institutions throughout Japan.[23] NACSIS developed and operates for Monbusho the Science Information Network (SINET) connecting major universities and research institutions via an Internet backbone and local area networks, interconnected with some 90 national universities, 40 municipal and prefecture universities, 240 private universities, and over 200 other types of research organizations. For a total of over 600 institutions, it also connects with international research networks in the United States, United Kingdom, and Thailand. NACSIS also conducts research and educational programs; many of the staff are also researchers and faculty members. Similarly, university faculty members often are associated with NACSIS for specific research projects as collaborators. In this sense, NACSIS can be compared to the NSF in its function of infrastructure creation and management, or to the National Library of Medicine (NLM) in that it provides content while participating in research with the community. In August 1998, after a series of committee studies, the Japanese government issued an interim report recommending the establishment of a core inter-university institute focusing on information research and education, by reorganizing and expanding the role of NACSIS.[24] Thus, like the changing roles of the two U.S. federal agencies, NSF and NLM, in the Internet era, Monbusho's NACSIS has also affected and been affected by the changing landscape of networked information resources for research and education in Japan.

The Broadband-ISDN Business Chance and Culture Creation (BBCC) Experiment. Among the many networking R&D projects, this project has the unique characteristic of being a collaborative effort in which the regional government plays a central role. BBCC is a non-profit organization established in December 1992 in Japan's new Kansai Science City in the Kyoto-Osaka-Nara region.[25] Away from Tokyo, though not totally free of its influence and participation, BBCC combines resources from local universities, industries, and governments (Kyoto, Osaka, Nara, and Hyogo prefectures) to develop multimedia and network-based experiments and explore their business opportunities (Matsumoto 1995). This is a university-based consortium, and faculties at Kyoto University, Osaka University, the new Nara Institute of Science and Technology, and others are involved, beginning with the development of an agenda, research, prototyping, and technology demonstration. On the private side, BBCC has the support of a wide range of industries as corporate partners, including not only the usual computer and telecommunications companies but also those industries from the electric, power, gas, steel, machinery, and construction sectors, as well as financial services seeking a new foothold in the information and network fields. Over 200 organizations are taking part in this experiment, which is designed to evaluate the feasibility and demonstrate the performance of new models of work, learning, and leisure activity in a networked society. The ultimate benefit is clearly to create commercial and business opportunities. In the era of the Internet, BBCC through its new action plans appears to be restructuring itself by expanding its regional emphasis and linking more closely to national goals and projects. Most notably, BBCC now stands for Broadband-network Business & Culture Creation, a clear break from the initial emphasis on ISDN. Secondly, BBCC is also part of the Japan Gigabit Network Initiative (JGN), which will establish three research and development facilities, one of them in the Kansai region. The latest in a series of Japan's efforts to stimulate the economy included a new R&D initiative, JGN. Similar to the NGI in the United States, JGN also focuses on advanced networking research and society applications. These unfolding events may significantly change the inter-relationships of university, industry, and government in Japan for years to come.

The WIDE (Widely Integrated Distributed Environment) Project. In discussing Japanese R&D, the common perception is that it is dominated by the central or regional governments; the WIDE project is a salient exception to this pattern.

Supported with funds primarily raised by private companies in Japan, WIDE is known as Japan's pioneer in the development of Internet-based technologies and services.[26] It is also a networked research consortium, with teams and individuals from both academia and industry engaged in a wide range of research projects contributing to all aspects of the Internet. WIDE's stated research goal is to establish a large-scale distributed computing environment by providing the wide area communication infrastructure with the best telecommunication and operating system technologies. It also lists "contributing to people/society" as an integral part of its goal. If there is one element missing from WIDE, it is a clear and explicit role for the government. In a recent membership list of more than 400 e-mail addresses, their affiliations are almost exclusively academic or industrial; none of the addresses ends in go.jp (for Japanese government).

More impressive is WIDE's research history and its technical achievement. In less than a decade, beginning in the late 1980s, this consortium project has contributed to nearly every major stage of Internet development and applications, from the early works of packet switching, ISDN, and directory services to more recent works of routing strategies, multicast protocols, network security, next generation Ipv6, and mobile and wireless computing. Working with a large number of both Japanese and international institutions or consortia, WIDE has established many cooperative arrangements for joint research or Internet-based applications. Of those arrangements, the most recent and perhaps best advertised and well-known is its "Asian Interconnection Initiatives," including a project named "Asian Internet University," which utilizes a satellite-cased system for education in Asian countries.

While WIDE is not aimed only at developing ultra high-speed connectivity among research institutions, as in the case of the U.S. Abilene project, its resemblance to Abilene is striking, in terms of academic-industry collaboration and their technical research and development agendas. Both were created and have progressed with little if any explicit government directions. These "government-independent" initiatives, however, cannot be totally independent of the government, nor, we argue, should they be completely without the support and participation of government in this interconnected society. Like Abilene in the United States, while WIDE will in all likelihood remain a private, research-oriented entity (because of its strong technical organization and independent-minded leadership), its overlapping objectives, constituency, and even funding sources may increasingly overlap with those of Japan's government-sponsored programs.

Analysis

We will now examine the ingredients of change in the context of the characteristic impacts of the Internet on university-based research, across the cases described in the previous section. Rather than attempting a quantitative comparison between the United States and Japan, we summarize the highlights in Tables 3 and 4,

Table 3 Highlights of Internet's Impacts—U.S. Activity

Activity	Accelerating Convergence of CCC	Creating and Using Social Capital	Achieving New Economies of Scale and Scope	Leveraging Global, National, and Regional Resources	Connecting to Society
CIC Program—The New HPCC	Moderate. Each of five program components contributed differently.	Yes, especially by the Large-Scale Networking (Networks) & Human Centered Systems (Organizations, Trusts).	Yes. High End Computing cuts across disciplines (Scope); Human Centered Systems scale across users, masses of information.	National resources highly leveraged, especially across federal agencies.	Moderate. Mainly applies to scientific communities, but new CIC work is to emphasize impact on general public.
Cross-agency Digital Library Initiatives	Strong impact. Research partners from content industry a primary factor; Computing and Communications industries Joint content-based work.	Yes. Consortia arrangements for collaborative research and strong emphasis on linkage to public/professional groups and local civic organizations.	Moderate. Phase 1 had limited impact; Phase 2 is to emphasize cross-discipline work (Scope).	National and regional (state-level) resources higly leveraged; Phase 2 is to have an international component for global content-based research.	Moderate for Phase 1. Projects linked to K12 education, media, or state-level missions. Phase 2 to emphasize benefits to society.
Partnerships for Advanced Computing Infrastructures (PACI)	Yes, though impact is primarily on computers and communications. Content research should increase in new partnership/consortia arrangements.	Very strong. Alliance to build a national R&D grid for next generation scientific computing (NCSA); new knowledge processing center (San Diego).	Strong in economies of scope (infrastructures for all sciences).	National and regional resources highly leveraged; The San Diego Center has recently formed international partnerships.	Mainly to scientific communities. Connection to society (education, health) should improve in new program.
vBNS, Internet 2 and Next Generation Internet (NGI)	Moderate so far, but new applications-focused research will enable increased acceleration.	Yes. Internet 2 (private) and NGI (federal) each provides a unique and complementary mechanism to bring about trust and new relationships.	Yes, as an enabler.	vBNS has an international component. Internet 2 and NGI are both domestic but have clear global implications.	Yes, through their applications research.
The Abilene Project	Yes, potentially as an enabler to fuel growth in computing and communications with content-based work.	Potentially high, due to complementary resources, expertise, and needs for consortia members—academic and industrial partners.	Yes, as an enabler.	National (government, industry) and regional (university, state) resources highly leveraged.	Potentially very high. Success depends on Abilene's ability to integrate its network infrastructure with applications research in NGI and Internet.

Table 4 Highlights of Internet's Impacts—Japan's Activity

Activity	Accelerating Convergence of CCC	Creating and Using Social Capital	Achieving New Economies of Scale and Scope	Leveraging Global, National, and Regional Resources	Connecting to Society
Real World Computing (RWC) in the Internet era	Initial research focused on new technologies for computing and communications. New research is to emphasize networked, content-based work.	Yes, through its strong consortia and partnerships among industry, universities, and government.	Moderate. Only through its strong partnerships across diverse industries (scope).	A long-term MITI project. National and regional resources highly leveraged. Has international research partners for selected areas.	Society benefits not a visible goal in initial research. New work to emphasize human-centered, networked computing.
Digital Contents and the Next Generation Digital Library	Strong impact. Content industries (media, publishing) highly engaged. Traditional electronics and computer industries also drawn into content-based research.	Yes, through strong consortia and partnerships between industry and universities. Presence of government support important.	Moderate. Fixed architecture tends to limit opportunities to achieve economies of scale or scope across multiple knowledge domains.	National and regional resources highly leveraged.	Research goals and outcomes well-linked to societal/economic benefits.
National Center for Science Information Systems (NACSIS)	Yes, due to its rich and highly developed information services; research in multimedia information systems also a factor.	Moderate. Has a unique role in support of research and higher education. Also links to industry and professional groups are helpful.	Yes, across multiple scientific disciplines (scope) and contents (scale across data).	National assets highly leveraged. Network links to international gateways (Thailand, UK, US).	Yes, but primarily to science communities.
Broadband-ISDN Business and Culture Creation (BBCC)	Very high impact. Well-selected experimental projects provide evidence for convergence of CCC.	Yes, through its very strong partnerships. Highly effective management for bringing together users, researchers, organizations.	Moderate. Resources and expertise shared through diverse fields and industries (scope).	Regional resources highly leveraged. Also connected to national resources via networks.	Very high. Societal and business benefits a top research goal.
The WIDE Project	Very effective in integrating computer and networking technologies. Content-based research has not been a research focus.	Yes, through its strong relationships with industries, universities, and professional groups. Very effective technical working group for targeted projects.	Moderate. Has a large number of well-connected participants (scale) in diverse fields (scope).	National and regional resources highly leveraged. Close association with most Internet-related technical working groups, domestic and international.	Yes, but primarily to science communities.

analyzing the perceived impacts of the Internet across roughly similar cases.

Because the Internet world has generally relied on the concepts and program models first developed and experimented with in the United States, Tables 3 and 4 sketch the similarities and parallel developments between the two countries. Japan, and much of the rest of the world, developed their own versions of the Internet road map and the result is an orderly march in synchronous, if not identical, steps. Yet there are cultural, political, and social factors that point us in different directions, now and in the future, in the ways we manage research and innovation in the networked environment. In Table 5, we contrast the different management "formulas" according to a set of five issues: How can we decide what information infrastructure to build? Where and how strongly do universities and industries link up in their research investment? What is the predominant process for innovation? How is the success of R&D measured? Finally, what is the role of government in supporting new research and advancing technology?

Conclusion—Emerging Issues and Outlook

The Internet has profoundly affected university-based research, both in the United States and in Japan. For those who are concerned with public policies in support of science and technology, the changing relationships between university and industry, with government as a catalytic player in the Internet era, is of utmost importance. Our discussion, while necessarily limited in scope, nevertheless offers a starting point for broader and more in-depth analyses. A natural question is, Where do we go from here? More specifically, What are the emerging issues that are facing both the United States and Japan? Are there things we can do together? Obviously, there are no simple and quick answers yet, but we offer several observations here.

First, it is clear that in both nations many of the government-sponsored research programs increasingly emphasize outcome-oriented results. The recent Government Performance and Results Act (GPRA) passed by the U.S. Congress is a prime example of this emphasis, and it is affecting the programs of all Federal agencies.

Table 5 Managing Networked Research

Issue	U.S. formula	Japanese formula
How to decide what infrastructure to build	Most innovative, advanced technology	Appropriate technology tailored to local or national needs
Where university and industry meet	University-based research essential to industrial innovation: ties are more formal	Ties less formal and not as active in the past; new laws render university-industry joint work increasingly important
Process of innovation	Entrepreneurship actively pursued and encouraged; strong collaborative research within disciplines	Team work; more and stronger cross-discipline research
How success is measured	Size of budget; new ideas, breakthroughs; connection to economy first, society second	Step-wise performance improvement; connection to society first, economy second
Roles of government	Catalyst; assessor; mediator	Catalyst; forecaster; broker

Japan's R&D programs have traditionally been more society-oriented, at least in their goals and stated justifications. Changes in the Japanese economy are sure to affect university-based research. Across the U.S. university community, historically focused on knowledge creation but not on utilization, coupling research with society is no longer just a visionary hope, but a goal to be achieved. How is the Internet going to help mediate this apparent conflict? We argue that the Internet has recently established a bridge connecting research and society, reaching far beyond the usual scientific community, a role that will continue to expand as the Internet moves into the next generation. These Internet-induced connections will most likely be a powerful source for technology innovation that leads to new products and services. A dynamic market is essential to technology innovation and the Internet will be increasingly a vital part of it.

Second, the Internet's most potent influence is not so much to provide world-wide connectivity as to put distributed "contents" of all kinds across a variety of digital media. The Japanese government, more than its U.S. counterpart, came to this realization in the early 1990s and initiated new research programs on multimedia and human-centered technologies, joined by industries as Japan's consumer electronics and media companies converged into networked businesses, while in the United States, research emphasis on "Content" and "Information" came in much later. The HPCC's transition to the Computing, Information and Communications program exemplified this new recognition. Between the two countries, the Digital Library research initiatives have had a major impact on our ability to create, access, use, and store information of all kinds anywhere and anytime. As citizens of the world, this is a worthy goal; as scientists, the Internet's ability to help create and propagate "content" and "knowledge" across cultural and national borders represents an enormous challenge. Among the pressing issues for making university-based research more effective for innovation, the question of Internet governance, especially in the areas of intellectual property rights and copyright laws when applied to digital content, is especially urgent. Is an Internet-based university-industry partnership less able to resolve such issues as ownership when collaborative research leads to innovative technology? Where do the United States and Japan stand on rights protection in the era of digital content?

Final answers to these questions, as in many other Internet governance issues, are not in sight, but new studies suggest where we might be heading. For example, in a study done by the World Technology Evaluation Center at Loyola College,[21] author Shamos makes an insightful observation about the relative developments of IPR/CR laws in Japan and the United States. Contrary to common belief, according to Shamos, Japan's legal and popular concepts of intellectual property protection are very compatible with those of the West, and its systems are both rich and advanced. His many studies of pertinent laws, digital economics, and recent R&D programs for Digital Libraries in Japan and the United States led him to conclude that Japan is better prepared as a society to harness "content"-oriented technologies and to deal with the ownership and use of digital information. In short, says Shamos, when it comes to information on the Internet, "Japan will get there first."

In this interconnected world, the real question is not who is ahead, but how to get there together. This brings us to the question of global partnerships. The Internet has fostered collaboration between university and industry and the sharing of resources at both national and local levels. However, with some exceptions in highly visible areas (human genome research and environmental science/global change, for example), collaborative research on a global scale is more theory than practice in most disciplines. This is especially true in the information technology field, as the question of collaboration vs. competition naturally takes precedence over the need for sharing resources between competing parties. While U.S. university-based research, even with industry participation, seems openly accessible, access to Japan's research, done mostly in industrial and proprietary settings, appears restricted, which has impeded collaborative efforts between the two countries. A recent NRC report (NAS 1997) on "Maximizing U.S. Interests in Science and Technology Relations with Japan" offers an authoritative, contemporary view of where we were and where we might be headed on this subject. As discussed earlier, Japan's 1997 Basic Plan for increased support of science and technology and other reforms to encourage university-based research for innovation are aimed at making Japan's pre-competitive research more accessible. The opportunity is therefore at hand for reinventing the ways the two countries do collaborative work. In this Internet world, there should be less chance that the two nations will let that opportunity slip away.

Notes

1. The Internet's origin and its subsequent development are not the main topics of discussion here. A concise and up-to-date treatment of these topics may be found in several articles posted in the Internet Society's Web Site: <http://www.isoc.org/> including "Hobbes' Internet Timeline" by Robert H. Zakon and "A Brief History of the Internet" by some of those who made that history.

2. A list of Internet survey organizations may be found in the Internet Society's Web Site: <http://www.isoc.org/>.

3. URL for Network Wizards: <http://www.nw.com/>.

4. For information about UARC, see the Web Site for the School of Information at the University of Michigan: <http://www.umich.edu>.

5. For comprehensive coverage of this subject, see the NSF workshop report on "Human-centered Systems—Information, Intelligence, and Interactivity," available from the Division of Information and Intelligent Systems: <http://www.cise.nsf.gov/iis>. Also available from the Beckman Institute of Advanced Science and Technology, University of Illinois, <http://www.beckman.uiuc.edu/>.

6. Library of Congress's Digital Library program (<http://www.loc.gov>).

7. More information may be found in National Library of Medicine's URL: <http://www.nlm.nih.gov>.

8. URL for the CIC program: <http://www.ccic.gov/>.

9. There is a National Synchronization Web Site for the joint NSF/DARPA/NASA Digital Library Initiative during Phase 1 (1994–1998): <http://dli.grainger.uiuc.edu/national.htm>.

10. Digital Library Initiative, Phase 2, Announcement of a Joint-agency program on Digital Libraries, Phase 2 by the NSF98-63. Information about this program and other details may be found on the home pages of the sponsoring agencies: National Science Foundation (<http://www.nsf.gov>), Defense Advanced Research Projects Agency (<http://www.darpa.mil/ito>), National Library of Medicine (<http://www.nlm.nih.gov/ep>), Library of Congress (<http://cweb2.loc.gov/ammem/dli2/>), National Aeronautics and Space Administration (<http://www.nasa.gov>), National Endowment for the Humanities (<http://www.neh.gov/htm/guidelin/dli2.html>).

11. NSF PACI URL: <http://www.cise.nsf.gov/acir/>.

12. NCSA Alliance URL: < http://alliance.ncsa.uiuc.edu/>.

13. SDSC NPACI URL: < http://www.npaci.edu/>.

14. <http://www.vbns.net>.

15. <http://www.internet2.edu>.

16. <http://www.ucaid.edu>.

17. <http://www.ngi.gov>.

18. Why the name Abilene? The Abilene Project is named after a railway built from Texas to Kansas during the 1860s. In its time the ambitious railroad project staked a claim on what was then the frontier of the United States; the present day Abilene project establishes a foothold from which to explore and develop pioneering network technology. The links of last century's railway changed the way people worked and lived. The new information superhighway provided by the new Abilene can potentially transform university-based research into industrial innovation for the twenty-first century.

19. URL for RWC: <http://www.rwcp.or.jp/>. For more information about the RWC, its accomplishments and next-phase research, see a recent report by the Asian Technology Information Program (<http://www.atip.org>), which gives a summary and commentary on the 1998 RWC Symposium held in Tokyo, 1998.

20. See "Industry-academia R&D wall crumbling," an article in *The Nikkei Weekly*, June 15, 1998. Also of interest is the article, "Legislative proposal offers chance

to bridge academia-industry gap," in *The Nikkei Weekly*, May 11, 1998.

21. The Workshop on Digital Information Organization in Japan, held by the World Technology Evaluation Center at Loyola College, May 12, 1998, received an initial report by a team of U.S. experts after a site visit to Japan sponsored by NSF and DARPA. Its full report is in preparation at this writing and will be available by contacting Loyola College's Center: <itri@loyola.edu>.

22. Proceedings of the Research and Development Conference for the Next Generation Digital Library System, March 13, 1998 in Tokyo, Japan. This material is mostly in Japanese, but much of the technical discussion in this review conference appeared in a paper by Hiroshi Mukaiyama of JIPDEC at the International Symposium on Digital Libraries, ISDL97, held in Tsukuba, November, 1997. A useful Web Site for this is: <http://www.dlib.jipdec.or.jp/>.

23. URL for NACSIS: <http://www.nacsis.ac.jp/>.

24. "The Interim Report of the Committee on an Institute of Informatics" (August 1998) is the culmination of a series of recent activities in behalf of Japan's intense planning work on information research, education, and infrastructures. This report was based initially on a recommendation of the Science Council of Japan in May 1997 and subsequent proposals by the Ministry of Education. More information, especially about the role of NACSIS in this new development, can be found in the March, 1998 NACSIS Newsletter.

25. Information about Kansai Science City may be obtained from Kansai Research Institute, Keihanna Plaza, Hikaridai, 1-7, Seika-cho, Sourakuun, Kyoto 619-02, Japan. The URL for BBCC is <http://www.bbcc.or.jp>.

26. URL for WIDE: <http://endo.wide.ad.jp/index.html>.

References

Branscomb, Lewis M. and James H. Keller, eds. 1998. *Investing in Innovation: Creating a Research and Innovation Policy that Works.* Cambridge, MA: MIT Press.

Brooks, Frederick and Ivan Sutherland. 1995. "Evolving the High Performance Computing and Communications Program to Support the National Infrastructure." Washington, D.C.: National Research Council.

Chien, Y.T. and Randy Katz, eds. 1994. "Information Infrastructure Technology and Applications for the High Performance Computing and Communications (HPCC) Program." Washington, DC: Report of the HPCC National Coordination Office.

Fountain, Jane. 1998. "Social Capital: A Key Enabler of Innovation." In Lewis Branscomb and James H. Keller, eds., *Investing in Innovation.* Cambridge, MA: MIT Press.

Matsumoto, Hiroshi. 1995. "High Speed Network for Digital Libraries." Proceedings of the 1995 International Symposium on Digital Libraries. Tsukuba, Japan.

Messerschmitt, David G. 1996. "The Convergence of Computing and Telecommunications: What are the Implications Today?" Proceedings of the IEEE. 84(8): 1167.

National Academy of Sciences (NAS). 1997. "Maximizing U.S. Interests in Science and Technology Relations with Japan." Report of the Competitiveness Task Force. Washington, D.C.: National Academy Press.

National Research Council (NRC). 1993. *National Collaboratories: Applying Information Technology for Scientific Research.* Washington, D.C.: National Academy Press.

Rayport, Jeffrey F. and John J. Sviokla. 1995. "Exploring the Virtual Value Chain." *Harvard Business Review* (November–December): 75.

Tanaka, Yasumasa. 1978. "Proliferating Technology and the Structure of Information Space." In Alex Edelstein et al. (eds.), *Information Societies: Comparing the Japanese and American Experiences.* Seattle, WA: International Communication Center, University of Washington.

V

Distributional Issues: University Roles in Regional and National Development

Having examined the broader outlines of university-industry interactions in Japan and the United States, we turn to the issue of how this dynamic affects, and is affected by, policy making at both the local and national levels. The papers in this part focus on specific cases in each country in which public policy has attempted to meet the economic, scientific, and educational needs of a region or an institution.

18

Why Older Regions Can't Generalize from Route 128 and Silicon Valley: University-Industry Relationships and Regional Innovation Systems

Michael S. Fogarty and Amit K. Sinha

Introduction

Faced with intimidating industrial decline during the last part of the 1970s and early 1980s, older U.S. industrial regions seized on university-industry relationships (UI) as a primary economic development tool (Coburn 1995). State science and technology (S&T) programs have sprung up everywhere, but disproportionately in industrial states. The stimulus for creating these programs comes from the successes of Boston's Route 128 and California's Silicon Valley. The reasoning was: It worked in these places, so why not here?

What is the economic rationale for state and local S&T programs' focus on UI? National studies provide evidence that the nation's investments in academic research produce a high social rate of return (perhaps 25 to 50%) (OTA 1986; Mansfield 1991, 1995; Henderson, Jaffe, and Trajtenberg 1995). R&D spillover networks associated with university research are a primary mechanism creating substantial public benefits. Evidence indicates that R&D spillovers cause a large gap between social and private rates of return—a difference on the order of 50 to 100%.[1]

However, a high national social rate of return is not sufficient to justify a state or region's investment in research or a federal agency's decision to fund R&D in a specific region.[2] The reason is straightforward: Knowledge spillovers are available to everyone, so benefits get widely dispersed (Jaffe 1989).[3] In fact, states and local areas face a dilemma, namely that top research programs are essential for producing globally competitive technology, but the best research with the greatest commercial potential will quickly become known throughout the world. At best, geographic proximity of research confers only a temporary advantage to a region and its industries. Capturing spillover benefits hinges on speedy diffusion of knowledge within local R&D networks.

Equally important, and usually overlooked, calculating a rate of return to state and local research investments requires tracing spillover benefits to local industries. By itself, a region's performance of R&D is not sufficient for producing stronger economic performance over the long run. If university research is to raise a particular region's productivity growth via technology, it must connect with local industry performance. Eventually, local gain requires that new technology be commercialized and take the form of investment in local facilities. This can occur directly through various UI mechanisms or paths, such as: startups, further development of the technology by local industry R&D labs, attraction of new industry labs or inducement of additional R&D by existing labs, or indirectly by raising the area's educational level and thus its support of important regional industries. The destination of graduates from local institutions will substantially affect any calculation of payoff from state and regional investments in research.

R&D spillovers associated with the new technology will become a source of long-run economic benefit only if the local industry R&D network draws from the technology, if commercialization occurs locally, and if the region's industries capture the technology through diffusion and investment.

This paper examines the experience of one older region, Cleveland, Ohio, in the context of the regional innovation system, focussing on the tension between national, state, and local objectives for encouraging university-industry relationships. We conclude that older industrial regions need a new framework for evaluating policy choices and considering S&T investments. State and local areas seek to capture economic benefits locally. For the most part they can improve their odds of success only if they (1) clarify objectives, particularly with respect to the geography of spillover benefits; (2) invest strategically by building the strongest UI paths from basic research to local industry; and (3) develop a new local S&T management system to ensure the integrity of the system. This requires becoming knowledgeable about technologies and economics of the local innovation system, structuring incentives for intermediaries, and insisting on evaluation based on an understanding of UI mechanisms linked to goals.

We developed a policy framework that raises several questions. How can a region maximize benefits for a region from university-industry relationships? What is the necessary decision process for investing in university research programs? What role should S&T intermediaries play? What should be national goals?

Finally, we offer a few suggestions based upon our study, as to how older U. S. industrial regions can develop appropriate UI structures.

Response to Industrial Decline: Building an S&T Infrastructure

Cleveland, Ohio includes three contiguous metropolitan areas: Cleveland, Akron, and Lorain-Elyria. It has been a major industrial center for a century and a half, specializing in steel, autos, metalworking, and chemicals. It still produces about 1% of U.S. GDP. The economy and population declined sharply in the late 1970s and early 1980s, and the region responded by developing an impressive S&T infrastructure, which includes a very active group of S&T intermediaries.

Fostering university-industry relationships has been a centerpiece of their efforts. Despite the large investment already made, many of the fundamentals remain weak, as suggested by the region's declining per capita income relative to U.S. metropolitan areas over the past twenty years.

Methodology

We take two related approaches in examining university-industry relationships in this setting. The first draws lessons from three Cleveland technology case studies: Functional Electrical Stimulation (FES), Polymer Displaced Liquid Crystals (PDLC), and Microelectro-Mechanical Systems (MEMS). In each case a wide range of Cleveland's S&T infrastructure institutions became engaged and a wide spectrum of university-industry relationships have been employed.

The second approach uses patent data and our new systems methodology to analyze regional implications of R&D spillovers associated with UI. This approach defines university-industry interactions with patent citations connecting universities and industry at two levels: (1) We analyze local and external R&D spillovers for all technologies associated with UI R&D networks, with the expectation that R&D spillover flows will be much less favorable toward older industrial regions; and (2) We also separately analyze R&D spillovers from microelectro-mechanical systems technology. We chose MEMS because the technology is globally significant and because it is a very important current Cleveland case—in fact, Ohio and Cleveland are increasing investment in MEMS, focused on Case Western Reserve University at the center of a statewide partnership called Ohio MEMSNet. Cleveland's current position suggests that it cannot reasonably expect to benefit from MEMS spillovers unless it has an industrial R&D network poised to capture them.

The problem of older industrial regions is complicated by regional growth patterns characterized by increasing returns: Regional growth tends to become self-reinforcing. Specifically, R&D spillovers favor newer, "knowledge" regions and work against older, lagging regions. Put simply, it is very hard for older industrial regions to get spillovers derived from university-industry interactions to stay there once established. Figure 1 suggests a regional pattern of increasing returns to R&D related to a region's share of a technology in period t and the strength of its industry R&D network. With a low share of the technology and weak R&D networks, region Cleveland is positioned to the left of X (below a critical mass); region Silicon Valley is positioned to the right of X with a high share of the technology and strong R&D networks.

Now imagine the same university technology emerging in each location. Our expectation is that Cleveland's R&D spillovers will largely be external, while Silicon Valley's spillovers will be localized. In fact, region Silicon Valley will absorb Cleveland's spillovers, further widening the Silicon Valley-Cleveland gap. By t+10, Cleveland's share of the technology has declined along A, B, C, D and Silicon Valley's share of the technology has increased along E, F, G, H.

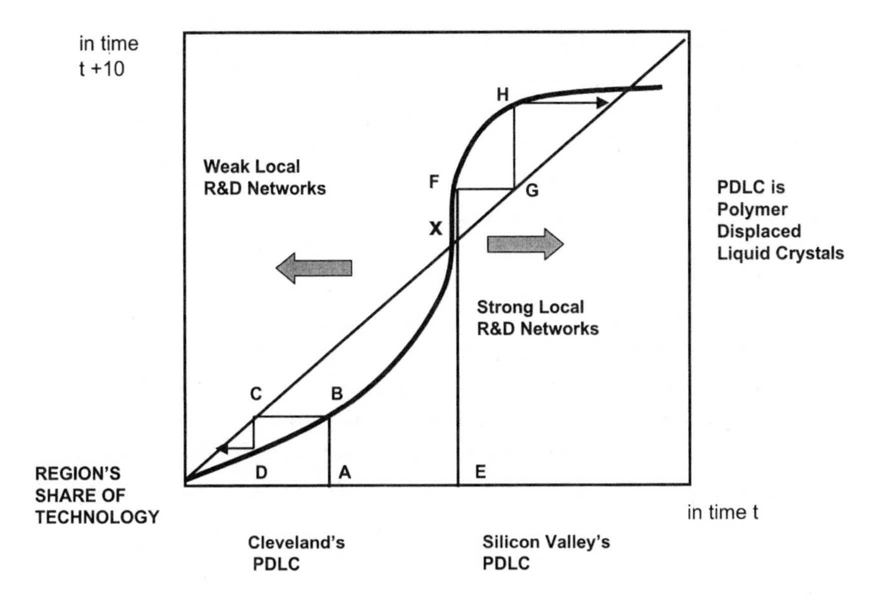

Figure 1 Hypothetical Regional Patterns of New Technology Development.

Background

The Cleveland region experienced over 130 years of continuous economic growth from 1840 into the early 1970s, spurred by the growth of several industry clusters based on the steel, metalworking, chemicals, and auto industries. Possibly as early as the 1950s, the region's industries entered a phase of decline, creating a local crisis in the late 1970s and early 1980s. Roughly one-fifth of all manufacturing jobs were lost from 1979 to 1982.

The magnitude of this shock propelled fifteen years of sustained efforts by Cleveland and the state of Ohio and Cleveland to reverse the decline.[4] University-industry relationships became the mantra of local decision makers.[5] Seeing Route 128 and Silicon Valley as models, Cleveland, the State of Ohio, and the federal government invested heavily in a new science and technology infrastructure. In effect, what Cleveland sought to do was to build both the technical as well as the social capital necessary to compensate for its disadvantaged market position.[6] The following summarizes the federal, state, and local S&T infrastructure.

Federal. The federal government has a strong regional presence in the Cleveland region, which includes the NASA-Lewis Research Center;[7] the Great Lakes Technology Center (GliTec), a federally-funded Regional Technology Transfer Center created by NASA to facilitate contact between business and the federal laboratory system; the Great Lakes Manufacturing Technology Center, which is funded by the National Institute of Standards and Technology (NIST) to assist in the deployment of technology; and the center for Advanced Liquid Crystalline and Optical Materials (ALCOM), a National Science Foundation (NSF)-funded S&T center focused on liquid crystals at Kent State University in partnership with Case Western Reserve University and the University of Akron.[8]

State. The state of Ohio is well established in the Cleveland region with three Edison Technology Centers: the Edison Biotechnology Center (EBTC), the Edison Polymer Innovation Corporation (EPIC), and the Cleveland Advanced Manufacturing Program (CAMP, Inc.). CAMP also administers the federally funded Great Lakes Manufacturing Technology Center. In addition, through the Ohio Board of Regents, higher education selectively funds academic research with an eye to commercialization through three research programs.[9] Acknowledging the need to strategically coordinate science and technology policy decisions, the Governor of Ohio created the Ohio Science & Technology Commission (OSTC) as a joint high-level industry-state body. Its purpose is to give science and technology greater visibility and to consider broad state S&T strategy issues. The Ohio Aerospace Institute (OAI) is a statewide consortium involving the NASA-Lewis Research Center in Cleveland, Wright Patterson Air Force Base in Dayton, nine Ohio universities, and private companies. Its mission is to be a collaborative network of government, university, and industry organizations for fostering aerospace technology development, commercialization, and application.[10] MEMS-Net is a statewide consortium of Ohio's universities, the Cleveland Clinic, Edison Technology Centers, two federal laboratories, and industry. Its purpose is to provide infrastructure support for fabrication, design, and testing of microelectro-mechanical systems technology.[11]

Local. The Cleveland region operates a range of S&T infrastructure institutions and programs. Most important, the Technology Leadership Council (TLC) is a CEO-level organization organized

Federal	State	Local
NASA-Lewis	Ohio Science & Technology Commission	Technology Leadership Council
ALCOM–NSF S&T Center (liquid crystals) —*PDLC case study*	Edison Biotechnology Center (EBTC)—*FES case study*	Enterprise Development Inc. (EDI) (incubators and entrepreneurs)
Great Lakes Manufacturing Technology Center (GLMTC-NIST)	Edison Polymer Innovation Corp. (EPIC)	Research Park (focus on biotechnology)
GliTec-RTT (technology transfer from Great Lakes federal labs)	Cleveland Advanced Manufacturing Program (CAMP)	Primus (venture capital fund)
		The Ohio Innovation Fund (seed capital)
	Ohio Aerospace Institute	

Figure 2 Cleveland's S&T Infrastructure.

to increase the region's research and commercialization. Its mission and activities are comparable to those of the Ohio Science & Technology Commission, but with a local focus. The activities of all three Edison Technology Centers are constantly monitored and shaped by the Technology Leadership Council. Because it was believed that availability of early-stage funding for startups was crucial, the Technology Leadership Council worked closely with the state to establish two funds: the Primus Venture Capital Fund was established in 1985 with both corporate and State of Ohio public pension funds; The Ohio Innovation Fund was created in 1998 to fill a perceived gap in funding for the very earliest commercialization stage. Enterprise Development Inc. (EDI) was also created to encourage entrepreneurship and startups, particularly startups originating in local universities. EDI is affiliated with Case Western Reserve University. Most recently, working with many partners, the Technology Leadership Council has been working to create a new research park. Its purpose would be to foster an agglomeration of new high-tech businesses emphasizing biotechnology within the geography surrounding Case Western Reserve University and the Cleveland Clinic.

The Technology Leadership Council first focused on strengthening academic research (Fogarty et al. 1991), leading the efforts to increase the region's scale of university research through selective investments in research and graduate programs. (Research scale is measured simply as the region's total research expenditures per capita relative to the median of the thirty-seven comparable metropolitan regions.) Relative to the set of thirty-seven comparable metropolitan regions, Cleveland's university research scale increased from 68% of the median in 1985 to 76% in 1994. The TLC also sponsored efforts to increase commercialization, including efforts to encourage commercialization through technology transfer offices and university-industry research centers (UIRCs). For example, CWRU has two prominent UIRCs: the Center for Automation and Intelligent Systems Research (CAISR), which is an affiliate of the Cleveland Advanced Manufacturing Program, an Ohio Edison Technology Center. A second is the National Center for Microgravity Research on Fluids and Combustion, which is sponsored by NASA and is co-located at Case Western Reserve University and NASA-Lewis in Cleveland.[12]

Why Intermediaries?

There are now numerous voices raised in the region's S&T decision making. The list of intermediary institutions is long and the outcomes are extraordinarily difficult to measure and evaluate.[13] What do they do? One question often posed is: Why should Cleveland have intermediaries when Silicon Valley doesn't seem to need them? One answer is simply that intermediaries have been created to raise the odds that S&T investments will work for the older region. The economic rationale for intermediaries is reasonably straightforward: The Cleveland region is still disadvantaged. The market, as indicated by flows of private investment, educated people, venture capital, R&D spillovers, and high-tech industry continues to favor newer regions; the market fails to produce, commercialize, and use the new technology necessary to improve Cleveland's relative economic performance.[14] New institutions may be necessary to bridge the gap, but it is not at all clear how intermediaries can overcome the market disadvantage.

One key is leadership. Cleveland's strategy has been led by the Technology Leadership Council, which takes its lead from CEOs of Cleveland's largest companies in partnership with government and nonprofit S&T organizations. Working closely with the state of

Ohio, the Council assumed responsibility for building and loosely managing the S&T infrastructure.[15] Its efforts are those of a high-level UI relationship: building the institutions, helping to increase the region's scale of university research (which was about 28% below average in 1985), commercialization, and, increasingly, technology deployment (Fogarty et al. 1991).

Has There Been a Turnaround?

Have the region's efforts reversed economic decline? Today the Cleveland regional economy accounts for 1% of US GDP. Although this share has been declining, there is clear evidence of a turnaround across a wide spectrum of economic conditions (Gottlieb and Bogart 1997). For example, the region's economy has added nearly 200,000 jobs over the past decade (a 15% gain), and its unemployment rate has dropped from about 7.5% in 1986 to 4.8% in April 1997—equal to the current U.S. rate.[16] Yet one troubling fact is that per capita income has declined relative to other metropolitan regions since 1969. The declining trend appears to have stabilized after 1990. The primary reason for Cleveland's declining relative per capita income is the poor productivity performance of the region's industries. Per capita income in real terms (adjusted for inflation) cannot increase over the long run without productivity growth (Fogarty 1983; Baumol, Blackman, and Wolff 1989, chap. 2).

Numerous economic studies have accumulated evidence that the lion's share of productivity growth is caused by improvements in two factors, education (including training) and technology.[17] Ideas from these studies increasingly influenced Cleveland's economic development strategy (Fogarty 1998). But they also present Cleveland with a big challenge. First, reflecting the "dead hand of the past," for two decades the region's educational level has declined relative to the median of a benchmark set of 37 comparable metropolitan regions (Fogarty 1998). Education is measured as the percent of persons over 25 with at least one year of college. Not surprisingly, the trend in per capita income parallels that of education. Per capita income declined from about 6% above the U.S. metropolitan median in 1969 to 4% below the median in 1990.

Second, although no comparable measure for technology exists, various measures suggest a weak technology base. One is that even though Cleveland's university research scale has grown as a result of investments in local universities since 1985, the scale is 24% below the median of the 37 metropolitan regions, with virtually all gains in the medical and biological sciences. A second is that industry is shifting R&D to other regions. (Industry R&D is estimated to be $2 billion, which is 20% above average per capita.)[18] From 1986 through 1995, the region's share of U.S. (domestic) patents fell sharply, from 1.72% to 1.12%. Also, Cleveland's share of patents produced locally by its top 52 corporate labs declined from 19% to 11% over the same period. Third, there are no data to show that long-term benefits of R&D are being incorporated into the region's industries. For example, no data exist on private investment in equipment and facilities at the metropolitan level. This is critical for evaluating the local payoff from R&D because investment is the mechanism for incorporating new technology into production. However, as a summary indication, one measure of the Cleveland region's overall productivity growth shows decline over this period relative to a set of fifty-six comparable U.S. regions.

Issues

The picture that emerges is complex. A big commitment made by many decision makers caused the infrastructure to be built. However, fundamental economic trends are weak. A few things stand out. First, even though the infrastructure has been assembled, it is not clear how it is supposed to function. How are decisions being made? Who is in charge? And to what purpose? On the surface, it seems that the Technology Leadership Council manages the local S&T system. Closer observation, however, reveals a very complex decision process that involves all levels of government, universities, industry, and numerous nonprofit intermediaries.

Second, intermediary funding sources affect outcomes. Table 1 shows the funding history of a subset of these S&T institutions. The figures represent the total annual budget and percent of funding by sources for Enterprise Development Inc., the Cleveland Advanced Manufacturing Program (CAMP), the Edison Polymer

Table 1 Sources of Funding for Selected Cleveland S&T Intermediaries

Source	1985	1990	1997
State	59%	51%	22%
Industry	41%	37%	18%
Federal	0%	12%	60%
Total Budget (in millions)	$0.9	$13.2	$57.9

Innovation Corporation (EPIC), the Edison Biotechnology Center (EBTC), the Ohio Aerospace Institute (OAI), the federal Great Lakes Technology Center (GliTec), and the center for Advanced Liquid Crystalline and Optical Materials (ALCOM).

Not shown, data for individual intermediaries reveal wide variations in funding level and sources of funds. State funding has become far less important and less predictable while federal funds have grown significantly. The specific funding sources for S&T intermediaries alter incentives and, consequently, shape the geography of R&D spillover benefits. Scale is also a major issue. For these organizations, the budget totals represent a scale roughly one-tenth of 1% the size of Cleveland's economy, which exceeds $60 billion annually. Given this scale, what is a reasonable expectation for the role of intermediaries?

UI Lessons from Three Case Studies

The best way to learn about university-industry relationships is to examine particular cases involving specific technologies in depth. This section draws on our Cleveland case studies of three technologies.

Functional Electrical Stimulation (FES). FES is a rehabilitative technology that restores motor function to paralyzed limbs by applying low-level stimulating current to the body (Rae 1994). One system consists of electrodes surgically implanted into the muscles. Potential applications now under clinical trial involve paralyzed limbs and bladder function.

FES, developed from research over a fifteen-year period at Case Western Reserve University, is a classic university startup example. Cleveland had invested in the traditional basic research and created an intermediary (the Edison Biotechnology Center, or EBTC) to help build a biomedical industry. No large industry R&D sector was involved in the technology either locally or nationally. Most local partners worked very hard to get the technology to lodge in the region.

A full-scale effort was undertaken to demonstrate that it could be done in Cleveland: Several companies with related technology sought licensing with CWRU; the FES partners made the purposeful decision to commercialize FES as a startup for the region. The result was the founding of a company called Neurocontrol in 1993.

The FES case shows that, with hard work and some luck, an intermediary operating in this setting can make a difference. Even when local partners are highly motivated to capture the technology locally, achieving this result for a region that was significantly behind the major biomedical device regions required vigorous intervention. The Edison Biotechnology Center was highly instrumental in getting FES to a commercialization stage for the Cleveland region. The company exists today, but its experience over the past five years reflects the absence of a critical mass of biomedical device companies in the region. The magnitude of the challenge is highlighted by our systems analysis of biomedical technology. (See below for a discussion of our R&D spillover systems methodology.) Cleveland was ranked fourteenth nationally in biomedical device technology with a system value (measure of system influence) only about one-fourth that of the fifth-ranked U.S. region (San Francisco). Cleveland has no critical mass on which to build. The challenge is: How can the region invest to build critical mass?

PDLC (Polymer Displaced Liquid Crystals). PDLC is an important technology that emerged at Kent State University (KSU) in the Cleveland region (Hoshiko 1994). This technology involves liquid crystals entrapped in a plastic medium, forming microdroplets surrounded by plastic, causing the material to become opaque. When voltage is applied, the crystals align and the material becomes transparent. The first application was privacy windows for cars.

This case is far more complicated than FES. Strong networks of highly influential companies exist outside the region, serving as powerful magnets for the technology. No clear path existed between the PDLC technology and local industry. Evidence based on patent citations shows that the technology was being quickly diffused to Japan,

California, and, to some extent, Texas. Although KSU's patents received an astounding 439 patent citations as of mid-1998 (the average patent receives three or four citations), not one citation to the PDLC technology came from an Ohio R&D lab, even though various labs' technologies were closely related. Clearly, there is no evidence of localized R&D spillovers. From our interviews, it appears that local R&D labs working in related technology decided it was not worth their investment because Japanese companies were too far ahead. As a result, all university-industry interactions associated with the university patents were external to Cleveland. In fact, it is likely that local university-industry relationships fostered by the center for Advanced Liquid Crystals and Optical Materials' industry board sped up commercialization of the technology for other regions. (ALCOM is a National Science Foundation S&T Center.) The intermediaries have been unsuccessful at commercializing the technology for the Cleveland region.

The main lesson from this case seems to be that when a highly important technology is produced locally, with few clear links to the region's industry R&D labs, extraordinary efforts are needed to capture economic benefits. PDLC is a case in which Cleveland's technology was highly influential worldwide but the local industry R&D network's R&D labs chose not to develop the technology.

MEMS (Microelectro-mechanical Systems). This technology merges computation with sensing and actuation, making integrated electromechanical systems and miniaturization possible. MEMS has a potentially large market in which a few universities are playing prominent roles—predominantly at Berkeley and MIT (Fogarty, Sinha, and Jaffe 1998; Fogarty 1998). Because MEMS is a story in the making, its lesson is not yet clear. However, Cleveland seems to be taking a lesson from the PDLC experience. If MEMS is to produce the hoped-for success, the region must employ a different strategy.

First, if a significant technology opportunity emerges from a local university, and the associated local industry R&D network is sparse, it may be possible to build an R&D network as a partnership of local companies working on closely related technologies. Our assessment of the Cleveland region's MEMS R&D network found that the region has no R&D labs producing technology that is core to MEMS. However, some local companies are producing related technology, and their labs interact with major university sources of MEMS in Silicon

Valley. Some of these companies who were unlikely to increase investment in MEMS individually do support a regional MEMS partnership. With aid from the state of Ohio and NASA, this partnership is being formalized as a statewide MEMS network involving the state's universities, NASA-Lewis, Wright Patterson Air Force Base, Battelle's Great Lakes Industrial Technology Center, and the Cleveland Clinic.

Second, the Cleveland region's scale of MEMS technologies puts Northeast Ohio at a disadvantage—most of the flows involving MEMS technologies are toward networks around Berkeley and MIT as well as Japan. To overcome the disadvantage, it would be essential to build critical mass quickly. As we discuss later, because these flows tend to be cumulative and self-reinforcing, delay in building critical mass to support emerging, enabling technologies like MEMS makes catching up far more difficult. The window of opportunity may be quite narrow.

Issues

A number of issues emerge from our three UI case studies. The prevalence of conflict in objectives is the first. For example, each participant represents different geographic interests. Most obvious, the objective of most federal R&D support is to commercialize for the United States while the state and local objective is to commercialize locally. Industry's objective is to commercialize wherever it can benefit the most, which may or may not correspond with the interests of the university, inventor, or region. Moreover, conflicting objectives are typically not made explicit in decisions concerning university-industry relationships or the technology. Usually these conflicts are glossed over, causing ambiguity about the extent to which private and social benefits coincide; the intermediaries and government both presume that the two are roughly equivalent and, therefore, rarely ask tough questions, such as: How do we know when goals are being achieved?

Second, UI takes many forms and involves complex partnerships with many governmental and nonprofit institutions at all levels. For example, the three technology cases all have involved active management by the Technology Leadership Council, which is headed by large Cleveland corporations. Nevertheless, all levels of government participated in all three cases in many different ways. The forms of university-industry relationships ranged widely: venture capital funding of startups, membership on industry boards of university centers, funding of applied research, hiring of graduates by R&D labs, licensing of patents, and membership on university boards of trustees.

Within this context, intermediaries can clearly make a difference. This was true in the case of FES but not for PDLC; intermediaries are actively engaged with MEMS. These relationships are complex and their effects are both subtle and not well understood. The question worth asking is: How do these relationships influence research, affect commercialization, and shape the geography of R&D spillovers?

Third, funding matters a lot—budgets drive most decisions by shaping incentives. In particular, the mix of public and industry funds alters the mix of social versus private benefits. Consequently, Ohio's diminished importance as a funder of Ohio's Edison Technology Centers is telling, raising the question: What should be the state's role? In all three cases, federal funding was the most important source of research support. However, state funding was also crucial; in one or two cases even local government dollars supported the technology.

Finally, the three case studies make it clear that the flow of investment, educated workers, and new technology strongly favors "knowledge" regions. R&D spillovers associated with university-industry relationships contribute to the pattern of increasing returns, placing older industrial regions in a disadvantaged position. Their challenge is to utilize university-industry relationships to maximize the payoff from their science and technology investments.

Regions and R&D Spillover Networks Based on University-Industry Interactions

This section focuses on one university-industry relationship form, namely R&D spillovers associated with university-industry relationships linked to patents. Current methods for analyzing R&D spillovers use patent citations to measure spillovers (citations are the references on a patent to earlier inventions). The term "spillover" captures the idea that economic benefits resulting from R&D are shared by organizations other than the R&D performer. For example, a citation on a patent by company X to university Y's patent suggests that X's technology builds on knowledge from Y.[19]

Our interviews and quantitative findings support the view that spillovers occur within a system or network of knowledge flows. The basic unit in our analysis is the R&D lab representing an organization (company, university, or government lab), working on a specific technology, located in a particular region/country. We interpret citations over a period of time as a form of communication among these R&D organizations.

We have developed a fuzzy systems methodology that reveals a hierarchical structure based on R&D spillover networks.[20] More specifically, company X and university Y above are components of a larger R&D spillover network, linking various companies, universities, and government labs both directly and indirectly. Indirect links occur when, for example, company Z learns from university Y through company X. In effect, X processes the knowledge for further diffusion. The methodology incorporates such second- as well as third-level diffusion.

This methodology differs significantly from the existing methodology, which treats all citations as equal (each citation gets a value of 1 in evaluating "importance") and overlooks the system.[21] In addition, the existing methods do not incorporate higher-level diffusion of spillovers that occur through R&D networks. We use the fuzzy methodology to analyze the intensity of interaction between individual company R&D labs and particular universities working on specific technologies, assigning a value between 0 and 1 to each company-university link. To illustrate, two R&D organizations, each with a MEMS patent getting 25 citations, would be treated as equally influential with the existing R&D spillover methodology. In contrast, our systems method assigns a unique value between 0 and 1 for each citation based on the intensity of interaction between the citing R&D Lab A and cited R&D lab B. This value also reflects the relative dependence of A on B as a user of A's technology and B on A as a source of B's technology.

We construct industry R&D networks by first identifying R&D companies whose R&D is predominantly focused on a particular industry using either a high-tech corporate directory (CorpTech)[22] or more specialized technology information, as in the case of MEMS.[23] The second step involves identifying both significant sources and users of the technology generated by the initial data set. This step is repeated up to the third level of diffusion. The system source and user interactions are then normalized with respect to all interactions contained in the specific industry R&D network.

We can aggregate these measures in various ways to analyze how regions, organizations, and technologies interact over time as sources and learners in developing new technology. A measure of technology-specific system influence of a region's R&D labs can be obtained by aggregating to get each region's system influence by industry technology.

The sum of these measures of interaction across the R&D network associated with university-industry interactions by technology is a measure of system influence or importance. For this paper we aggregate these values across R&D labs and technologies for U.S. metropolitan regions to indicate the relative importance of university-industry relationships by region. A strong UI region has many intense UI links (that is, a larger number of UI links with values close to 1). Summing the value of many individual strong links (UI interactions) produces a large system value.

We also separate the UI value into two parts: UI local, which represents the portion of UI influence that is due to within-region university-industry interactions; and the portion of UI influence that is external, occurring when a region's industry R&D labs interact with and therefore benefit from universities outside the region.

Data

The data are drawn from the universe of patents granted by the U.S. Patent Office between 1963 and 1995; information on citations made to these patents covers the period 1977–1995 (approximately 3 million patents).[24] Information on patent citations begins in 1977; electronic data on assignees is available beginning in 1969. We geographically locate patents using the inventor's address, which means that location in our analysis is the R&D lab's location based on country as well as county within the United States. Sorting inventors into counties permits us to aggregate by metropolitan area. Consequently, regions are the sources of the technology rather than headquarters (assignee) locations.

University-Industry Interactions for Selected US Metropolitan Regions

We focus our attention on nine U.S. metropolitan regions (Consolidated Metropolitan Statistical Areas, or CMSAs). Our objective was to select places with universities producing influential patents. We chose nine places with universities receiving the largest number of patent citations from individual university patents receiving at least twenty citations—university patents that passed a reasonably high minimum threshold level of "importance."[25] The period is 1985–1995. The question we ask is: What are the geographic R&D spillover implications of influential university technology?

The resulting R&D spillover systems analysis is summarized in Figure 3. The size of each circle represents a region's total learning from universities, both within the region as well as externally.[26] For example, with system values of 838 and 821, San Francisco and New York–New Jersey possess R&D networks that draw most heavily from university technology. The vertical position of each circle measures the percent of learning from universities that stems from local interactions. Boston, for example, draws the largest share from local universities; Washington-Baltimore draws the least. Regions are ranked along the horizontal axis by total system value from high to low. A region's position reflects the influence of each region based on all interactions and all technologies—not just those derived from university-industry interactions. For instance, of the selected regions, New York–New Jersey possesses the most influential R&D network.

The picture that emerges is very suggestive of the underlying hypothesis. In general, it shows two types of places. The first represents strong, influential R&D centers which tend to be characterized by R&D networks drawing extensively from university technology—both locally and externally (strong R&D regions interact most intensely with other strong regions and learn extensively from the local R&D network).[27] We expect these centers to increase their share of new technology over time, possibly because a strong UI component helps

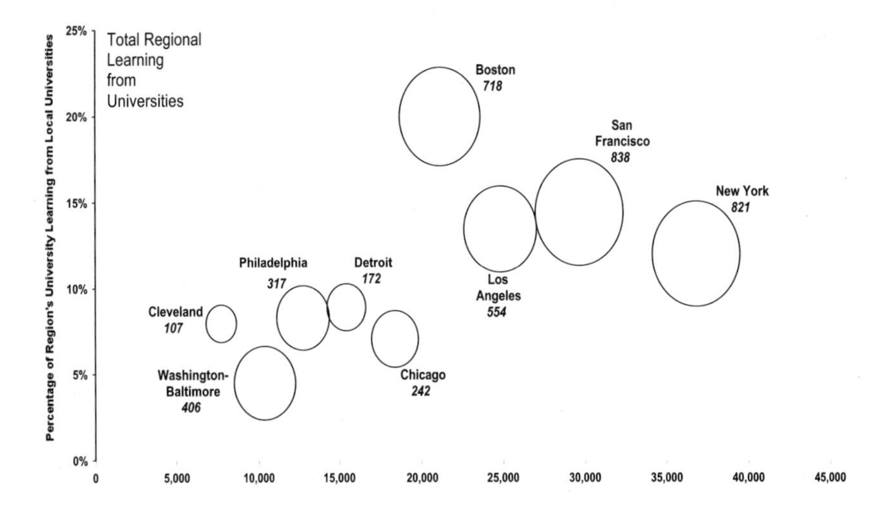

Figure 3 Regions as Learners from Local and External University Sources.

to maintain the region's technology leadership. The second type of place is represented by weaker, less influential and older R&D centers which have R&D networks minimally drawing from university technology. The upward sloping pattern, coupled with smaller circle size, indicates that less influential R&D regions possess weak UI, locally as well as externally.

A region is hampered from becoming an important source of new technology unless it is a good learner—that is, it learns extensively from other important regional sources as well as locally. Among the nine, Cleveland learns the least from universities: Cleveland's UI scale is roughly one-eighth that of the San Francisco region. Further, like the other older industrial regions, the percentage of UI spillovers drawn from local university-industry interactions is low. A large part of the difference between the two types of regions reflects the maturity of the technologies—because of history, older industrial regions specialize in more mature technologies, limiting their connection to universities. One example is the increasing importance of electronics and computer technology relative to other technologies in the automobile. As a result, Cleveland's R&D is becoming less important to its auto industry while Silicon Valley's role is expanding.

The Case of Microelectro-Mechanical Systems (MEMS)

An important question is: Given the development of an important specific university technology, what is the likelihood that such a region will capture a significant share of R&D spillovers? To explore this issue, we next analyzed MEMS R&D networks using the systems methodology over the period 1985–1995. We chose MEMS for two reasons:[28]

(1) MEMS is an emerging technology worldwide, with about 200 firms actively engaged in MEMS R&D—about 80 are US firms; Japan is the second major player. According to the Department of Defense, the MEMS market was $1 billion in 1994; projections for the year 2000 range from $8 to 14 billion. The technology cuts across a number of manufacturing industries, including sensors, industrial and residential controls, electronic components, computer peripherals, automotive and aerospace electronics, analytical instruments, and office equipment. The industry list suggests a potential for generating a large volume of technology spillovers.

(2) MEMS exhibits characteristics of an emerging technology involving substantial university-industry interactions. About 30 universities and government labs are actively pursuing MEMS technologies. U.S. industry investment in MEMS so far has been fairly modest (about $120 million in 1995). In contrast, in the same year federal R&D support of MEMS was a large component (about $35 million), $30 million of which came from the Department of Defense (mainly Defense Advanced Research Projects, or DARPA).[29] The National Science Foundation's MEMS support was $3 million. Labs contributed about $2 million. Since 1989 the NSF has sponsored 124 MEMS-related projects at 61 organizations (mainly universities), with funding of about $25 million. About $1.4 million consists of Small Business Innovation Research (SBIR) grants.[30]

MEMS Patents and the MEMS R&D Network

Our analysis of MEMS starts with a core database of about 1,200 MEMS patents which we identified beginning with a short list of key inventors and federally-funded MEMS projects. Citations to these initial patents were used to identify additional MEMS candidate patents. Each candidate patent abstract and exemplary claims was read to ensure that the patent was a MEMS technology.[31] This analysis identifies a set of core MEMS R&D organizations which was used to construct the MEMS network. Our analysis identifies the MEMS R&D network by technology (patent class), organization (assignee), and region.

We find that MEMS is highly geographically concentrated in five top U.S. regions (San Francisco, Boston, Los Angeles, New York–New Jersey, and Chicago), and that universities and government labs play a major role. MEMS R&D spillovers disproportionately involve interactions among these higher-order regions, of which Cleveland is not even in the top twenty-five. There is no evidence that Cleveland's existing R&D network interacts in an important way with the dominant MEMS regions. Given the cumulative nature of new technology development, this condition suggests that Cleveland would be in an increasingly disadvantageous position, particularly with respect to the absence of critical mass and presence of a weak R&D network for capturing MEMS spillovers.

Despite this condition, as discussed earlier, Ohio is making a major investment in MEMS. If our hypothesis is correct, then left to the marketplace, Ohio's investments in MEMS will most likely create R&D spillovers for the dominant MEMS regions. However, Ohio can increase the odds of capturing the benefits of its investments if its intermediaries succeed in building a partnership of industry R&D labs with a stake in the development of MEMS technology—or at least a strong specific (niche) MEMS R&D network. The broader question is: What strategy will maximize the local payoff from an older industrial region's investments in MEMS?

A Policy Framework to Maximize Regional Economic Benefits from University-Industry Interactions

This section develops a framework for examining the regional implications of science and technology investment decisions. MEMS provides a particularly interesting case because it is an important enabling technology with a potentially large market and significant involvement of university and government labs, coupled with high geographic concentration of R&D networks in Japan, San Francisco, and Boston. MEMS may represent a case where S&T investments that strengthen university-industry relationships yield a high social rate of return for the country but a highly problematic return for Ohio. Under what circumstances can Cleveland expect its investments in MEMS to move the region from the left to the right side of X in Figure 1?

S&T intermediaries are necessary but not sufficient in regionalizing a technology. Our case studies suggest several observations, reflected in Figure 4. Do intermediaries compete or collaborate as a system? Do intermediaries work for individual firms or the region? First, it is crystal clear that intermediaries do not work as a system. With few exceptions, they are competitive rather than cooperative organizations. Most obvious, their core budgets are not predictable and, therefore, they compete aggressively for funding from various sources, whatever the implications for local benefits.

Second, despite their origins, the intermediaries do not have clear economic development objectives. This ambiguous purpose is underscored by an emphasis on individual firms as clients—S&T intermediaries operating more as private consultants than as economic development institutions.[32] Decision-makers typically think, "If I work with a firm, then it's economic development!"[33]

Third, as is evident from the three case studies, the wide mix of participants, coupled with an absence of clear goals, produces highly conflicted objectives—especially concerning the geography of benefits. The federal government's objective is to commercialize for the United States, the state's objective is to commercialize for Ohio, and Cleveland's is to commercialize for Cleveland.

Finally, S&T intermediaries operate with a nearly universal absence of any clear, consistent, established link between the services they provide and regional economic development (Fogarty and Lerner 1993). The task of uncovering such a link is not simple, because intermediaries provide a wide range of services to companies (including research support, technology development, training assistance, commercialization, and technology deployment) and the evaluation methodologies for each service are inadequately developed.

Moreover, time lags present a special problem because the current process emphasizes early payoff. Estimates of time lags

Client Focus

Minimum Economic Impact

Operational Approach	Individual Firm	Regional Economy
Individual Intermediary	Competitive intermediaries targeting individual firms	Competitive intermediaries targeting industry clusters
System of Intermediaries	Collaborative intermediaries targeting individual firms	Collaborative intermediaries targeting industry clusters

Maximum Economic Impact

Figure 4 Maximizing Economic Benefits Requires New Incentives to Encourage Cooperation and a Regional Focus.

between basic research and economic impact range from five to fifteen years.[34] The time frame depends on the mechanism for turning research into productivity. For example, if the effect occurs when graduates take jobs with local companies, then it will take the six or seven years required to earn a science or engineering doctorate; if it occurs through faculty consulting with local firms, the lag may be considerably less. In the latter case, timing hinges on how long it takes for the research to reach a point where commercialization opportunities exist. If a university technology is transferred through patent licensing to a local company, it may take longer and require additional industry R&D to commercialize the technology (Pressman et al. 1995).

We examined the academic and policy literature for lessons and models that could guide investment in S&T infrastructure and provide a framework for making decisions like Cleveland's MEMS, and found that it yielded few insights. It is extensive but highly fragmented on many important topics that bear on the effectiveness of university-industry relationships as a strategy for improving older industrial regions. The topics range widely: cooperation, entrepreneurship, seed and venture capital, university research, technology transfer, licensing, research parks, market assessment of technology, and metrics. However, the literature provides very little help on a number of critical issues, such as: conflicting objectives among university, state, federal, and industry partners; effects of funding level, source, and predictability on behavior of intermediaries with respect to regional objectives; incentives affecting the intermediary decision-making process; appropriate scale of investment; the process of identifying a state or region's strengths; the relationship of technology to other factors, such as education; managing a local S&T system; and evaluation metrics focused on a regional economy.[35]

Maximizing Economic Benefits from R&D Spillovers by Building UI Paths

If state and local policy makers cannot trace the effects of S&T investments to better technology incorporated into local industries or to a better-educated regional workforce, then there does not exist an economic development justification for the investments. The Cleveland region will reap the economic benefits of its investments in academic research only if (a) there is a strong path between the research and local industry and (b) it organizes to capture the benefits (Jaffe, Trajtenberg, and Henderson 1993; Jaffe and Trajtenberg 1997; Jaffe, Fogarty, and Banks 1998). This section proposes a framework for considering S&T investments to increase commercialization and raise the odds of capturing local economic benefits.

The S&T System's Four Components

Figure 5 represents our adaptation of Don Stokes' "Pasteur's Quadrant" framework to regional science and technology systems (Stokes 1997). The system is represented by four major components that describe a region's interactions and UI paths. It should be noted that the interactions are not limited to the region. Interactions among the four parts shape the development and commercialization of new technology as well as the geography of

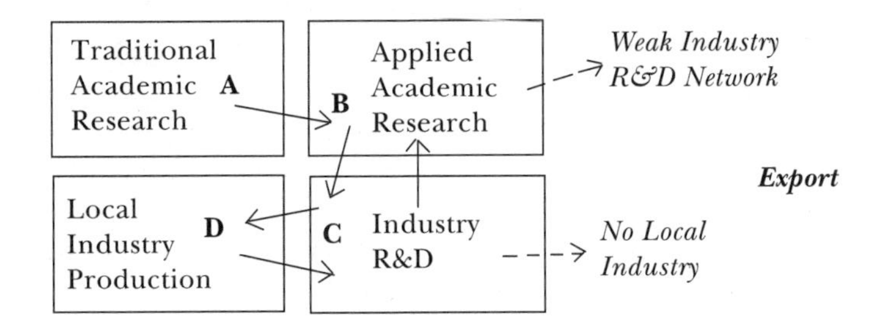

Figure 5 Maximizing Local Spillovers from University-Industry Relationships Requires Science and Technology Investments That Build Paths from Research to the Region's Industries.

economic benefits. We can characterize the four S&T capabilities and the paths that connect them. The top two quadrants are housed in universities while the bottom two quadrants belong to industry. (To keep the picture simple, research hospitals and government R&D labs are excluded, although both are important sources of new technology and play important roles in the Cleveland region.)

Although the sharp division shown in the chart between traditional basic research and interactive research does not in fact exist, it is helpful to draw this distinction for several reasons. First, the quality of a university's applied research leading to significant commercialization opportunities ultimately depends on the quality of basic research. Most of a region's attention is drawn to the applied areas of research because the products are more visible and thus closer to having commercial value. It is in this quadrant that we find university-industry research centers (UIRCs), industry funding of university research, technology transfer efforts, and patents (Cohen, Florida, and Goe 1994). From the outsider's perspective, this is where the action happens. State and local policy makers commonly focus on the tip of the iceberg—the most obvious emergent university technologies, such as FES, PDLC, or MEMS (that is, the potential new products and processes, or a start-up company).

In reality, there exist multiple interactions between universities and industry that affect industry R&D and performance: graduates take jobs, faculty consult, companies fund academic research and serve on boards of UIRCs, and academic papers influence corporate invention. For example, a recent study of citations to academic papers on corporate patents shows increasingly close links between basic science and corporate patenting (Narin, Hamilton, and Olivastro 1996).

Given the visibility of university-industry interactions involving specific technologies, it is easy for funding decisions to give greater weight to activities located in the second quadrant while neglecting earlier-stage basic research investments. For example, an important question is: Does Ohio's investment in MEMS include sufficient support for quadrant 1 to lay a foundation for future technologies related to MEMS so that MEMS is not just a one-shot technology?

Second, sources of funding for each quadrant differ and affect outcomes. Most direct funding for traditional basic research comes from the federal government. In the case of public institutions, states fund basic research by paying for buildings, equipment, chaired professorships, and graduate programs. This situation highlights the federal government's role as primary funder of basic research and the necessity for states to separately evaluate implications of federal R&D policy. For example, roughly two-thirds of the growth in federal research support to universities over the past decade has supported research in the medical and biological sciences (Bania et al. 1991). Research funding by the National Institutes of Health (NIH) heavily influenced the priority Cleveland has given to cultivating a local biomedical instruments and devices industry.

The bottom two quadrants belong to industry—the lower right-hand component is industry R&D (which is about $2 billion in the Cleveland region). Localized R&D spillovers from patents involve interactions between the applied university research quadrant and local industry R&D. For the most part, state and local decisions concerning investments in university research programs largely overlook the region's industry R&D network. Is the local network strong enough to capture R&D spillovers? For Cleveland's PDLC technology, the external university-industry interaction component was very high while the local component was zero. (As shown in the previous section for nine prominent U.S. metropolitan regions, the bulk of local industry's UI is external—a region's companies draw mostly from universities located elsewhere.)

The second industry component represents industry production. This is the local destination for technology. It is essential to ask: Is there a path from industry R&D to local production? If either or both of industry's quadrants are missing or weak, the possibilities are very limited for localization of R&D spillovers that will affect long-run economic performance. The point about local production is simply that regional productivity gains from local S&T investments require that the technology be incorporated into local production facilities. The presence of an industry cluster connected to the region's R&D substantially raises the odds that R&D spillovers will produce long-run economic benefits. National economic benefits occur if the geographic association of a complete path from research to industry production increases the speed of commercialization and diffusion, causing increasing returns, not just of R&D but also of industry output.

The Importance of Investing in the System

Seeing these components as part of a highly interactive system is essential for developing sound science and technology policy for older industrial regions. The framework suggests why local knowledge and management of the system matters. For example:

The second quadrant (applied, interactive academic) may produce important technology but its effects can be short-lived. This may occur if the first quadrant (traditional basic research) is weak, eventually causing quadrant 2 to dry up as an important source of innovation.[36] It is important to ask: Does the plan include sufficient investment in basic research?

The first quadrant may produce nationally competitive basic research but, because quadrant 2 is weak, basic research findings produce little of potential commercial value locally. The results of the region's basic research may be used by institutions in other regions. However, it is possible that this may be partially mitigated if knowledge is transferred by graduates taking positions in the region or through faculty consulting with local industry.

The two university quadrants could produce excellent research but the technology is exported because (a) there is no connected industry R&D network, or the R&D network is weak, as with PDLC, and/or (b) there is no local industry destination for the technology. This should cause local decision makers to ask: Do university-industry interactions associated with specific S&T investments contain a strong link to local R&D networks?

Given this framework, why would Ohio and Cleveland think MEMS is a good investment? What assets does Cleveland have that suggest MEMS could be a good investment? Among the reasons are the presence of: (1) one major university MEMS inventor with an entrepreneurial approach; (2) actively engaged intermediaries, including the statewide MEMS-Net and state support; (3) NASA's commitment to the partnership and funding of MEMS research through the Glennan Microsystems Initiative (see http://www.glennan.org) ; (4) several local companies' expressing interest in being partners; and (5) a large local industry customer base.

While important questions are unanswered, the region's selection of MEMS is at least loosely guided by the innovation system framework. In particular, the region has worked to develop an R&D network based on a government-university-industry partnership, suggesting that several lessons may have been gleaned from the PDLC experience. The R&D partnership is critical because local companies were unwilling to undertake the R&D individually; however, they appear to be more enthusiastic about joining a partnership. The strategy appears to consist of investing in research capacity, using the intermediary system to cultivate a MEMS R&D network, and working extraordinarily hard with local industry to commercialize and capture the technology.

Conclusions

Using university-industry relationships, older industrial regions are working hard to generate and capture R&D spillovers, increase commercialization, and convert new technology into local economic benefits. Our findings based on our three case studies and a systems analysis of university-industry interactions are consistent with Figure 1: A pattern of increasing returns and the system of R&D spillover more specifically favors "knowledge" regions. The

FES, PDLC, and MEMS case studies clearly reveal the difficulty that one older industrial region faces in getting R&D spillovers to remain in the region. Systems analysis of university-industry relationships using patents shows that regions like Cleveland operate at a significant disadvantage: They possess weak R&D networks that draw very modestly from university R&D spillovers, whether local or external to the region. In fact, the more influential the technology, the quicker knowledge get diffused across the world's R&D networks and regions. We believe that other older industrial regions are confronted with the same disadvantage (see Detroit, Chicago, Philadelphia, and Baltimore–Washington DC in Figure 2.)

The Cleveland region's approach to MEMS loosely approximates our framework. Despite this, the MEMS partnership does not solve underlying problems, such as conflicting objectives, mixed incentives to commercialize locally, or lack of a framework to manage the local innovation system. How can such an older industrial region increase the likelihood that S&T investments will produce long-term economic benefits for the region? While improving understanding of the economics of regional innovation systems is integral to success, the biggest challenge will be to develop a well-managed S&T infrastructure—one that invests in the best UI paths from basic research to industry. That S&T intermediaries are necessary institutions but clearly not sufficient, is amply illustrated by the three case studies.[37] At the strategic level, both Cleveland's Technology Leadership Council and the Ohio Science and Technology Commission exist to create a setting for considering the long-run consequences of S&T investments. The idea is to compensate at least partially for politics' short time-horizons. Unfortunately, the payoff from S&T investments in the form of higher productivity and per capita income occurs too late for most officeholders to get credit, so an alternative management structure makes sense even though it is still very difficult.

Decisions affecting the status of each component of Figure 4 are determined by the interplay of many autonomous institutions operating at all three levels—federal, state, and local. One approach to consider would be to create an advisory board whose purpose it is to gather information regularly about the four quadrants using new case studies and a systems analysis of the region's R&D networks based on patents. The board must be highly respected for its knowledge, integrity and objectivity, and long-run perspective. One purpose would be to foster discussion of conflicting objectives. The new board should also look for opportunities to make local objectives consistent with national objectives. Local university research must be nationally competitive for R&D spillover to produce important local benefits.

S&T investments should build the best UI paths from basic research to industry. Most decisions affecting a region's capabilities tend to be piecemeal. For example, the Ohio Board of Regents' decision to dramatically increase investment in computer science and MEMS and the Technology Leadership Council's creation of a Corporate Technology Mining program and a new research park are essentially separate initiatives. One reason interventions are piecemeal is that no one has developed sufficient knowledge of the complex ways in which the system incorporates university-industry relationships. An important step is to commit to developing that knowledge as a foundation for better decisions. One approach is to use in-depth case studies of interventions such as FES, PDLC and MEMS as an educational device widely shared and used to raise issues, clarify goals, and encourage mutual learning among decision makers.

The board should also formally analyze R&D networks so that the region bases its investments on knowledge of its own networks. A competitive S&T system implies both strong internal capabilities and paths and strong external networks. For the country's top R&D centers, internal and external university-industry interactions go hand-in-hand; regions with strong external UI R&D networks also tend to have strong local UI networks. Ideally, the region develops and utilizes technology originating locally and draws on the world's best sources of technology for local industry. One purpose of S&T intermediaries should be to use university-industry relationships to foster both local as well as external learning (Lynn, Aram, and Reddy 1997). Universities play a unique role as conduits for knowledge spillovers, which if connected through UI to local industries will confer both state and local benefits.

The board should also consider developing a formal decision tool to weigh the complex set of factors that underlie an investment decision in a specific technology. The tool would be quantitative and qualitative, drawing on experts in specific areas of technology and the systems analysis of R&D networks to weigh the potential of a technology, the region's local capability in that technology, and its human as well as production capacity against its potential impact. Together these components would provide estimates of expected economic benefit, which could then be compared with expected costs.

Finally, as is evident from the three case studies, the federal government plays a critical role in shaping the future of the country's older industrial regions. Federal policy should explicitly incorporate national as well as state and local objectives, even if the goal is only to build a strong national innovation system. Equally important, states have an opportunity to shape their future economic and social health through investments that utilize university-industry relationships. However, given their unique circumstances, they cannot generalize from the experience of Route 128 and Silicon Valley. To overcome a disadvantaged status, they must take responsibility for developing knowledge of their part of the innovation system. This knowledge is an essential foundation for guiding investments in science and technology. Each state is choosing its future by the weight it gives these issues.

Notes

1. Two types of R&D spillovers are particularly relevant here: (1) "knowledge spillovers"—these occur because knowledge created by one firm is typically not contained within that firm, and thereby creates value for other firms and other firms' customers; and (2) "network spillovers"—these result from the profitability of a set of interrelated and interdependent technologies that may depend on achieving a critical mass of success so that each firm pursuing one or more of these related technologies creates economic benefits for other firms and their customers (see Jaffe 1996). Griliches (1992, Table 1) summarizes the R&D spillover evidence: "R&D spillovers are present, their magnitude may be quite large, and social rates of return remain significantly above private rates."

2. Federal policy could explicitly choose to leverage its R&D funding by favoring the R&D projects and institutions in regions with clearly strong increasing returns to R&D, such as Silicon Valley. The argument would be that other regions would then gain by building new technology from a larger pool of R&D spillover benefits.

3. Industrial consulting is an important mechanism for transferring the benefits of university research to firms. By one estimate, firms in high-technology indus-

tries spend an additional 20% on academic consulting above industry support of university research (see Mansfield 1996).

4. Investing in S&T infrastructure was one of several major initiatives, which included major downtown investment, manufacturing, and neighborhoods. See Harvard Business School (1996).

5. For a discussion and national assessment of Ohio's Thomas Edison Centers, see NRC 1990.

6. For a very interesting discussion of social capital in support of innovation, see Fountain (1998: 112–142).

7. NASA-Lewis' research is focused on new propulsion, power, and communications technologies. It has special capabilities in the microgravity science disciplines of fluid physics, combustion science, and some materials science. See <http://www.lerc.nasa.gov/WWW/PAO/aboutlew.html>.

8. See <http://alcom.kent.edu/ALCOM/ALCOM.html>.

9. See <http://www.regents.state.oh.us/rsch/rschsupport.html>.

10. See <http://www.oai.org/CCRP.html>.

11. See <http://mems.eeap.cwru.edu/memsnet>.

12. See <http://www.ncmr.cwru.edu/about/whoweare.html>.

13. These intermediary institutions are equivalent to what Harvey Brooks calls "buffer institutions" (Brooks and Randazzese 1998: 383–386).

14. To use the terminology of externalities, each firm's performance depends on the performance of an industry cluster (industry neighbors). But individual firms ignore the effect of their investment decisions on the health of the industry cluster neighbors. This interdependence is increasingly based on an industry's knowledge networks, a large part of which is regional. These networks become a major source of productivity growth and competitiveness. If individual decisions are predominantly to disinvest, the market undermines the competitiveness of the industry cluster. The outcome can be cumulative disinvestment. Both "macro" and "micro" (operational) intermediaries are necessary to offset the market failure.

15. The Technology Leadership Council has no authority to manage anything but its membership. However, because its members are CEOs of very large companies, with close connections to government at all levels, its influence on funding of the region's science and technology infrastructure and operation of intermediaries is extensive. In fact, the TLC created most intermediaries.

16. The 7.5% unemployment rate is the 1986 Cleveland PMSA rate. The Akron MSA unemployment rate for that year was 8.0%.

17. According to Denison's estimates, approximately half of the growth rate in real GDP per capita from 1929 to 1982 was due to education plus R&D. (Denison 1985).

18. Most of the estimated $2 billion industry R&D is unaffected by government-funded defense work. Although we do not have local data on DOD support of the

region's industry R&D, the statewide figure was 16% in 1995. This amount would include R&D associated with Wright Patterson Air Force Base in Dayton. The percentage for the Cleveland region is likely to be less than 16%.

19. See Jaffe, Trajtenberg, and Henderson (1993) for a more thorough discussion of patents and the use of patent citations to measure knowledge flows.

20. For a more thorough discussion of our fuzzy R&D spillover systems methodology, see Fogarty, Sinha, and Jaffe (1998).

21. For example, see Jaffe, Trajtenberg, and Henderson (1993: 577–598).

22. The CorpTech database is a directory of about 50,000 high-tech companies provided by Corporate Technology Services, Inc. The file used here is for 1996. CorpTech estimates that the file contains 99% of companies over 1,000 employees, 75% of companies with 250–1,000 employees, and 65% of companies with fewer than 250 employees.

23. MEMS patents were analyzed by Dave Hochfelder, who holds a Masters degree in electrical engineering and the Ph.D. from CWRU's History of Technology program.

24. The study uses the comprehensive patent database developed by NBER and Case Western Reserve University, with support from the National Science Foundation. The specific variables are: 400 patent classes; name of the organization to which the patent is assigned; the inventor(s) with address, allowing for identification of R&D location, including county; claims; country; type of organization (corporation, university, hospital, government lab, or independent inventor); and cited patents and other cited sources, such as academic publications.

25. New York–Northern New Jersey–Long Island; San Francisco–Oakland–San Jose; Los Angeles–Riverside–Orange County; Boston–Worcester–Lawrence; Chicago–Gary–Kenosha; Detroit–Ann Arbor–Flint; Philadelphia–Wilmington–Atlantic City; Washington–Baltimore; and Cleveland–Akron CMSAs.

26. UI firms may be members of more complex R&D networks that form agglomerations of highly interacting companies, universities, and government laboratories.

27. This conclusion holds for five technologies analyzed in a separate study for the Public Policy Institute of California: aerospace, information technology, automotive, advanced materials, and biomedical devices.

28. For a description of MEMS, see Department of Defense (1995). For an interesting description of MEMS, see *Discover* (March 1998). MEMS combines computation, sensing, and actuation with miniaturization to make mechanical and electrical components. The bulk of applications are pressure sensors, optical switching, inertial sensors, fluid regulation and control, and mass data storage. Our systems analysis of MEMS technologies ranks the United States as first, with Japan second, followed by Germany, France, and Great Britain. Ranks are based on each country's systems value as a MEMS source (i.e., our fuzzy estimate of each country's contribution to MEMS technologies).

29. Our research shows that DARPA presently funds 62 projects at 48 organizations (17 universities, 5 government labs, 18 large companies, and 8 small firms). DARPA is currently funding 5 SBIR projects at 4 companies; they previously funded an additional 5 SBIR projects. The Army has funded 17 MEMS-related projects at 14 firms through its SBIR program. The projects amount to nearly $2 million. NASA has sponsored 20 MEMS-related SBIR projects. (No dollar amount was available.) However, the MEMS working group at NASA-Lewis in Cleveland supported $2.5 million MEMS R&D by 17 S&Es. Moreover, Ohio MEMS-Net has funded $2.4 million for capital investments in 1995 and 1996.

30. Ranked by total NSF support of MEMS projects, the top ten institutions include: Stanford, UC Berkeley, University of Michigan, Cornell, University of Utah, University of Pennsylvania, University of Illinois Chicago, Case Western Reserve University, University of Minnesota, and University of Hawaii. Most of the MEMS university projects are associated with fairly extensive patenting. The 61 MEMS universities currently account for 312 MEMS patents.

31. The analysis of MEMS patents was partly supported by the Advanced Technology Program, National Institute of Standards through a grant to the National Bureau of Economic Research. The top five MEMS technologies are: Semiconductor Device Manufacturing Processes; Metal Working; Electricity, including Electrical Systems & Devices; Incremental Printing of Symbolic Information; and Optics Systems, including Communication and Elements.

32. A shift to the regional economy as a client requires: intermediaries that become experts on knowledge networks within industry clusters, a focus on the "right" firms, evaluation metrics that reward activities leading to long-term economic impact through improvements in productivity, and the right balance between government- and firm-based funding.

33. The percentage of an intermediary's budget coming from industry is usually considered the single best measure of intermediary performance. Carried to an extreme, where fees represent 100% of the budget, the intermediary is a purely private consulting firm.

34. See Mansfield (1996) and Adams (1990). Adams estimates a lagged effect of up to fifteen years. Jaffe (1996) estimates a lag of six years from R&D spillover to expected effects on Massachusetts' Gross State Product.

35. For good examples of the literature, see the following: Anderson (1994); Kassicieh and Radosevich (1994: 183–194); Carr (1992, 1995); Lynn, Aram, and Reddy (1997: 129–145); Lynn, Reddy, and Aram (1996: 91–106); Reddy, Aram, and Lynn (1991: 295–304); and Fogarty and Lerner (1993).

36. Our terminology and use of quadrants differs from Stokes' framework. Stokes develops a 2x2 matrix whose two dimensions are: Quest for Fundamental Understanding (yes, no) and Consideration of Use (no, yes). For example, Stokes' quadrant 1, which he calls "Bohr's Quadrant," contains "pure basic research." His quadrant 2 is "Pasteur's Quadrant," consisting of "use-inspired basic research." Quadrant 3 is "Edison's Quadrant." This is purely applied research.

37. A shift to a system of intermediaries requires agreement on a mission, more coordination and focus to gain critical mass and greater expertise, shared resources with less duplication of effort, increased communication and knowledge flow, evaluation metrics that reward collaboration, and a better-informed state partner.

References

Adams, James D. 1990. "Fundamental Stocks of Knowledge and Productivity Growth," *Journal of Political Economy* 98: 673–702.

Anderson, Lawrence K. 1994. "Technology Transfer from Federal Labs: The Role of Intermediaries," in S.K. Kassicieh and H.R. Radosevich, eds., *From Lab to Market: Commercialization of Public Sector Technology.* New York: Plenum Press.

Bania, Neil, Michael S. Fogarty, Mohan Reddy, and Michael Ginzberg. 1991. "University Research Performance." Report. Cleveland: Center for Regional Economic Issues, Case Western Reserve University.

Baumol, William J., Sue Anne Batey Blackman, and Edward N. Wolff. 1989. *Productivity and American Leadership: The Long View.* Cambridge, MA: MIT Press.

Brooks, Harvey and Lucien P. Randazzese. 1998. "University-Industry Relations: The Next Four Years and Beyond," in Lewis M. Branscomb and James Keller, eds., *Investing in Innovation: Creating a Research and Innovation Policy that Works.* Cambridge, MA: MIT Press.

Carr, Robert K. 1992. "Doing Technology Transfer in Federal Laboratories: A Survey of Federal Laboratories and Research Universities."<http//millkern.com/rkcarr/flpart1.html>.

Carr, Robert K. 1995. "Measurement and Evaluation of Federal Technology Transfer." Proceedings of the 20th Annual Meeting of the Technology Transfer Society, Washington, DC (July).

Coburn, Christopher. 1995. "Partnership: A Compendium of State and Federal Cooperative Technology Programs." Columbus: Battelle Memorial Institute.

Cohen, Wesley M., Richard Florida, and Richard W. Goe. 1994. "University-Industry Research Centers in the United States." Pittsburgh: Carnegie Mellon University.

Denison, Edward F. 1985. *Trends in American Economic Growth, 1929–82.* Washington, DC: The Brookings Institution.

Department of Defense. 1995. Microelectromechanical Systems: A DOD Dual Use Technology Industrial Assessment (Final Report)." Washington, DC: U.S. Department of Defense.

Fogarty, Michael S. 1983. "The Decline of the Metropolitan Economy: A New Interpretation." *Urban Studies* 20: 495–497.

Fogarty, Michael S. 1998. "Cleveland's Emerging Economy: A Framework for Investing in Education, Science, and Technology." Working paper. Cleveland: Center for Regional Economic Issues, Case Western Reserve University.

Fogarty, Michael S., Neil Bania, Mohan Reddy, and Michael Ginzberg. 1991. "University Research Performance." Cleveland: Center for Regional Economic Issues, Case Western Reserve University.

Fogarty, Michael S., and Joshua Lerner. 1993. "An Evaluation Methodology for the Edison Technology Centers." Cleveland: Center for Regional Economic Issues, Case Western Reserve University.

Fogarty, Michael S., Amit K. Sinha, and Adam B. Jaffe. 1998. "ATP and the U.S. Innovation System: A Preliminary Report." Report. NIST Advanced Technology Program.

Fountain, Jane E. 1998. "Social Capital: A Key Enabler of Innovation," in Lewis M. Branscomb and James Keller, eds., *Investing in Innovation: Creating a Research and Innovation Policy that Works.* Cambridge MA: MIT Press.

Gottlieb, Paul D. and William T. Bogart, 1997. "The Downtown's Economic Revival: Cleveland's Recent Success and Next Steps," Paper prepared for *Cleveland Tomorrow* as part of the planning process for Civic Vision 2000 and Beyond. (August).

Griliches, Zvi. 1992. "The Search for R&D Spillovers." *Scandinavian Journal of Economics,* Supplement, 29–47.

Harvard Business School. 1996. "Leadership in Action." Boston: Harvard Business School.

Henderson, Rebecca, Adam Jaffe, and Manuel Trajtenberg. 1995. "Universities as a Source of Commercial Technology: A Detailed Analysis of University Patenting, 1965–88." NBER Working Paper No. 5068 (March).

Hoshiko, Jim. 1994. "Polymer Displaced Liquid Crystals: A Case Study." Cleveland: Center for Regional Economic Issues, Case Western Reserve University.

Jaffe, Adam B. 1989. "Real Effects of Academic Research." *The American Economic Review* (December): 957–970.

Jaffe, Adam B. 1996. "Economic Analysis of Research Spillovers: Implications for the Advanced Technology Program." Economic Assessment Office, The Advanced Technology Program, National Institutes of Standards and Technology, U.S. Department of Commerce (November).

Jaffe, Adam B., Michael S. Fogarty, and Bruce Banks. 1998. "Evidence from Patents and Patent Citations on the Impact of NASA and other Federal Labs on Commercial Innovation." *Journal of Industrial Economics* XVLI(2): 183–205.

Jaffe, Adam B., and Manuel Trajtenberg. 1997. "Knowledge Flows across Time and Space as Evidenced by Patent Citations." Paper presented at NBER Science and Technology Policy Meeting, July 14.

Jaffe, Adam B., M. Trajtenberg, and R. Henderson. 1993. "Geographic Localization of Knowledge Spillovers as Evidenced by Patent Citations." *Quarterly Journal of Economics* CVIII(3): 577–598.

Kassicieh, S. K. and H. R. Radosevich, eds. 1994. *From Lab to Market: Commercialization of Public Sector Technology.* New York: Plenum Press.

Lynn, Leonard H., John D. Aram, and N. Mohan Reddy. 1997. "Technology Communities and Innovation Communities." *Journal of Engineering and Technology Management* 14: 129–145.

Lynn, Leonard H., Mohan Reddy, and John D. Aram. 1996. "Linking Technology and Institutions: The Innovation Community Framework. *Research Policy* 25: 91–106.

Mansfield, Edwin. 1991. "Academic Research and Industrial Innovation." *Research Policy* 20: 1–12.

Mansfield, E. 1995. "Academic Research Underlying Industrial Innovations: Sources, Characteristics, and Financing." *The Review of Economics and Statistics* (February).

Mansfield, Edwin. 1996. "Industry-University R&D Linkages and Technological Innovation." Paper presented at the annual meeting of the American Economic Association, January.

Narin, Francis, Kimberly Hamilton, and Dominic Olivastro. 1996. "Linkage Between Agency-Supported Research and Patented Industrial Technology." Paper presented at the National Bureau of Economic Research (March 25).

NRC. 1990. National Research Council, Commission on Engineering and Technical Systems, "Ohio's Thomas Edison Centers: A 1990 Review." Washington, D.C.: National Academic Press.

OTA. 1986. Office of Technology Assessment. "Research Funding as an Investment: Can We Measure the Returns?" Washington, DC: Office of Technology Assessment.

Pressman, Lori et al. 1995. "Pre-Production Investment and Jobs Induced by MIT Exclusive Patent Licenses: A Preliminary Model to Measure the Economic Impact of University Licensing." Cambridge, MA: MIT Technology Licensing Office.

Rae, Linda. 1994. "The Transfer of Functional Electrical Stimulation Technology From University to Industry: A Case Study." Cleveland: Center for Regional Economic Issues, Case Western Reserve University.

Reddy, N. Mohan, John D. Aram, and Leonard H. Lynn. 1991. "The Institutional Domain of Technology Diffusion." *Journal of Product Innovation Management* 8: 295–304.

Stokes, Donald E. 1997. *Pasteur's Quadrant: Basic Science and Technological Innovation.* Washington, DC: The Brookings Institution.

19

The Regional Economic Impact of Public Research Funding: A Case Study of Massachusetts

Amy B. Candell and Adam B. Jaffe

Introduction

Most of the long-run economic effects of research activities are indirect and difficult to quantify. Nonetheless, there is a substantial body of research that identifies and quantifies or partially quantifies important pathways by which government research support leads to increased innovation, birth of new firms, creation of jobs, and increases in sales and profits for high-technology firms. The available research varies in the degree to which these effects are analyzed in systematic and quantified ways. At one end of a continuum is qualitative and anecdotal information about the role of research institutions and their interactions with other parts of the economy; at the other are quantitative statistical estimates of the ultimate economic impact of particular kinds of research activities. Between these two extremes lies a variety of quantitative data that cover certain aspects of the impact of research activities, but which cannot be translated directly into economic measures such as sales or employment.

This paper attempts to bring together these different forms of empirical information and explores their implications in the context of a quantitative model of regional economic development. The result is a set of quantitative estimates of the effect of public research on employment and income in Massachusetts. Many judgments and assumptions were needed in carrying out this exercise; the results should be viewed as illustrative of the measurement of the effects of public spending on research, without undue weight being placed on the precise quantification described.

Because the interconnections among research institutions and firms are complex and dynamic, the effect of public funding cannot be quantified in the abstract. Rather, we model the impact of public funding by examining a very specific funding scenario, and compare economic growth under that scenario to a baseline scenario. The funding scenario we examine was constructed originally in the context of political concerns in the state that pressures in the U.S. Congress for a balanced federal budget would lead to significant funding cuts for Massachusetts. We believe, however, that this scenario—embodying significant cuts in research spending of about 35% in real terms over five years—is a useful illustration of the magnitude of impacts associated with public research support. Although cuts of this magnitude were embodied in the 1995 Concurrent Budget Resolution, such a drastic reduction in funding has not been enacted.[1]

In order to quantify the impact of these reductions on the economy, one must postulate what the level of spending would have been under the baseline scenario. For the baseline scenario, we used the recent spending level extrapolated into the future based on an existing economic forecast. The baseline forecast has civilian federal R&D growing slightly in real terms and defense-related R&D declining in real terms, resulting in an overall baseline forecast for federal research support in Massachusetts which is very close to a constant level after adjustment for inflation.

The model underlying our analysis is illustrated in Figure 1. The public sector provides research funding both to nonprofit entities such as universities and hospitals, and to for-profit firms. The regional economic impacts of this funding operate through three primary mechanisms:

- a direct demand or Keynesian multiplier effect, whereby the spending of federal dollars in the state stimulates the local economy, in the same way that federal spending on highways or schools would;
- an induced demand effect, which occurs because spillovers of research results from the nonprofit sector induce additional spending by firms on research to exploit commercially the fruits of public-sector research; and

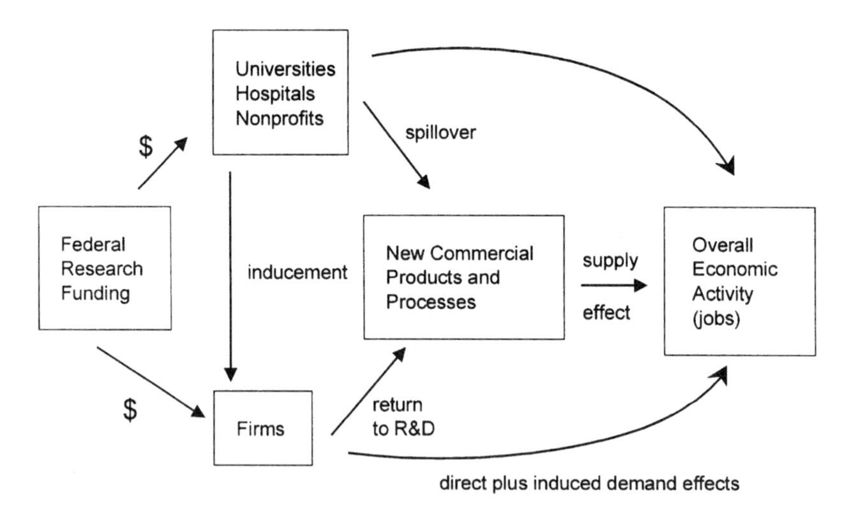

Figure 1 The Innovation System.

• a supply or productivity effect derived from the new products and processes brought about by the combined effects of public and induced private research.

The paper provides a quantitative overview of these different elements of the Massachusetts innovation system, then describes what is known about the linkages among the different components of the system illustrated in Figure 1 and presents the results of an exercise in which existing empirical evidence about these linkages is combined with a regional economic model in order to estimate the consequences of changes in federal research funding. Finally, we draw some conclusions from our study and point out their implications for Massachusetts state policy.

The Role of Science and Technology in the Massachusetts Economy

The research-intensive sectors of the Massachusetts economy are shown in Table 1. For most of our analysis, we will distinguish what can be broadly labeled the nonprofit research sector (including universities, hospitals, and other nonprofit research laboratories) from the for-profit research sector, principally comprising firms in high-technology industries.

Table 1 R&D-Intensive Sectors in Massachusetts

Industry	1994 Employment (000s)	% of Private Sector Employment	1993 Total R&D Expenditures (Millions)*
Computer Equipment	28.1	1.1%	$2,095
Engineering & Research Services	98.7	3.9%	1,964 **
Colleges & Universities	70.9	2.8%	1,076
Communication Equipment	15.8	0.6%	876
Electronic Components	25.4	1.0%	680
Hospitals & Other Medical Research Institutions	133.3	5.3%	628
Aircraft & Missiles	18.3	0.7%	568
Optical, Surgical, Photographic & Other Instruments	24.8	1.0%	294
Measuring Instruments	22.5	0.9%	230
Chemicals & Allied Products (excluding Drugs)	13.0	0.5%	204
Drugs	4.9	0.2%	57
Total	455.7	18.1%	$8,673

*ERG estimates for 1993 from latest available data.
**Includes non-manufacturing industrial R&D, MIT Lincoln Laboratories, and non-medical, not-for-profit research institutes.
Source: NSF, Massachusetts Dept. of Employment and Training.

In the nonprofit sector, universities and colleges, hospitals, and other nonprofit organizations are both major performers of research and key employers of researchers. Approximately 120 Massachusetts colleges and universities employ over 70,000 people, nearly 3% of total Massachusetts employment. In 1993, 31 of these institutions performed $770 million in federally funded research. Hospitals and associated medical laboratories are also significant employers and performers of research. In 1993, for example, it is estimated that hospitals conducted over $600 million in research, sponsored both by the federal government and by industry and other sources. Other nonprofit research organizations such as Draper Laboratory, MIT-affiliated Lincoln Laboratory, and MITRE Corporation conduct research, much of it financed by the federal government.[2]

In the for-profit sector, most of the industrial R&D in Massachusetts is performed in several key industries: chemicals and drugs;

Table 2 Summary of 1993 R&D Spending by Funding Source for Massachusetts (Millions of Dollars)

R&D Performer	Source of Funds Total	Federal	Other
Industrial R&D	$6,952	$1,878	$5,074
Academic R&D	1,076	770	306
Other Nonprofit R&D*			
Hospitals & Medical Research Institutes	628	502	126 ***
Other Research Institutes**	554	554	
Federal Intramural	384	384	
Total R&D Funding	$9,594	$4,088	$5,506

*Breakdown of nonprofit institutions by type not available from NSF.
**Includes MIT Lincoln Laboratories, Draper Laboratory, MITRE, and non-medical research institutes.
***ERG estimates.
Source: NSF, ERG estimates.

computer equipment; electronic components and communication equipment; aerospace (including guided missiles); and professional and scientific instruments. Although these industries represent only about 6% of total employment, they account for 70% of the industrial R&D. Thus, along with the nonprofit research sector, these industries are the locus of technological innovation and technology-driven economic growth in Massachusetts.

Table 2 provides an overall summary of performers and sources of funds for research in Massachusetts. Research expenditures are large in overall scale and in importance to the state economy. In 1993, $9.6 billion was spent on research and development in all types of organizations in Massachusetts, nearly 6% of Gross State Product (GSP). The majority of research activity in Massachusetts takes place in firms, with industrial R&D accounting for nearly 72% of the total, while academic institutions and nonprofit organizations account for about 23% of the research spending.

Federally funded research and development is an important component of Massachusetts R&D. According to NSF surveys, 43% of the total research and 72% of academic research in the state are funded by the federal government. For industrial firms, the federal government funds about 27% of total research efforts. Figure 2

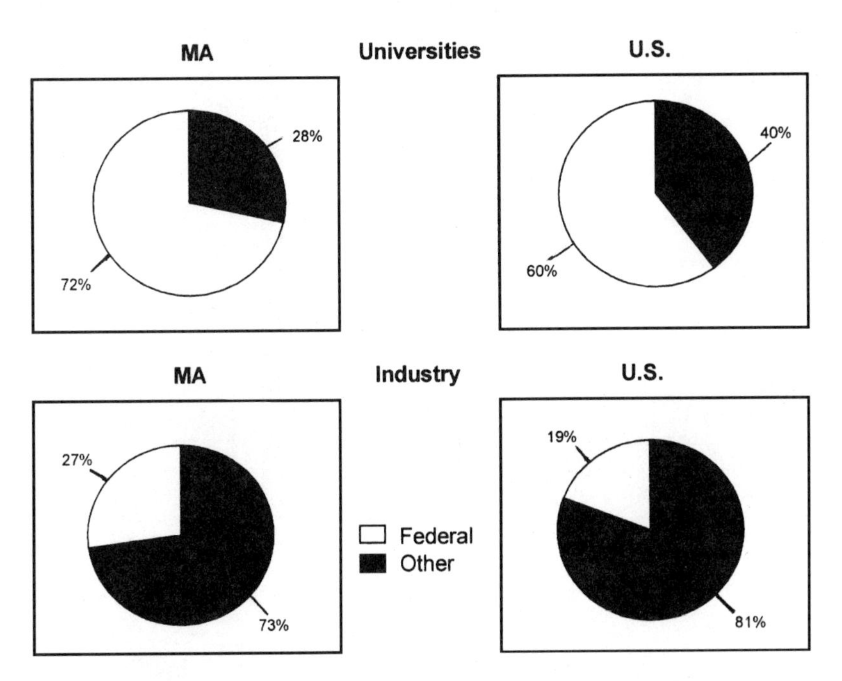

Figure 2 Importance of Federal Support in Massachusetts and the United States. Source: NSF.

shows that federal research funding in Massachusetts is more important than for the United States as whole, in both the academic and the industrial sectors.

In addition to receiving funding from the federal government, support for R&D comes from other sources as well—principally the industrial sector. For example, the largest support for industrial R&D is companies' own funds, accounting for 53% of total research effort. Approximately 20% of all sponsored hospital research comes from industry.[3] Other forms of support for research and development include state and local funding, foundations, and institutions' own funds.

Research spending is a significant component of economic activity in the state, whether measured by the magnitude of the expenditures or by their intensity (R&D as a fraction of GSP). Massachusetts ranks fourth in overall research expenditures, behind California, New York, and Michigan. Measured as a share of the state economy, however, Massachusetts ranks first among the major industrial states. Figure 3

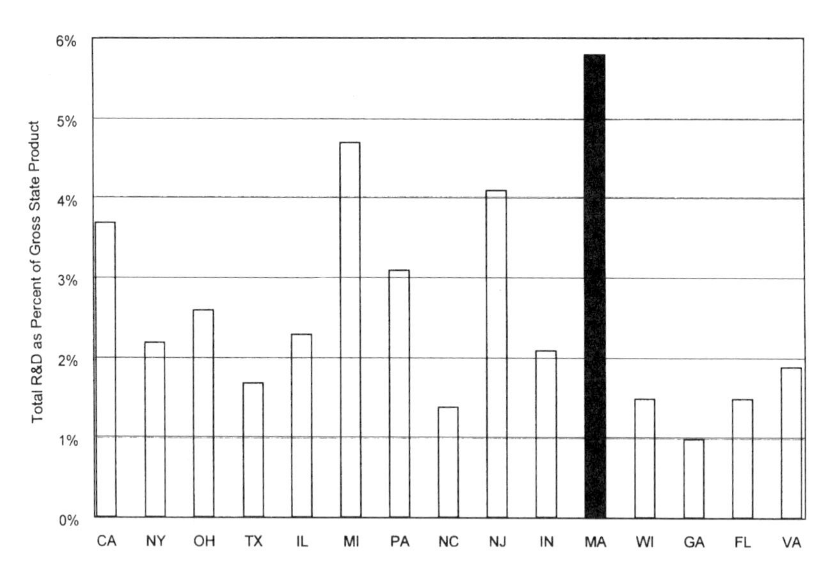

Figure 3 Research Intensity in Largest Manufacturing States. Source: NSF.

shows the ratio of total R&D to GSP for the 15 largest industrial states. After Massachusetts at 6%, the next highest is Michigan at about 4.7%, followed by New Jersey and California at around 4%.[4]

Estimating the Regional Economic Impact of Public Research Funding

Overview of Pathways of Economic Impact of Research Cuts

The level of external funding of research activity within the state affects the level of economic activity through several direct and indirect mechanisms, as illustrated in Figure 1. The direct, short-run effect is that research funding is used to pay employees and purchase supplies and equipment; these wages and purchases have further "ripple" effects through the local economy, creating a multiplier effect that is standard in modeling the regional economic impact of spending changes. The impact of change in federal funding of research in Massachusetts is qualitatively similar to the impact of changes in any other form of federal spending.

What makes research spending different is that research activities, particularly in the nonprofit sector, generate "spillovers." The economic consequence of these spillovers can be broken into two components. First, because firms undertake R&D of their own in order to capture and exploit the spillovers from the nonprofit sector, a reduction in nonprofit R&D induces a reduction in industrial R&D, independent of any changes in direct federal support of industrial research. We call this the "inducement" effect. Because the overall scale of R&D in industry is much larger than in the nonprofit sector, the inducement effect can be large. This induced reduction in industry R&D has a multiplier effect just like the reduction in federally supported R&D; thus the arrow in Figure 1 labeled "direct plus induced demand effects" captures the combined multiplier effect of changes in industry R&D due to changes in federal industry support, and the indirect or induced change in industry R&D due to a change in federal support of nonprofit research.

In addition to generating this secondary or induced demand side effect, a change in university research, combined with the induced change in industry research, has an effect on the supply side of the economy. Fewer new products and processes are introduced, slowing the growth in sales of the affected industries. This supply effect will, in general, take much longer to manifest itself than the demand-side effects; however, it will also last much longer, whereas a change in demand generated by a change in spending on wages and supplies (including the "multiplier effect") in both the nonprofit and the for-profit sectors will affect the local economy almost immediately, but this effect will diminish over time. A change in the introduction of new products and processes, on the other hand, will take much longer to have any impact, but this supply-side effect will continue to affect the growth of the relevant industries for a long time.

There are a variety of pathways through which nonprofit research benefits the for-profit sector, including explicit technology transfer, the startup of new firms, joint university/industry research centers, incubators, and industry councils. Clearly some of these pathways have more impact on the state's economy than others. For example, explicit technology transfer through universities, hospitals, and joint research centers has a larger impact than incubators and industry councils in terms of economic activity generated. However, there is some tradeoff between scale and the likelihood of localization: While industry councils and incubators are likely to generate less activity, the spillovers from these activities are more likely to be located within the state. In examining some of the spillover mechanisms, we found evidence of some degree of localization. For example, we examined two institutions—MIT and Massachusetts General Hospital—and discovered that their technology licenses are disproportionately granted to local companies. In our modeling, described in the next section, we have not tried to separate effects from various pathways. Instead, we use statistical estimates of the overall magnitude of spillovers to model quantitatively the economic effects of cuts in research activities.

Quantitative Estimates of the Cumulative Impact of Federal Research Support

As discussed above, federal research support affects the Massachusetts economy through direct demand effects, induced demand effects, and supply or innovation effects. In order to estimate the magnitude of these effects, we use a computer-based model of the Massachusetts economy developed by Regional Economic Models, Inc. (REMI), which provides a baseline forecast of economic growth in Massachusetts. We have estimated the impact of federal research spending by examining how economic activity in this baseline forecast differs from the level of such activity in a forecast in which federal research support is hypothetically reduced. The reduction that we model, which was based on budget proposals discussed in the U.S. Congress in 1995, corresponds to a significant reduction in such support after allowance for inflation.

The REMI model is a widely used, commercially available tool for understanding the economic impacts of events such as changes in U.S. government policy. Its core is an input-output structure for distributing final demand among various sectors of the Massachusetts economy. The core of the model estimates transactions among industries, and uses these data for analyzing the interdependence between sectors and the impact which changes in one sector have on another. The REMI model utilizes standard statistical techniques for interconnecting the various sectors of the model. (This particular project utilized a 53-sector version of the model of the state's economy, which was specifically calibrated to examine

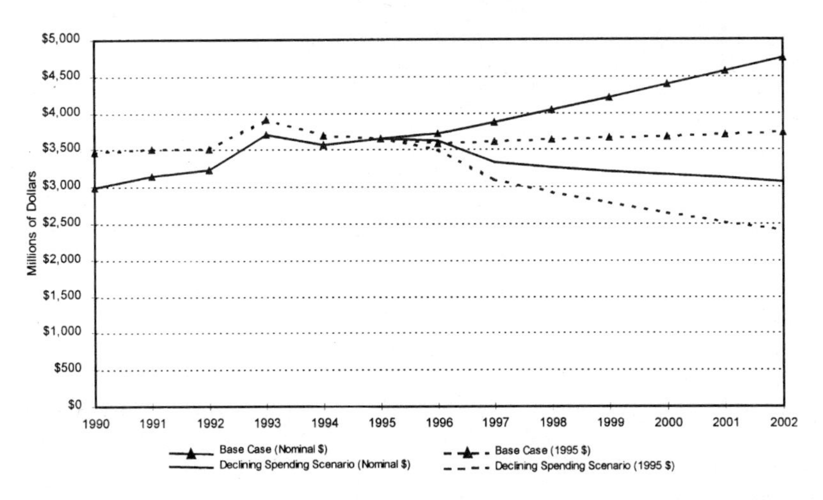

Figure 4 Federal R&D Spending in Massachusetts: Baseline Forecast versus Declining Spending Scenario. Source: ERG calculations.

the reduction in the flow of federal expenditures.) The model also takes into account the extent to which industries sell to other firms or households within the state and the extent to which they export to the rest of the United States or to the rest of the world.

In order to estimate the economic impact of reductions in federal research expenditures, it is not sufficient merely to identify what the spending levels will be; we must also specify what spending levels would have been had the policy change being modeled not occurred. It must be kept in mind that such an economic impact is only revealed by the differences between the funding levels in a given scenario and the funding levels in a baseline forecast. This baseline forecast for federal research support must be consistent with the baseline economic forecast embodied in the REMI model. The model does not contain explicit data or assumptions regarding research expenditures; rather, we have developed baseline projections for federal research support in Massachusetts from the model's estimates of overall federal spending in the state, divided between military and civilian expenditures.

Figure 4 shows the overall federal R&D spending assumed under both the baseline and declining-spending scenarios. The baseline scenario has real spending holding approximately constant between 1996 and 2002, while the declining-spending scenario has

Table 3 Projected Federal Research Funding Cuts by Sector, Relative to Baseline, Declining-Spending Scenario (Millions of Dollars)

	Total Federal Funding Cuts by Sector							
Industry	1995	1996	1997	1998	1999	2000	2001	2002
Hospitals & Health Care	$0	$3	–$74	–$109	–$145	–$181	–$220	–$258
Education	0	–23	–172	–227	–282	–336	–389	–443
Chemicals & Drugs	0	0	0	0	0	0	0	–1
Machinery & Computers	0	–1	–3	–5	–6	–8	–9	–11
Electric & Electronic Equipment	0	–22	–89	–134	–174	–215	–255	–296
Transportation & Aerospace	0	–13	–54	–81	–105	–130	–154	–179
Professional & Scientific Instruments	0	0	–1	–2	–2	–3	–3	–4
Miscellaneous Professional Services	0	–39	–155	–228	–295	–362	–428	–495
Total	$0	–$96	–$549	–$785	–$1,010	–$1,234	–$1,459	–$1,687
Total in 1995 dollars	$0	–$93	–$511	–$705	–$876	–$1,034	–$1,182	–$1,321

actual expenditures reduced each year between 1996 and 2002. Using the inflation rates forecast by the model, overall federal support for research corresponds by the year 2002 to a reduction of 34% after adjusting for inflation, approximately 35% less than is projected to occur if there were no change in federal policy.

Because it incorporates the linkages among industries, the REMI model quantifies the overall effect of a spending reduction, including "multiplier" effects that occur as reduced spending leads to lower income and hence to further reduced spending by downstream firms and industries. For example, decreases in R&D spending reduce salaries and purchases of other goods and services. The decrease in salaries means that dollars will not be spent again on the goods and services that those employees demand. In addition, the firms that provide goods and services to the R&D-performing sector will experience a decrease in demand.

Direct demand effects. Direct demand effects are simply the difference in spending between the baseline scenario and the declining-spending scenario. The latter consists of estimates of future funding levels that were likely to occur under the balanced-budget scenario that existed in late 1995, as estimated by the American Association for the Advancement of Science (AAAS 1996) and by the Department of Defense.[5] Estimates of R&D spending in the baseline are extrapolated from the REMI model's estimates of overall federal spending.

The cuts are summarized by sector from 1996 to 2002 for each scenario in Table 3.[6] Modeling the direct effects of reductions in federal research support is straightforward. After inputting reductions in R&D by sector, the model then adjusts employment and output of other industries in response to this reduction in spending.[7]

Spillover effects: Induced demand and supply/innovation effects. What makes research spending different from other forms of federal spending is that it generates spillovers. As discussed above, spillover effects have two components: an induced demand effect and a supply/innovation effect. In order to quantify the impact on the Massachusetts economy of a potential loss of research funding, we need quantitative measures of the inducement and the supply/innovation effects and a measure of the distribution of these effects across industries. To estimate spillover effects, we proceed in three steps. First, we use information regarding the rate of citation by firms in different industries to university and other nonprofit patents to estimate the pattern of spillover influences of nonprofit research across industries. Second, we estimate the reduction in company-funded research that is likely to occur in each of these industries as a result of the decreased flow of new science and technology out of Massachusetts nonprofit research institutions. This "induced" reduction in research effort has demand-side effects similar to those resulting from reductions in federal expenditures. Finally, we take the reductions in federal nonprofit research, federal industry research, and induced industry research and attribute a corresponding reduction in the rate at which new commercial products and processes are introduced. This is the supply/innovation effect, which we model as a reduction in the growth rate of output in the affected industries relative to the baseline.

Allocating spillover impacts across industries. There are estimates in the economic literature of the overall impact of university research on firms in the private sector, but there are no estimates showing how these impacts are distributed across sectors. In order to fill this gap, we analyzed the citations made by private firms to patents

Table 4 Linkages between Massachusetts Universities, Hospitals, Other Nonprofits and Industry (Based on Industry Citations to Patents from Massachusetts Institutions)

	Fractions of Citations to Patenting Entity by Industry							
Patenting Entity	C&D	M&C	E&E	T&A	P&S	E&R	Other	Total
Hospitals	49%	0%	2%	2%	21%	12%	14%	100%
Other Nonprofit	7%	25%	13%	28%	16%	4%	8%	100%
Lincoln Labs	1%	27%	22%	19%	17%	5%	9%	100%
University Departments								
Engineering	17%	12%	14%	13%	20%	5%	19%	100%
Physical Sciences	56%	3%	1%	5%	8%	7%	20%	100%
Geosciences	0%	14%	14%	29%	0%	0%	43%	100%
Math & Computer Sciences	0%	39%	22%	26%	3%	0%	10%	100%
Life Sciences	44%	5%	0%	5%	11%	5%	29%	100%
Other Science Depts.	14%	0%	5%	5%	14%	55%	9%	100%

Note: University department data is based on citations to MIT patents.
C&D: Chemicals and Drugs. M&C: Machinery and Computers. E&E: Electric and Electronic Equipment. T&A: Transportation and Aerospace. P&S: Professional and Scientific Instruments. E&R: Engineering and Research.

Table 5 Cuts in Industry R&D Induced by Cuts to Universities, Hospitals, and Other Nonprofits, Declining–Spending Scenario (Millions of Dollars)

SIC	Industry	1996	1997	1998	1999	2000	2001	2002
28	Chemicals & Drugs	$0	–$9	–$13	–$17	–$21	–$25	–$29
35	Machinery & Computers	–19	–86	–122	–155	–190	–224	–260
36	Electric & Electronic Equipment	–8	–39	–55	–71	–88	–105	–122
37	Transportation & Aerospace	–2	–8	–10	–13	–15	–18	–20
38	Professional & Scientific Instruments	–2	–18	–25	–32	–40	–47	–55
87	Engineering & Research	–2	–29	–40	–52	–65	–77	–91

Source: ERG calculations as described in text.

received by universities, hospitals, and other nonprofits in Massachusetts. Our assumption is that these citations are a proxy for technological impact, hence that the fraction of citations coming from a given industry will be indicative of the fraction of technological impact enjoyed by that industry. Because the industrial location of this spillover will obviously vary with the nature of the research, we looked separately at citations from different kinds of institutions and different university departments. The pattern of citations allowed us to allocate overall impacts to particular industries.[8] Table 4 shows the breakdown of citations to a particular discipline or nonprofit institution by companies in different industries. These results provide a "weighting matrix" that is used to map spillover effects into specific industries resulting from cuts in funding to particular parts of the nonprofit research sector.

Estimating induced industry R&D. To quantify the spillover effects, we used statistical estimates from the academic economics literature. A statistical analysis estimated that, for every 10% increase in university R&D spending within a state, firms within the state increased their own R&D spending by about 7% (Jaffe 1989). This estimate, however, implies constant university research in other states, but we have assumed throughout that the Massachusetts reduction is proportional to the nationwide reduction. To correct for this, we utilized an inducement ratio of one-fourth the effect found in the literature: For each 1% reduction in university research relevant to a particular industry, we reduced that industry's R&D spending by .175%.

This inducement parameter was combined with the distribution of impacts by industry (the "weighting matrix") to estimate reductions in industry R&D funding generated by reductions in federal support for universities, hospitals, and other nonprofits. First, we took the reductions in federal support shown in Table 3 and allocated them across institutions and fields of science (for university research) based on historical spending patterns. Then we used the "weighting matrix" to transform spending cuts by discipline and year to induced reductions by industry and year. The magnitude of induced demand effects is summarized in Table 5.

Supply/innovation effects. The reductions in federal nonprofit research, federal industry research, and induced company-funded industry research will each have an impact in the form of a reduced flow of new commercial technology. We calculate these types of effects for all industries except the engineering and research industry, whose output is, in some sense, research. To estimate the size of these impacts, we utilized estimates from the economic literature on the rate of return to different forms of R&D. For the induced reduction in company-funded industry R&D, we used an estimate of the gross rate of return to private R&D of 40%, roughly the mid-point of published estimates that range from about 25% to 60%. For the impact of university and nonprofit research, we used an estimate from the Jaffe (1989) study on the state-level impact of university research, combined with an assumed return on private

Table 6 Output Reductions Due to Supply/Innovation Effect, by Industry, Declining-Spending Scenario

Industry	2002	2003	2004	2005	2006	2007	2008	2009	2010
C&D	–0.01%	–0.11%	–0.31%	–0.61%	–1.00%	–1.46%	–1.99%	–2.33%	–2.52%
M&C	–0.05%	–0.30%	–0.77%	–1.44%	–2.30%	–3.32%	–4.49%	–5.23%	–5.64%
E&E	–0.05%	–0.25%	–0.63%	–1.18%	–1.89%	–2.75%	–3.74%	–4.37%	–4.72%
T&A	–0.08%	–0.39%	–0.98%	–1.84%	–2.98%	–4.37%	–6.02%	–7.07%	–7.66%
P&S	–0.01%	–0.08%	–0.22%	–0.43%	–0.70%	–1.03%	–1.41%	–1.65%	–1.78%

C&D: Chemicals and Drugs. M&C: Machinery and Computers. E&E: Electrical and Electronic Equipment. T&A: Transportation and Aerospace. P&S: Professional and Scientific Instruments.

R&D of 40%. Finally, for the reduction in federally funded industry research, we used an estimate that federally funded industry research has an effect on industry sales 35% as large as the effect of company-funded research; combined with our estimate of 40% for the rate of return to company-funded research, this yields a return to federally funded industry research of 14%.[9]

Economic studies imply that reductions in R&D reduce the growth rate of the industry's sales over time. Because R&D affects the growth rate, the effect of reductions is cumulative over time; on the other hand, as time passes, the effect of old R&D diminishes because of obsolescence as new technology becomes more relevant. In order to capture this effect, we assumed that the impact of R&D erodes at a rate of 15% per year.

A final issue with the supply/innovation effects is the time lags that are likely to occur before reductions in research affect the rate at which new commercial technologies are introduced. Because next year's products and processes are likely to be based on the results of research from several years ago, reductions in research today will reduce the flow of new products and processes at some unknown future date. These lags are likely to be particularly significant with respect to the commercialization of university technologies. We have incorporated a lag of six years into our analysis, based on survey results.[10]

Table 6 summarizes the decline in industry output that occurs because of reduced research spending. While the direct as well as the induced demand effects occur in the near term (concurrent with the changes in federal funding), the supply effects do not

Table 7 Summary of Impacts of Federal Research Spending Cuts on the Massachusetts Economy, Declining-Spending Scenario

	1997		2002		2010	
Source of Impact	GSP	Emp.	GSP	Emp.	GSP	Emp.
Reduction in Federal Support of Research at Universities & Other Nonprofits						
Direct Demand Effect	–$0.440	–10.0	–$0.966	–21.5	–$0.647	–13.5
Induced Industry Research Demand Effect	–$0.200	–2.5	–$0.461	–5.1	–$0.306	–2.8
Supply/Innovation Effect	N/A	N/A	–$0.021	–0.2	–$2.708	–20.4
Subtotal—Universities & Other Nonprofit R&D	–$0.640	–12.5	–$1.448	–26.8	–$3.661	–36.7
Reduction in Federal Support of Research in Industry						
Direct Demand Effect	–$0.277	–4.4	–$0.681	–10.2	–$0.440	–6.1
Supply/Innovation Effect	N/A	N/A	–$0.008	–0.1	–$0.668	–6.7
Subtotal—Industry R&D	–$0.277	–4.4	–$0.689	–10.3	–$1.108	–12.8
Overall Impact	–$0.917	–16.9	–$2.137	–37.1	–$4.769	–49.6

GSP: Gross State Product in Billions of 1995 Dollars. Emp.: Employment in Thousands.

begin to affect the economy until 2002. These effects are initially extremely small, less than one-tenth of a percent of baseline output in these industries. The effects grow over time, however, both because the reductions in research increase over time and also because they work by slowing the industry's growth.

Overall Impacts

Table 7 presents snapshots of cumulative impact of the federal spending reductions taken in 1997, 2002, and 2010. In 1997, there are no supply/innovation effects, so the only impact is the Keynesian multiplier effect, which reduces GSP by about $900 million and employment by about 17,000; the bulk of these effects occurs in the nonprofit sector, where the cuts are the largest. Over time, however, these demand-side effects come to be dominated by the supply/innovation effects. By the year 2010, the freeze in nominal spending is estimated to reduce GSP by about $4.8 billion (1.8%

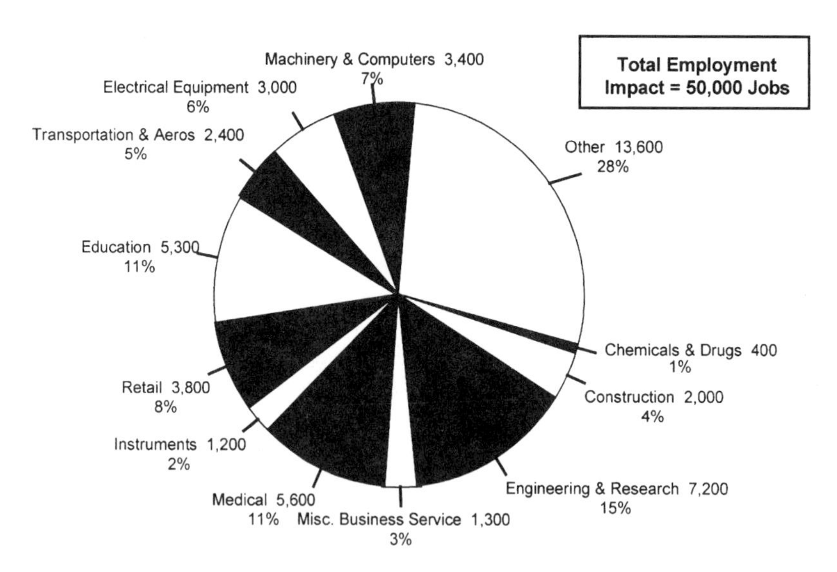

Figure 5 Distribution of Employment Impacts across Industry Sectors, Declining-Spending Scenario: 2010. Source: ERG calculations using REMI model.

relative to baseline GSP) and employment by 50,000 (1.4%). The fraction of losses due to the supply/innovation effect are larger for Gross State Product than for employment, reflecting the high labor intensity of the education and medical sectors where the direct demand effects originate. One characteristic of these high-quality jobs is that they pay far more than the average job in the state; in the past, they have been an important component of the state's continued growth. The magnitude of these losses reflects the vital role that research and innovation play in the state economy.

Figure 5 shows the distribution of long-run employment impacts across industrial sectors. The largest are in the directly affected sectors of education, medicine, and engineering and research services. Because of the multiplier effects, however, there are also significant impacts in such sectors as retail trade and construction. Again, these numbers all represent reductions in employment relative to the baseline forecast. In the medical sector, for example, the REMI model forecasts an increase in employment of about 33% in the declining-spending scenario between 1995 and 2010, despite the reduction relative to the baseline of about 5,600 jobs. Only in transportation and aerospace would employment in 2010 be lower than it was in 1995.

Conclusions and Policy Implications

Federal research support is an important component of the innovation system in a state like Massachusetts: It provides the predominant source of funding for the nonprofit sector and a significant source of funding for firms. Modeling a hypothetical reduction in the level of such support suggests that the reduction in federally sponsored R&D has a modest but significant effect on the overall level of economic activity in Massachusetts. Because of the nature of the innovation process, these effects are dynamic and cumulative.

Massachusetts, like many other states, increasingly views the innovation system as a target for regional economic policy. Our analysis suggests a number of observations regarding the possibilities for regional or state-level policy intervention designed to foster economic development via innovation.

At a broad conceptual level, there are three possible leverage points at which the contribution of science and technology to economic growth within a state or region can be enhanced: (1) the scale of research activity in either the nonprofit or the industrial sectors can be increased; (2) the rate or effectiveness of transfer of technology from the nonprofit sector to the industrial sector can be enhanced; and (3) the likelihood that spillovers from nonprofit research are captured within the region can be increased.

As we move from the nonprofit research sector, to research and development efforts in the for-profit sector, to commercial implementation of a new technology, the scale of possible impacts and the likelihood of localization within a region generally move in opposite directions. Nonprofit research institutions are almost entirely infrastate institutions, and the need for communication at the applied research stage makes conducting that research nearby desirable. However, there may be a tradeoff between scale and localization: If and when a technology moves to the production stage, such active communication may be less important, so that a product conceived in a local university and nurtured in a local corporate laboratory may be manufactured in another state or another country. Thus the probability of "capture" of benefits as a consequence of geographic localization falls as we proceed from university research through industry research to commercialization. On the other hand, the scale of potential economic impact will generally increase at the same time the likelihood of localization is decreasing: If we trace the trajectory of a particular technology, the number of people employed will typically be very small at the university, somewhat more at the industry R&D phase, and larger still at the production stage. As a result, even though a significant fraction of the industry R&D, and much or most of the commercial production, induced by nonprofit research may be outside the state, it is possible that the employment impact on the state may be larger at the industry R&D stage, and larger still at the production stage. This suggests that increasing the probability that new technologies stay within the region all the way through to the production stage may be an important aspect of maximizing the regional economic return from research.

We also found that start-ups or "spin-offs" are a crucial part of the process: A climate conducive to the formation of new firms in general, and academic spin-offs in particular, is an important component of an approach to technology-based regional economic growth. This observation is not new (Governor's Council 1995), but it is reinforced by several aspects of our analysis of the spillover process. New firms are more likely to be local than other academic licensees, and they are more likely to keep the higher-employment, larger-scale production phase of a product within the state. In addition, a climate conducive to formation of new firms will help to keep entrepreneurially minded graduates of local institutions living within the region.

Another factor is the match between public research strengths and industrial base. Spillovers from nonprofit research are more likely to be localized to the extent that the particular industries most likely to develop the new technologies are located within the state. Thus the extent of match between the areas of scientific strength of a region's universities and the areas of technological strength of the region's industrial base will be an important factor in the ability to capture the spillovers. Specific efforts to support technology development and/or transfer are most likely to be successful in areas where a region has strength in both the nonprofit research and the industrial sectors.

Both broad statistical studies and the detailed examination of licensees by MIT suggest that the amount of money industry is induced to spend in developing university technology may be large relative to any amount of direct funding from local public initiatives. Similarly, the induced impact on industry or hospital R&D of government procurement or health care funding may be large relative to actual government funding of research itself. This suggests that these inducement mechanisms are at least as important as funding mechanisms.

Acknowledgments

This paper is based on a study performed by The Economics Resource Group, Inc., for the Massachusetts Technology Collaborative. The results of that study are contained in a report, *Planning for Change, Preparing for Growth: Implications of Reduced Federal Research Spending for Massachusetts,* 1996. The views expressed in this paper, as well as any errors it may contain, are the responsibility of the authors.

Notes

1. See Massachusetts Technology Collaborative (1998) for discussion of the implications of current federal research funding proposals in Massachusetts.

2. For the purpose of the economic impact analysis, the nonprofit research institutions other than hospitals (plus Lincoln Laboratory) are grouped together with the private firms in the Engineering and Research Services sector.

3. AUTM Survey and ERG estimates, based on conversations with hospital research directors.

4. Overall, Massachusetts ranks fourth in research intensity, behind New Mexico, Maryland, and Delaware, small states in which a very small number of large research institutions produce a high ratio.

5. Projections for non-defense R&D came from AAAS (1996); projections for defense R&D came from U.S. Budget 1996. Federal support for R&D in Massachusetts was estimated for each sector using historical funding levels, historical funding sources, and projected funding by agency.

6. After 2002, we assume that there are no further reductions in R&D spending relative to the baseline.

7. There are no budget projections beyond the year 2002. Because we wish to model long-run effects, we must make some assumptions about what happens to spending levels beyond 2002. For simplicity, we assume that spending after 2002 resumes the rate of increase in the REMI baseline forecast: The reductions of 1996–2002 represent a permanent reduction in the level of support, but after 2002 there are no further reductions relative to the baseline projection.

8. We linked Massachusetts nonprofit patents from 1963 to 1993 with subsequent industrial citations to these patents. We then assigned a standardized industrial classification (SIC) code to the industrial citer and created a matrix of nonprofit and industrial categories.

9. For a brief discussion of studies on the rate of return of R&D, see Tassey (1995). For evidence on the rate of return on federally funded private research, see Griliches (1986).

10. The most detailed analysis of these lags has been done by Mansfield (1996). He estimates that the lag between university research and its commercial impacts varies from 5 to 9 years, with a mean of 6 years.

References

AAAS. 1996. Report 20: Research and Development Fiscal Year 1995 and FY96 R&D Appropriation Update, Table A, (January 10).

Economics Resource Group, Inc. 1996. "Planning for Change, Preparing for Growth: Implications of Reduced Federal Research Spending for Massachusetts." Report to The Massachusetts Technology Collaborative.

Governor's Council. 1995. "Renewing the Venture Capital: Improving the Environment for Financing Emerging Companies in Massachusetts." Governor's Council on Economic Growth and Technology Task Force on Financing Emerging Companies (July).

Griliches, Zvi. 1986. "Productivity, R&D, and Basic Research at the Firm Level." *The American Economic Review* 76(1): 141–154.

Jaffe, Adam. 1989. "Real Effects of Academic Research." *The American Economic Review* 79(5): 957–970.

Mansfield, Edwin. 1996. "Industry-University R&D Linkages and Technological Innovation." Presented at the annual meeting of the American Economic Association (January).

Massachusetts Technology Collaborative. 1998. Analysis of the Impact of Federal R&D Investment Scenarios on Economic Growth (July).

Tassey, Gregory. 1995. "Technology and Economic Growth: Implications for Federal Policy." National Institute of Standards and Technology (October), pp. 15–18.

U.S. Budget. 1996. Historical Tables: Budget of the United States Government, Fiscal Year 1996. Table 5.1. Washington DC: U.S. Government Printing Office.

20

The Growing Sophistication of Research at a Time of Broadened Participation in Higher Education

Shinichi Yamamoto

Introduction

Universities and colleges have always trained researchers, teachers, and other professionals, performed research and development, and identified the talented young people who would later lead Japanese society. This traditional role, however, is now changing greatly because of the broadening of access to higher education, and because of the sophistication of modern research in science and technology. The research and research training functions of universities need to be re-evaluated and improved, at the same time that the universities respond to the various educational demands in a higher education system that includes more than 47% of the 18-year-old population.

In response, gradual changes and reforms are taking place in the university research environment. Selective allocation of research funds, expansion of graduate training accompanied by new financial aid programs, encouragement of research cooperation with industry, restructuring of research units at major universities, and so on, are some of the trends observed; "university reform" has become a key word in describing the current Japanese higher education system. The new Basic Plan for Promotion of Science and Technology is expected to play a major role in the reform of university research.

In my short essay, therefore, following a brief introduction of the Japanese higher education systems' relations with research, I will present some facts and data on these trends, emphasizing the conflict between research and teaching, for both the institutions and their faculties. Growth of enrollment in graduate programs in Japan has been the highest among major advanced countries for two decades, and nearly 60% of the 18-year-old population is expected to seek a university education within the next 10 years.

A recent survey, in which I was involved, shows how faculties allocate their time between research, teaching, administration, and so on. While a strong preference for research over other activities is frequent, much has been changing as more students enroll in higher education and postgraduate programs. This essay will conclude by suggesting the new direction of the relationship between university and research.

Features of the Japanese University System

The first modern university established in Japan was the University of Tokyo in 1877, which was re-organized as the Imperial University in 1886. The missions of the Imperial University were to train future elites and to introduce or interpret Western science to Japanese society, both essential to the modernization of Japan. Engineering, taught outside the university in most Western countries, was regarded as an essential part of the newly created university system. In the mid-Meiji era, around the 1900s, when Japan established its hierarchical higher education system, from the imperial universities down to various types of private schools, it was realized that a university education would guarantee a good job and social prestige: Even a student from a poor family could move up to a higher social status.

The secondary educational system, however, was the "European-style double-track" system, in which a student was prepared either for the university or for employment or vocational training. In addition, university tuition was expensive, further limiting access to higher education. The postwar introduction of the "American-style single-track" secondary education system, followed by the enormous growth of the Japanese economy in 1960s and 1970s, triggered a rapid increase of enrollment in higher education.

Table 1 Numbers Enrolled in Higher Education, 1997

	Number of Institutions	Enrollment
Universities	586	2.633.790
Junior Colleges	595	446,750
Colleges of Technology	62	56,294
Special Training Colleges	2,981	652,072

Source: Monbusho 1997b, Statistical Abstract.

The Current Higher Education System in Japan

Japan's higher education system in May 1997 comprised 586 universities, 595 junior colleges, and 62 colleges of technology. In addition to these universities and colleges, nearly three thousand special training colleges, offering a variety of practical vocational and technical education programs, were regarded as a part of the non-university sector of the higher education system. They admit graduates of senior high schools; there are also colleges of technology, which accept junior high school graduates.

As shown in Table 1, universities enroll the majority of the students. Their function, according to the School Education Law, is to conduct in-depth teaching and research in specialized academic disciplines and to provide students with advanced knowledge. They are expected to conduct both advanced research and teaching, and it is important to remember that all universities without exception must provide equally for both research and teaching. Although the reality is somewhat different from the letter of the law, this universal requirement is a unique characteristic of the Japanese university system, whereas in the United States institutions with different missions may compete with each other for students and research funds.

This brings us to the Japanese system of funding research. In Japan, a general university fund is provided to each national university according to some standard formula, say, the size of the faculty or the number of students. The portion of these funds allocated to research recently has been far exceeded by individual and competitive research funds. This is a second different characteristic of Japanese higher education system. A study in the late

1980s shows that nearly three-fourths of research funds were classified as general university funds in Japan, while only one-fourth were so categorized in the United States, indicating that the research environment in Japanese universities was much the less competitive (Irvine, Martin, and Isarde 1990). Improvements in the funding system have been discussed that would more effectively allocate limited resources into active and prospective research projects.

A third different characteristic is that Japanese universities provide undergraduates with specialized education, such as law, psychology, civil engineering, computer science, or medicine, rather than concentrating on liberal arts and sciences; when they finish undergraduate programs, the students already have a specialty. Advancing to graduate programs in humanities and social sciences, therefore, has been regarded as appropriate training only for future academics. Engineering, however, has attracted many graduate students, especially in master's degree programs.

A further difference is the smaller scale of graduate education than in the United States, in terms of the ratio of graduate to undergraduate students or the number of graduate students in the population compared to other countries. While Japan had only 6.1 graduate students per 100 undergraduates in 1994, the United States had 16.4 graduate students in the same year (Monbusho 1995). This is a great incentive for both academics and the Monbusho to call for the expansion of graduate education in Japan.

Different Characteristics of Groups

Not only does the Japanese university system have several specific characteristics as described above, there are also some differences in relation to graduate education within each level. When we divide universities into three groups, national, local public, and private, they differ greatly with each other (see Table 2). The majority of students take their undergraduate courses at private institutions, while national universities exceed others in graduate education. We can also find the same differences among the various disciplines (see Table 3 and 4). In the humanities and social sciences, most students leave their institutions with bachelor's degrees and

Table 2 Number of Students Completing Programs in 1997 (by type of institution)

	Undergraduate		Master's Program		Doctoral Program	
Type of Institution	#	%	#	%	#	%
National	104,100	19.8	31,025	61.5	7,024	71.2
Local public	15,808	3.0	2,000	4.0	446	4.5
Private	404,604	77.1	17,405	34.5	2,390	24.2
Total	524,512	100.0	50,430	100.0	9,860	100.0

#: Number of students completing the program.
Source: Monbusho 1997b, Statistical Abstract

Table 3 Number of Students Completing Programs in 1997 (by major field)

	Undergraduate		Master's Program		Doctoral Program		Doctoral Degrees	
Field	#	%	#	%	#	%	#	%
Humanities and Social Sciences	299,324	57.1	4,234	8.4	1,570	15.9	522	4.3
Science and Engineering	136,773	26.1	26,393	52.3	4,359	44.2	4,663	38.8
Health	23,571	4.5	2,033	4.0	3,370	34.2	6,480	53.9
Others	64,844	12.4	17,770	35.2	561	5.7	366	3.0
Total	524,512	100.0	50,430	100.0	9,860	100.0	12,031	100.0

Source: Monbusho 1997b, Statistical Abstract

Table 4 Number of Degrees Awarded to Undergraduate and Graduate Students, 1994

	Japan	United States	United Kingdom
Social sciences			
(1) undergraduate	185,027	413,008	56,100
(2) graduate	4,470	188,447	32,400
(2)/(1)	0.02	0.46	0.58
Science & Engineering			
(1) undergraduate	122,132	210,738	63,000
(2) graduate	32,932	82,737	19,200
(2)/(1)	0.27	0.39	0.30

Source: Monbusho

the number of graduate students is very small compared to the number of undergraduates. Advanced research activities in this group are highly concentrated in a few institutions and the rest are facing a massive expansion at the undergraduate level.

In science and engineering, students advance to graduate programs more often than do those in the humanities and social sciences; in fact, masters' programs in engineering are regarded as the most successful case of graduate education in Japan. Doctoral enrollment in this field is also much larger than in the humanities and social sciences.

The health sciences (medicine, dentistry, and pharmacy) present a different picture. In Japan, medical schools provide undergraduate programs to train physicians, surgeons, or others, who may go on to graduate programs in this field. The ratio of doctoral programs to undergraduate programs in the health sciences is greater than in other fields, as is the number of doctoral degrees awarded each year. This implies that, in Japan, training of medical doctors and Ph.D. researchers in medicine are combined in a single system.

The Changing University Environment in Japan

Broadening of Higher Education and the Decline of the 18-Year-Old Population

The university research environment has changed rapidly over the past few years. The most significant factor is the expansion of higher education, which has made the traditional notion of the "unity of research and teaching" difficult to maintain (Clark 1993).

Japan experienced its first rapid growth of higher education in the 1960s and early 1970s. Due to various causes (Yamamoto 1997: 294–307), the entry of 18-year-olds into higher education grew to 38.6% in 1976 from 10.3% in 1960. This broadening of participation meant not only a quantitative growth of higher education, but also a radical change in its character. Higher education is no longer for elite students, but for everyone who needs higher education. The increased demand for education brought about diversification in offerings, from purely academic to practical instruction.

To respond to this rapid quantitative and qualitative change, in the mid-1970s the Monbusho (Ministry of Education, Science, Sports and Culture) initiated a new policy intended to control the quantity and improve the quality of university education; this policy, however, became unworkable when the 18-year-old population again began to grow.

The second stage of expansion began in the early 1990s, in response to a surge in the 18-year-old population in the late 1980s. The participation ratio of the 18-year-old population grew from 36.3% in 1990 to 47.3% in 1997. But this was short-lived, and has since been followed by a steady decline in the 18-year-old population, which, it is estimated, will drop from 2.05 million in 1992 to 1.20 million in 2009. Except for a few prestigious universities, the colleges and universities must therefore consider how to deal with a future shortage of applicants and how to attract students.

Along with the broadening of higher education, a growing number of people complained about the content of education, saying that teachers spent too much time on academic matters while many students prefer to take practical courses that will be useful for their future jobs outside academia. Another problem is the students' declining incentives to learn. Because many students who might not have enrolled in higher education two decades ago, are not accustomed to study abstract ideas taught in academic language, universities find themselves forced to change their way of teaching and the structure of their curriculum. "Faculty development"(FD) has become a fashionable phrase in Japan in discussions of improvement in teaching. It is one way universities must reform themselves, in a situation where an inadequate response will mean they can no longer attract students.

Expansion of Graduate Education

The current system of graduate education in Japan, introduced after the Second World War, aimed at carrying out basic research activities in all academic disciplines to provide a sound basis for the development of scientific research of all types, producing highly qualified researchers and professionals. However, until the late

1970s, the system mainly provided for research training for future academics. In some areas such as engineering, growing enrollment had gradually changed the character of graduate education from academic research training toward professional training. Reform of the graduate school system was undertaken by the Monbusho in the 1970s and 1980s, followed by the introduction of systemic flexibility and the expansion of functions.

Although graduate education aims at both academic research training and professional training, it is also regarded as an important place for research activities (Yamamoto 1996). Due to the broadening of university education, concerns about university research have shifted from undergraduate education toward graduate education and training. Graduate school seems to be a sanctuary not only for university faculty members who wish to unite research and teaching but also for policy makers who regard university research as an engine for economic growth and technological innovation. Attitudes toward research as a primary focus are shown, worldwide and in Japan, in Table 5.

The growing number of graduate students, especially in engineering, reflects new expectations from the industrial sector. Master's degree programs grew far more rapidly than those at the undergraduate level. The proportion of students who advanced from undergraduate to master's degree courses was low in the engineering field during the 1960s and early 1970s, but it had reached nearly a quarter by 1996; at the University of Tokyo, for example, 69% of undergraduate students at the School of Engineering advanced to graduate courses in that year. On the other hand, in the social sciences, this ratio has remained low.

Although enrollment is different among disciplines, financing of graduate education has been closely connected to research intensity at Japanese national universities. The level of general university funds allocated to each national university from the Monbusho is greatly different according to whether the university has doctoral programs or master's programs. The size of the general university fund allocated for a research unit that is connected to a doctoral program is at least twice that of a research unit that is not. For private universities and also for local public universities, having doctoral programs confers prestige among neighbor institutions, even if they do not attract enough students.

Table 5 Research or Teaching: Responses to the Question, "Regarding your own preferences, do your interests lie primarily in teaching or in research?"

	Primarily in Teaching (%)	Leaning to Teaching (%)	Leaning to Research (%)	Primarily in Research (%)
Australia	13	35	43	9
Brazil	20	42	36	3
Chile	18	49	28	5
Germany	8	27	47	19
Hong Kong	11	35	46	8
Israel	11	27	48	14
Japan	4	24	55	17
Korea	5	40	50	6
Mexico	22	43	31	4
The Netherlands	7	18	46	30
Russia	18	50	29	3
Sweden	12	21	44	23
United Kingdom	12	32	40	15
United States	27	36	30	7

Source: Boyer (1994)

Graduate education has thus been expanding by responding not only to the growing needs of society but also to faculty insistence that today all the national universities have at least a master's program, and that 80% of them have doctoral programs. Additionally, 47% of private universities have doctoral programs, 19% have master's degree programs, while 34% have only undergraduate programs. The annual growth of graduate enrollment in Japan was the highest among major countries in the world: While the United States experienced about 1.8% percent of annual growth during the 1980s, Japan had 5.6%.

Selective Allocation of Resources

The university budget fell on hard times in the 1980s, when due to the government's budget deficit the growth of general university funds was nearly frozen. (A detailed discussion of this fiscal problem may be found in chapter 6 of this volume.) However, because the revitalization of university research remained urgent in order for Japan to promote advanced research and secure economic competitiveness, the Monbusho increased other types of research funds. These funds were not formula-based, but competitive. This led to a noticeable change in the structure of university research funding in the 1990s. The establishment of special funds for graduate schools and individual fellowships for doctoral students and postdoctoral researchers became typical.

The aim of these new policies, along with the growing amount of competitive grants-in-aid programs, is to give additional resources not to all institutions and students, but to selected schools and scholars whose research quality and performance are outstanding. A new funding program, "Research for the Future," established in 1996 by the Japan Society for the Promotion of Science (JSPS), is funded through capital investment made by the Japanese government to promote and expand the frontiers of scientific research. Awards are decided by JSPS's committees after designation of the specific research fields to be funded.

In addition, some universities, such as the University of Tokyo, have recently reformed their research units to place more emphasis on graduate education. This has helped them to get more research funds from the Monbusho. A "Center of Excellence" (COE) program is also an example of the selective allocation of resources. This program aims at the establishment of superior research bases, to which the Monbusho provides active support (Monbusho 1997a).

Financial support for graduate students and post-doctoral researchers is important for research training. In this individual domain, selective allocation of resources has also been promoted. For graduate students, scholarship loans provided by the Japan Scholarship Foundation (JSF) has played the biggest role, enabling competent students who lack financial resources to attend graduate schools. More than 40% of master's program students and more than 60% of doctoral program students took advantage of these loans in the 1970s. Although the growth of JSF scholarships does not follow the expansion of student population (the figures have declined to 30% and 50%, respectively), they provide basic financial support for graduate students. Students who after graduation serve as researchers for universities or related institutions for a certain period do not need to repay this scholarship loan.

In 1985 the Monbusho established a new, more competitive fellowship program for young researchers, called "Fellowships for

Young Japanese Scientists." With the aim of cultivating young researchers who will conduct innovative and trail-blazing research, this program provides a limited number of promising young researchers with fellowships and research grants so as to allow them to concentrate on their research, which they conduct in laboratories or under supervising researchers of their choice for a specified period (two to three years). This new fellowship, which is administered by the JSPS, is provided for graduate students and post-doctoral researchers on a highly competitive base. In fiscal 1997, 2,420 doctoral students and 1,070 post-doctoral researchers were granted the fellowships.

Along with the promotion of "The Program to Support 10,000 Post-doctorals," which is included in the Science and Technology Basic Plan of 1996, these new competitive support devices will be expanded not only by the Monbusho but also by other governmental agencies, including the Science and Technology Agency (Monbusho 1997b).

Cooperation with Industry

Although there is severe criticism of the current role of universities and university research, there are also diverse expectations and demands from industry, and from other sectors of society, for scientific research activities at universities which will solve specific and practical problems. It is important for universities to respond to these social demands as much as possible, while, of course, retaining their initiative and their original mission of scientific research.

In their cooperation with industry, universities utilize their accumulated achievements and research abilities, and by doing so they are able to make their contributions to society on the one hand, and to obtain useful stimuli for their research activities on the other. From this point of view, universities should make efforts to engage in cooperative research with industry according to their stated goals.

The relationship between universities and industry, however, has only recently come to be regarded as desirable by university faculties, especially in the social sciences and humanities. Previously it had been argued that cooperation between these two sectors would endanger the freedom of research and the university's autonomy. Some faculties thought that research for the sake of research, or purely "curiosity-oriented" research, was fundamental for a university. The popularization of mass university education and the advancement of basic science and technology gradually moderated such objections while increasing demand for accountability of universities, and the worsening of the research environment after the 1980s forced universities to change their role, structure, and relations with other social sectors.

The government has encouraged universities to introduce the system of joint research between university and industry, and also the system of contract research, by which university researchers can be engaged in specific research projects commissioned by industrial enterprises on a contract basis. Since 1987 the Monbusho has been helping national universities to establish joint research centers aimed at contributing to the promotion of research cooperation with industrial firms in the in-service training of engineers and researchers. Furthermore, in 1998 a new system of technology transfer related to patents was introduced.

Concentration of Resources: Research Universities vs. Others

As we have seen, the funding structure for university research has been changing. Universities and their faculties now increasingly expect extra funding outside of general university funds, from private companies or competitive governmental grant-in-aid programs (see Table 6). At some laboratories within research-intensive universities, in engineering, for example, more than a third of research funds come from the private sector. These external funds have become essential for the maintenance of the laboratories. Under these circumstances and for the vigorous development of research activities corresponding to social demands, the system of introducing financial resources from outside sources has become increasingly simple and more flexible.

While every university has the same mission in Japan, in reality each has a different function. If we categorize Japanese universities by the U.S. model, we can recognize 24 "research universities" out of more than 400 universities, as shown in an earlier scholarly work

Table 6 Major Research Funds for National Universities (¥ million)

	1987	1992	1997
General University Fund	97,824	117,873	154,052
Grants-in-aid	45,080	64,600	112,200
Donation from Industry	22,361	48,184	52,783
Contract Research	5,451	9,449	41,853

Source: Monbusho

(Keii 1984). When we compare the growth of enrollment in "research universities" and others, the growth of graduate enrollment exceeds that of undergraduates in research university groups, while a much larger increase occurred in undergraduate programs in other university groups (see Table 7). This means that as research universities have become more research-oriented and non-research universities have become more teaching-oriented, the difference between the two becomes more pronounced.

Allocation of research grants is tremendously different among institutions: the University of Tokyo received 11.5% of the Monbusho's Grants-in-aid in 1998; the top 15 universities, among more than a thousand higher education institutions, received more than half (Table 8).

Conclusion

The 1990s in Japan were the decade of university reform, intended to make the university more accountable to students, funding institutions, and the general public, in contradiction of the traditional notion of university or faculty autonomy. It has been realized that expanded, broadened higher education and the progress of scientific research have made Japan's university system no longer fit the new requirements of its people.

In teaching, students prefer more practical to academic courses, and they require that the curriculum be more systematically organized; in research, the outcome, it is said, should be more applicable to actual use. The government also states that basic research at academic institutions is essential for the maintenance and improvement of the national economy and international competitiveness.

Table 7 Research Universities vs. Other Universities (Enrollment)

	Research Universities	Other Universities	Total
1975			
Undergraduate	40,607	228,996	269,603
Master's program	10,417	12,315	22,732
Doctoral program	5,458	3,360	8,818
1997			
Undergraduate	51,465	454,496	505,961
Master's program	18,073	33,532	51,605
Doctoral program	9,098	8,732	17,830
Increase, 1975–1997			
Undergraduate	10,858	225,500	236,358
Master's program	7,656	21,217	28,873
Doctoral program	3,640	5,372	9,012

Source: Yamamoto (1996)

Table 8 Allocation of Grants-in-Aid to Universities, April 1998

Institution	Number of projects	Amount of grant (¥ thousands)	%
1 Tokyo	2,261	8,994,600	11.5
2 Kyoto	1,729	6,051,300	7.7
3 Osaka	1,548	5,097,800	6.5
4 Tohoku	1,441	4,824,000	6.1
5 Hokkaido	1,114	3,110,600	4.0
6 Kyushu	1,189	3,038,100	3.9
7 Nagoya	968	2,988,000	3.8
8 Tokyo Inst. of Tech.	680	2,376,000	3.0
9 Hiroshima	692	1,624,700	2.1
10 Tsukuba	630	1,351,700	1.7
11 Kumamoto	380	1,108,000	1.4
12 Kobe	480	988,300	1.3
13 Keio *	491	961,400	1.2
14 Okayama	458	895,300	1.1
15 Chiba	393	879,700	1.1
Top 15 total	14,454	44,289,500	56.4
Total	34,104	78,554,920	100.0

* Private institution. Others are national universities.
Source: Monbusho

According to the Monbusho, there are three reasons for current university reform: first, to respond to progress in scientific research and changes in human resource development; second, to respond to both the rise in the percentage of students continuing to higher education and their diversification; third, to respond to the growing need for lifelong learning and the rising expectations of society for universities (Monbusho 1995). In this situation, each institution should identify its own mission, whether it is research-intensive or not.

An important fact is that nearly three-quarters of the faculties in Japan show more interest in research than in teaching, although the difference is small when spread among universities, junior colleges, and colleges of technology. This implies that faculties have resisted and will continue to resist diversification of the university mission and wish their affiliated institution were research-intensive. Although the realities of the university environment may make it more difficult to maintain a single-mission system of higher education, if more resources are allocated to a limited number of research-intensive universities, the ability of university faculty and researchers to move among their institutions should be taken into consideration.

References

Boyer, Ernest L. 1994. "The Academic Profession, An International Perspective." Princeton NJ: The Carnegie Foundation.

Clark, Burton, ed. 1993. *The Research Foundations of Graduate Education: Germany, Britain, France, United States, Japan.* Berkeley: University of California Press.

Irvine, John, Ben Martin, and Phoebe Isarde. 1990. *Investing in the Future: An International Comparison of Government Funding of Academic and Related Research.* Worcester, UK: Billing and Sons.

Keii, Tominaga, ed. 1984. *Study on University Evaluation.* Tokyo: University of Tokyo Press.

Monbusho (Ministry of Education, Science, Sports and Culture). 1995. "Japanese Government Policies in Education, Science, Sports and Culture." Tokyo: Printing Bureau, Ministry of Finance.

Monbusho. 1997a. "Statistical Abstract of Education, Science, Sports and Culture." Tokyo: Printing Office, Ministry of Finance.

Monbusho. 1997b. "The University Research System in Japan." Tokyo: Printing Office, Ministry of Finance.

Yamamoto, Shinichi. 1996. "Graduate Schools in Japan, from the Perspective of Academic Research." *University Studies* No.15: 1–287.

Yamamoto, Shinichi. 1997. "The Role of the Japanese Higher Education System in Relation to Industry," in Akira Goto and Hiroyuki Odagiri, eds., *Innovation in Japan.* Tokyo: Oxford University Press.

21

Public Policies for Japanese Universities and the Job Market for Engineers

Seiritsu Ogura and Hiroto Kotake

Introduction

In view of Japan's enormous public and private investment in the last thirty years, it is surprising that there is still an excess demand in the higher education market.[1] Why has the supply never caught up with the demand in this particular market? For so long, the Japanese public has been conditioned to believe that higher education should be rationed, that the selection should be done through uniform examinations, and that the price mechanism in higher education only invites corruption. Due to these screening beliefs, the Japanese government and the universities believe the real problem in the higher education market is not so much excess demand as excess supply.

The national government has been playing two separate but related roles in the higher education market. On the one hand, it has been its dominant producer: Prior to and during the Second World War, it was clearly the military importance of the sciences and technologies that convinced the government to develop and maintain its strong imperial university system; after the war, the national government has not only kept these universities, but has also added a national network of "new" universities to guarantee equal access to higher education. All of them have been supported by the general tax revenue, their tuition set at only a fraction of the costs.

On the other hand, the national university has been the omnipotent regulator of private universities. Because the budget of the Ministry of Education (Monbusho) has limited the capacity of national universities to accept applicants, the rest have been left for the private universities to accommodate. Although private universities charge many times more than national universities in tuition and other fees, their classrooms have remained severely overcrowded and their curriculum weak. The Ministry of Education has been trying to regulate this problem by issuing "Minimum Standards for New Universities" (*Daigaku Setchi Kijun*),[2] numerical standards governing almost every conceivable aspect of university education. The real source of the academic weaknesses of private universities, and of national universities to some degree, has been the national universities' low tuition and the absence of competition, which have imposed low ceilings on the tuition of private universities and lower ceilings on the quality of their education (Ogura and Kotake 1998). The absence of competition has protected many substandard universities and professors.

No one, including the officials of the Ministry, has been aware of the full consequences of these policies, which in the long run can seriously hurt the economy. There are three possible problems in these policies. First, they are like manpower planning in a command economy, where the government decides how many to train for each type of work and how to train them. If the government makes a mistake in forecasting the sizes or compositions of future needs, firms will not able to hire the kinds of workers in the numbers they need. Second, centralizing resource allocation in research and technological innovation can be very risky, since no one can always pick the winners of research races. In practice, almost always, the government has chosen the status quo, which has kept the allocation of university's research resources rigidly fixed over the years.[3] Finally, there is nothing in the regulatory system that can dissolve those universities whose education or training programs have become obsolete or ineffective. They would be driven out of a competitive market by new or efficient ones, but the restrictive policies on new entries and the low tuition of national universities have been extremely effective in protecting the weak ones.

In short, as far as higher education is concerned, the Japanese economy has been subjected to the same risks that have plagued centralized economies, and cannot take advantage of a diversification of risks through a decentralized market mechanism or efficiency resulting from competition.

These shortcomings can be either magnified or partially neutralized by the flexibility of the particular regulations. If they are flexible enough to accommodate the adjustments universities need to meet changes in the markets, these shortcomings may not be too serious. If, however, the regulations are very specific and leave little room for adjustments by the universities, these shortcomings may become quite apparent and harm the health of the economy, as has happened in the case of Japan's schools of engineering.

In what follows, we will try to identify the resource allocation problems of schools of engineering since 1970, by examining the relationship between what the government decided, what the schools have produced, and what industry has needed. We begin with an overview of the government's regulations governing Japanese universities, then we present the results of our study of those governing admission quotas of major fields of specialization in engineering and their outcome, and examine the available evidence on the relationship between training and jobs for the graduate. Finally, we summarize our findings and offer a few suggestions.

Public Regulations on Universities and Schools of Engineering

Minimum Standards for New Universities

Under the present law, a university can add a new department if it obtains an approval from the Ministry of Education, by proving that it will meet all the minimum requirements on staff and physical resources of that specialization and will offer an appropriate curriculum. Likewise, it may not increase or decrease the annual student admission of an existing department unless approved by the Ministry.

The present regulations, including Minimum Standards, were originally introduced by the Ministry of Education in 1956 as the new university system replaced the old one, and a large number of new universities were created. During the 1960s, however, these regulations had only a limited effect on private universities, which regularly admitted twice their quotas or more to accommodate the first baby-boom generation. But when the baby-boomers left the market in the early 1970s, inflation and economic stagnation following the first oil-shock hit the debt-ridden private universities hard, and they had to turn to the national government for financial help.

In addition to a new subsidy program in 1976, the Ministry of Education decided to freeze application quotas[4] for ten years. In return, the private universities promised to comply with the Ministry's regulations. Even after the freeze had been formally lifted in 1986, the newly created University Council, in charge of making policy recommendations on higher education, declared the Ministry should continue its restraint in approving expansion of universities. The recommendation is not surprising, as most of its members are selected from university communities.

Even under the present regulations, private universities are generally very sensitive to developments in the job market, as they have to charge several times more for tuition and fees than the national universities. Thus for private universities to remain competitive, it is essential to offer education in the fields with better job prospects, but it has been difficult for them to obtain approval from the Ministry of Education. The process is very time-consuming: It takes at a minimum two years to start a new department, up to five years or longer.

It can be far more difficult for a national university to start a new department than for a private university. As a new program involves new costs to the government, under the present system, the faculty or the university cannot make the decisions; it is up to the Ministry to decide, and up to the Ministry of Finance to ask for additional funds for a new department to be included in the national budget.

Regulation and Fragmentation: Observations

A department is the standard unit the Ministry regulates. The smallest departments, with an annual admission quota of ten students or fewer, can be found in medical schools, and the largest departments, with annual admission quotas of five hundred or more, can be found in schools of social science. The size of a department, under the present regulatory system, reflects the scope of the training or education of the department. In particular, if the education or training is very specific, its resource requirements are usually strict, and demand for it will be limited as well.

In this regard, engineering schools are a good example. Engineering departments bear names that suggest a very close relationship with a particular industry or with a group of industries. They are small in size. For instance, in the engineering faculty of the

Table 1 Quotas for Engineering Departments, University of Tokyo, 1995

Department	Quota
Civil Engineering	45
Architecture	65
Urban Engineering	56
Mechanical Engineering	50
Engineering Synthesis	43
Mechano-Informatics	42
Precision Machinery Engineering	53
Shipbuilding	55
Aeronautics	57
Electrical Engineering	43
Information and Communication Engineering	40
Electronic Engineering	48
Applied Physics	55
Mathematical Engineering and Information Physics	60
Quantum Engineering and Systems Science	40
Geosystem Engineering	33
Materials Science	38
Metallurgy	43
Applied Chemistry	57
Chemical System Engineering	55
Chemistry and Biotechnology	55

University of Tokyo in 1995, there were twenty-one such departments (Table 1).[5] Even in private universities, departments of engineering are relatively small, as their admission quota is usually somewhere between 100 and 200, a third or a fourth that of other schools in these universities.

These specific regulations may well have been effective at the start. Unlike other schools of Japanese universities, many still believe, most of the engineering schools have effective education and training programs—a reputation that no doubt reflects the successful history of firms in Japanese manufacturing industries where their graduates have found jobs. Recently, however, the schools have been having difficulty in attracting good students, and the popularity of the field continues to fall among college applicants. At the same time, it is frequently pointed out that schools of engineering have not kept up with the state of the art in technologies, unlike the research laboratories of large private firms, particularly in their physical equipment. In other words, the schools are losing ground in competition both within the universities and with other private research organizations.

The experience of the schools of engineering in Japanese universities illustrates a fundamental problem in the public regulation of higher education. These regulations may have had their desired effects in the short run, but they did so at the cost of badly fragmenting engineering schools into many small independent departments, guaranteeing the departments their virtual independence from either the school or the university, and even from the market. Furthermore, once created they became independent of the Ministry of Education itself through the constitutional protection of academic freedom. In the real world, this means that it would be next to impossible to transfer resources across these fragmented units, even when there is an overwhelming economic justification for it. Thus, in the long run, the fragmentation caused by regulation cost the schools their relevance in the marketplace, and cost the economy the benefits of competition.

Finally, in the early 1990s the Ministry of Education, seemingly realizing its mistake, reorganized the schools of engineering in the old imperial universities. These universities are now treated primarily as institutions of graduate education, where departments are no longer used as units of regulation. But even this fundamental change in policy has yet to reverse their long history of fragmentation; furthermore, it is not clear how far the Ministry is willing to go in deregulating the other universities or the other schools.

Undergraduate Admission Quotas by Major Field since 1970

Quantifying Policies

Ideally we would like to have three sets of data. The first is detailed data on the amount of resources that have been allocated to individual departments and their sources in all the schools of engineering since 1970. Resources should include not only the money allocated for their operating costs, but also for capital equipment and research activities. The second is information on students and their jobs after completing their education. The third is information on faculty and their research outputs.

Unfortunately, the Ministry of Education does not publish operating budget figures of individual departments or universities. Nor does it publish comprehensive statistics on the distribution of the Scientific Research Funds (*Kagaku Kenkyu Hi*) they manage. And there is no comprehensive database available on the physical characteristics of individual private universities.

We have two sets of sample data for our analysis. One is the admission quota data for individual departments. The annual admission quotas are set when the departments are first approved, and determine the sizes of their annual budgets and the faculty numbers in the national universities. Even in private universities, admission quotas practically determine the faculty size and composition of a department, under an elaborate formula devised by of the Ministry of Education. The other data set is on the number of faculty and their specialization for each department, in the same universities. We have compiled both these data sets for this research.

Sample Data for Three Groups of Universities

In reviewing the admission quota policy of the Ministry of Education since 1970, we find it convenient to classify all the universities into three relatively homogeneous groups, based on their ownership and history. The first group consists of just seven major national universities that used to be members of the imperial university system.[6] The second group consists of the other national universities,[7] most of which were created after the Second World War. The third group consists of all private universities.[8]

Construction of Sample Data. The admission quota of each department is public information found in the annual "List of All Japanese Universities" (*Zenkoku Daigaku Ichiran*) published by the Ministry of Education (Monbusho 1970–1995). Due to time constraints and limited resources, we constructed a sample of admission quota data since 1970, from a reasonably comprehensive sample of universities, but omitting MS programs of schools of engineering. In particular, we used the following three rules to select our samples:

1. Admission quota data of every department in the engineering schools of the old imperial universities were collected. There are

seven such universities with a total of 119 different departments. Our sample coverage for this group is 100% of its population.

2. As to the national universities in our second group, at least one national university is selected from each prefecture, including those in (1). This resulted in our selecting 32 additional national universities with a total of 203 departments. For 1995, our sample coverage is slightly less than 80% of the quotas of the second group.

3. Fifty private universities were selected, first the five largest—Nihon University, Tokai University, Waseda University, Keio University, and Tokyo Science University. Then the other 45 universities were selected at random. These 50 private universities had 370 different departments in 1995. Coverage is more than 75% of the total admission quotas.

The summary statistics of annual admission quotas are given in Table 2.[9] In order to translate these admission quota statistics of the sample universities to those of national populations, in each year, we simply multiplied the sums of the second and the third groups by constant factors respectively, to cover the universities in each group that are not included in our sample.[10] These multipliers are then used to obtain population totals of each field in each group used in the following analyses.

Admission Quotas of the Old Imperial Universities

All the universities in this group have their origin in the old imperial university system and are still considered the best research universities in Japan in most disciplines, including engineering. Ever since the old universities formally lost their privileged status, the government has been trying to decide how much to spend on these top universities at the expense of "average" national universities, or vice versa. The experience of the schools of engineering in national universities since 1970 is an interesting case in point.

In 1976 a formal freeze was placed on the expansion of all universities, which was to last until 1986. But from Figure 1, it is clear that for a decade literally nothing changed in the old imperial university group. Even before the freeze, admission quotas had not been increased, in fact the freeze seemed to have started two years earlier in 1974 and to have lasted until 1983. In 1986 and 1987, a significant

Table 2 Summary Statistics of Three University Groups (Old Imperial, New National, and Private)

	mean	s.d.	max.	min.
1970s				
Old Imperial	51	28.78	235	20
National	71	34.72	240	25
Private	94	48.67	440	20
1980s				
Old Imperial	53	31.81	260	20
National	79	36.21	250	30
Private	109	50.59	440	20
1990s				
Old Imperial	60	37.63	260	25
National	92	37.43	240	35
Private	125	46.78	440	20

expansion took place, which seemed to have ended in 1994. Figure 1 shows that although many fields benefited from this expansion, clearly electrical engineering benefited the most. Even after this expansion, however, most of the departments in this group remained small. For example, there are three departments in the field of electrical engineering in the University of Tokyo: departments of electrical engineering (with an annual admission quota of 43), electronic information engineering (40), and electronic engineering (48).[11] Chemical engineering has almost the same share as mechanical engineering and civil engineering in this group.

Admission Quotas of National Universities in the Second Group

From Figure 2, it is clear that the experience of the national universities in the second group has been very different. This group has experienced two waves of expansion instead of one, and in between, the freeze came and slowed down their expansion. In the first wave before the freeze, their quota was growing moderately so that, by 1984, it was more than 30% larger than in 1970.

In 1985, the second wave of expansion began, and by 1995 this group had gained another 30% in its total admission quota over the end of the freeze. Since 1970, electrical and communication engineering has accounted for about 40% of the expansion in

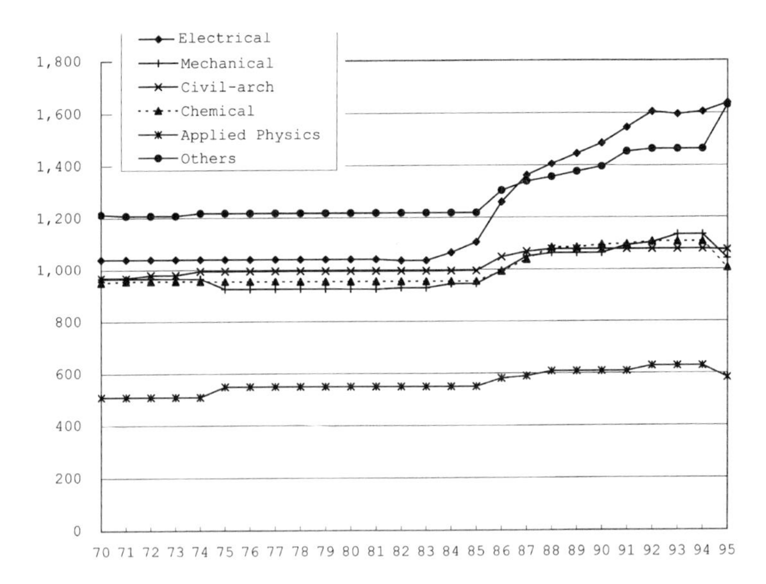

Figure 1 Student Quotas in Old Imperial University Group.

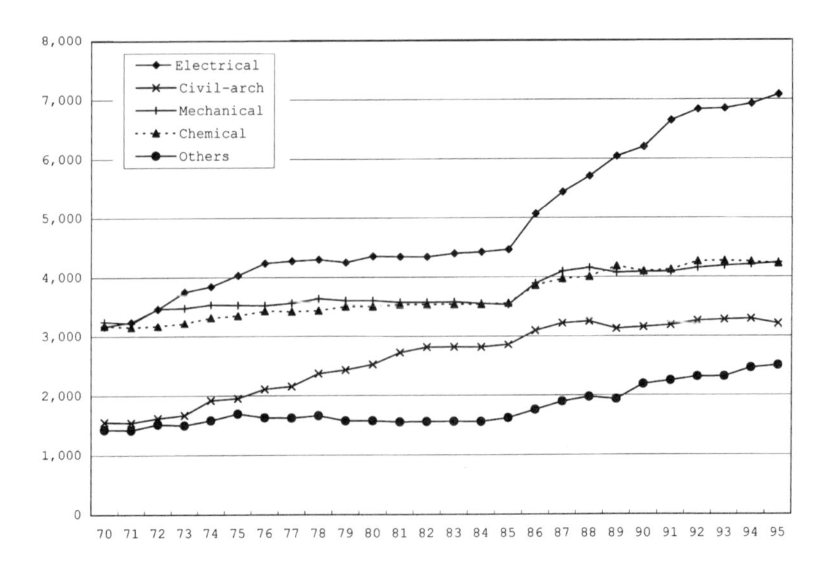

Figure 2 Student Quotas in New National University Group.

admission quotas of this group. Two out of three new departments in the field created since 1970 are for "information engineering," but they tend to be small, with only 40 students or so; only about 20% of the new departments in the field are for hardware electrical engineering. Almost all hardware departments in electrical engineering experienced a 10% or greater increase in the second wave of expansion, in 1986 and 1987. Most of them now have annual admission quotas of 80 students or more. Chemical engineering and mechanical engineering each grew at about half the rate of electrical engineering: In 1970 civil engineering was only half as large as these two, but in 1995 it was two-thirds their size.

Admission Quotas at Private Universities

Between 1970 and 1995, private universities would appear to have experienced three waves of expansion in their admission quotas, instead of one or two for national universities (Figure 3). But the first wave of expansion shown as having occurred in 1976—the very first year of the freeze—was only apparent;[12] the first actual wave came in 1986–1987, the second in 1992 and 1993.

By the end of 1995, compared with 1984, their quota had increased by more than 46%, compared with 34% for the second group and 22% for the first. In terms of absolute numbers, these differences are quite substantial. For private universities the increase was 18,995, while for the second group it was 5,369 and for old imperial universities it was only 1,246. The increase in the number of admitted students in private universities may be 50% larger than this quota, as will be explained shortly.

In 1970, private universities' quota in electrical and communication engineering was 7,057, but in 1995 it was almost 17,597. Many new departments were created during this period, but it is again "information engineering" that appears most frequently as the name of new departments in this field since 1986. In hardware engineering, many electronic engineering departments expanded twice, and most electrical engineering departments have also expanded, albeit at lower rates. Departments of private universities in this field tend to be larger than in the first or second groups.[13]

Currently, the second largest field is civil engineering and architecture. In 1970, their quota was almost 6,300 and in 1995 it

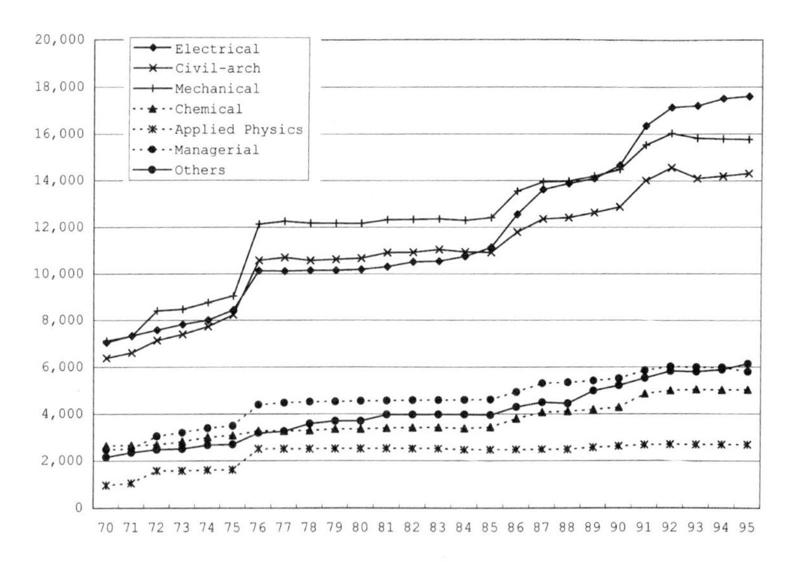

Figure 3 Student Quotas in Private University Group.

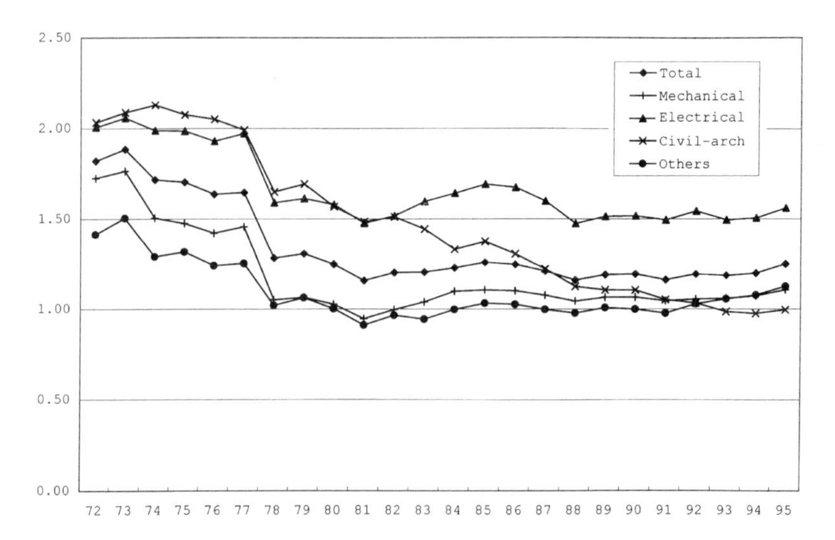

Figure 4 Estimated Excess Admission Ratios of Major Engineering Fields in Private Universities.

Table 3 Actual Number of Third-Year Engineering Students across Fields

	1970	1975	1980	1985	1990	1995
Mechanical	11,336	13,503	16,688	16,887	19,626	21,024
Electrical	11,259	13,514	15,589	16,700	22,330	26,274
Civil-arch	8,911	11,181	14,183	14,766	17,100	18,591
Chemical	6,759	7,385	7,815	7,953	9,463	10,242
Applied Physics	1,523	2,217	3,112	3,074	3,312	3,400
Nuclear	236	333	335	333	363	366
Metal	456	510	513	510	565	642
Aero-Ship	165	169	291	300	346	365
Managerial	2,529	3,544	4,553	4,616	5,517	5,799
Others	3,838	4,488	5,299	5,598	7,437	8,756
Total	47,013	56,844	68,379	70,737	86,061	95,459

reached almost 14,000, or about 23% of the total. The third largest field is mechanical engineering. In 1970, the field's quota was more than 6,100 and in 1995 it was more than 12,000, or about 23% of the total. In contrast, chemical engineering has been in a distant fourth place with a less than 5,600 admission quota in 1995.

Measuring Outcomes of Regulations

Undergraduate Education

National universities strictly observe their quotas, but in actual admissions private universities have a tendency to exceed their quotas for additional revenue.[14] We can compute the actual admission ratio for private universities as the ratio of the number of students they have actually admitted to that of their quota.[15] This is shown in Figure 4. In 1970, the private universities admitted close to twice their departmental quotas in their schools of engineering. But in 1976, the figure shows a sharp drop in these ratios, to around 1.5. As we pointed out earlier, this sudden drop happened on paper only, as the Ministry approved increased quotas without any real changes taking place. In 1985 a decline began that ended in 1988. In the 1990s, the ratio seems to be generally stable with a very slight upward trend.[16]

The numbers following the names of the fields in Table 3 are the actual numbers of third year students[17] in 1960, 1970, 1980, 1990, and 1995.[18] In Figure 5, we show how the numbers of third-year

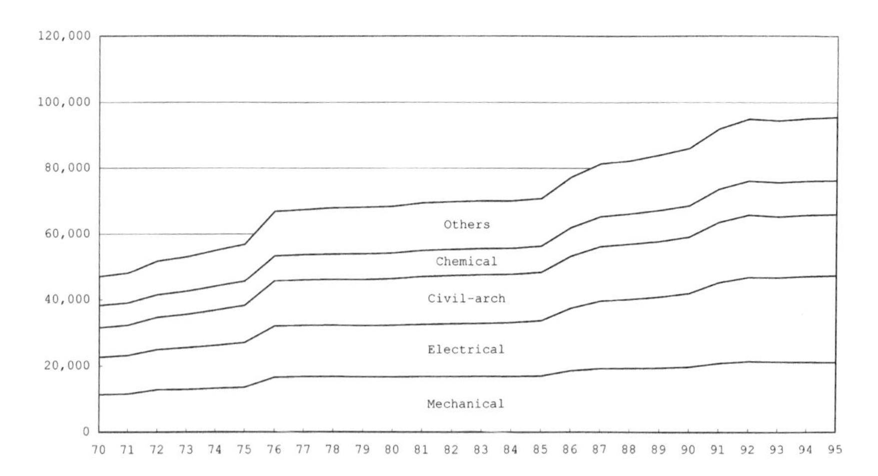

Figure 5 Number of Major Fields of Third-year Undergraduate Students in the Schools of Engineering.

students in these fields and their shares changed between 1970 and 1989. All through this period, the three top fields have remained the same, consistently accounting for 75% of the engineering students. In particular, from the very beginning the top field in the rankings has been "electrical, electronic, and communication engineering."[19] Until 1971, mechanical engineering was ranked a close second in the number of students and civil engineering, with architecture in third place. Although, by and large, the movements of these three fields have been very similar to each other, between 1971 and 1985 mechanical engineering fell to third place and civil engineering moved up to second place. In the mid-1980s, however, mechanical engineering regained second place, where it has remained. The only other field of specialization that came close to these three was chemical engineering around 1961 and 1962, but since then it has grown at less than half the rates of the three big fields.

Graduate Education

In contrast to undergraduate education, the supply of graduate education has been increasing at a fairly rapid rate (Figure 6). In 1976, 4,475 students graduated with MS degrees and found jobs immediately, but in 1996 19,494 students graduated with MS

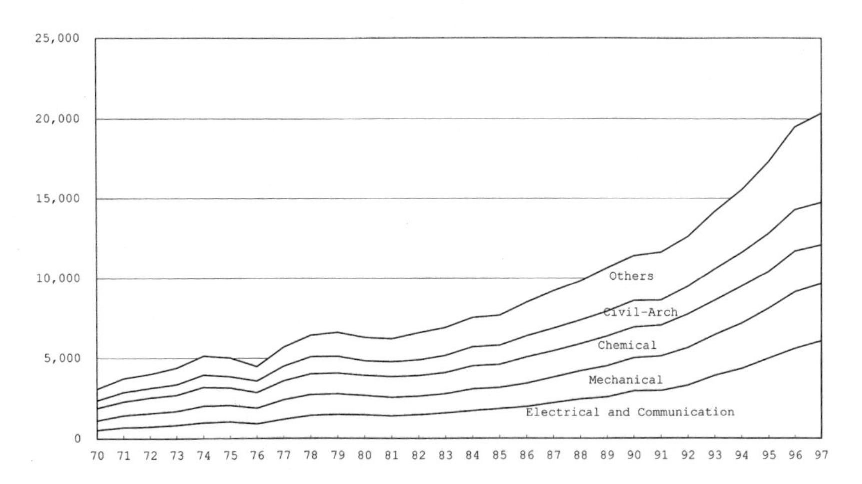

Figure 6 Fields of Specialization of New Engineers with MS Degrees (Numbers of Persons).

degrees and found jobs. Thus in the last twenty years, the absolute numbers of students graduating with MS degrees and finding jobs has quadrupled.

In Figure 7 we show the shares of these graduates' fields of specialization. Unlike the shares of the undergraduate students, the shares of graduate students have small short-term ups and downs as well as noticeable trends. The upward trend of electrical and communication engineering is fairly pronounced, as its share increased from 17% in 1970 to 26% in 1997. In contrast, the downward trend in chemical engineering is also clear, as its share decreased from 25% in 1970 to 17% in 1997. Throughout the period, civil engineering and mechanical engineering more or less maintained constant shares.

Faculty and Their Specialization

In Table 4, using the Basic School Statistics published by the Ministry of Education, we show the size of teaching staffs of the schools of engineering every five years since 1970. For example, in 1970, there were 3,633 professors and 2,517 associate professors.[20] The number of professors exceeded 4,800 in 1980 and 6,900 in 1995, and the number of assistant professors increased as well.

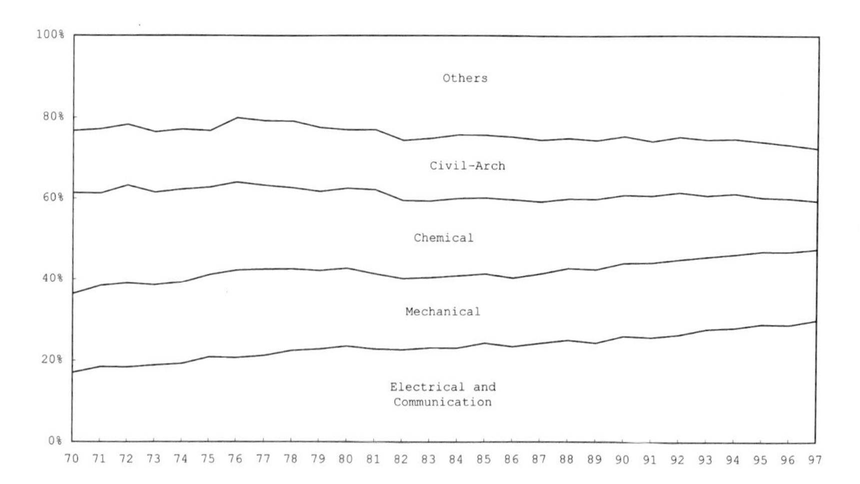

Figure 7 Share of Fields of Specialization of New Engineers with MS Degrees (%).

Table 4 Number of Teaching Staff

	1970	1975	1980	1985	1990	1995
Professors						
National University	1,806	2,063	2,412	2,604	2,900	3,012
Municipal University	156	156	159	152	225	254
Private University	1,671	1,857	2,236	2,448	3,087	3,571
Total	3,633	4,076	4,807	5,204	6,212	6,837
Assistant Professors						
National University	1,545	1,837	2,153	2,342	2,523	2,480
Municipal University	115	130	124	130	307	203
Private University	857	1,145	1,452	1,577	1,792	1,707
Total	2,517	3,112	3,729	4,049	4,622	4,390

It is important to determine how many of these professors are working in a given field of specialization. We hypothesize that the size of teaching staffs of individual departments should reflect their quotas of undergraduate and graduate students. The allocation of faculty resources across the major fields of specialization is itself important information, as engineering schools are expected to constitute an important part of the innovation system in the Japanese economy.

Data Sources. For our source of information on the faculty fields of specialization, we used the National List of University Staff

(*Zenkoku Daigaku Shokuin Ichiran*), an annual commercial publication. In particular, we have simply counted into our sample the number of professors, associate professors, instructors, and technical staff members whose names are listed in each department, counting at five-year intervals since 1980. (Due to anomalies in the data for 1970 and 1975, these years have been omitted pending further study.)

Faculty Size and Admission Quota: Regression Analysis. We represented the faculty sizes of individual departments by their number of professors and assistant professors, and took the sum of the index over all the departments in a given field for each university. We have also taken the sums of explanatory variables such as the undergraduate admission quota, MS admission quota, and Ph.D. admission quota over all departments in a given field for each university. We then regressed the faculty size on the undergraduate admission quota, MS admission quota, Ph.D. admission quota, major field dummies, year dummies, and university dummies. More specifically, we have pooled all the data of each major field of each university for 1980, 1985, 1990, and 1995 and run separate regressions for each group of universities using the ordinary least-squares method. The regression results for each group are shown in Table 5. For all three groups, the coefficients are for the year 1980 and for mechanical engineering.[21]

In the old imperial universities, for every additional person in their undergraduate quota the number of professors or assistant professors increases by almost .15, or one faculty for every seven extra students. Apparently MS students do not count, but each additional doctoral student counts as much as an undergraduate. Controlling for the number of students, all the other major fields except civil engineering are relatively smaller than mechanical engineering. There is a very small upward trend in faculty size of this group. This equation explains about 70% of the total variations in the faculty size within this group of universities.

For the second group of national universities, an additional undergraduate increases the number of faculty by 0.16, or by almost the same magnitude as in the old imperial universities. The coefficient of the quota of MS students, on the other hand, is puzzling, as it has a significant negative coefficient equal to –0.19, but the quota of doctoral students apparently does not matter.

Table 5 Independent Variable: Number of Professors and Assistant Professors

	Old Imperial University		New National University		Private University	
	Coef.	t-value	Coef.	t-value	Coef.	t-value
Faculty	0.146	12.083	0.156	8.635	0.047	9.295
Master	0.034	0.713	−0.193	−2.523	0.099	2.874
Doctoral	0.167	2.021	0.302	1.119	1.511	7.276
Chemical Eng.	6.253	2.192	−14.296	−6.938	−6.206	−2.933
Applied Physics	−0.854	−0.269	−11.995	−2.138	−2.819	−0.744
Teaching	−1.523	−0.248	—	—	−7.848	−2.581
General Studies	0.944	0.195	−7.607	−1.157	17.207	8.371
Metallurgy	−1.600	−0.518	−9.041	−2.345	−9.387	−1.932
Management	—	—	−26.869	−2.906	−7.049	−3.357
Nuclear Sci.	0.708	0.219	—	—	−5.867	−0.714
Industrial Arts	—	—	−14.033	−1.581	−8.970	−1.372
Aero-Ship	0.686	0.191	—	—	−17.774	−2.919
Mining	−2.710	−0.803	−6.160	−1.148	—	—
Textiles	—	—	−12.543	−2.558	—	—
Naval Eng.	−5.024	−1.263	−26.329	−3.436	0.246	0.029
Electric	0.987	0.339	−13.711	−6.330	−4.015	−2.424
Civil-architectural	8.705	2.669	−16.332	−6.761	2.602	1.398
Applied Science	—	—	−6.397	−0.633	—	—
Applied Biology	—	—	−8.064	−0.798	—	—
Others	8.512	2.749	−9.778	−4.037	−3.198	−1.384
year 1985	0.215	0.121	−0.329	−0.186	0.137	0.097
year 1990	1.621	0.773	−1.368	−0.756	2.535	1.768
year 1995	2.416	1.333	1.814	0.968	1.501	1.042
Constant	11.414	3.247	17.968	4.120	7.472	1.993
Obs./Adj.R2	284	0.7108	672	0.2641	951	0.5876
SS (Residual/Total)	29128.5	110051.8	157785.2	232045.8	202021.0	527654.8
df (Residual/Total)	259	283	620	671	882	950
MS	112.5	388.9	254.5	345.8	229.0	555.4

Again, all the other fields are smaller than mechanical engineering. The fit of this equation is relatively weak, as its adjusted R^2 (0.26) shows.

For private universities, an additional undergraduate will increase the number of faculty by only 0.05, which is only one-third the increase in national universities. Furthermore, if we consider that private universities actually admit about 50% more students than their quota, the increase in the number of faculty for each

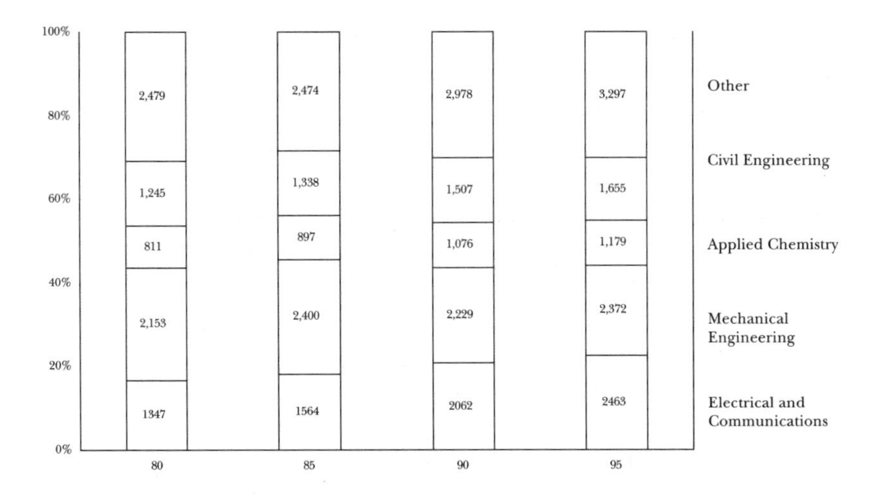

Figure 8 Number of Professors/Assistant Professors by Major Fields.

Table 6 Summary of Numbers, Professors/Assistant Professors

Year	Observations	Mean	S.D.	Min	Max
1980	18	446.39	609.43	7	2153
1985	18	481.83	680.07	6	2400
1990	16	615.75	769.06	0	2229
1995	18	609.22	850.43	0	2463

actual additional student is around 0.0333, or one-fifth of what it is in national universities. An extra MS quota will increase the faculty size by 0.1, and a doctoral student by 1.5, which is an unlikely magnitude and is probably a result of spurious correlation. The fit of this equation is relatively good (adjusted R^2=0.59).

Fields of Specialization. Table 6 shows the distribution of the professors and assistant professors in major fields of specialization in our sample universities for these four years. The numbers are simple sums of our sample observations and they are not adjusted to reflect those not in our sample.

The distribution in Figure 8 seems strikingly stable, with the top four engineering fields, electrical and communications, mechanical, civil, and chemical, accounting for 70% of the total. This stability should not be surprising, as almost all professors and assistant professors are tenured in Japanese universities, and as the mobility between industry and the universities has been very low.

The most significant change in this period is the exchange of shares between mechanical engineering and electrical and communication engineering. In 1980, the share of the former was 27% and that of the latter was 17%, but in 1995 both were almost equal at 22%. In fact, 35–40% of the net hiring between 1985 and 1995 was in electrical and communication engineering, while there was a net decrease in mechanical engineering. In contrast, the shares of civil engineering (about 15%) and of the others taken together (about 30%) have stayed the same in this fifteen-year period.

Evaluating the Outcomes of Regulations

Training and Jobs

Since 1970, the Japanese economy has experienced a number of shocks, such as the end of the high-growth era, trade frictions, oil shocks, and the end of the cold war, even as science produced two technological innovations of historical importance, one in microelectronics and the other in biochemistry. Each of these shocks forced a serious adjustment process on the economy as firms have tried to reorganize their product lines and production activities. When the firms made their necessary adjustments, their demand for engineers —in both kind and number—changed substantially as well.

On the other hand, as we have seen in the preceding analysis, the supply was rigidly fixed until the mid-1980s in terms of both quantity and training, and we should find at least some evidence of a mismatch between supply and demand during this period. As there are no micro-data available, in what follows we will examine two macro-statistics that reflect on the relationship. One will be the industry distribution, the other the professional job distribution, of the graduates of the schools of engineering.

Industry Distribution of Jobs: First Evidence

Figure 9 gives the industry distribution of those with new Bachelor's or Master's degrees who found jobs since 1970; Figure 10 gives their proportions.

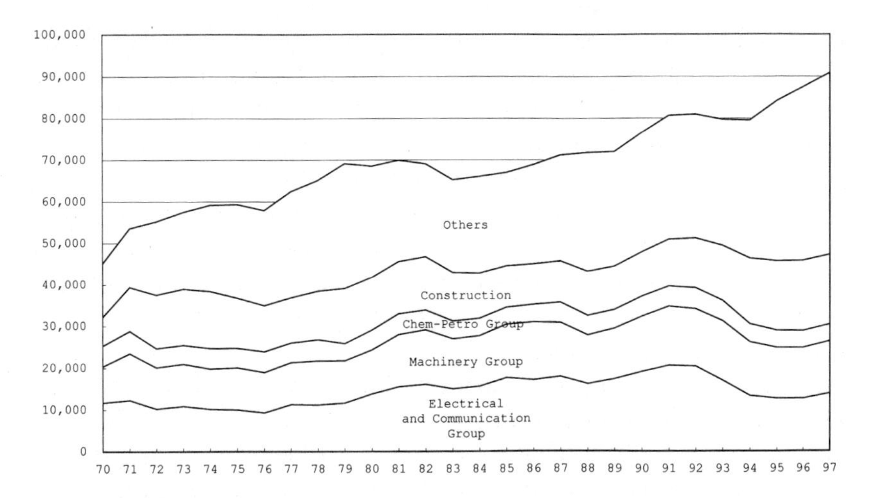

Figure 9 Jobs for New Engineers by Industry Groups (BS and MS).

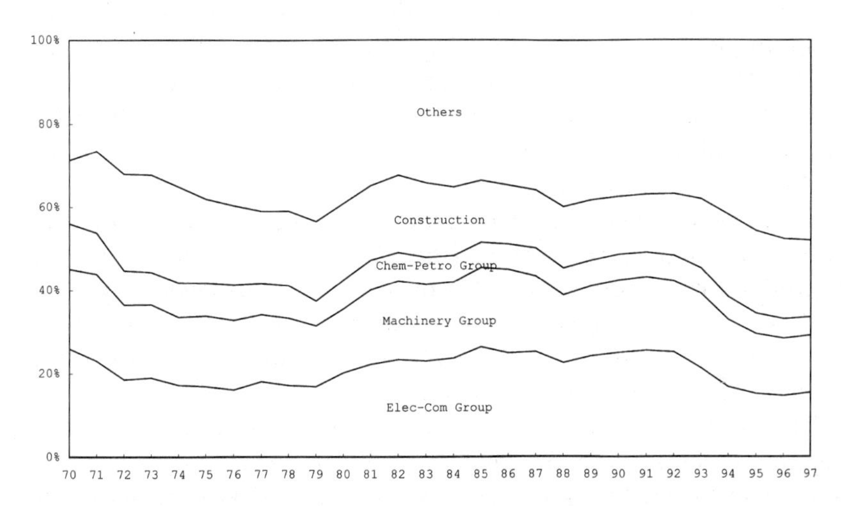

Figure 10 Share of Jobs for New Engineers by Industry Groups (BS and MS).

Declining Shares of Manufacturing. In 1970, almost 70% of these students were hired by the manufacturing sector, but the proportion fell sharply in the first half of 1970s, to about 50% in 1980. In the mid-1980s the proportion recovered somewhat, but again it started to decline and in 1995 it stood at 43%.

In the manufacturing sector, the firms in electrical machinery have hired the most but, at the same time, these firms have been the most volatile employers. In 1970, the industry absorbed 24% of all the new engineers; subsequently its share declined until it reached 14% in 1979. Again in the first half of 1980s, the industry's share started to recover, reaching the 25% mark in 1985, but in 1996 it dropped to 12%, the lowest level in the last 25 years.

There are two industries that have ranked next to electrical machinery in hiring new graduates: transportation machinery and general machinery. In 1970, the transportation machinery industry hired about 6% of these students, and its share has been fairly stable. In 1970, the general machinery industry hired 11% of them, but in the 1980s its share was around 7%, and in 1995 it was still 7%. Firms in the chemical industry hired almost 10% of the new engineers in 1970, but its share seems to be in a long, gradual decline: In the 1980s, the chemical industry's share was around 5% and in 1995 it was 3%.

Construction Industry. Firms in the construction industry have been important employers of new graduates, hiring 15% of them in 1970 but since maintaining an average proportion of close to 20%. The industry's demand seems to move so as to partially offset the cyclical changes caused by electrical machinery and other industries in manufacturing.

The Increasing Share of the Service Industry. It is generally agreed that the traditional market for engineering graduates has been the manufacturing sector and such industries as construction, public utilities, and transportation and communication; in 1970 these firms hired almost 90% of these new engineers, but the ratio declined steadily until it reached about 70% in 1995.

The decline in this section reflects the structural changes in Japan's economy that in turn resulted in the expansion of the service industry. Among these graduates, in 1970 only 3% of new graduates found jobs in the service sector, but by 1980 the proportion was up to 9%, and by 1995 it grew to 15%; in 1997, its share was more than 20%.

Distribution of Career Choices: More Evidence

In the Basic School Statistics, a limited description of the jobs obtained by graduating students is available for mechanical engineers, electrical engineers, chemical engineers, civil engineers,

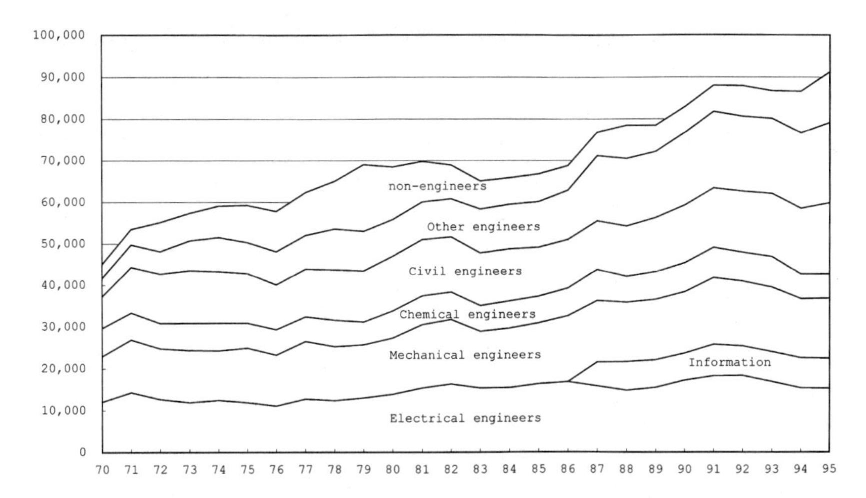

Figure 11 Distribution of Fields Chosen by New Engineers (BS and MS Degrees).

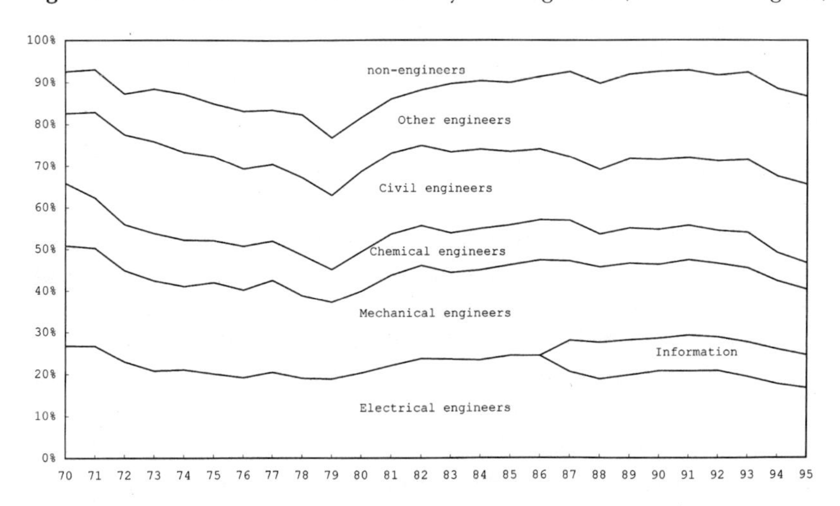

Figure 12 Share of Fields Chosen by New Engineers (BS and MS Degrees).

information engineers, and several other kinds of engineers.[22] In Figure 11, we show the distribution of these five major categories as well as of all the other engineers added together, including those who are hired explicitly as non-engineers. Comparing this figure with Figure 12, we can make the following observations:

If we interpret the number of graduating students taking non-engineering jobs as an index of excess-market supply, it reached its peak in 1979, with more than 20% of the graduates, and subsequently shrank to 8% in 1991. Reflecting the deepening recession, however, in 1995 it stood at 17%.

If we take the number of students in the most popular field of specialization, namely electrical and communication engineering, as the supply of electrical engineers, supply exceeded demand very substantially in the 1970s, although in the 1980s demand began to catch up. If we add information engineers to electrical engineers in the demand for these students, supply and demand seemed to have been roughly equal between 1987 and 1991, but after 1993, the excess of supply began to widen again. In mechanical engineering, up to 1987 supply and demand seem to have been more or less equal; since 1988, however, an excess supply seems to be emerging. In civil engineering and architecture, there was an excess supply throughout the 1970s and most of the 1980s. Quite recently, demand seems to be catching up with supply, which possibly reflects the massive counter-cyclical public works projects of the last few years.

University Research and the Needs of Industry

Under the present regulations, controlling the annual admission quota has been almost equivalent to controlling the size and distribution of faculty, and if a mistake is made in the manpower program, it may lead to the mistakes in the allocation of the universities' research capacity. To prove that such a misallocation has actually occurred is beyond the scope of the present analysis, and we will simply compare the allocations of faculty resources within the universities and two different allocations of resources, either of which can be justified as a reasonable basis for comparison. One is the size of the GDP of the relevant industries and the other is the net investment of those industries.

We will be looking at the four big fields in the schools of engineering. For electrical and communications engineering, the relevant industries are taken to be the electrical machinery industry and the transportation and communication industry; for chemical engineering, they are the chemicals and petrochemicals industry; for mechanical engineering, they are general machinery, precision

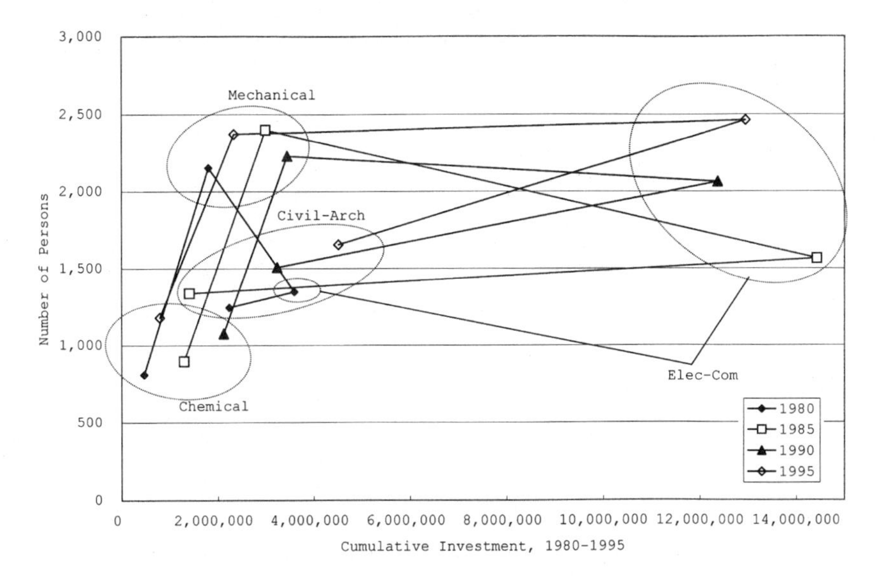

Figure 13 Investment 1980–1995 and Number of Professors/Assistant Professors.

machinery, and transportation machinery; for civil engineering and architecture, simply the construction industry.

Figure 13 shows the correlation between the number of professors or assistant professors and the cumulative investment of the relevant industries during the preceding five years for each of the four years. It suggests that in 1985 and the following years universities were unable to increase their faculty in electric and communication engineering in sufficient numbers when investments in electrical machinery started to rise, sometime between 1980 and 1985. Clearly they had failed to take advantage of the demand for knowledge in the field for many years. They seem to have had far more faculty in mechanical engineering than is justified by this criterion. If one draws an imaginary line connecting the chemical (lower left) and electrical (upper right) engineering fields, the points above the line belong to mechanical engineering. The distance between the points and the imaginary line seems to be close to 1000, or about 40% of their sizes.

Figure 14 shows the correlation between the number of professors or assistant professors and the cumulative GDP of the relevant industries during the preceding five years for each of the four years.

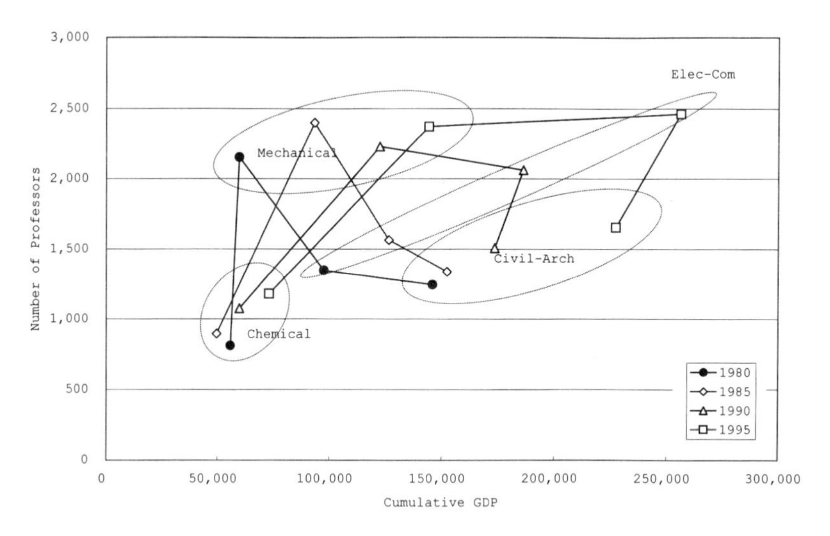

Figure 14 Industry GDP and Number of Professors, 1980–1995.

Again the story is similar to the preceding one. Here the breaks appear after 1985, due to a large jump between 1985 and 1990 in the cumulative GDP figures for the electrical machinery industry. If we draw a similar line connecting the chemical and electrical engineering fields, the distance between mechanical engineering and the line is about 700 or 800.

Conclusion

For historical reasons, in most universities most departments in their schools of engineering have been defined according to specific industries or groups of closely related industries. In the last twenty years, however, such a structure has proved to be a serious handicap for the schools, when the relative importance of the industries began to fluctuate widely. Given the walls between the departments, it became increasingly hard for schools to keep track of industry needs both in education and research. There is a substantial amount of evidence indicating that the system has been creating a serious mismatch between what the schools have been producing and what the industries outside have needed.

Many fundamental changes will be necessary in the higher education market if we are to remain competitive in the world market. It is clear that two changes in government policies should be included in any serious reform package. First, the government must end its restrictive policies on new entrants and make the market for higher education as competitive as possible. Secondly, the government should stop making decisions for individual national universities, which should run their own universities and should also be financially responsible for their decisions. They should start charging full costs to their students, and compete with private universities on a level playing field. These are theoretically correct changes any sensible economist would recommend.

There are other important problems specific to the field of engineering education. It seems evident that we must abandon the industry-specific model of schools, because we live in a world where the relative importance of technologies can change abruptly. If a school needs such flexibility in resource allocation, how can it be organized? How should we finance it? What kind of people should be there? What is the optimal mix of science and technology for an engineer? These are much harder questions for an economist to answer, and we must leave them for engineers to answer.

Notes

1. Due to a rapid decline in Japanese birth rate, however, the number of 18-year-olds has decreased from more than 2 million in 1991 to 1.6 million in 1998, and it will drop to 1.2 million around 2010. This drastic change in demography is expected to bring about a fundamental change in the higher education market in Japan.

2. When the new university system was established in 1949, the Japanese University Accreditation Association, a non-government organization, was to perform that function, but in 1957, the Ministry decided to perform the function itself, by claiming that a non-government voluntary organization is not fit to set compulsory standards for universities.

3. In some instances, the Ministry was willing to assert its preference, but this was synonymous to putting everyone's eggs in the same basket.

4. Formally, there are exceptions for universities located outside metropolitan areas and in disciplines with insufficient supply to meet the regional demand.

5. Recently the Ministry of Education changed its policy toward some national universities (particularly old imperial universities) and encouraged them to form less formal groups of faculty resources, rather than departments. But it is not clear whether or not their policy has changed toward other universities.

6. The seven universities are University of Tokyo, University of Kyoto, Osaka University, Nagoya University, Tohoku University, Hokkaido University, and Kyushu University.

7. There are several major national research universities in the second group that are closer to the first group than the rest, such as Hitotsubashi University, Tokyo Institute of Technology, Tsukuba University, and Kobe University. In this paper, we have drawn the line rather arbitrarily.

8. We will ignore public universities as they account for less than 2% of engineering students.

9. The statistics of the first group are population statistics. In 1970, the total quota of this group amounted to less than 5,700, or about 12% of the national total. In 1995, the total of this group was 6,974, or only 8% of the national total. The data of our second and third groups are sample data. In 1995, for example, the simple sum of admission quotas of universities in the second group amounted to 16,728, whereas the population total was estimated to be 21,250. Also in 1995, in the third group, the simple sum of admission quota of universities in (s_3) amounted to 45,955, whereas the population sum was estimated to be 59,855.

10. For this purpose, we used the total admission quota of national universities (X) and that of private universities (Y) published by the Ministry of Education for each year (Monbusho 1947–1987). We then proceeded in the following manner: (a) From the total admission quota of national universities, we subtracted the sum of the first group (x_1) and treated the residual as the "population" total of the second group ($X - x_1$). (b) We then obtained a multiplier for the second group by dividing this estimated population total ($X - x_1$) by the sample total (s_2) for each year, or $(X - x_1)/s_2$. (c) For the third group, we computed a multiplier by dividing the Ministry's total for private universities (Y) by the sample total of the third group (s_3) for each year, or Y/s_3.

11. The exceptions are Nagoya University with its single electrical engineering department (217), as well as Kyoto University with two departments (140 for hardware and 101 for software). But there were only five departments with admission quotas of 80 or more in 1995 among 24 departments in this group.

12. When the government decided to start subsidizing the private universities, the Ministry of Education wanted them to reduce excess admissions by providing additional quotas in return for their promises to reduce congestion on their campuses in future.

13. There are 104 individual departments in this field in our sample, but only 13 have an admission quota lower than 100.

14. In the absence of any "penalties," admission quotas may become more or less a moral guideline as far as private universities are concerned. In fact, the freeze that had started in 1976 worked well because the Ministry packaged it with sizeable subsidies, which would have been reduced or cut off if they had exceeded their quota intolerably. The freeze worked well because the subsidy was quite sizeable at that time, and the market demand was weak for demographic reasons.

15. We can compute the ratio from the two sets of numbers: The numerator is the Ministry's tabulation of the number of third-year students in each field, and the denominator is our estimates of admission quotas in each field. Since we know the admission quota of our three groups of universities, and since national universities would never overshoot their targets, we can compute actual admission ratios of private universities for each field.

16. Another interesting inference one can make from these figures is the relative popularity of these four major fields, as the relative popularity of each major field has moved in the same way as its excess admission ratio and its job prospects. In the early 1970s, the excess ratio was the highest in civil engineering, and the lowest in chemical engineering. Chemical engineering remained low throughout this period, but subsequently civil engineering gave way to electrical engineering and mechanical engineering. The two remain very close to each other in their excess ratios.

17. Since in some universities, including the University of Tokyo, engineering students do not select their field of specialization until their third year, an accurate tabulation of students' specialization can be made only in their third year.

18. Since 1959, the Ministry's annual report has published the number of students by their fields of specialization and their school year. While their classification has changed a little over the years, it is still possible to consolidate the changes and obtain the following list of fields for the years since 1959.

19. Throughout this paper, we simply refer to this field as "electrical engineering," but it is far more comprehensive that this shorthand name suggests. For example, the growth in the number of students in electrical engineering in the second half of the 1980s is due to the expansion of computer software engineering.

20. In addition, there were almost 6,100 instructors and lab-technicians called "assistants," but their numbers are not shown here.

21. For the first group, the standard university is Osaka University, for the second group, Gunma University, and for the third group, Hosei University.

22. The classification of information engineering engineers has been available only since 1987.

References

Amano, I. 1986. *Japanese Higher Education and Its Structure.* Tokyo: Tamagawa University Press (in Japanese).

Amano, I. 1988. *Age of Trial for Universities.* Tokyo: Tokyo University Press (in Japanese).

Monbusho. 1947–1987. Annual Report of the Ministry of Education. Tokyo: Ministry of Education (in Japanese).

Monbusho. 1970–1995. List of All Japanese Universities (*Zenkoku Daigaku Ichiran*). Tokyo: Ministry of Education (in Japanese).

Ogura, S. and H. Kotake. 1998. "A Simulation of Japanese Higher Education Market using an Econometric Model," in *Proposals on the Reform of Japanese Higher Education System by Economists.* Tokyo: Keizai-Kikakuchou (in Japanese).

VI

Public Policies for the Global Economy

After exploring the nature of university-industry activities, the cultural and historical constraints, and the incentives and barriers to successful industrialization of knowledge, we conclude with questions about how to think about what advanced societies, such as the United States and Japan, should ask of this relationship. The universities must ask themselves, of course, how advantages to the university might compromise the academic values they must protect. And firms must ask themselves, Is it realistic to expect university research to be sufficiently focused and timely as to constitute a useful basis for commercial innovation? In this final part we first take a snapshot of some informed industrial views about universities as technical partners, and then we step back from the issue of university-industry transaction and ask two larger questions: What are the demands on universities by the evolving global economy? Does this perspective give us a more balanced view of the university as a node in a modern economy? Our conclusion is that any examination of the economic role of the university must include all of the university's contributions to the social capital of the community, the region and the world, a context in which a collection of licensed patents may not be the most important contribution the university can make.

22

The Industrial Perspective on University-Industry Relationships in Japan and the United States

Sheryl Winston Smith

Industrial perspectives on university-industry relationships in Japan and the United States reveal a complex but important relationship. In both nations, industry leaders recognize the necessity of effective university-industry collaboration in the face of global competition, as well as the inherent constraints associated with transforming academic research into successful market products. At the same time, distinctive structural and policy environments in the two nations lead to important differences in both the arrangement and the success of such relationships. In Japan, while significant structural and historical impediments stand as barriers to formal university-industry relationships, informal connections are important conduits for information flow; in the United States the interchange between university and industry is more open, but intellectual property agreements are a troublesome area. Despite these differences, the industrial perspectives from both countries point to surprisingly similar problems related to crafting effective collaborations between these distinctive sectors.

University research is a potentially significant resource for companies in both the United States and Japan. Companies in both countries are reacting to the pressure of global competition by seeking knowledge from a greater variety of sources. As Charles Hora noted, for the Lord Corporation "global competition has caused a fantastic rate of obsolescence of product lines. Hardly have you initiated a new product than a number of imitative products have come out from competitors, and the only thing that protects us anymore from bankruptcy or from disappearing from the marketplace is the velocity of our technological change and our ability to innovate in the marketplace." Developing working relationships with university researchers and establishing programs geared toward this, he continued, enable the company to gain insights into fundamental knowledge and thus gain "an advantage and an edge on our competitors who find it difficult to do so." In Japan, the current economic difficulties heighten this pressure of global competition, leading Japanese companies to outsource innovation in order to respond to market demands for rapid innovation, as both Tetsuhiko Ikegami and Michiyuki Uenohara observed. As global competition increases, companies continuously need new sources of innovation and new ways to deliver value, as Fujio Niwa points out in chapter 5 of the present volume.

While both the United States and Japan experienced close university-industry linkages in the prewar era, the war and postwar developments led formal university-industry relationships in different directions. In Japan, the disconnect that developed between universities and industry after the war has made it difficult to reestablish formal ties between the two sectors. In contrast, in the United States the wartime involvement of universities in the national research effort solidified and shaped the formal connections that evolved in the postwar period. These divergent paths in the postwar period shape the nature and ease of current university-industry relationships in both nations.

The dynamics of current university-industry relationships in the United States are rooted in the continuous evolution of these two sectors since the mid-19th century. The incorporation of universities and academic researchers into the industrial effort of the Second World War formalized mechanisms for contracting and highlighted the relationship of academic research to national goals. In the postwar era, these relationships continued to develop (Brooks and Randazzese 1998; Mowery and Rosenberg 1993). In particular, most U.S. research universities have established technology licensing offices to encourage and oversee relationships with industry, offices that have proliferated since the mid-1970s in response to the emergence of new science-based industries and facilitated by the Bayh-Dole Act. Their role has also expanded, particularly in biomedical and engineering research, as universities increasingly attempt to derive revenue from these relationships (see Mowery, Nelson, Sampat, and Ziedonis, in this volume). From the industrial perspective, these offices serve as a liaison and facilitate the contracting and intellectual property aspects of the relationship.

The current relationship between industry and universities in Japan was shaped by the emergence of Japanese industry as a strong global presence in the postwar period. However, in Japan the predominant postwar trend was toward greater separation between universities and industry, as described by Hashimoto in chapter 9 of this volume. Ikegami, during panel discussions, mentioned that the relationship was "very good and in some sense very natural" before the Second World War, but "collapsed" after the war when a large gap developed between university and industry researchers. As industry developed into a global presence, it was increasingly successful with its own innovation and had less need for universities as a source of innovation. Industry invested heavily in R&D, and many companies created corporate laboratories. Further distance was placed between universities and companies as these companies developed their own in-house educational facilities (Odagiri and Goto 1993). For several decades Japanese companies followed American industry, as Japanese industry was engaged in technological catch-up. Japanese industry primarily focused on reverse engineering innovations introduced by U. S. industry to obtain market share, an activity that did not rely on basic research and thus reduced the need to utilize basic research from the universities (Odagiri and Goto 1993). However, as Uenohara pointed out, Japanese industry can no longer rely on "following industry, not following the market." Instead, industry in Japan is changing, learning to get new business ideas from the market, and developing closer contacts with universities in order to use new knowledge to generate new products. The gap between universities and industry was also fostered by the attitude of university researchers, who sought autonomy from industry. Despite changes in government regulations, the consequences of this persist; Uenohara noted that "most Japanese professors still think they have to keep sufficient distance from industrial problems." Relationships were further hindered by a lack of mobility between industry and academic employment (Ishii and Hirano 1993). In particular, Uenohara noted that mobility fosters networks of effective human relationships, such as his own experience at Bell Laboratories, which later helped him to support NEC's research.

Industry encounters many difficulties in trying to establish working relationships with Japanese universities. Given the poor connections between universities and industry, Japanese universities have not developed formal structures to coordinate university-industry relationships, although this is starting to change. The lack of formal structures is an impediment for Japanese industry seeking collaborative relationships with a university. Ikegami identified several key barriers. First, there is no clear path for initiating a relationship. Contact offices responsible for industry interactions rarely exist at Japanese universities, and when they do they are ineffective relative to their counterparts in American research universities. Moreover, universities and industry have no common language for contracting. Ikegami also pointed out that the ultimate return is often unclear to the sponsoring company, as Japanese universities tend to produce scanty documentation, in contrast to the final reports produced by their American counterparts. As a result of barriers such as these, he noted, Japanese industry invests about three times more in university research abroad than in Japan.

While formal structures for collaboration are weak, and despite the arms-length relationships between the universities and industry, informal relationships among individuals provide important avenues for the exchange of ideas and the formation of ongoing relationships between academic researchers and those in industry. Ikegami described a relationship that, while "very degraded," nonetheless exists to some extent. In particular, he noted that academic societies, such as the Japan Society of Applied Physics or the Institute of Electronics Information, enable informal technological interaction between industry, universities, and the national laboratories, particularly through papers given at the yearly meetings. Importantly, these papers are not reviewed and can be submitted by industry researchers as well, resulting in "a kind of free zone in terms of exchanging technological information."

Uenohara explains that the informal network created through the ongoing relationship between professors and former students leads to joint authorship, as "students often visit their professors in universities and most joint research begins based on private discussions"; furthermore, the "thesis doctor" system in Japan fosters relationships between industry and university professors (Uenohara, pers. com. 1999; NSF 1997). In this system, industrial researchers receive their doctorate after doing course work in graduate school, based on a dissertation topic worked out between the company research director and a professor who acts as an informal thesis advisor. The process takes about two years, and frequently includes research at the university with the thesis advisor, and may result in joint authorship of papers (Uenohara, pers. com. 1999).

Government policies also play direct and indirect roles in the university-industry relationship. In Japan, government regulations have limited the ability of researchers in the national universities to work directly with industry (NSF 1997). As civil servants, university faculty have been highly proscribed with regard to direct consultation with industry. Recently, this has been relaxed to an extent, but academic researchers remain constrained, as Uenohara observes (pers. com. 1999) and as Yoshihara and Tamai note in chapter 13. Uenohara noted the negative consequences of this, observing that the government restrictions regarding full-time employees still limit consulting with companies, and have resulted in a situation where "the majority of faculty do not know critical industrial problems and their research subjects are based on technical journals and conference digests." (Uenohara, pers. com. 1999). The Japanese government has started taking steps toward fostering university-industry relationships (Yoshihara and Tamai, chapter 13 in this volume). As Ikegami noted, the Japanese government is focusing on the central importance of science and technology as engines of growth, and to that end is constructing policies to encourage collaboration between industry and universities, and to foster horizontal collaborations between agencies and universities in place of vertical ones. In the United States, to a large extent, the role of the government has been to promulgate policies that push aggressive diffusion (Branscomb 1993; Branscomb and Florida 1998).

At a fundamental level, the most significant issue in both countries has more to do with the distinctive nature and needs of academic and industrial research, which creates inherent barriers to technology transfer from universities to industry, than with different national conditions. Industry and universities approach knowledge differently. Hora described the "fundamental difference between industry-driven research and the kind of research that goes on in academia." As he explained, industry research must be market- and product-driven. It is "of necessity highly focused, frequently governed with market inputs and needs, customer complaints about products that currently are inefficient or that need to transcend certain capabilities that present product lines do not have." This contrasts with the nature of university research, which is "knowledge-driven, blue sky, and very much more with an idea toward generating new information." Moreover, university and industry research operate in different time-frames, as industry research is characterized by urgency in the face of the need to commercialize. Similarly, Ikegami suggested that to work together effectively industries and universities must bridge a multidimensional "perception gap" in terms of time constant, agility, urgency, and the role of the market. Increasingly, information technology and networking may help to overcome this perception gap arising from the different senses of time and agility in academic and industrial research.

In order to overcome these dissimilarities, university and industry must work to achieve mutual understanding. Hora identified the "very significant impedance mismatch between university needs and requirements and industrial needs and requirements." Many difficulties involve differences in culture between academics and industry, and thus each side must strive to understand the culture of the other. Jim Foley observed that academic researchers need to grasp the strategies of the sponsoring companies, while the company needs realistic expectations of what university research can do for the company. As he noted, "it also takes a belief that it's the right thing to do. It takes a belief that there's value to be received, and certainly in my own lab at Georgia Tech we were very much wanting to understand the issues of industry because all of us wanted our research to have an impact." The researchers' having a specific understanding of the company's product strategies and long-term goals increases the likelihood that this will occur. As an academic researcher in computer science, Foley noted that he took steps to ensure that the development environment in his laboratory was the same as his industrial partners.'

In Japan, this gap is bridged somewhat through the thesis doctorate. As Uenohara noted, "In the company laboratory we encourage excellent young researchers to acquire the doctorate and allow them to perform basic research. Most of those subjects are based on key technological problems and hence the professor learns indirectly industrial basic technological problems" (Uenohara, pers. com. 1999).

Even in industry, translating corporate research into business outcomes is difficult. As Uenohara underscored, in over fifteen years of managing NEC corporate research, he had found that the "most headaches, nightmares, were how to transfer corporate research outcomes to the business division—very similar to university and industry relationships." While the gaps between universities and industry in Japan are compounded by the fact that industry previously relied on research from foreign sources, Uenohara suggested that industry in Japan needs better overall strategies for using university research to meet new business ideas from the market. He suggested that industry should adopt an active approach to university research, asking university researchers to learn more about the market and industry and helping them to explain research outcomes in industrial language.

Fundamentally, university and industry collaborations involve creating working relationships between people. Uenohara observed that a significant rationale for funding basic research at NEC is fostering closer contact with academic researchers and enabling corporate researchers to understand new knowledge that translates into industrial knowledge. Likewise, Foley noted that effective technology transfer between university and industry is "a contact sport, requiring people, not papers and reports." Importantly, this is based on recurrent relationships, which are nurtured over time, not through "one-year feel-good funding." Hora echoed this belief, criticizing industry approaches to university collaboration for being too often perfunctory. In contrast to this, he cited a successful program at his own company, the Lord Corporation, which works to create R&D programs more comparable with those at a university. By generating excitement in the company and with the board of directors tempered by the long time horizon, this program fosters interactive growth that enables industrial needs to evolve in the face of new and unexpected information. Collaborations at Lord focus on selecting certain universities with specific expertise in areas the company wants to develop and through ongoing contact between the academic researchers and Lord's scientists, as well as through summer internships and exchanges. Hora noted that such relationships serve as a powerful recruiting mechanism and create the foundation for future collaborations.

Intellectual property rights issues also are a major concern in university-industry relationships, especially in the United States, where intellectual property issues often impede effective relationships because both sides have unreasonable expectations. Foley suggested that companies fail to recognize that their sponsorship does not cover the real cost of the research, while academic researchers and the university often have unrealistic expectations of royalty income. He described mutually beneficial intellectual property strategies, such as the university's placing research results in the public domain through publication in technical journals, giving the sponsoring company a head start over competitors. In this case the company wins by being first to market while the university wins by securing research support. Alternatively, he suggested that the university and the company share ownership and licensing revenue: Either can license out and share fees, with a larger percentage to the broker, and the company pays no fee for its own use of the technology. Since neither party knows in advance what intellectual properties will result, Hora suggested that both parties need to be flexible initially and to have faith that any unique ideas will be worked out in conjunction with the university technology office, with the company given first refusal on co-patenting or licensing. Ikegami discussed the need for "good faith" in patent arrangements between universities and industry, noting that this would engender a "bargaining power [on the part of the universities] over the academic societies and [that] the university should be respected by industry in terms of research and producing trained students. Such functions are not that good in Japan."

An important subtext has been the university-industry relationships between the United States and Japan. Japanese industry sought relationships with U.S. universities because of the difficulties in approaching and working with Japanese universities, which made U.S. universities appear to be better collaborative partners. From his perspective as an academic researcher, Foley suggested that Japanese companies make better use of their university relationships through greater investment of time and genuine effort than many U.S. companies. As he has observed in his experience at Georgia Tech and other schools, it was "much more the case that Japanese companies have been willing to make that investment than have American companies." Interestingly, survey evidence suggests that Japanese companies make greater use of U.S. research universities than U.S.-owned firms (National Academy of Engineering, 1996). In some sense, a set of complementarities emerges from these comparative discussions. A paradigm takes shape in which U.S. universities relate to industry along the lines of structured relationships and career fluidity between academic and industrial paths. At the same time, it seems that U.S. industry might be better able to utilize its relationships with academia by incorporating lessons from Japanese industry, investing real effort and time in these relationships and perhaps by looking beyond a limited region or even nation for potential ones.

Acknowledgments

This chapter builds on discussions during a September 12, 1998, workshop by distinguished panelists from industry: Michiyuki Uenohara, Chairman, NEC Research Institute, Inc.; Tetsuhiko Ikegami, President, NTT Advanced Technology Corporation; Charles Hora, President, Lord Corporation; and James Foley, President, Mitsubishi Electric Information Technology Center America. The workshop was part of a conference held at the John F. Kennedy School of Government, Harvard University, on Universities and Science-Based Industrial Development. Unless otherwise noted, all quotations from the participants are taken from notes and transcripts of the panel discussion.

References

Branscomb, Lewis M., ed. 1993. *Empowering Technology: Implementing a U.S. Strategy.* Cambridge, MA: MIT Press.

Branscomb, Lewis M., and Richard Florida. 1998. "Challenges to Technology Policy in a Changing World Economy." In *Investing in Innovation: Creating a Research and Innovation Policy That Works,* eds. Lewis M. Branscomb and James H. Keller. Cambridge, MA: MIT Press.

Brooks, Harvey, and Lucien P. Randazzese. 1998. "University-Industry Relations: The Next Four Years and Beyond." In *Investing in Innovation: Creating a Research and Innovation Policy That Works,* eds. Lewis M. Branscomb and James H. Keller. Cambridge, MA: MIT Press.

Ishii, Masamichi and Yukihiro Hirano. 1993. "Comparative Study on Career Distribution and Job Consciousness of Engineering Graduates in Japan and the U.S." *NISTEP Study Material No. 28.* Japan: National Institute of Science and Technology Policy, Science and Technology Agency.

Mowery, David C. and Nathan Rosenberg. 1993. "The U.S. National Innovation System." In Richard R. Nelson, ed. *National Innovation Systems: A Comparative Analysis.* New York and Oxford: Oxford University Press.

National Academy of Engineering. 1996. *Foreign Participation in U.S. R&D: Asset or Liability?* eds., Proctor P. Reid and Alan Schrieshiem. Washington, DC: National Academy Press.

NSF. 1997. "The Science and Technology Resources of Japan: A Comparison with the United States." Special Report NSF 97(324, by Jean M. Johnson. Arlington VA: National Science Foundation.

Odagiri, Hiroyuki, and Akira Goto. 1993. "The Japanese System of Innovation: Past, Present, and Future." In Richard R. Nelson, ed. *National Innovation Systems: A Comparative Analysis.* New York and Oxford: Oxford University Press.

Uenohara, Michiyuki. 1999. E-mail communication to author, February 10, 1999.

23

Engine or Infrastructure? The University Role in Economic Development

Richard Florida and Wesley M. Cohen

Introduction

While a growing number of academics are interested in technological change, the innovation process, organizational transformation, and economic development, few have considered the role of the university in this context. Surprisingly, the literature lacks an adequate conceptual understanding of the role of the research university in contemporary capitalism. This gap is significant since, as numerous scholars have noted, capitalism is changing. A recent article in *The Economist* characterized the university as a "knowledge factory." As observers increasingly note, knowledge has replaced natural resources and labor-intensive industry as a primary source of wealth creation and economic growth (Drucker 1993; Nonaka and Takeuchi 1995; Florida 1995; Romer 1993, 1995; Leonard-Barton 1995). In this new economy, knowledge and ideas are a critical component of economic advantage, with intellectual capital being a pivotal resource (Stewart 1997; Edvinsson and Malone 1997). Taken in the context of this broader economic transformation, it stands to reason that the university's role is becoming increasingly important as an economic and social institution.

What is the role of universities in knowledge-based capitalism? Should we be surprised that universities are increasingly involved in areas of direct relevance to industry? Answers to these questions require a better theoretical understanding than we now have. This chapter draws from the previous chapters in this book, from our survey research on university-industry research centers, and from other sources to lay out the underpinnings of a better conceptual understanding or theory of the university's role in knowledge-based capitalism. It also deals with the tensions that new role is generating and reconsiders the notion of the university as an "engine" of economic development. While we do not expect to fill the wide theoretical gap, we do hope to place the debate in a richer conceptual context.

First, we want to emphasize that the university is involved in the knowledge economy in many complicated ways, direct and indirect, formal and informal, that are not yet clearly articulated, identified, or understood. Second, we want to offer a new way of thinking about what the university does. Drawing on our joint work, we emphasize the notion that universities act to optimize eminence and we highlight the tensions between the quest for eminence and the pursuit of research support from industry. We conclude that the university functions less as a direct engine of economic development than as an actor fulfilling an even more important role: that of an enabling infrastructure for technological and economic development.

Toward a Theory of the University

Conceptual foundations for understanding the role of the university are weak. The most important work includes that of Merton (1973), the more recent work of Dasgupta and David (1987, 1994), and of Rosenberg and Nelson (1994); from an historical perspective there is the work of Noble (1977) warning of corporate manipulation, Leslie's work (1987, 1990, 1993) on Stanford and MIT during the era of "cold war science," Etzkowitz's (1988, 1989, 1990) notion of the entrepreneurial university, and the work of Geiger (1986, 1993) and Veysey (1965) on the historical development of the research university.

Robert Merton (1973) argued that academic science should be an open project. While this view has often been understood as a normative prescription, Merton's own justification was grounded in efficiency. Firms are motivated to undertake scientific and technical advance by their quest for profit and intellectual property. Academic science has its own motivations that are centered on the efficient creation of knowledge and advance of scientific frontiers. The quest to discover and publish early creates a productive competition; information is quickly disseminated, as openness leads researchers to write their results on the "blackboard of science" promptly.

Building upon Merton's view, Dasgupta and David (1987, 1994) have presented an economic argument for keeping university and industry research separate. Academic science is a quest for fundamental discovery; industry research focuses on profit motives and proprietary access, they argue, and any intermingling of these functions would have negative social welfare implications, hence only a strong separation will optimize resource allocation and social welfare. Other economists, such as Mansfield (1991) and Jaffe (1990), have probed the relationships between university and industrial R&D and the effects of those relationships.

In contrast, Nathan Rosenberg (1982) has argued that the divide between science and technology is difficult or impossible to discern. Applied work often begets fundamental work and vice versa. Rosenberg and Nelson (1994) trace the ways in which university science contributes to technical advance in industry, and the ways in which technical advance in industry contributes to fundamental understanding. While such large-scale theories help illuminate the interaction of science and technology generally, they tell us little about the specific role of the university. How do the individual and organizational incentives of the university affect collaboration with industry and the government?

Two theories examine the university more specifically. One, associated with David Noble (1977) can be referred to as the "corporate manipulation" thesis, essentially arguing that corporations interfere with the normal pursuit of academic science and seek to control relevant university research for their own ends. A second theory, espoused by Etzkowitz (1988, 1990), Geiger (1986, 1993), and Slaughter and Leslie (1997), is that of "academic entrepreneurs." These scholars argue that university faculty members and administrators act as entrepreneurs, cultivating opportunities for federal and industry funding to advance their own agendas. Despite these important advances in our understanding of the role of the university in capitalism, they fail fully to grasp the objective function of the university, the intricate and complex ways in which the university is embedded within economy and society, and the full nature of the tensions thereby generated.

Let us outline what some elements of a better theory of the university might do. It would begin by identifying what a university does and what it wants to do. We find that the university is an institution that generates and disseminates knowledge. It competes with other institutions, and the nature of this competition is around eminence, which the university seeks to optimize along with reputation and prestige. In this regard, the university engages in a productive competition for highly regarded faculty, who attract outstanding graduate students, with the university's increased reputation in turn attracting leading undergraduates, and so on. The pursuit of eminence is reflected in contributions to new knowledge, typically embodied in academic publications. Universities, however, like all social and economic institutions, require funding to pursue their objectives. This gives rise to a fundamental tension which underscores the nature and history of interactions and relationships between the university and industry—a tension between the pursuit of eminence and the need for funding support. Today's debate over university-industry relations is the most current manifestation of this underlying tension.

To elaborate this argument we note three points. First, the university is fundamentally engaged in knowledge production or knowledge creation, but the nature of that production has changed over time. In general, we have seen a shift in emphasis at universities from knowledge transfer in the 19th century, with an emphasis on training students who then go out into the world, to research and knowledge creation in the mid- and late-20th century. Furthermore, this shift in university activities was related to the evolution of science-based industry in the late 19th and 20th centuries, in particular the rise of the industrial R&D laboratory (Hounshell 1996; Servos 1994). Today the life sciences and molecular biology best represent the contemporary profile of science-based industry (Blumenthal et al. 1986a,b). The growing emphasis on the role of knowledge in production has focused attention on the important contribution that universities might make to industry. In an era of knowledge-based capitalism, the capacity to combine diverse approaches to research makes the research university a particularly good place to pursue knowledge creation. Today's research university has the advantage of being able to cultivate and incubate a wide range of research approaches and strategies that are potentially relevant to industrial R&D and commercial technology.

Second, the university is becoming much more important as an economic and social institution than it has ever been. The reason for this is basic. The shift from industrial capitalism to knowledge-based capitalism makes the university ever more critical as a provider of critical resources such as talent, knowledge, and innovation. The university, however, is embedded and enmeshed in this system of knowledge-based capitalism in subtle, nuanced, and complicated ways we must better understand if we are fully to comprehend the broader processes of innovation, value creation, and economic growth. And, given this, the university is a very useful laboratory for understanding the broader dimensions of knowledge-based capitalism and of regional and national development.

Third, universities strive to increase their eminence, prestige, and reputation: The pursuit of external research support from industry and other sources essentially involves balancing new financial support against eminence. Generally speaking, attracting corporate funds does not hinder the quest for eminence, but industry funds may at times come with too many restrictions: control over publishing, or excessive secrecy. Furthermore, strategies for attaining eminence have changed over time. In some periods, eminence dictated a focus on teaching; at other times it has advocated work to enrich our stock of knowledge. There are tensions embedded across the entire historical evolution of university-industry interactions; so it should come as no surprise that they are evident today in issues involving the "skewing" of academic research from basic toward applied research, or in growing concerns about the increased "secrecy" of academic research as discussed in chapter 11. The initial wave of industrial support of and involvement in academic research appeared at the turn of the century in conjunction with the rise of industrial R&D. Chemistry and engineering departments at the time were host to a deep struggle between faculty who wanted to pursue applied, industry-oriented research, and other faculty who wanted to study anything so long as it was basic research. This tension ran particularly deep at MIT (see Servos 1980), where departments that became dependent on industry funds lost eminence as prestigious faculty members moved away. One goal of postwar government funding for university research (Brooks 1993) was to counteract this negative impact of industrial support by creating "steeples of excellence."

University-Industry Research Centers in the United States

To shed some additional light on this general argument, let me now turn to empirical evidence provided in a detailed Carnegie Mellon survey study of what we have called university-industry research centers (see Cohen, Florida, and Goe 1994). The study indicates that university-industry ties in the United States are quite extensive, identifying 1,056 university-industry research centers as of 1990. Moreover, the magnitude of spending by these joint research centers is substantial: a total of $4.12 billion in 1990, with $2.9 billion spent directly on R&D. For comparison's sake, this is more than double the National Science Foundation's $1.3 billion of support for all academic R&D in 1990 and almost one-fifth of all government expenditures in science and engineering. Between 1970 and 1990, it should also be pointed out, the share of industry funding of academic R&D more than doubled, rising from 2.6 to 6.9%.

These university-industry research centers involve not only a lot of money, but also a large number of faculty and students. They include, according to the CMU survey, roughly 12,000 university faculty members, 22,300 doctoral-level researchers (15% of the total), and 16,800 graduate students. These people do not necessarily work for university-industry research centers full-time. Indeed, one advantage these centers enjoy is their ability to leverage resources, including faculty time.

Another indicator of deepening university-industry ties in the United States is academic patenting. In 1974, 177 patents were awarded to the top 100 research universities. In 1984, this number increased to 408; in 1994, it jumped dramatically to 1,486 (Cohen et al. 1998: 182). In 1997, the 158 universities in the survey conducted by the Association of University Technology Managers applied for more than 6,000 patents. Universities granted roughly 3,000 licenses based on these patents to industry in 1998, generating roughly $500 million in royalty income, up from 1,000 in 1991. Furthermore, a growing number of universities, such as Carnegie Mellon and the University of Texas at Austin, have become directly involved in the incubation of spin-off companies, sometimes with great success as in the case of Carnegie Mellon and Lycos. And, as Josh Lerner discusses in chapter 15, a growing number of universities have sought to develop ties to venture capital funds, to encourage venture capitalists to open offices, and in some controversial cases, such as Boston University and Seragen, to make direct venture investments themselves.

Academic Entrepreneurs? University Initiative and Federal Science Policy

A theme gaining increasing attention in the debate over university-industry ties revolves around the concept of academic entrepreneurship. This view stands in some contrast to the notion that universities are more or less unwitting pawns of corporate manipulation. The question becomes: To what extent do universities actively cultivate and forge ties to industry?

The findings of the CMU survey of university-industry research centers indicate that universities, rather than industry, were the prime movers in the drive to develop closer academic-industrial ties. This contradicts the corporate manipulation thesis and tends to support the argument advanced by Henry Etzkowitz that "entrepreneurism" has permeated U.S. universities. The findings of the CMU survey clearly indicate that the main initiative for university-industry research centers originated with universities. More than two-thirds (73%) of university-industry research centers in the CMU survey report that the main impetus for their establishment came from the entrepreneurial efforts of university faculty and administration. For comparison's sake, it is useful to note that just 11% of centers reported that the main impetus for their establishment came from industry.

This university initiative did not occur in a vacuum; it was in many respects prompted and conditioned by shifts in federal science and technology policy. Here it is important to point our that more than half of all funding for university-industry research centers comes from government. Of the university-industry research centers that participated in the CMU survey, 86% received government support, 71% were established based on government support, and 40% reported they could not continue without this support (Cohen, Florida, and Goe, 1994).

Three specific policies conditioned the move among universities toward university-industry research centers. First, the Economic Recovery Tax Act of 1981 extended industrial R&D tax breaks to support research at universities. Second, the Patent and Trademark Act of 1980, otherwise known as the Bayh-Dole Act, permitted universities to take patent and other intellectual property rights on products of federally funded research. This allowed universities both to take patent rights and to assign or license those rights to others, frequently industrial corporations. Third, government agencies began funding a relatively small number of research centers such as the NSF Engineering Research Centers and Science and Technology Centers, both of which required industry participation, creating the perception that government resources would in the future be tied to joint university-industry research initiatives.

The push for linkages to industry had its roots in a combination of perceived declining government research funds in the 1970s and in the debate over U.S. competitiveness in the 1980s. A number of university leaders, including Derek Bok (1982), then President of Harvard University, posed the university as a potentially potent and under-used weapon in the battle for global industrial competitiveness. The National Science Foundation promoted joint centers as a way of encouraging closer ties between universities and industries and improving the transfer of academic research to industry. Although less than a hundred centers were originally funded under various NSF programs, these initiatives encouraged universities to seek closer research ties to private industry, by creating the perception that future competitions for federal funds would require demonstrated links to industry, thus prompting more and more universities to establish such centers, sometimes with their own funds. Universities, for example, provide 18% of the total support for the centers in the CMU survey, much of it coming in the form of cash support.

The Reaction from Industry

Industry's views of growing university-industry research ties are decidedly mixed. A first approximation comes from the findings of the CMU survey. A significant portion of university-industry research center funding, $800 million, comes from industry, representing 50% of industry's total contribution to academic research, and 1.5% of industry's own R&D budget. This funding takes the form of grants, dues, equipment, and even some endowed chairs.

A richer perspective comes from our interviews with corporate leaders. Almost every company we interviewed thinks that universities are doing what they do very well. Cutting-edge academic research is superb, and students are being well educated. What companies are concerned about is the move into applied research that university industry research centers represent. This concern has three causes. First, students are the most important product that universities produce, and industry is worried that a focus on profit will hurt the education function. Second, industry now feels that it can get better research results out of one-on-one interactions with professors. University-industry research centers offer the advantage of strong government support, allowing firms to leverage their investments; but overhead in research centers is often high, and the results often not directly relevant to the interests of the participating company. Faced with increasing pressure to achieve results, company research divisions are resorting to smaller contracts with individual faculty members which last several years. This ensures them faculty commitment, a counterpart with aptitude for the business culture, and a check on overhead costs. One vice-president for research summarized industry's feelings about university-industry research centers: "The university takes this money, then guts the relationship." Third, firms are concerned with university wrangling over intellectual property rights. They are particularly concerned with the time delays this may cause. They are also concerned that even though they fund research up front, they are forced into unfavorable negotiations over intellectual property when something of value emerges. Furthermore, some companies are concerned that the centers they support will share vital information with their competitors. Because several firms normally participate in a single research center, faculty members may inadvertently make public vital information. For industry, this risk of information leakage is significant.

Implications for the Research University

What do closer ties to industry mean for the university and its traditional missions of research and teaching? Where are the tensions manifesting themselves? This is a question that is on the minds of many inside the university as well as in government. Closer ties between university and industry clearly pose important implications for the research university. For universities, the key issue has to do with the tradeoff between the quest for eminence and the pursuit of funding support from industry. The CMU survey indicates that industry is still capable of affecting the direction of research agendas, their policies on information disclosure and publication, and, perhaps most troubling, the amount of communication within the center itself. Of the centers surveyed, 65% indicated that industry exerts a "moderate to strong influence" over the direction of their research. Furthermore, it is important to distinguish between two distinct issues facing research universities. The first can be referred to as the "skewing problem"—the alleged shift in research effort from basic to applied research. The second is the "secrecy problem" and involves the rise in restrictions on publication of research findings.

Many contend that growing ties to industry tend to skew or shift the academic research agenda from basic toward applied research. The evidence here is mixed. David Blumenthal and others (1996) found that industry-supported research in biotechnology tended to be "short term." Surveys by Rahm (1994) and Morgan (1993, 1994) found some empirical association between greater faculty involvement with industry and more applied research. The CMU survey found that the research direction of centers is associated with the extent to which they expressly take on the mission of improving industrial products and processes. Centers that view this mission as important devote 29% of their R&D activities to basic research, while centers that do not consider it important devote 61% of their R&D activities to basic research (Cohen, Florida, and Goe 1994). While this evidence is interesting, it remains unclear whether or not industry funding is causing academics to shift their research agendas. We may simply be observing a "selection effect," in the sense that centers that are already oriented toward more applied research are more likely to obtain industry funding and to embrace a more commercial orientation. In fact, the National Science Board data show the composition of academic R&D between basic and applied research has remained relatively stable since 1980, at about 66%, though this is down from 77% in the early 1970s (Brooks and Randazzese 1998).

Complicating the matter, the findings of the CMU study of university-industry research centers appear to indicate that such centers are able to achieve significant gains in industrially relevant technology. Furthermore, the CMU study and related research (Cohen, Nelson, and Walsh 1996; Cohen et al. 1998: 171–199) clearly shows that the process of knowledge or technology transfer from university to industry occurs through multiple channels, such as publications, students, informal discussions, consulting relationships, intellectual property, spin-off companies, and so on (see also Faulkner and Senker 1995). Policies and programs that seek to tie university and industry through more formal systems for technology transfer and commercialization appear to strengthen some channels, while weakening others. This is particularly true of financing agreements, which include disclosure restrictions. The same study, however, finds that these channels often operate in synergistic ways. In this regard, it may be a mistake to attempt to alter the system in ways we do not yet fully understand.

A larger and more pressing issue appears to revolve around growing "secrecy" in academic research. Most commentators have posed this as an ethical issue, suggesting that increased secrecy contradicts the norm of open dissemination of scientific knowledge. The real problem is not simply this normative and ethical challenge, but that academic secrecy may threaten the efficient advance of scientific frontiers. One dimension of this issue revolves around the nature and extent of so-called disclosure restrictions, that is, restrictions on what can be published and when it can be submitted for publication. Here again, the findings of the CMU survey are illustrative. Over half the centers in the CMU survey said that industry could force a delay in publication, and over a third reported that industry could have information deleted from papers prior to publication. Even though some have argued that these delays are for relatively short time periods, and the information deleted tends to be of marginal value, the issue of disclosure restrictions opens up a veritable Pandora's box. Blumenthal and his collaborators (1997) report that 82% of companies they surveyed, which support academic research in the life sciences, require academic researchers to keep information confidential to allow for filing a patent application, and that 47% of the firms report that their agreements with universities occasionally require academic institutions to keep results confidential for longer than is necessary to file a patent. The study concludes that participation with industry in the commercialization of research is "associated with both delays in publication and refusal to share research results upon request" (Blumethal et al. 1997). Furthermore, a survey (Rahm 1994) of more than a thousand technology managers and faculty at the top 100 R&D-performing universities in the United States found that 39% of technology managers reported that firms place restriction on information sharing by faculty, and that 79% of them and 53% of faculty members reported that firms had asked for research findings to be delayed or kept from publication. A 1996 article in *The Wall Street Journal* (King 1996) reported that a major drug company, by disallowing publication of research it had funded in a major scientific journal after that article had been accepted, suppressed findings of sponsored research at the University of California San Francisco, when the researchers found that cheaper drugs made by other manufacturers were therapeutically effective substitutes for its drug, Synthroid, which dominates the $600 million market for drugs to control hypothyroidism. While prestigious universities with strong federal funding are often able to avoid the deleterious impact of industry investment, less prestigious research universities are not (Randazzese 1996). Furthermore, as Mowery et al. point out in chapter 11, there is concern that growing secrecy in biotechnology research tools and techniques may be holding back advances in that field.

There are also growing concerns both among university faculty and industry that U.S. universities may have become "overzealous" in the pursuit of revenues from technology transfer. There is mounting concern over the practice and policies of technology transfer offices, and university intellectual property staffs in particular. Industry is increasingly nervous about disclosure restrictions and intellectual property policies at universities, and particularly the increased legal wrangling that occurs. They are concerned both about legal wrangling over intellectual property and the time delays it may cause, as noted earlier. This perception of overzealousness on the part university technology transfer operations may in some cases be weakening the relationship between the university and industry. It also appears to be provoking some negative reactions on the part of at least some faculty members. While such a negative reaction is not yet pervasive, it continues to bubble under the surface as a general sentiment, voiced variously as "Why are we doing this, what does it mean, why are we compromising ourselves?" Among university faculty in the United States, it is safe to say that there is some aversion to technology transfer offices.

A related tension revolves around the impact on university finances of increasing internal university funding of university-industry research centers. According to Irwin Feller (chapter 3), the most rapidly increasing source of academic research funding comes from the university itself. While federal funds are holding constant or increasing slightly and industrial funds for research are increasing somewhat, the fastest-growing segment of research funds is internal funds. Universities increasingly believe, according to Feller, that they need to make investments in internal research capabilities, by funding centers and laboratories for example, in order to compete for federal funds down the road. The need for internal funding is an important motivator for university technology transfer efforts. The revenue from these efforts is in the form of discretionary funds which can be invested in new activities such as closer research ties to industry, which it is hoped will someday lead to greater revenue streams in the form of larger federal grants.

Carnegie Mellon, for example, generated more than $20 million from its initial equity stake in Lycos, which it is using among other things to finance endowed chairs in computer science and the construction of a new building for computer science and multimedia research. It is the quest for these sorts of discretionary funds, which are growing (in some case rapidly) at the margin, that motivates increased university interest in revenues from technology transfer.

Finally, closer ties to industry are helping to bring about a change in personnel at research universities. Whereas universities used to comprise only faculty and students, university-industry research centers are creating a new faction within the university, the research scientist. These research scientists work primarily on sponsored research and outside of the realm of graduate education. They consider themselves a different group, and some universities have created new career tracks for them. This means that research scientists have personal and institutional goals that differ from those of faculty and students. These divergent goals may create distortions when universities make important decisions affecting their trade-off between eminence and cash flow. Interestingly, as the interests of research scientists become better represented, a university's eminence may suffer.

University-Industry Relations in Japan

Japan is also moving to a knowledge economy, and university and industry are working more closely together, as well. In fact, as the chapters in this volume indicate, university-industry ties in Japan are much more extensive than most U.S. scholars, analysts, and policymakers have typically thought. Furthermore, as the chapters in this volume document, the nature of university-industry ties in Japan differs considerably from those in the United States. The simplest way of saying this is that university ties in Japan are considerably less formal than such ties in the United States, and depend on an informal network structure of relationships as opposed to formal contractual relationships.

University-industry ties in Japan are quite extensive, as previous chapters amply demonstrate. As Odagiri documents in chapter 10, there were 1,448 joint research projects between university and industry in Japan in 1994, involving 89 universities and 883 firms. As Pechter and Kakinuma show in chapter 4, nearly half (46%) of all publications emanating from an industrial corporation in Japan in 1996 had a university co-author, up from 23% in 1981. As Kneller (chapter 12) and Yoshihara and Tamai (chapter 13) demonstrate, Japanese patent statistics sorely understate the role of university research in patentable innovations in Japan. Chapter 13 shows that Japan's Patent Agency and Monbusho reported only 129 university patent applications in 1994, less than one percent of all patent applications. Furthermore, as that chapter documents, while the Japan Patent Office reported only two patent filings by University of Tokyo faculty in 1994, a survey by the university's Department of Engineering reported that there were as many as 150 inventions made by faculty members, suggesting that 148 of these academic inventions were filed by the private sector rather than the by university faculty members. A study of Japanese patent applications in genetic engineering cited by Kneller in chapter 16 found that half of the approximately 600 patent applications in that field listed a Japanese university scientist as co-inventor.

Taken as a whole, the chapters in this volume on university-industry ties in Japan provide the outlines of a broad model of how the process of technology transfer from university to industry works in that nation. The defining principle of this model is the mobilization of knowledge through informal but well-articulated networks. In its simple (and most over-simplified) form, the model works like this: In return for intellectual property emanating from academic laboratories, industry tends to compensate the academic inventors in the form of donations. The use of donations as the preferred form of industrial support is a product of Japanese law, which since the Second World War has prohibited direct industrial support of university research. Faculty members use these donations to conduct research work in their laboratories. When the research leads to something of relevance to industry, that research is informally transferred to industrial partners who patent the discovery. These same industrial sponsors also tend to hire the graduate students from the university laboratory. In this way, the system tends to create a productive cycle. Large R&D-intensive

firms sponsor academic research of direct relevance to them through donations, which in turn support relevant academic research. The results of the research are transferred to those sponsors via intellectual property ownership and transfers of human capital. More donations come in, more research get done, more graduate students are trained, more patents go to the firms supporting the research, more graduate students get hired by those same firms, and so on. It is important to point out that this informal system appears to be an effective way of cross-pollinating the channels for technology transfer, in that it involves the transfer of formal codified knowledge along with human capital (in the form of graduate students familiar with the technology).

Rethinking the University's Role in Economic Development

We now turn to a key issue: the role of the university in economic development. Here, in particular, we want to suggest that the conventional metaphor of the university as an "engine" of regional economic development is misapplied. The university's economic role is much more complicated, subtle, nuanced, and complex than such mechanistic thinking allows. Instead of thinking of the university as an engine of economic development, it is more appropriate to conceptualize it as a pivotal component of an underlying infrastructure for innovation on which the system of knowledge-based capitalism draws.

The role of the university in economic development has captured the fancy of business leaders, policy-makers, and academics as they have looked at the examples of technology-based regions like Silicon Valley and the Route 128 region surrounding Boston and Cambridge. They have concluded that the university has played a fundamental role in developing the technological innovations and technologies that power those regional economic models. A theory of sorts has been handed down based mainly on anecdotes and so-called success stories of the university as "engine" of regional economic development. This view is similar in many respects to the now widely criticized "linear model of innovation" which rests on the assumption that there is a linear pathway from university science and research, to commercial innovation, and on to regional development in the form of ever-expanding networks and genealogies of newly formed companies. This model is in turn reflected in a wide variety of university-based and publicly supported technology transfer programs which aim to increase the output of university "products" that are of value to industry.

There are self-evident reasons to question that view. It is quite clear that Silicon Valley or the Cambridge/Boston regions are not the only places with excellent universities working in areas of potential commercial importance. One way to begin to structure the problem is to think of the relationship between the university and the economy as composing a simple system, in which the university transmits a signal, which the regional economic environment must absorb. Increasing the volume of the signal need not result in effective transmission or absorption if the region's receivers, so to speak, are not turned on or are functioning ineffectively. In short, the university appears to be a necessary but insufficient condition for regional technological and economic development. To borrow a phrase from Cohen and Levinthal (1990), what appears to matter here—and what is too often neglected in policy circles—is what we might call "regional absorptive capacity," the ability of a region to absorb the science, innovation, and technologies which universities generate. Another way of saying this is that regions need to capture the "spillovers" of the technologies and innovations they generate.

In chapter 18, Michael Fogarty and Amit Sinha examine the flow of intellectual property (in this case patents) from universities to other universities and to firms around the nation. They identify a simple but illuminating pattern: a significant outward flow of intellectual property from universities in older industrial regions such as Detroit and Cleveland to high-technology regions such as the Boston/Cambridge region, the California Bay Area, and the greater New York metropolitan area. Their work suggests that even if the ability to generate new ideas and new knowledge exists in many places, it is those places having the ability to use and absorb those ideas that are able to turn them into economic wealth.

This brings us to the most critical contribution of the university to economic development; this lies in the domain of talent. As is increasingly noted, talent is the key resource of the knowledge economy. As a factor of production, it has a number of critical features. First, talent is highly mobile. Second, the distribution of talent in scientific and technical fields is highly skewed. Finally, the labor market for knowledge workers is different from the general labor market: Smart people do not necessarily respond to monetary incentives alone; they want to be around other smart people. In this regard, talent tends to attract talent, which is why universities tend to compete to attract the best talent, the so-called academic stars, and why they do so by leveraging the reputations of the talent they already have, for example by highlighting the number of Nobel laureates on their faculty.

The university plays a magnetic role in the attraction of talent—a classic increasing returns phenomenon. The fact is that good people attract other good people, and places with lots of good people attract firms that want access to talent, creating a self-reinforcing cycle of growth. According to Dr. Uenohara of NEC, one of the most significant corporate impacts of NEC's Research Institute in Princeton was that it helped the firm to attract better talent in Japan—having a basic research faculty with Nobel Prize-caliber talent was an important factor in attracting bachelor's level engineers. The need to attract talent is also one of the key reasons why firms organize their internal research units in ways that emulate university research laboratories, with investigator autonomy and the ability to publish, hold seminars, invite visitors, and so on. A key role of the university in the knowledge economy then is as a collector of talent—a growth pole which attracts eminent scientists and engineers (see Zucker, Darby, and Brewer 1998) who attract graduate students, who in turn create spin-off companies, and eventually encourage other companies to locate nearby.

Furthermore, it is important to recognize the dynamic nature of this system for attracting talent. Over time, any university or growth region will be constantly re-populated with new talent. Leading universities constantly replenish their stock of talent, with professors and graduate students moving in and out. In short, universities—and the labor market for knowledge workers more broadly—are distinguished by high degrees of "churning." It is not simply capturing any given stock at any given moment that matters: What matters is the ability to attract and replenish that stock. This is particularly true in advanced scientific and technical fields, where "learned skills" (such as engineering degrees) tend to depreciate rather quickly. So growth regions benefit from this dynamic process of talent creation and attraction.

This has important implications for public policy. Consider the fact that virtually all public policy in this area, whether it is national, state, or local, has been organized as a giant "technology push experiment." The basic logic goes like this: If the university can just push more innovations out the door, those innovations will somehow magically turn into economic growth. Avoiding a lengthy critique of the naive assumptions made here about the "localized" nature of spillovers, it can simply be said that the process for turning academic research into companies that create regional growth is long and complex.

The key, then, is to move away from the limited concept of the university as an engine of economic development, and to begin to view the university as a complicated institutional underpinning of regional and national growth. If nations and regions are really serious about building the capability to survive and prosper in the knowledge economy and in the area of talent, they will have to do much more than simply enhance the ability of the university to transfer and commercialize technology. They will have to act on this infrastructure both inside and surrounding the university in ways that make places more attractive to and conducive to talent. And, it is here—in the attraction of talent—that national and regional policymakers have a great deal to learn from the universities, who have been doing just this—creating organizational and institutional environments appropriate to knowledge workers—for a very long time.

Acknowledgment

This chapter summarizes key themes that will appear in the book *For Knowledge and Profit: University-Industry Research Centers in the United States* (New York: Oxford University Press, forthcoming). The Ford Foundation and National Science Foundation provided research support.

References

Blumenthal, David, Michael Gluck, Karen Seashore Louis, and David Wise. 1986a. "Industrial Support of University Research in Biotechnology," *Science* 231: 242–246.

Blumenthal, David, Michael Gluck, Karen Seashore Louis, Michael Soto and David Wise. 1986b. "University-Industry Research Relationships In Biotechnology: Implications for the University," *Science* 232: 361–366.

Blumenthal, David, et al. 1996. "Relationships between Academic Institutions and Industry in the Life Sciences: An Industry Survey," *New England Journal of Medicine* 334 (6): 368–373.

Blumenthal, David, et al. 1997. "Withholding Research Results in Academic Life Sciences: Evidence from a National Survey of Faculty," *Journal of the American Medical Association* 227: 1224–1228.

Bok, Derek. 1982. *Beyond the Ivory Tower.* Cambridge, MA: Harvard University Press.

Brooks, Harvey. 1993. "Research Universities and the Social Contract for Science," in Lewis Branscomb (ed.), *Empowering Technology.* Cambridge, MA: MIT Press.

Brooks, Harvey, and Lucien Randazzese. 1998. "University-Industry Relations: The Next Four Years and Beyond," in Lewis Branscomb and James Keller (eds.) *Investing in Innovation: Creating a Research and Innovation Policy that Works.* Cambridge, MA: MIT Press.

Cohen, Wesley, Richard Florida, and W. Richard Goe. 1994. *University-Industry Research Centers in the United States.* Pittsburgh: Carnegie Mellon University.

Cohen, Wesley, Richard Florida, and Lucien Randazzese. Forthcoming. *For Knowledge and Profit: University-Industry Research Centers in the United States.* New York: Oxford University Press.

Cohen, Wesley, Richard Florida, Lucien Randazzese, and John Walsh. 1998. "Industry and the Academy: Uneasy Partners in the Cause of Technological Advance," in Roger Noll (ed.), *Challenges to Research Universities.* Washington DC: Brookings Institution Press.

Cohen, Wesley, and Daniel Levinthal. 1990. "Absorptive Capacity: A New Perspective on Learning and Innovation," *Administrative Science Quarterly* 35: 128–152.

Cohen, Wesley, Richard Nelson, and John Walsh. 1996. "Links and Impacts: New Survey Results on the Influence of University Research on Industrial R&D," unpublished paper, Carnegie Mellon University.

Dasgupta, Partha, and Paul David. 1987. "Information Disclosure and the Economics of Science and Technology," in G. Feiwel (ed.), *Arrow and the Ascent of Modern Economic Theory.* New York: New York University Press.

Dasgupta, Partha, and Paul David. 1994. "Toward a New Economics of Science," *Research Policy* 23 (3): 487–521.

Drucker, Peter. 1993. *Post-capitalist Society*. New York: HarperBusiness.

Edvinsson, Leif, and Michael S. Malone. 1997. *Intellectual Capital*. New York: HarperBusiness.

Etzkowitz, Henry. 1989. "Entrepreneurial Science in the Academy: A Case for the Transformation of Norms," *Social Problems* 36 (February): 14–29.

Etzkowitz, Henry. 1990. "MIT's Relations With Industry: Origins of the Venture Capital Firm," *Minerva* 21(2–3): 198–233.

Etzkowitz, Henry. 1998. "Making of an Entrepreneurial University: The Traffic Among MIT, Industry and the Military, 1860–1960," in E. Mendelsohn, M.R. Smith, and P. Weingart, (eds.), *Science, Technology and the Military* 12. Boston: Kluwer Academic Publishers.

Faulkner, Wendy, and Jacqueline Senker. 1995. *Knowledge Frontiers: Public sector Research and Industrial Innovation in Biotechnology, Engineering Ceramics and Parallel Computing*. New York: Oxford University Press.

Florida, Richard. 1995. "Toward the Learning Region," *Futures* 27 (5): 527–536.

Geiger, Roger. 1986. *To Advance Knowledge: The Growth of American Research Universities, 1900–1940*. New York: Oxford University Press.

Geiger, Roger. 1993. *Research and Relevant Knowledge*. New York: Oxford University Press.

Hounshell, David. 1996. "The Evolution of Industrial Research in The United States," in Richard Rosenbloom and William Spencer (eds.), *Engines of Innovation*. Boston: Harvard Business School Pres).

Jaffe, Adam. 1990. "The Real Effects of Academic Research," *American Economic Review* 79(5): 957–978.

King, Ralph. 1996. "Bitter Pill: How a Drug Firm Paid for University Study Then Undermined It," *Wall Street Journal* (25 April): A1, A13.

Leonard-Barton, Dorothy. 1995. *Wellsprings of Knowledge: Building and Sustaining the Sources of Innovation*. Boston: Harvard Business School Press.

Leslie, Stuart. 1987. "Playing the Education Game to Win: The Military and Interdisciplinary Research at Stanford," *Historical Studies in The Physical and Biological Sciences* 18: 55–88.

Leslie, Stuart. 1990. "Profit and Loss: The Military and MIT in the Postwar Era," *Historical Studies in The Physical and Biological Sciences* 21(1): 59–86.

Leslie, Stuart. 1993. *The Cold War and American Science*. New York: Columbia University Press.

Mansfield, Edwin. 1991. "Academic Research and Industrial Innovation," *Research Policy* 20: 1–12.

Merton, Robert. 1973. *The Sociology of Science*. Chicago: University of Chicago Press.

Noble, David. 1977. *America by Design: Science, Technology and the Rise of Corporate Capitalism.* New York: Oxford University Press.

Morgan, Robert. 1993. "Engineering Research at U.S. Universities: How Engineering Faculty View It," paper prepared for the IEEE-ASEE Frontiers in Education Conference.

Morgan, Robert. 1994. "Engineering Research at U.S. Universities: How University-Based Research Directors View It," paper prepared for the ASEE Annual Meeting.

Nonaka, Ikujuro, and Hirotaka Takeuchi. 1995. *The Knowledge-Creating Company.* New York: Oxford University Press.

Rahm, Diane. 1994. "University-Firm Linkages for Industrial Innovation," paper prepared for the Center for Economic Policy Research/AAAS Conference on University Goals, Institutional Mechanisms and the Industrial Transferability of Research.

Randazzese, Lucien. 1996. *Profit and the Academic Ethos.* Ph.D. Dissertation, Department of Engineering and Public Policy, Carnegie Mellon University.

Romer, Paul. 1995. "Beyond the Knowledge Worker," *World Link* (January–February): 56–60.

Romer, Paul. 1993. "Ideas and Things," *The Economist* (11 September).

Rosenberg, Nathan. 1982. *Inside the Black Box.* New York: Cambridge University Press.

Rosenberg, Nathan, and Richard Nelson. 1994. "American Universities and Technical Advance in Industry," *Research Policy* 23: 323–348.

Servos, John. 1980. "The Industrial Relations of Science: Chemical Engineering at MIT," *ISIS* (December).

Servos, John. 1994. "Changing Partners: The Mellon Institute, Private Industry and the Federal Patron," *Technology and Culture* 35(2): 221–257.

Slaughter, Sheila, and Larry Leslie. 1997. *Academic Capitalism: Politics, Policies and the Entrepreneurial University.* Baltimore: Johns Hopkins University Press.

Stewart, Thomas. 1997. *Intellectual Capital: The New Wealth of Organizations.* New York: Currency/Doubleday.

Veysey, Laurence. 1965. *The Emergence of the American University.* Chicago: University of Chicago Press.

Zucker, Lynne, Michael Darby, and Marilyn Brewer. 1998. "Intellectual Human Capital and the Birth of U.S. Biotechnology Enterprises," *American Economic Review* (March): 290–306.

Index